Stingless Bees
神奇的无刺蜂

Their Behaviour, Ecology and Evolution

［瑞士］克里斯托弗·格鲁特　著

王　凯　译著

中国农业科学技术出版社

著作权合同登记号：01-2023-0734

图书在版编目（CIP）数据

神奇的无刺蜂/(瑞士)克里斯托弗·格鲁特著;王凯译著. --北京：中国农业科学技术出版社，2022.12

ISBN 978-7-5116-6059-6

Ⅰ.①神… Ⅱ.①克…②王… Ⅲ.①蜜蜂—研究 Ⅳ.①S893

中国版本图书馆CIP数据核字（2022）第225447号

First published in English under the title
Stingless Bees: Their Behaviour, Ecology and Evolution by Christoph Grüter
Copyright © Springer Nature Switzerland AG, 2020
This edition has been translated and published under licence
from Springer Nature Switzerland AG.

责任编辑　白姗姗
责任校对　马广洋
责任印制　姜义伟　王思文

出 版 者	中国农业科学技术出版社
	北京市中关村南大街12号　邮编：100081
电　　话	（010）82106638（编辑室）　　（010）82109704（发行部）
	（010）82109709（读者服务部）
网　　址	https://castp.caas.cn
经 销 者	各地新华书店
印 刷 者	北京建宏印刷有限公司
开　　本	170 mm×240 mm　1/16
印　　张	31.25
字　　数	580千字
版　　次	2022年12月第1版　2022年12月第1次印刷
定　　价	258.00元

◀━━ 版权所有·侵权必究 ━━▶

序　言

　　近年来，蜜蜂越来越受到人们的关注，这主要也是因为我们很清楚现今蜜蜂的生存环境越发严峻，传粉蜂数量、多样性不断下降。作为造成这一现象的直接原因，人类可能因为授粉不足而影响自身生存。而且蜂类不仅是高效的授粉昆虫，它们在很多方面也都值得研究。在过去几十年来发表的成千上万篇关于蜂类的研究文章中可以看出，研究者也确实从多方面对传粉蜂展开了相关研究。这些研究中大多数聚焦在温带蜜蜂和熊蜂，往往不太关注热带蜂。关于热带无刺蜂的研究也就相对较少，而热带无刺蜂却是社会性蜂中最大的、最古老的以及最多样化的类群，因此亟待研究者加大关注力度。当前，无刺蜂的相关信息往往不像蜜蜂那样容易获得。实际上，许多关于无刺蜂的重要文献都是以非英语语言出版的，有些甚至没有在科学杂志上发表。事实上，大量的有关无刺蜂的信息只存在于未发表的硕士和博士论文中，例如，我在本书中就提到过其中一些硕博士论文。因此，本书的主要出版动机之一是想要将这些不同的信息汇集在一起，对无刺蜂这个大群体的行为、生态及进化相关的内容进行总结，希望能给读者提供一些这方面的信息。David Roubik 所著 *Ecology and Natural History of Tropical Bees*（1989）以及 Schwarz（1948）、Michener（1974）、Sakagami（1982）和 Wille（1983）等发表的文章也提供了关于无刺蜂的重要信息，在此我也推荐读者去阅读，希望想要了解热带蜂的读者可以有所收获。

　　此外，我觉得当代研究者逐渐关注无刺蜂研究（Hrncir et al., 2016），他们越来越重视无刺蜂养殖、无刺蜂文化以及无刺蜂的保护（Vit et al., 2013, 2018; Heard，2016; Quezada-Euán，2018; Quezada-Euán and Alves，2020），这也是促使我写这本书的原因之一。最近开始有研究者发表关于无刺蜂基因组的论文（Kapheim et al., 2015; Freitas et al., 2020），这将进一步推动无刺蜂生物学研究。我们也可以重新认识无刺蜂自然史和生态学的重要性，这将有望提高我们对无刺蜂在热带生态系统中发挥的不同作用的认识。

近年来，无刺蜂研究取得了很大进展的一个重要领域是无刺蜂系统发育学和分类学（Rasmussen and Cameron，2007，2010）。了解无刺蜂和其他具花粉筐蜂的系统发育关系（Romiguier et al., 2016; Bossert et al., 2017, 2019）将成为我们理解这些蜂类表型特征进化和表达的关键。说到这里，关于蜂种鉴定仍然存在许多不确定性，一些类群间的内部关系仍然存疑（第2章）。例如，新热带大属 *Plebeia* 属无刺蜂仍然是一个无刺蜂系统发育的难点（Rasmussen and Cameron, 2010）。读者可能会对本书中使用的一些分类学名称有异议，但是面对这个庞大而多样化的蜂类群时，分类学上的分歧是无法避免的。我希望本书能够激励新一代的蜂分类学家对无刺蜂进行分类学修订，特别是新热带区的 *Plebeia* 属和 *Trigonisca* 类群、非洲热带的麦蜂动物群及许多其他类群的分类学修订，这项工作本早该进行。我们需要更好地了解无刺蜂的生物多样性，这将大大有利于我们研究和保护无刺蜂。

读者在阅读本书时可能也会好奇我为什么经常使用西方蜜蜂（*Apis mellifera*）作为参照物，而不是与无刺蜂亲缘关系更密切的熊蜂（第1章，第2章）。这是因为尽管无刺蜂和蜜蜂在较长的时间内处于不同的进化轨迹上（第2章），无刺蜂和蜜蜂的生活方式却更为相似。无刺蜂和蜜蜂都是多年生高度社会化蜂类，它们的生存面临着类似的挑战，但有趣的是它们为了应对这些挑战采用了不同的解决方案（例如，蜂巢结构，第3章；蜂群迁徙，第4章；繁育幼虫的协作配合，第5章；劳动分工，第6章；蜂群防御，第7章；招募交流，第10章）。而我经常用蜜蜂作为参照的另一个原因是，我是一名资深的蜜蜂生物学家，因此我常常忍不住通过蜜蜂研究者的视角来研究无刺蜂（反之亦然）。事实上，我第一次接触到无刺蜂是在阿根廷布宜诺斯艾利斯大学 Walter Farina 的实验室攻读博士学位期间，那时我正在研究蜜蜂。当时我正在尝试训练我的蜜蜂去糖水喂食器上采集，而 *Tetragonisca fiebrigi* 这种好斗的无刺蜂的小守卫蜂攻击了我的蜜蜂，因此，那时的我自然不会对无刺蜂产生太多亲近之感。

而当我在英国萨塞克斯大学（University of Sussex）的实验室做博士后时，Francis Ratnieks 才真正地将我带入了无刺蜂的世界，我也深深感谢他的引导。在我们去巴西的旅行中，Francis 不断给予我知识、兴趣和灵感。这些旅行也让我遇到很多有趣的人，达成了许多奇妙的合作。我也想对 Cristiano Menezes、Denise Alves 和 Ayrton Vollet-Neto 表示深深的感谢，感谢他们分享给我很多关于无刺蜂的知识，感谢他们友情相助，也感谢他们对本书部分章节的改进。Cristiano Menezes 还慷慨地为本书提供了许多照片素材。此外，我还要感谢圣保罗大学的 Fabio Nascimento，我在他的实验室做博士后研究时以及后来的访问中，如果没有他的支持，我对无刺蜂的大部分研究都不可能完成。同样，这些年来 Sidnei Mateus 也给予我极大的帮助，他热情地分享了他关于无刺蜂及其习性的知识。我还要感谢 Eduardo Almeida，在他的专业帮助下我解决了许多关于无刺蜂系统发育和分类学的棘手问题，他还为本书的某些部分提供了重要的反馈意见。感谢 Lucas von Zuben、Jamille Veiga、Ricardo Oliveira、Luana Santos、Túlio Nunes 和 Vera Imperatriz-Fonseca 在我们的合作项目中慷慨热情地分享了许多他们的专业知识。特别值得一提的是，感谢 Jamille Veiga 为收集表 1.3 中总结的蜂群规模大小数据做了大量的工作。感谢 Michael Hrncir 和 Robbie I'Anson Price 认真地阅读本书并对其中几个章节进行了改进。我也非常感谢在巴西之行中 Robbie I'Anson Price、Tianfei Peng 和 Simone Glaser 与我进行了多次讨论。感谢许多人善意地允许我使用他们的照片和插图，具体使用情况在文中均有说明。感谢 Nadja Stadelmann 非常专业地为本书绘制了图 1.4 和图 1.9。

我还要感谢 Tomer Czaczkes、Jelle van Zweden、Sam Jones 和 Patrícia Nunes-Silva，感谢他们在 São Simão 的 Fazenda Aretuzina（Paulo Nogueira Neto 的农场）时和我一起并肩作战。而农场主 Paulo 博士是无刺蜂研究的先驱，也是巴西久负盛名的环保主义者，他的热情好客也使我们的研究成为可能。在其农场开展工作让我得以与他讨论无刺蜂，他也一直鼓励我不断前行。最后，我要感谢我最重要的同事和合作者 Francisca Segers。她的建设性意见大大改进了本书的几个章节。

我们能够进行无刺蜂研究要感谢多个组织提供的经费支持。我在巴西圣保罗大学进行的研究得到了圣保罗州立研究基金（*Fundação de Amparo à Pesquisa do Estado de São Paulo*，FAPESP）、瑞士国家科学基金（*Swiss National Science Foundation*，SNSF）、Ethologische Gesellschaft 和德国研究基金（*German Research Foundation*，DFG）的资助。特别感谢由国家科技发展委员会（*Conselho Nacional de Desenvolvimento Científico e Tecnológico*，CNPq）和 Coordenação de Aperfeiçoamento de Pessoal de Nível Superior（Capes）资助的 Ciência sem fronteiras（"科学无国界"）博士后奖学金，在它的支持下我在圣保罗大学的 Fabio Nascimento 实验室度过一段漫长的时光。这使我能够更多地了解无刺蜂，使我不断加强现有的合作并发展新的合作契机。这项奖学金也为本书的出版奠定了基础。

Christoph Grüter
克里斯托弗·格鲁特
2020 年 6 月于英国布里斯托

参考文献

Bossert S, Murray EA, Almeida EAB, Brady SG, Blaimer BB, Danforth BN（2019）Combining transcriptomes and ultraconserved elements to illuminate the phylogeny of Apidae. Molecular Phylogenetics and Evolution 130:121–131

Bossert S, Murray EA, Blaimer BB, Danforth BN（2017）The impact of GC bias on phylogenetic accuracy using targeted enrichment phylogenomic data. Molecular Phylogenetics and Evolution 111:149–157

Freitas FCP, Lourenço AP, Nunes FMF et al.（2020）The nuclear and mitochondrial genomes of *Frieseomelitta varia* – a highly eusocial stingless bee（Meliponini）with a permanently sterile worker caste. BMC Genomics 21:386

Heard T（2016）The Australian Native Bee Book: keeping stingless bee hives for pets, pollination and sugarbag honey. Sugarbag Bees, Brisbane, Australia

Hrncir M, Jarau S, Barth FG（2016）Stingless bees（Meliponini）: senses and behavior. Journal of Comparative Physiology A 202:597–601

Kapheim KM, Pan H, Li C, Salzberg SL et al.（2015）Genomic signatures of evolutionary transitions from solitary to group living. Science 348:1139–1143

Michener CD（1974）The Social Behavior of the Bees. Harvard University Press, Cambridge

Quezada-Euán JJG（2018）Stingless Bees of Mexico: The Biology, Management and Conservation of an Ancient Heritage. Springer, Cham

Quezada-Euán JJG, Alves DA（2020）Meliponiculture. In: Starr C（ed）Encyclopedia of Social Insects. Springer International Publishing, Cham, pp 1–6

Rasmussen C, Cameron SA（2007）A molecular phylogeny of the Old World stingless bees（Hymenoptera: Apidae: Meliponini）and the non-monophyly of the large genus *Trigona*. Systematic Entomology 32:26–39

Rasmussen C, Cameron S（2010）Global stingless bee phylogeny supports ancient divergence, vicariance, and long distance dispersal. Biological Journal of the Linnean Society 99:206–232

Romiguier J, Cameron SA, Woodard SH, Fischman BJ, Keller L, Praz CJ (2016) Phylogenomics Controlling for Base Compositional Bias Reveals a Single Origin of Eusociality in Corbiculate Bees. Molecular Biology and Evolution 33:670–678

Roubik DW (1989) Ecology and Natural History of Tropical Bees. Cambridge University Press, New York

Sakagami SF (1982) Stingless bees. In: Hermann HR (ed) Social Insects Ⅲ. Academic Press, New York, pp 361–423

Schwarz HF (1948) Stingless Bees (Meliponidae) of the Western Hemisphere. Bulletin of the American Museum of Natural History 90:1–546

Vit P, Pedro SR, Roubik D (2013) Pot-honey: a legacy of stingless bees. Springer, New York

Vit P, Pedro SR, Roubik D (eds) (2018) Pot-Pollen in Stingless Bee Melittology. Springer International, Cham

Wille A (1983) Biology of the stingless bees. Annual Review of Entomology 28:41–64

译者序

无刺蜂（Stingless bees）隶属于膜翅目（Hymenoptera）、蜜蜂科（Apidae）、蜜蜂亚科（Apinae）、麦蜂族（Meliponini）昆虫，在麦蜂族下已报道的无刺蜂属超过 50 个，目前，世界上已发现无刺蜂蜂种超过 500 种。它们是一类营群体生活并能酿蜜的昆虫，因其无螫针，而命名为无刺蜂。主要分布在热带和亚热带地区，如东南亚地区、非洲、澳大利亚、中美洲和南美洲等。无刺蜂体小灵活，是热带地区植物的主要授粉者，在生态环境和经济效益方面担当重要角色。我国无刺蜂主要分布在云南、海南、广西、西藏和台湾地区等热带亚热带区域。目前，我国无刺蜂相关研究仍处于初始阶段。

2016 年，本人应马来西亚常青集团执行董事蔡天佑先生的邀请，同孙丽萍研究员一同前往马来西亚砂拉越州泗务市开展蜂胶成分和生物活性的合作研究，也在那时我第一次近距离接触到神奇的无刺蜂，当地称之为"银蜂"。从那一刻起，我就对这种神奇的小生物产生了极大的兴趣。随后本团队在海南、云南、西藏等地调研发现，我国存在数量庞大的无刺蜂野生和养殖种群，我国无刺蜂和无刺蜂蜂产品同样具有极大的开发潜力和研究价值，发展我国无刺蜂产业，大有可为。

2020 年，我在查阅无刺蜂相关文献时，发现施普林格出版社刚出版的《Stingless Bees: Their Behaviour, Ecology and Evolution》一书，阅读后就被此书深深地吸引住了。从那一刻起，就想把它翻译出来，献给那些渴望深入了解无刺蜂神奇世界的我国读者，我也第一时间同此书作者英国布里斯托大学克里斯托弗·格鲁特博士进行了邮件联系，表达了我想翻译此书的意愿。格鲁特博士热情洋溢的回复，以及他所指导的博士彭天飞老师在后续翻译过程中对我的悉心帮助，既极大地丰富了本人对无刺蜂的知识，也是本书最终能够得以出版的重要保障。

此书出版颇费周折，但幸运的是本书出版得到多位领导、同事的关心与支持，在中国农业科学院蜜蜂研究所彭文君所长、吴黎明副所长、孙丽萍研究员、薛晓锋研究员和浙江大学胡福良教授、中国科学院西双版纳热带植物园汪正威副研究员、昆明理工大学郭军副教授、西藏自治区高原生物研究所达娃副研究员、西双版纳云蜂古山生物科技有限公司潘鹏先生等同仁的大力支持下，本书得以付梓问世。同时也向国内外无刺蜂有关科研机构和国际上从事无刺蜂研究的友好人士致以衷心的谢意。

在本书付梓之际，谨对本书出版和本人在蜂学研究提供资助的国家自然科学基金（批准号：32172791）、北京市科技新星计划（批准号：20220484101）、中国农业科学院"青年创新专项"（批准号：Y2022QC09）、质兰公益基金会（批准号：2022040591B）表示衷心感谢，同时也对中国农业科学技术出版社各位编辑的辛勤工作表示诚挚的感谢！

最后还要感谢我的妻子，也是我最重要的科研战友——金晓露博士（中国农业大学动物科学学院），逐页反复核对译文，往往因一字而反复推敲、斟酌并眷正译文，直到该译著既清晰流畅易懂，又符合科学严谨性。

大力推广开发无刺蜂产业，将有助于推动蜂业科技创新与乡村振兴战略深度融合，同时亦可为蜂业科技助力精准扶贫提供一种新途径、新思路。我相信，本书的出版不仅可以让读者更好地了解神奇的无刺蜂世界，而且可以使我国蜂业同仁能借鉴国外无刺蜂研究中的一些宝贵经验，最终推动我国蜂业的提质增效。

由于译者的能力和水平有限，加之时间紧张，成书过程仓促，书中可能会有一些翻译不准确的地方，本人的联系邮箱是：kaiwang628@gmail.com，敬请各位读者给予批评指正！

王凯　博士　副研究员

中国农业科学院蜜蜂研究所

2022.12.15 于北京香山

目 录

1 无刺蜂：概述 ··· 1
    1.1 蜂王 ·· 5
    1.2 工蜂 ·· 8
    1.3 雄蜂 ·· 11
    1.4 无刺蜂蜂产品 ·· 12
        1.4.1 蜂蜜 ·· 12
        1.4.2 巢质和巢脂质 ··· 15
    1.5 无刺蜂与蜜蜂之间的异同 ··· 16
    1.6 蜂群规模 ·· 18
    1.7 蜂群寿命 ·· 26
    1.8 无刺蜂对于人类的重要性 ··· 27
        1.8.1 无刺蜂养殖 ·· 28
        1.8.2 无刺蜂蜂产品的药用价值 ···································· 30
        1.8.3 无刺蜂的精神和宗教重要性 ································· 30
        1.8.4 无刺蜂蜂产品的其他用途 ···································· 31
    1.9 人类活动对无刺蜂的新挑战 ·· 31
        1.9.1 杀虫剂 ·· 33
    参考文献 ·· 34

2 无刺蜂的进化和物种多样性 ·· 55
    2.1 当今无刺蜂的多样性和分布 ·· 57
        2.1.1 当今有多少种无刺蜂？ ······································· 57
        2.1.2 哪里可以找到无刺蜂？ ······································· 59
    2.2 无刺蜂的起源 ·· 60
    2.3 无刺蜂的生物地理学 ··· 64
        2.3.1 无刺蜂生物地理分布的相关设想 ···························· 65
    2.4 无刺蜂的种属清单 ··· 72

2

2.4.1　新热带区无刺蜂蜂种 ······ 72
2.4.2　非洲热带区无刺蜂蜂种 ······ 92
2.4.3　印尼-马来半岛生物地理区域和澳大拉西亚区的无刺蜂蜂种 ······ 95

参考文献 ······ 100

3　筑巢生物学 ······ 113

3.1　筑巢地点 ······ 113
    3.1.1　地上筑巢型无刺蜂 ······ 115
    3.1.2　地下筑巢型无刺蜂 ······ 118
    3.1.3　在活蚂蚁或白蚁巢穴中筑巢 ······ 119
    3.1.4　暴露型巢穴 ······ 123
    3.1.5　筑巢习性的种内差异 ······ 124

3.2　无刺蜂蜂巢巢体建筑结构、建筑材料及无刺蜂的建筑行为 ······ 124
    3.2.1　入口 ······ 126
    3.2.2　产卵育虫区 ······ 129
    3.2.3　食物储存区 ······ 137
    3.2.4　监禁室 ······ 138

3.3　无刺蜂对蜂巢内部气候条件控制 ······ 139
    3.3.1　主动气候控制 ······ 140
    3.3.2　被动气候控制 ······ 142

3.4　巢穴结构的演变 ······ 142

3.5　无刺蜂群落及其与其他生物体的联系 ······ 143
    3.5.1　蜂群群落与蜂群密度 ······ 143
    3.5.2　无刺蜂蜂群间的紧密联系 ······ 146
    3.5.3　寄食昆虫 ······ 147
    3.5.4　微生物 ······ 149

参考文献 ······ 150

4　无刺蜂分蜂及婚飞 ······ 169

4.1　无刺蜂分蜂 ······ 169
    4.1.1　分蜂阶段 ······ 171

- 4.2 分蜂距离 ·· 181
- 4.3 分蜂时期 ·· 182
- 4.4 换王 ·· 183
- 4.5 蜂王的产生和选择 ·································· 184
- 4.6 无刺蜂中的单雄交配 ································ 188
- 4.7 无刺蜂的一夫一妻制 ································ 192
- 参考文献 ·· 193

5 无刺蜂幼蜂繁育 ·· 206
- 5.1 食物供给和产卵过程 ································ 206
  - 5.1.1 幼虫食物供给前阶段 ··························· 207
  - 5.1.2 幼虫食物供给和产卵阶段 ······················· 207
- 5.2 工蜂产卵 ·· 213
- 5.3 幼蜂发育阶段 ······································ 220
- 5.4 幼蜂的繁育 ·· 221
  - 5.4.1 工蜂幼蜂的繁育 ······························· 221
  - 5.4.2 雄蜂幼蜂的繁育 ······························· 224
  - 5.4.3 蜂王幼蜂的繁育 ······························· 228
- 5.5 无刺蜂的蜂王级型确立机制 ·························· 229
  - 5.5.1 营养确立蜂王级型 ····························· 229
  - 5.5.2 遗传确立蜂王级型 ····························· 232
  - 5.5.3 非 *Melipona* 属无刺蜂的微小型蜂王 ············ 234
- 5.6 无刺蜂的繁殖矛盾 ·································· 236
  - 5.6.1 繁育雄蜂的矛盾冲突 ··························· 237
  - 5.6.2 雌蜂级型分化的矛盾冲突 ······················· 240
- 参考文献 ·· 241

6 无刺蜂蜂群的劳动分工 ···································· 261
- 6.1 无刺蜂的时间阶段性职级分工 ························ 262
- 6.2 无刺蜂主要工作任务的顺序和工作内容 ················ 263
- 6.3 无刺蜂的生理性次级分工 ···························· 275
- 6.4 工蜂工作活动的分布情况 ···························· 281

6.5　工作的无刺蜂雄蜂和蜂王 ··································· 283
参考文献 ··· 285

7　无刺蜂的天敌、威胁及蜂群防御 ··································· 300
7.1　无刺蜂的天敌 ··· 302
7.1.1　小型天敌 ··· 302
7.1.2　大型天敌 ··· 305
7.1.3　疾病 ··· 307
7.2　抢劫蜂和盗窃蜂 ··· 309
7.2.1　无刺蜂攻击的组织安排 ······································· 312
7.2.2　影响无刺蜂蜂群易受袭击的因素 ······························· 314
7.2.3　防御反应的不同 ··· 316
7.3　无刺蜂的防御策略 ··· 317
7.3.1　避免被攻击的策略 ··· 317
7.3.2　蜂巢入口守卫策略 ··· 318
7.3.3　警报信息素 ··· 320
7.3.4　无刺蜂蜂巢的入口封锁和树脂的使用 ··························· 322
7.3.5　同巢与非同巢蜂的识别 ······································· 324
7.3.6　形态适应：大型守卫蜂和兵蜂 ································· 328
7.3.7　建筑防御 ··· 330
参考文献 ··· 331

8　无刺蜂的采集 ··· 353
8.1　无刺蜂采集的物质 ··· 353
8.1.1　蛋白质 ··· 353
8.1.2　碳水化合物 ··· 357
8.1.3　树脂和其他黏性植物材料 ····································· 364
8.1.4　其他资源 ··· 365
8.2　无刺蜂采集的资源时空分布 ······································· 366
8.3　无刺蜂采集的专业化 ··· 367
8.4　无刺蜂的采集活动力 ··· 368
8.4.1　季节性采集活动 ··· 368

        8.4.2 日常活动 ·· 369
    8.5 无刺蜂的采集范围和采集旅程时长 ····················· 373
    8.6 无刺蜂的资源竞争和资源分配 ·························· 377
        8.6.1 无刺蜂之间的竞争 ·································· 377
        8.6.2 无刺蜂与蜜蜂之间的竞争 ························· 382
        8.6.3 无刺蜂与其他动物之间的竞争 ··················· 384
    8.7 无刺蜂采集过程中的学习 ································· 385
        8.7.1 关联学习 ·· 386
        8.7.2 无刺蜂对采集地点的忠诚性 ······················ 389
        8.7.3 无刺蜂的访花恒定性 ································ 390
    参考文献 ·· 391

9 无刺蜂授粉 ·· 419
    9.1 作物授粉 ·· 426
    参考文献 ·· 433

10 无刺蜂采集活动中的招募和交流 ························· 442
    10.1 基于蜂巢的招募交流 ····································· 444
        10.1.1 "之"字形或推挤式前进 ························· 444
        10.1.2 胸部振动 ·· 446
        10.1.3 食物气味的社会性学习 ·························· 449
        10.1.4 有采集经验的采集蜂的重新激活 ············· 450
    10.2 位置特异性招募 ··· 452
        10.2.1 信息素轨迹 ··· 456
        10.2.2 无刺蜂在食物源上留下的化学标记 ·········· 461
        10.2.3 无刺蜂的引导飞行（引路）行为 ············· 464
    10.3 无刺蜂的"窃听"行为 ··································· 465
    10.4 局部增强学习和刺激增强学习 ························ 467
    10.5 招募交流的进化 ··· 468
    参考文献 ·· 469

1 无刺蜂：概述

热带和亚热带有成千上万种不同类型的蜂。其中无刺蜂（Stingless bees, Meliponini 麦蜂族）经常引起观察者注意。在新热带植物区花朵上看到的蜂中约有一半就属于无刺蜂类（第9章）。有些无刺蜂比果蝇还小，有些却像巨型蜜蜂——大蜜蜂（*Apis dorsata*）一样大。同蜜蜂族（Apini）类似，无刺蜂也是群体生活并能酿蜜。它们是数千种植物的潜在传粉媒介，并在人类文化中发挥了重要作用。在当今时代，无刺蜂同其他野生昆虫一样也面临着许多新的挑战，特别是栖息地大规模丧失、农用化学品的广泛使用、气候变化和外来物种入侵等，这些挑战都给无刺蜂种群的生存带来了巨大压力（第1.9节）。

同人们熟知的熊蜂（Bombini 熊蜂族）、蜜蜂（Apini 蜜蜂族）、兰蜂（Euglossini 兰蜂族）一样，无刺蜂也属于具花粉筐蜂（膜翅目，蜜蜂科）。这4种蜂的共同特点是都会习惯于将花粉放在后足上的"花粉筐"中（图1.1）。无刺蜂类群涵盖了几十个属约550个种，所含物种数量多于其他三类蜂（图1.2；第2章）（Eardley，2004; Rasmussen and Cameron，2010; Camargo and Pedro，2013; Rasmussen et al., 2017）。由于无刺蜂没有螫针，且翅脉结构较为简单，因此可以很容易将其同其他蜂区分开来（图1.3）。无刺蜂翅脉退化的原因可能是无刺蜂祖先的体型较小（Melo，2020）。无刺蜂的其他典型形态学特征是具有"刚簇"，即后足胫节根部坚硬且弯曲的刚毛簇（图1.1），而且无刺蜂后足基跗节外角没有花粉廓（压实花粉）结构，它是后基跗节底部外角的一种小突起结构（Michener,2007, 图102-2; 更多细节参见 Quezada-Euán 2018年对无刺蜂的形态学描述）。

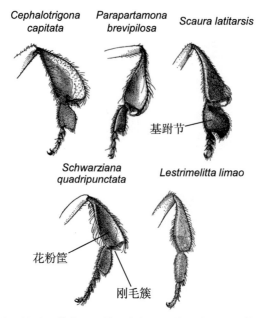

图 1.1　无刺蜂工蜂后胫节变宽，其上有长且厚的毛或刚毛（花粉篮或花粉筐）

注：*Scaura latitarsus* 无刺蜂基跗节具有囊状结构［该蜂种名不确定。该蜂种目前被认为是 *Scaura amazonica* 或 *S. aspera*（Nogueira et al., 2019）］（也见图 8.1 中的 *S. longula*），而 *Lestrimelitta limao* 无刺蜂的花粉筐退化且无刚毛簇（原图来自 Schwarz, 1948）。

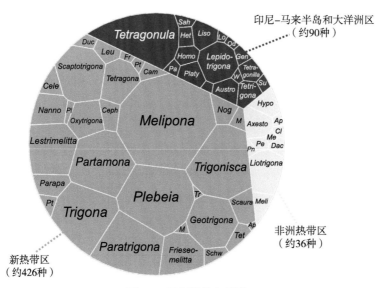

图 1.2　无刺蜂的多样性

注：每个属区块的大小与其物种数量成正比（关于每个属的物种数量和名称的详细信息见第 2 章）。

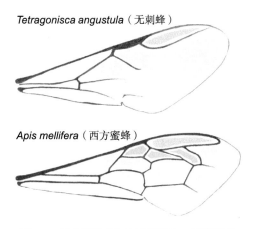

图 1.3　无刺蜂与西方蜜蜂相比，翅脉退化

注：无刺蜂的边缘翅脉室（橙色）通常是开放的，而且其翅脉边缘下翅室难以辨认或根本看不见（图中西方蜜蜂的翅脉边缘下翅室标记为绿色，从左到右依次为边缘下细胞 1～3）。更多无刺蜂翅脉情况参见 Michener（2007，图 120-1）。

大部分无刺蜂蜂种都是在新热带植物地区（77%）被发现的，其次是印尼-马来半岛/澳大拉西亚区（16%）和非洲地区（7%）。无刺蜂在热带生态系统中发挥了重要作用，然而令人惊讶的是同其他真社会性的蜜蜂、熊蜂相比，我们对大多数无刺蜂物种知之甚少。例如，尽管无刺蜂在所有真社会性蜂种中占 70% 左右，但关于社会性蜂的科学文献中只有约 6% 涉及无刺蜂[1,2]（Hrncir et al., 2016）。

无刺蜂是所有社会性蜂中最古老的类群。它们出现在公元前 8 700 万—公元前 7 000 万年的白垩纪晚期，它们甚至与恐龙共存生活了数百万年（第 2 章）。在那个时候无刺蜂的祖先就已经没有了功能性的螯刺。但是，"无刺蜂"这个名字存在一些歧义，因为无刺蜂的尾部残存了一些螯针的残肢（图 1.4）（von Ihering, 1886; Schwarz, 1948; Kerr and de Lello, 1962; Michener, 2007）。麦蜂族无刺蜂像许多蚂蚁一样失去了看似强大的武器，这十分令人困惑，最可能的解释是无刺蜂更适合采用其他防御手段来对付它们的敌人（第 7 章）（Kerr and de Lello, 1962; Sakagami, 1971）。早期无刺蜂的体型较小

---

［1］根据 2020 年 1 月在 Web of Science 的搜索结果所示，约 78% 的文献是研究蜜蜂的，约 16% 是研究熊蜂的。检索词为：① Apini/honeybees/honey bees；② Bombini/bumblebees/bumble bees；③ Meliponini/stingless bees/stingless bee。

［2］无刺蜂研究简史见 Quintal 和 Roubik（2013）。

可能是其防御手段发生变化的原因之一，因为小型蜂的刺击对阻止其敌人入侵作用并不大（Melo，2020）。无刺蜂采用的替代螯针的防御手段就包括强大的上颚、腐蚀性分泌物、难闻的气味或应用黏性物质固定敌人等。相关报道指出多种无刺蜂（*Lestrimelitta, Melipona, Oxytrigona and Tetragonisca*）会偶尔攻击并消灭蜜蜂，在其攻击过程中，上述这些防御性手段显得十分有效（Nogueira-Neto, 1970; Sakagami, 1971; Grüter et al.,2016）。

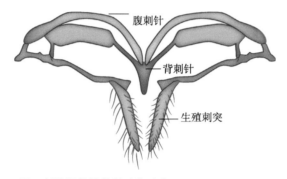

图 1.4 *Melipona* 属无刺蜂退化的螯针（修改自 Michener,2007;Quezada-Euán,2018）

注：不同的无刺蜂蜂种的腹刺针（也称为第一产卵器）和背刺针（第二产卵器）都有不同程度的退化（von Ihering,1886；参见 Michener,2007，图 120–6）。而西方蜜蜂的腹刺针和背刺针结合在一起形成一个功能性螯刺针（摄影：Nadja Stadelmann）。

无刺蜂许多性状与蜜蜂相同。它们都能生活在长年蜂群中（第 1.7 节），它们都有明显的蜂王和工蜂的级型分化（图 1.5），它们都会建造精细的蜂巢来储存大量食物，并且它们都会使用复杂且交互的通信系统来协调日常活动。同蜜蜂一样，无刺蜂蜂王也不能独自建立新蜂群（第 4 章，第 5 章，第 10 章）[有关无刺蜂生物学的早期综述，请参见 Schwarz（1948）；Michener（1974）；Sakagami（1982）；Wille（1983）；Roubik（1989）；Heard（2016）；Quezada-Euán（2018）]。但是，无刺蜂与蜜蜂也有很大的不同，这些不同有些甚至令人惊讶（第 1.5 节）（Sakagami, 1971; Michener, 1974），这意味着无刺蜂有别于蜜蜂进化出了一种高度真社会化生活方式（第 2 章）。

图 1.5 *Trigonisca mepecheu* 无刺蜂的蜂王和工蜂

注：*Trigonisca mepecheu* 无刺蜂是体型最小的新热带无刺蜂之一（改绘自 Engel et al., 2019）。

1.1 蜂王

蜂王司职产卵，它们的体型和外貌同工蜂存在显著不同（图1.5）。无刺蜂蜂王体重比工蜂更重（为工蜂的2～6倍），身长更长，其中 *Melipona* 属无刺蜂是个例外，该属无刺蜂蜂王和工蜂是用相同量的食物喂养而成的，因此二者大小十分相近（图1.5，图1.6）（Tóth et al., 2004; Grüter et al., 2017; Luna-Lucena et al., 2019）。无刺蜂蜂王在交配之后腹部膨大程度远超蜜蜂蜂王，即使是 *Melipona* 属无刺蜂的蜂王交配后腹部也会膨大，人们可以通过腹部膨大特征轻易地将无刺蜂蜂王与工蜂区分开来（图4.1）。与无刺蜂工蜂相比，蜂王的头部和眼睛相对更小（图1.7）、翅膀更短，但通常蜂王的胸廓更宽，且具有更长的颚眼距和更长的触角（Schwarz, 1948; Sakagami, 1982）[3]。

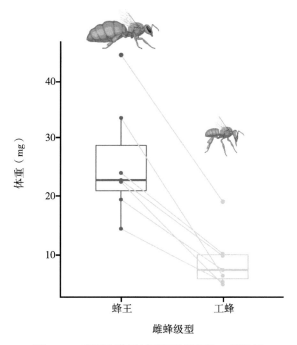

图1.6　7种新热带无刺蜂蜂种蜂王和工蜂体重

注：*Frieseomelitta varia*（Baptistella et al., 2012），*Nannotrigona testaceicornis*（Ribeiro et al., 2006; Grüter et al., 2017），*Plebeia pugnax* 和 *P. remota*（Ribeiro et al., 2006），*Scaptotrigona postica*（Velthuis, 1976），*Schwarziana quadripunctata*（Ribeiro et al., 2006），*Tetragonisca angustula*（Kerr et al.,1962; Segers et al., 2015）（摄影：Nadja Stadelmann）。

---

[3] 关于无刺蜂不同属的形态学详尽描述参见第2章第4节的列表。

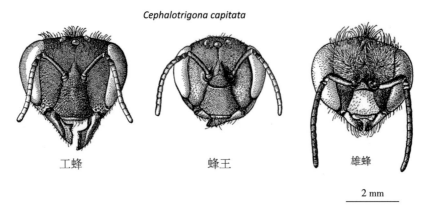

图 1.7 *Cephalotrigona capitata* 无刺蜂的工蜂、蜂王和雄蜂的头部

[源自 Schwarz（1948）的原图]

注：蜂王的眼睛小于工蜂和雄蜂。雄蜂的触角更长但鞭节更短。

无刺蜂蜂群通常孤雌繁育，即只有一只产卵蜂王，但 *Melipona bicolor* 无刺蜂是目前已知的唯一例外，因为 *Melipona bicolor* 蜂群通常是多雌繁育的（第 5 章）。不同无刺蜂蜂种的蜂王每天产卵数也有所不同，产卵数量从数十个（例如 *Plebeia juliani*）到数百个（例如 *Trigona recursa*）不等（表 5.1）。无刺蜂每个卵巢中的卵巢管数量远低于蜜蜂蜂王（100～200），大多数无刺蜂蜂种的蜂王每个卵巢中的卵巢管数为 4～8 个（Camargo, 1974; Cruz-Landim, 2000; Luna-Lucena et al., 2019），但一些 *Trigona* 属（如 *T. spinipes*）和 *Lestrimelitta* 属无刺蜂蜂王每个卵巢中的卵巢管数可高达 10～15 个（Sakagami, 1982; Cruz-Landim, 2000）。但是，蜂王生产力的提高与卵巢的延长有关，与卵巢管数量的增加无关（图 1.8）（Velthuis, 1976; Sakagami, 1982）。有研究指出，无刺蜂蜂王卵巢管数量与蜂群规模存在相关性，但这需要进一步研究（Sakagami, 1982）。无刺蜂蜂王同工蜂一样没有螯针，但其残留的螯针结构似乎比工蜂的更为明显（Schwarz, 1948; Sakagami, 1982）。

无刺蜂蜂王没有花粉筐结构，此外据 Sakagami（1982）报道，一些无刺蜂蜂种的蜂王也没有蜡腺。但有人已经观察到新热带区的一些无刺蜂属的处女蜂王存在泌蜡现象（第 7 章）。即使我们没有关于无刺蜂蜂王寿命的数据，但现有信息表明无刺蜂蜂王的寿命比工蜂要长得多，通常为 1～3 年，最高可达 7 年（表 1.1）（Kerr et al., 1962; da Silva et al., 1972; Darchen, 1977; Imperatriz-Fonseca, 1978; Imperatriz-Fonseca and Zucchi, 1995; van Veen et al., 2004; Ribeiro et al., 2006）。

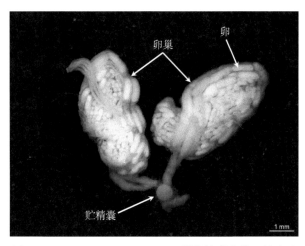

图 1.8 *Tetragonisca angustula* 无刺蜂的膨腹蜂王的生殖系统（修改自 Santos，2012）

表 1.1 无刺蜂工蜂、蜂王和雄蜂的平均寿命和最长寿命

无刺蜂蜂种	平均寿命	最长寿命	参考文献
工蜂	天数（d）		
Austroplebeia australis		240	Halcroft et al.（2013 a）
Friesellaschrottkyi	30.1		Giannini（1997）
Frieseomelitta languida	33.3		Giannini（1997）
Frieseomelitta varia	42	82	Cardoso（2010）
Meliplebeia beccarii	52.7	73	Njoya and Wittmann（2013）
Melipona beecheii	51	101	Biesmeijer and Tóth（1998）
Melipona bicolor	44	68	Giannini（1997）；Bego（1983）
Melipona eburnea	36～43	66	Bustamante（2006）
Melipona fasciculata	42.5	80	Giannini（1997）；Gomes et al.（2015）
Melipona favosa	40.5	108	Sommeijer（1984）；Roubik（1982）
Melipona fulva	34	84	Roubik（1982）
Melipona lateralis	35～54	73	Bustamante（2006）
Melipona marginata	41.1	70	Mateus et al.（2019）

续表

无刺蜂蜂种	平均寿命	最长寿命	参考文献
Melipona seminigra	30	54	Bustamante（2006）
Melipona scutellaris	43.8	75	Santos（2013）
Plebeia droryana	41.7	75	Terada et al.（1975）
Plebeia emerina		63	dos Santos et al.（2010）
Plebeia remota	67.7	96	van Benthem et al.（1995）；Grosso and Bego（2002）
Scaptotrigona postica	33.4～39.5	60	Simões and Bego（1979）
Scaptotrigona xanthotricha	94	97	Hebling et al.（1964）
Tetragonisca angustula	24	56	Grosso and Bego（2002）；Hammel et al.（2016）
Tetragonula minangkabau	37	60	Inoue et al.（1996）
Trigona pallens	29.4		Cardoso（2010）
Trigona pellucida	41.5	79	Cardoso（2010）
Tetragonula laeviceps	40～50		Inoue et al.（1984）
雄蜂	天数（d）		
Melipona eburnea	16	18	Bustamante（2006）
Melipona seminigra	21	23	Bustamante（2006）
蜂王	年		
Hypotrigona sp.		>4	Darchen（1977）
Melipona beecheii	>3		van Veen et al.（2004）
Melipona bicolor		7	Imperatriz-Fonseca and Zucchi（1995）
Melipona favosa	>3		Sommeijer et al.（2003）
Melipona quadrifasciata		～3	da Silva et al.（1972）
Paratrigona subnuda	3		Imperatriz-Fonseca（1978）
Plebeia remota	～1.5	>4	Ribeiro et al.（2006）

1.2 工蜂

在无刺蜂群中绝大多数都是工蜂，顾名思义，它们负责执行蜂群中大部分任务。具体包括筑巢（第3章，第4章）、哺育幼虫（第5章）、防御（第7章）和采集（第8章至第10章）。工蜂的形态特征也反映了它们执行的不同任务的需要（图1.9）。例如，蜂王大部分时间都在黑暗巢中生活，工蜂则需要外出从事采集工作，因而工蜂的眼睛相对蜂王要大一些（图1.7）。多种无刺蜂的工蜂都有强壮的上颚，可以有效地进行防御（Shackleton et al., 2015）。与蜜蜂不同，无刺蜂蜂蜡是从位于工蜂腹顶部的蜡腺向背侧分泌的（图3.7）（Müller，1874）。

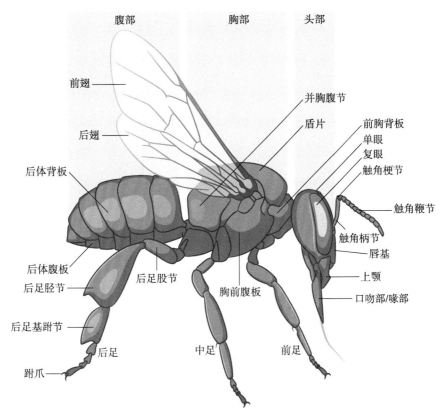

图1.9 无刺蜂工蜂的形态 [修改自 Camargo 和 Posey（1990），插图由 Nadja Stadelmann 绘制]

无刺蜂与许多其他社会性昆虫类群相比，其蜂群中即便存在蜂王，蜂群中的工蜂也有高度发育的卵巢（其内通常有4个卵巢管），而且多种无刺蜂蜂群中的雄蜂大多是工蜂的后代（第5章）（Cruz-Landim，2000; Tóth et al., 2004; Grüter，2018; Luna-Lucena et al.,2019）。无刺蜂的工蜂可以通过产下未受精的卵来繁殖雄蜂。这种情况的存在是合理的，因为雄蜂是单倍体，这意味着雄蜂只有一套染色体[4]。无刺蜂工蜂本身是二倍体，由于不能交配，所以它们不能生产二倍体雌性卵[5]。这种性别决定系统被称为单倍二倍性（haplodiploidy），所有膜翅目昆虫（蚂蚁、蜜蜂、胡蜂和切叶蜂）都使用这种性别决定系统。单倍二倍性造就了膜翅目昆虫特殊的家族群体和相互关联的结构，例如雄性个体没有父亲，工蜂/工蚁与子侄间的关系比与兄弟之间的亲缘关系更为紧密，这也对整个群体产生了深远的影响（第4章，第5章）。

无刺蜂工蜂利用后足胫节上的花粉筐来搬运花粉、树脂、蜂蜡和泥土等物质（图1.1，图8.1）。无刺蜂花粉筐较宽，凹陷处覆盖着长而浓密的绒毛或刚毛。其前缘和后缘都有长刚毛，使花粉附着在腿上。不同无刺蜂蜂种的花粉筐形状有很大的不同。盗食寄生性 *Lestrimelitta* 属无刺蜂的花粉筐退化且无刚簇，而 *Scaura* 属无刺蜂的花粉筐还额外具有囊样结构（图1.1，图8.1）。这种结构的功能可能是帮助采集蜂收集掉落到地上的花粉（第8章）（Roubik，2018）。

各种无刺蜂在体型、颜色、体毛、外形上具有很大差异。已知的最小的无刺蜂蜂种可能是源自马达加斯加岛上的 *Liotrigona bitika* 无刺蜂（它也是世界最小蜂的有力角逐者[6]，它的体型甚至比果蝇还小（Koch，2010），而体型最大的无刺蜂种是 *Melipona fuliginosa* 无刺蜂（单蜂体重约为125 mg），其体型与大蜜蜂（*Apis dorsata*）体型类似（Roubik，1989；Camargo and Pedro，2008）。无刺蜂工蜂平均寿命最短低至3周（数据来源于 *Tetragonisca angustula*, Grosso and Bego，2002；Hammel et al., 2016），最长达10周（数据来源于 *Plebeia remota*, van Benthem et al., 1995）（表1.1）。目前发现 *Austroplebeia australis* 无刺蜂的工蜂是研究者记录到的无刺蜂工蜂最长寿命保持者，这种无刺蜂在其澳大利亚栖息地的花朵短缺期间最多可以存活270 d（Halcroft et al., 2013 a）。随着工蜂日龄的增长以及可供使用的食物源的增多，

---

[4] 无刺蜂的染色体数量表现出相当大的变异性，数量从8到18不等（在工蜂和蜂王中是成对的，$2n=16\sim36$）（Tavares et al., 2017；Travenzoli et al., 2019）。

[5] 在非洲有一种特殊的蜜蜂叫海角蜜蜂 *Apis mellifera capensis*，它已经成功攻克了这个难题。这种蜜蜂亚种的工蜂能够用一种名叫孤雌生殖的方式无性繁殖其他雌性蜂（Goudie and Old royd，2014）。

[6] 该蜂种工蜂的平均体长为1.97 mm（Koch，2010），与 *Perdita minima* 和 *Quasihesma clypearis* 的雌性蜂（Exley，1980）相比，体长相似甚至略小，而 *Perdita minima* 和 *Quasihesma clypearis* 也是已知"最小蜂"称号的有力竞争者。

工蜂的采集活动也随之增加，相应的工蜂死亡率也随之上升（图 6.4）。例如，*Melipona fasciculata* 无刺蜂工蜂在雨季晚期才开始采集活动，因此它们的寿命比干旱季节的工蜂寿命长 50% 以上（Gomes et al., 2015）。对 *M. favosa* 和 *M. fulva* 无刺蜂而言，当旱季的采集环境也相对较好时，雨季工蜂的寿命大约是旱季工蜂寿命的 2 倍（Roubik，1982）。研究者观察发现，在蜂群失王后蜂群中工蜂的寿命也会变得更长，这种现象的产生也可以用无刺蜂工蜂失王后采集活动减少来解释（Lopes et al., 2020）。*Plebeia remota* 无刺蜂主要生活在亚热带地区且会季节性繁育幼虫，这种无刺蜂工蜂的寿命就与季节相关，van Benthem 等（1995）观察到该蜂种工蜂冬季的寿命比夏季长 25%～100%（Terada et al., 1975）。对于西方蜜蜂（*Apis mellifera*）而言，冬季蜜蜂和夏季蜜蜂之间寿命的差异与血淋巴和脂肪体中信息素和卵黄蛋白原滴度有关（Bitondi and Simões，1996; Smedal et al., 2009）。卵黄蛋白原会影响蜜蜂的采集活动、先天免疫，并缓解蜜蜂的氧化应激（Amdam et al., 2004, 2005; Seehuus et al., 2006），但对无刺蜂寿命影响的生理、分子机制则仍有待研究。

无刺蜂工蜂体型大小也可能会影响其寿命长短。*Melipona fasciculata* 无刺蜂生活在巴西亚马孙地区，其蜂群中体型越大的工蜂寿命越长，特别是在干旱季节，亚马孙地区开花植物越多，无刺蜂采集活动竞争就更为激烈（Gomes et al., 2015）。相反，*Tetragonisca angustula* 无刺蜂中体型较大的工蜂寿命同体型较小的工蜂之间并没有显著差异（Hammel et al., 2016）。

1.3 雄蜂

雄蜂的主要职责是同蜂王交配（第 4 章）。雄蜂在其出生蜂巢中几乎不承担任何工作（但无刺蜂雄蜂承担的工作多于蜜蜂雄蜂，见第 6 章），它们在 2～3 周龄时就离开蜂巢且不会再回到蜂巢中（van Veen et al., 1997; Sommeijer et al., 2003）。无刺蜂雄蜂和工蜂从外表上看非常相似，只有经验丰富的观察者才能在没有放大镜等工具帮助的情况下将它们区分开。这种相似性的根本原因是无刺蜂为雄蜂和工蜂建立的育虫巢室是同一类型的。而且由于无刺蜂雄蜂和工蜂幼虫在巢室内食用的幼虫食物也是等量的，因此二者的大小和外观极为相似。无刺蜂雄蜂和工蜂间较明显的可辨别的外部特征之一是雄蜂的眼睛更大（图 1.7）且其后足胫节退化。然而也存在例外，一些亚洲无刺蜂（如 *Tetragonula fuscobalteata* 等）蜂种的后足胫节同工蜂类似（但无花粉筐）。已经有人观察发现这些雄蜂的后足会携带花粉和蜂胶（第 8 章）

（Boongird and Michener，2010）。无刺蜂雄蜂和工蜂的另一个区别是触角的外观和位置。雄性触角通常保持在特定的"V"形位置（图1.10）。无刺蜂雄蜂的触角较长，由11个鞭节组成（而工蜂和蜂王则为10个），但是无刺蜂雄蜂的鞭节长度要短于工蜂（图1.7）（Schwarz, 1948; Carvalho et al., 2017; Month-Juris et al., 2020）。

图1.10　*Melipona flavolineata* 无刺蜂雄蜂

注：无刺蜂雄蜂的眼睛比工蜂更大，且它们的触角多呈"V"形（摄影：Cristiano Menezes）。

目前关于无刺蜂成年雄蜂寿命的信息很少，但是观察表明无刺蜂雄蜂的寿命要短于工蜂。

1.4　无刺蜂蜂产品

1.4.1　蜂蜜

无刺蜂的多年生生活方式意味着蜂群需要在觅食条件恶劣的时期储存蜂蜜和花粉。位于在巴西东北部卡廷加地区热带干燥森林或澳大利亚北部半干旱地区的无刺蜂蜂群就是这一情况的最极端例子。这些栖息地每年的干旱时期长达数月，长期干旱迫使蜂群进入长时间的活动受限状态（Halcroft et al., 2013 b; Maia-Silva et al., 2015; Hrncir et al., 2019）。无刺蜂蜂蜜是由无刺蜂工蜂加工花蜜、蜜露和果汁制成的，无刺蜂将其保存在"蜜罐"中（第8章）（图1.11）。由于无刺蜂采用这种蜂蜜储存方式，人们常常称无刺蜂蜂蜜为"罐蜜"（pot-honey）（Vit et al., 2013）。不同蜂种的无刺蜂生产的蜂蜜在颜色、质地和口味上都有差异，但总的来说，无刺蜂蜂蜜比蜜蜂蜂蜜流动性更强。无刺蜂蜂蜜含水量约为30%（20%～42%），而蜜蜂蜂蜜含水量约18%[7]（Roubik, 1983; Souza et al., 2006; Bijlsma et al., 2006; Dardón et al., 2013; Ferrufino and Vit, 2013; Fuenmayor et al., 2013; Biluca et al., 2016; Nordin et al., 2018; Ávila et al., 2018）。但Nweze等（2017）报道尼日利亚地区的蜜蜂蜂蜜和无刺蜂蜂蜜样本的含水量基本一致。

---

[7] 无刺蜂蜂蜜的高含水量等特性意味着罐蜜在一些国家不符合官方定义"蜂蜜"的标准（见粮农组织/世卫组织食品标准计划的食品法典）。而制定法律标准时通常只考虑到了西方蜜蜂蜂蜜（Vit et al., 2004; Alves, 2013; Vit, 2013）。

图 1.11 巴西帕拉州的 *Duckeola ghilianii* 无刺蜂蜂巢的内部

注：图中无刺蜂蜂巢里的蜜罐比育虫巢室要大得多，通常位于蜂巢的外围。该蜂种不在育虫巢室周围建造保护层结构（"包壳"）。

蜂蜜中报道的最多的糖是果糖、葡萄糖、蔗糖和麦芽糖。但是，据 Fletcher 等（2020）的最新分析结果表明，无刺蜂蜂蜜中常被鉴定为麦芽糖的二糖实际上是海藻酮糖（trehalulose），这是一种蔗糖的异构体。这是一个有趣的发现，因为海藻酮糖对人类具有潜在的健康益处（Fletcher et al., 2020）。新热带无刺蜂蜂蜜与来自亚洲和澳大利亚的无刺蜂蜂蜜相比在麦芽糖/海藻酮糖含量上表现出显著差异。例如，来自哥伦比亚的 15 种无刺蜂蜂蜜中果糖含量为 17%～40%，葡萄糖含量为 9%～38%，而蔗糖和麦芽糖/海藻酮糖的总含量仅为 1%～13%（图 1.12）（Fuenmayor et al., 2013）。这与其他来自新热带无刺蜂蜂种的蜂蜜相近（Biluca et al., 2016）。但是在许多亚洲和澳大利亚无刺蜂蜂蜜中麦芽糖/海藻酮糖含量则更为突出。澳大利亚 *Tetragonula carbonaria* 无刺蜂生产的蜂蜜中含约 20% 麦芽糖/海藻酮糖（Oddo et al., 2008），而 *Geniotrigona thoracica*、*Heterotrigona itama* 和 *Lepidotrigona terminata* 无刺蜂蜂蜜则分别约含 35%、34% 和 53% 麦芽糖/海藻酮糖（Nordin et al., 2018; Se et al., 2018）。花蜜中发现的主要糖分为蔗糖（Chalcoff et al., 2017），而蜂蜜中的蔗糖被消化酶分解，因此蔗糖在蜂蜜中含量很低。无刺蜂蜂蜜中的总糖含量通常为 68%～73%（图 1.12）（Roubik, 1983; Souza et al., 2006）。

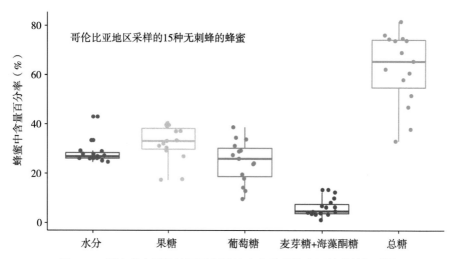

图 1.12　研究者在哥伦比亚无刺蜂蜂蜜中发现的主要糖类的相对量
（Fuenmayor et al., 2013）

注：研究者认定为麦芽糖的糖类可能是海藻酮糖（Fletcher et al., 2020）。

无刺蜂蜂蜜中还含有多种矿物质，尤其是钾。此外，无刺蜂蜂蜜中还含有氨基酸、灰分（Vit et al., 2013; Biluca et al., 2016, 2019）、黄酮和酚酸等（Vit et al., 2013; Tuksitha et al., 2018）[8]。应当注意的是，即使是相同蜂种的无刺蜂蜂蜜其物质组成（包括糖含量）在不同蜂群中也有所不同，并且无刺蜂蜂蜜的物质组成取决于采收季节、栖息地和蜜粉收集源。例如，Fuenmayor 等（2013）发现哥伦比亚的 *Tetragonisca angustula* 无刺蜂蜂蜜的糖含量不到 60%，而 Dardón 等（2013）报道发现危地马拉的 *Tetragonisca angustula* 无刺蜂蜂蜜的糖含量则达 70% 以上（Roubik, 1983; Souza et al., 2006）。

无刺蜂将花蜜、蜜露或果汁加工成蜂蜜的过程包括以下步骤：①水分的蒸发；②酵母和细菌的活性发酵；③添加工蜂头部腺体分泌的酶及其他物质（Menezes et al., 2013）。其中发酵过程产生的无刺蜂蜂蜜酸度高于蜜蜂蜂蜜（pH 值 3～5）（Dardón et al., 2013; Deliza and Vit, 2013; Vit, 2013; Heard, 2016; Nordin et al., 2018; Ávila et al., 2018）。无刺蜂蜂蜜的发酵过程可能是其产生抗氧化性和药用特性的原因之一（第 1.8 节）（Rodríguez-Malaver, 2013）。

人类通常是用注射器或抽气泵刺穿无刺蜂的密闭蜜罐来收获无刺蜂蜂蜜

---

[8] 这些黄酮和酚酸可能是从巢质罐体中泄漏到蜂蜜中的（Heard, 2016）。

的（Nogueira-Neto，1997; Souza et al., 2006; Almeida-Muradian, 2013; Vit，2013; Heard，2016; Quezada-Euán, 2018）。大多数无刺蜂蜂种的蜂蜜年产量与西方蜜蜂（*Apis mellifera*）相比相对较低，通常为 1～3 kg（Cortopassi-Laurino et al., 2006; Kumar et al., 2012; Alves，2013; Ferrufino and Vit，2013; Jaffé et al., 2015; Heard，2016）。但是有一些无刺蜂蜂种，如 *Melipona subnitida* 无刺蜂，每年可以生产 5 kg 以上的无刺蜂蜂蜜，而安哥拉人工饲养的 *Meliponula bocandei* 无刺蜂每年甚至可以收获 10～15 kg 的蜂蜜（Cortopassi-Laurino et al., 2006）。更有研究惊奇地发现 *Melipona seminigra* 或 *M. scutellaris* 无刺蜂蜂群每年可生产多达 15～25 kg 的蜂蜜（Kerr et al., 1967；另见 Alves, 2013）。有趣的是，无刺蜂蜂蜜的年产量似乎主要取决于无刺蜂体型的大小，而不是蜂群规模（Roubik，1983）。最广为人知的无刺蜂蜂蜜主要源自新热带植物区的 *Melipona*、*Scaptotrigona* 和 *Tetragonisca* 属无刺蜂和亚洲及澳洲的 *Tetragonula* 属无刺蜂（Alves，2013; Heard，2016）。由于无刺蜂的生产能力不及蜜蜂，因而无刺蜂蜂蜜的价格往往比蜜蜂蜂蜜要贵得多（Kumar et al., 2012; Alves，2013; Dardón et al., 2013; Ferrufino and Vit，2013; Fuenmayor et al., 2013）。

并非所有无刺蜂都能生产可供人食用的蜂蜜。人们认为 *Trigonisca mepecheu* 无刺蜂蜂蜜（图 1.5）就对人体有害，当地人都不食用这种蜂蜜（Engel et al., 2019）。同样的，据研究者记载"盗蜂"（*Lestrimelitta* 属）的蜂蜜和花粉也是有毒的（Schwarz，1948; Nogueira-Neto，1997）。Von Ihering（1903）报道人类在进食盗蜂蜂蜜[9]和花粉后，即使食用量很少，也会在 15 min 之内发生抽搐和呕吐，也有其他人报道发现人类食用盗蜂蜂蜜会导致暂时性麻痹（Nogueira-Neto，1997）。考虑到这种无刺蜂已分化出异常的采集觅食习性，包括专营盗食、从死牲畜中收集动物粪便和肉类等，因此人类最好避免食用此种蜂的蜂蜜（及其蜂花粉）（第 8 章）。此外，如果无刺蜂蜂蜜和花粉是源自某些植物物种，亦可能对人类有毒（如同期大量开花型的 *Luetzelburgia auriculata* 属植物，Nogueira-Neto，1997）。

1.4.2 巢质和巢脂质

无刺蜂是勤劳的建造者，它们会将各种物质（主要是蜂蜡和树脂）混

---

[9] von Ihering（1903, pp. 271-272）并没有明确这是盗蜂，但是他猜测这最可能是 *Lestrimelitta* 属蜂（另见 Nogueira-Neto，1997, pp. 289，其文中有更典型的案例）。

合在一起来建筑蜂巢（第3章）（Roubik，2006）。与蜜蜂不同的是，无刺蜂通常不使用纯蜂蜡作为建筑材料，而是将其与树脂材料混合创造出巢质（cerumen）。无刺蜂会使用巢质来建造大多数的蜂巢结构，包括育虫巢室、食物罐和包壳（involucrum）。包壳是一种保护性结构，用以包住无刺蜂蜂巢内的子区（图3.3）。这种树脂和蜂蜡的结合形成了一种更坚固的建筑材料，并且研究者认为无刺蜂所用巢质中的树脂由于其抗菌性可有效抑制蜂巢内真菌和细菌的繁殖（第3章）（Massaro et al., 2011; Çelemli, 2013）。而无刺蜂还创造出了一种更为坚固的名叫"巢脂质（batumen）"的建巢材料，无刺蜂创造巢脂质时会在巢质中混入更多的树脂、泥浆、植物材料，甚至某些蜂种会混入动物粪便（Roubik，2006）。无刺蜂主要将巢脂质覆盖在蜂巢腔的内壁，为无刺蜂蜂群创建出一个受到良好保护的嵌套空间。

1.5　无刺蜂与蜜蜂之间的异同

无刺蜂和蜜蜂具有许多相同的重要特征，但它们在许多方面也有所不同（表1.2）（综述参见 Sakagami，1971；Michener，1974）。与蜜蜂蜂王不同，无刺蜂蜂王只会与一只雄蜂交配（有个别例外，详情参阅第4章和第5章）。另一个重要的区别在于育子方式不同：蜜蜂蜂王产卵后，工蜂会不断给幼虫供应食物直到幼虫长到其最终大小，然后蜜蜂会对巢室进行封盖直到成年蜜蜂从巢室中化蛹而出；而无刺蜂工蜂则会先将全部幼虫食物反刍到一个空的幼虫巢室中，之后蜂王会在这些幼虫食物上产一枚卵，工蜂则立即将其封盖（第5章）。在幼虫食物上产卵的特征存在于无刺蜂与许多独居蜂种中（Michener，2007）。这种育子方式意味着蜂巢内成年无刺蜂不会与蜂卵和发育中的无刺蜂幼虫产生直接接触，这种情况很可能对无刺蜂生物学产生深远的影响。例如，这样会影响疾病传播的可能性，缩小了工蜂或蜂王监管的范围（即选择性地清除工蜂产下的卵），并阻止了工蜂去除二倍体雄性幼虫的行为，这也可能解释了为什么无刺蜂会进行单雄交配（第4章）。

表1.2　无刺蜂和蜜蜂的一些主要区别的总结

特征	无刺蜂	蜜蜂（*Apis*）	对应章节
物种多样性	目前记录了约550个种	约10个种	2
蜂巢建巢材料	巢质和巢脂质：蜂蜡与不同量的树脂及其他材料混合而成	巢室用的是纯蜂蜡，而蜂胶则接近开口处	3

续表

特征	无刺蜂	蜜蜂 (*Apis*)	对应章节
幼虫巢室和食物储存室	各种巢室的大小、形状和位置都有不同。工蜂和雄蜂巢室是一样的	工蜂巢室和食物储存室没有区别。在较大的巢室中养育雄蜂	3
分蜂时的离巢蜂王	处女蜂王飞往新巢,而母蜂王则待在旧巢中	母蜂王飞往新巢,而处女王则待在旧巢中	4
迁移情况	膨腹蜂王无法飞行,因而蜂群不能迁徙	交配过的蜂王能够飞行,且蜂群在应激等情况时可以迁徙到新位置	4
交配位置	多数在新建立的蜂巢处	在雄蜂聚集处	4
分蜂的持续时间	几天到数月不等	几小时到数天不等	4
交配雄蜂数	蜂王是单雄交配	蜂王与多个雄蜂交配	4
蜂王决定系统	命定的蜂王幼虫会获得更多的幼虫食物,*Melipona* 属无刺蜂中遗传因素似乎十分重要	养育蜂王需要质量和数量上都不同的食物	5
幼虫食物供应模式	莛饲型	零饲型	5
育虫巢室	开口向上且工蜂为每个卵建造新的巢室	侧开口,重复利用旧巢室来养育下一代幼虫	5
工蜂繁殖	多种无刺蜂工蜂有发育的卵巢且能生产雄蜂	工蜂卵巢未发育且在有王情况下保持不育状态	5
废弃物处理	蜂在巢内排泄,废弃物堆存在蜂巢内而后由其他蜂将其倾倒出蜂巢	工蜂在巢外排泄,内勤蜂会移除死蜂	6
防御	用上颚咬击,用树脂来控制住小型敌人,有兵蜂	大部分采用刺击,大蜜蜂会采用群体蜂浪进行视觉威慑	7
收集的资源	花蜜、花粉及各种资源	大部分是花蜜和花粉	8
招募交流	各有不同:有的独行,有的则采用化学轨迹	摆尾舞	10

　　无刺蜂和蜜蜂的蜂群内的社会分工总体表现出许多相似之处,但也有一些值得注意的差异。无刺蜂和蜜蜂的工蜂都会随日龄切换工作。年轻工蜂主要参与哺育幼虫过程和其他巢内内勤任务,而年长的工蜂则负责保卫蜂巢和

收集资源（第6章）。对于无刺蜂工蜂来说，建筑活动在其生命中占据了重要位置，远超蜜蜂工蜂。出现这种情况的原因之一是无刺蜂从不重复使用育虫巢室，而是为每个新卵从头建造新巢室。此外，无刺蜂工蜂还会用巢质和巢脂质来构建保护层，并不断进行维修来保护蜂群（第3章）。

无刺蜂和蜜蜂之间的另一个值得注意的区别是它们保卫蜂巢的方式不同。虽然无刺蜂守卫蜂没有螫针，但有几种无刺蜂的守卫蜂在行为上表现出更深度的专业化分工，即这种专业化守卫蜂可以长时间从事守卫工作，并且通常体型大于其他无刺蜂工蜂（第7章）（Grüter et al., 2011, 2017; Wittwer and Elgar, 2018; Baudier et al., 2019）。

蜜蜂会使用摆尾舞向蜂群传达可用食物源的位置和气味（von Frisch, 1967; Couvillon, 2012; I'Anson Price and Grüter, 2015）。无刺蜂在其蜂巢内也会表现出值得关注的运动方式，有人认为无刺蜂的旋转和"之"字形行进方式可能编码了食物源的位置信息，但目前缺乏令人信服的证据（第10章）（Nieh, 2004; Hrncir, 2009）。另外，一些无刺蜂蜂种会使用信息素轨迹来招募同伴准确抵达可用的食物源（Nieh, 2004; Jarau, 2009; Leonhardt, 2017）。

1.6 蜂群规模

蜂群规模是蜂类的一种基本的社会特性，在社会生活的许多方面都起着重要作用（Bourke, 1999; Dornhaus et al., 2012）。例如，蜂群规模同劳动分工（Oster and Wilson, 1978）、蜂群生殖力（Karsai and Wenzel, 1998）、繁殖冲突（Bourke, 1999）、招募的有益性和方法（第10章）（Beckers et al., 1989）以及竞争性互动（Hölldobler, 1976; Lichtenberg et al., 2010; Hrncir and Maia-Silva, 2013）都密切相关。不同蜂种的无刺蜂的蜂群规模（成年蜂的数量）差异很大，有些蜂群规模低至不到100只（*Melipona phenax*），有些则可最高达100 000只以上（*Trigona amazonensis*）（表1.3）。表1.3中所示的104种无刺蜂蜂种的平均蜂群规模约为5 200只，中位数为1 500只（图1.13）。多数建巢簇型蜂巢的无刺蜂蜂种（第3章）蜂群内的工蜂数量低于1 000只，而大多数建巢脾型蜂巢的无刺蜂蜂种的蜂群规模则大于1 000只（图1.13）。

表 1.3 无刺蜂蜂群规模估值和育虫巢室排布特征

区域	蜂种	蜂群规模范围	平均蜂群规模	育虫巢室排布	参考文献
非洲热带区	*Apotrigona nebulata*	195 ~ 2 000	1 700	巢脾型	Darchen（1969）; Kajobe（2007）
	Hypotrigona araujoi	2 000 ~ 2 500	2 250	巢脾型	Portugal-Araújo and Kerr（1959），同"landula"
	Hypotrigona gribodoi	100 ~ 750	450	巢簇型	Bassindale（1955）
	Meliponula bocandei	1 000 ~ 1 300	1 170	巢簇型	Kajobe（2007）; Njoya et al.（2018）
	Plebeina armata (=*hildebrandti*)		3 091	巢脾型/螺旋型	Namu and Wittmann（2016）
印尼-马来半岛/澳大拉西亚区	*Austroplebeia australis*		2 000	巢簇型	Hammond and Keller（2004）
	Austroplebeia cassiae		2 000	巢簇型	Hammond and Keller（2004）
	Lepidotrigona hoozana		10 000	巢脾型	Santos et al.（2014）
	Lepidotrigona ventralis	258 ~ 12 167	4 221	巢脾型	Chinh et al.（2005）
	Lisotrigona carpenteri	50 ~ 375	144	巢簇型	Chinh et al.（2005）
	Tetragonula carbonaria	2 500 ~ 11 000	6 750	螺旋型	Tóth et al.（2004）; Halcroft et al.（2013 a）
	Tetragonula hockingsi	3 000 ~ 10 000	7 000	半巢脾型	Tóth et al.（2004）
	Tetragonula iridipennis		2 500	巢簇型/半巢脾型	Schwarz（1939）
	Tetragonula laeviceps	487 ~ 2 800	747	巢簇型	Sakagami et al.（1983）; Chinh et al.（2005）
	Tetragonula mellipes	1 000 ~ 5 000	2 000	巢簇型	Tóth et al.（2004）
	Tetragonula minangkabau	300 ~ 2 600	750	巢簇型	Inoue et al.（1996）; Tóth et al.（2004）
新热带区	*Aparatrigona isopteraphila*	80 ~ 350	185	巢脾型/螺旋型	Roubik（1983）
	Cephalotrigona capitata	1 000 ~ 1 500	1 250	巢脾型	Michener（1974）
	Cephalotrigona zexmeniae	400 ~ 1 000	700	巢脾型	Roubik（1983）
	Duckeola ghilianii		10 000	半巢脾型	Tóth et al.（2004）

续表

区域	蜂种	蜂群规模范围	平均蜂群规模	育虫巢室排布	参考文献
新热带区	*Friesella schrottkyi*	300～2 500	1 400	半巢脾型	Sakagami et al.（1973）；Michener（1974）
	Frieseomelitta flavicornis		375	巢簇型	Roubik（1979），同 *T. savannensis*
	Frieseomelitta longipes	1 051～4 393	2 635	巢簇型	Leão（2019）
	Frieseomelitta nigra	400～1 500	950	巢簇型	Tóth et al.（2004）
	Frieseomelitta paupera	500～700	600	巢簇型	Sommeijer et al.（1984）
	Frieseomelitta silvestrii	400～600	500	巢簇型	Michener（1974）
	Frieseomelitta varia	800～1 600	1 200	巢簇型	Tóth et al.（2004）
	Geotrigona leucogastra	200～450	325	巢脾型	Roubik（1983）
	Geotrigona mombuca	2 000～3 000	2 500	巢脾型	Tóth et al.（2004）
	Geotrigona subterranea	2 726～11 074	7 485	巢脾型	Barbosa et al.（2013）
	Lestrimelitta danuncia		900	巢脾型	Roubik（1983）
	Lestrimelitta guayanensis		2 890	巢脾型	Roubik（1979）
	Lestrimelitta limao	2 000～7 000	4 500	巢脾型	Grüter et al.（2016）
	Lestrimelitta niitkib	3 000～5 000	4 000	巢脾型	Quezada-Euán and González-Acereto（2002）
	Melipona beecheii	300～2 000	800	巢脾型	van Veen et al.（2004）
	Melipona bicolor	150～800	425	巢脾型	Tóth et al.（2004）
	Melipona carrikeri		210	巢脾型	Wille and Michener（1973）
	Melipona costaricensis		2 000	巢脾型	Wille and Michener（1973）
	Melipona crinita	180～450	315	巢脾型	Roubik（1983）
	Melipona fasciata	200～2 500	1 000	巢脾型	Roubik（1983）；Tóth et al.（2004）
	Melipona fasciculata	300～1 000	600	巢脾型	Gomes et al.（2015）；Leão（2019）
	Melipona favosa	60～700	400	巢脾型	Sommeijer（1984）；Roubik（1982）

续表

区域	蜂种	蜂群规模范围	平均蜂群规模	育虫巢室排布	参考文献
新热带区	*Melipona flavolineata*	770 ~ 3 000	1 500	巢脾型	Grüter et al.（2016）；Leão（2019）
	Melipona fuliginosa	250 ~ 600	383	巢脾型	Roubik（1983）
	Melipona fulva	300 ~ 500	400	巢脾型	Roubik（1982）
	Melipona marginata	160 ~ 2 500	1 330	巢脾型	Tóth et al.（2004）
	Melipona micheneri	50 ~ 120	85	巢脾型	Roubik（1983）
	Melipona phenax	60 ~ 120	90	巢脾型	Roubik（1983）
	Melipona quadrifasciata	300 ~ 1 500	900	巢脾型	Michener（1974）；Tóth et al.（2004）
	Melipona rufiventris	500 ~ 700	600	巢脾型	Nieh et al.（2005）
	Melipona scutellaris	1 000 ~ 2 000	1 500	巢脾型	Tóth et al.（2004）
	Melipona seminigra	1 000 ~ 3 000	2 000	巢脾型	Grüter et al.（2016）
	Melipona trinitatis	1 000 ~ 2 000	1 500	巢脾型	Bijlsma et al.（2006）
	Melipona triplaridis	70 ~ 550	340	巢脾型	Roubik（1983）
	Nannotrigona mellaria		1 000	巢脾型	Roubik（1983）
	Nannotrigona perilampoides	700 ~ 1 200	950	巢脾型/螺旋型	Quezada-Euán and González-Acereto（2002）
	Nannotrigona testaceicornis	2 000 ~ 3 000	2 500	巢脾型/螺旋型	Michener（1974）
	Nogueirapis mirandula	2 281 ~ 4 076	3 180	巢脾型	Wille（1966）
	Oxytrigona mellicolor		5 442	巢脾型	Roubik（1983）
	Oxytrigona obscura		1 900	巢脾型	Roubik（1983）
	Paratrigona ornaticeps	1 100 ~ 2 400	1 710	巢脾型	Roubik（1983）
	Paratrigona opaca		250	巢脾型	Roubik（1983）
	Partamona aff. *cupira*	280 ~ 6 457	2 084	巢脾型	Wille and Michener（1973）；Roubik（1983）
	Plebeia droryana	1 070 ~ 3 000	2 400	巢脾型	Tóth et al.（2004）；Roldão-Sbordoni et al.（2018）

续表

区域	蜂种	蜂群规模范围	平均蜂群规模	育虫巢室排布	参考文献
新热带区	*Plebeia minima*	200～900	350	巢簇型	Leão（2019）
	Plebeia franki	70～250	113	巢簇型	Roubik（1983）
	Plebeia frontalis	100～1 900	1 000	巢脾型	Wille and Michener（1973）; Roubik（1983）
	Plebeia jatiformis	650～950	800	巢脾型	Roubik（1983）
	Plebeia juliani		300	半巢脾型	Drumond et al.（1998）
	Plebeia mosquito		1 175	巢脾型	von Ihering（1903）; Wille and Michener（1973）
	Plebeia nigriceps	100～200	150	巢脾型	Witter et al.（2007）
	Plebeia remota	800～5 000	2 900	巢脾型	van Benthem et al.（1995）; Tóth et al.（2004）
	Plebeia saiqui		7 000	巢脾型	Witter et al.（2007）
	Plebeia tica		612	巢簇型	Wille（1969）
	Plebeia tobagoensis	40～1 500	1 200	—	Bijlsma et al.（2006）; Hofstede and Sommeijer（2006）
	Scaptotrigona postica	6 000～10 000	7 400	巢脾型	Leão（2019）
	Scaptotrigona barrocoloradensis	3 000～7 827	5 400	巢脾型	Roubik（1983）; Tóth et al.（2004）
	Scaptotrigona luteipennis	3 000～10 000	5 600	巢脾型	Roubik（1983），同 *S. pachysoma*
	Scaptotrigona panamensis		5 201	巢脾型	Roubik（1983）
	Scaptotrigona pectoralis	2 000～5 200	3 600	巢脾型	Roubik（1983）; Quezada-Euán（2018）
	Scaptotrigona tubiba		1 764	巢脾型	Roubik（1979）
	Scaura latitarsis	387～450	400	巢脾型/螺旋型	Wille and Michener（1973）; Roubik（1983）
	Schwarziana quadripunctata	500～2 500	1 500	巢脾型	Tóth et al.（2003）

续表

区域	蜂种	蜂群规模范围	平均蜂群规模	育虫巢室排布	参考文献
新热带区	*Tetragona beebei*	1 500～2 000	1 750	螺旋型	Roubik（1983），同 *T. dorsalis beebei*
	Tetragona clavipes	5 383～29 000	7 343	巢脾型/螺旋型	Roubik（1979）；Duarte et al.（2016）
	Tetragona dorsalis	1 500～75 000	27 774	巢脾型/螺旋型	Roubik（1983）；Tóth et al.（2004）
	Tetragonisca angustula	2 000～8 000	5 000	巢脾型	Tóth et al.（2004）
	Tetragonisca buchwaldi	1 326～2 979	2 028	巢脾型	Wille（1966）；Wille and Michener（1973）
	Tetragonisca weyrauchi	2 000～3 000	2 500	巢脾型	Cortopassi-Laurino and Nogueira-Neto（2003）
	Trichotrigona camargoiana		200	巢簇型	Pedro and Cordeiro（2015）
	Trichotrigona extranea	43～163	107	巢簇型	Camargo and Pedro（2007）
	Trigona amazonensis	100 000～200 000	150 000	巢脾型	Roubik（1989, pp. 109），Inoue et al.（1996）
	Trigona branneri	43 758～59 820	51 789	巢脾型	Roubik（1979）
	Trigona cilipes	400～3 204	1 800	巢脾型	Roubik（1979, 1983）
	Trigona corvina	7 000～25 000	16 000	巢脾型	Michener（1974）；Roubik（1983）
	Trigona ferricauda		1 000	巢脾型	Roubik（1983）
	Trigona fulviventris	2 000～10 000	5 750	巢脾型	Roubik（1983）
	Trigona fuscipennis	5 000～10 000	7 500	巢脾型	Roubik（1983）
	Trigona hyalinata	最高至 40 000		巢脾型	Nieh et al.（2003）
	Trigona hypogea	1 200～5 000	1 750	巢脾型/螺旋型	Roubik（1983）；Camargo and Roubik（1991）
	Trigona necrophaga	1 000～3 000	2 000	巢脾型/螺旋型	Camargo and Roubik（1991）
	Trigona nigerrima	700～1 800	1 398	巢脾型	Roubik（1983）

续表

区域	蜂种	蜂群规模范围	平均蜂群规模	育虫巢室排布	参考文献
新热带区	*Trigona pallens*	658～8 170	4 400	巢脾型/半巢脾型	Roubik（1979）
	Trigona silvestriana		3 000	巢脾型	Roubik（1983）
	Trigona spinipes	5 000～180 000	92 500	巢脾型	Michener（1974）
	Trigona williana		2 492	半巢脾型	Roubik（1979）
	Trigonisca atomaria	160～500	330	巢簇型	Wille and Michener（1973）；Roubik（1983）
	Trigonisca buyssoni	136～600	368	巢簇型	Wille and Michener（1973）；Roubik（1983）

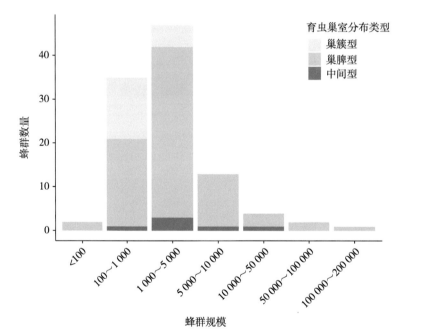

图 1.13　104 种无刺蜂的蜂群规模（表 1.3）分布

注：图中所示为不同蜂群规模大小和育虫巢室排布类型（第 3 章）的蜂种数量分布。其中 "combs"（"巢脾型"）包括了建造螺旋型巢脾的蜂种。"intermediate"（"中间型"）是指半巢脾型。

研究者在新热带 *Tetragona* 和 *Trigona* 属无刺蜂中，经常发现蜂群规模非常大。许多 Trigona 属无刺蜂会建立外部蜂巢，不受巢洞大小的限制。例如，Trigona corvina 无刺蜂建造的外部蜂巢重量就可高达 100 kg 以上（Roubik and Patiño，2009）。有人估算 Trigona amazonensis 和 T. spinipes 无刺蜂的蜂群规模有 18 万～20 万只，但一些研究者对此有所怀疑（Wille and Michener，1973; Roubik，1989），而且表 1.3 中的估计值应该被认为是对大多数无刺蜂蜂种的蜂群规模值的初步数据。这些蜂群规模值通常仅基于一两个蜂群，并且这些研究中估测蜂群规模的方法通常也不够精确，例如，有些蜂群规模是研究者通过飞行工蜂观察数结合育虫巢室计数估算而来的。一些研究甚至根本没有提供关于评估蜂群规模方法的信息。例如，Lindauer 和 Kerr 在其 1958 年发表的一篇无刺蜂研究领域有影响力的论文中指出，Trigona spinipes（即 "T. ruficrus"）蜂群中含 100 000～150 000 只无刺蜂，然后在其 1960 年发表的论文中[10]，此数据被更改为 5 000～180 000 只无刺蜂，但在这两篇论文中作者都并未提及蜂群规模的估算方法。

准确估测蜂群规模值的一种方法是，在所有采集蜂都在蜂巢时关闭蜂巢，并杀死或麻醉所有蜂，对其计数或称重（Roldão-Sbordoni et al.，2018）。但是，这种方法通常是不可行的，尤其是野外进行实地研究时。因此，研究人员试图根据更容易测量的特征（例如育虫巢室的数量）来估测蜂群规模（von Ihering，1903; Roubik，1979）。Roubik（1979）提出一个计算公式 $B_c = \frac{1}{4} B_r + \frac{1}{2} B_e$，$B_c$ 指成年蜂总数的估值，B_r 指幼虫总数，B_e 指成年蜂总数的区域估值。成年蜂总数的区域估值是指在开放状态的蜂巢附近飞行的蜂和在子脾上的蜂的总数量。Von Ihering［1930，由 Duarte 等（2016）引用］则利用公式 $x + x/2$ 估算蜂群总蜂数（包括所有幼虫），其中 x 指子脾巢室数。该公式意味着蜂群内成年蜂的数量约为子脾巢室数的 50%。而 Roubik（1983）则估算成年蜂数量为所有子脾巢室数的 1/3。Roldão-Sbordoni 等（2018）对蜂群内幼虫和成年蜂进行了计数，发现成年蜂数量（即此处使用的蜂群规模）约占幼虫数量的 40%。当然，任何基于子脾巢室数量的经验法则都应谨慎使用，因为成年蜂与未成年蜂的比例可能会随季节和蜂种而发生变化（第 5 章）。对于某些生物学问题，评估蜂群间的相对蜂群规模大小（"蜂群 A 大于蜂群 B"）就足够了。在这些情况下，研究者仅需对蜂群子脾巢室数进行计数或分析采集活动流量，这可能是较为直接的解决方案。

---

[10] 他们 1960 年发表的论文大部分是 1958 年德文论文的翻译。

1.7 蜂群寿命

理论上讲，无刺蜂蜂群可以长时间存在，因为衰老或死去的蜂王可以被新的蜂王代替（第4章）。一些无刺蜂蜂群可能是预料到蜂群存在失王风险，它们甚至会在"监禁室"中保留"备用"处女王（第4章）。实际上已经有多个报道发现有无刺蜂蜂群在一个洞穴中生活超过10年的案例（Schwarz，1948）。Kerr等（1967）研究过一个 *Frieseomelitta flavicornis* 蜂群，他研究时该蜂群存在时间已经达到了12年，而Starr和Sakagami（1987）所研究的无刺蜂（*Tetragonula sapiens*）蜂巢则至少存在超过10年。在笔者撰写这本书的时候，巴西圣保罗大学一个 *Trigona hyalinata* 无刺蜂蜂群至少已经存在了18年（图1.14），*T. hyalinata* 无刺蜂蜂巢存在时间高达20～30年的例子也并不少见（Sidnei Mateus）。Murillo［1984，引自Roubik（1989）］对35个在墨西哥南部人工饲养的 *Melipona beecheii* 无刺蜂蜂群进行了统计，这些蜂群存在寿命普遍为10～19年。更令人惊讶的是，有人报道有一个人工饲养的 *M. beecheii* 无刺蜂蜂群竟然存活了至少61年（Slaa，2006），而Schwarz（1948）提到，根据其私人关系得知有一个人工饲养的 *M. beecheii* 蜂群存在时间已高达一个多世纪了。但是需要注意的是，我们很难区分这种连续存在情况是同一蜂群的连续存活还是同一蜂种但不同蜂群的连续存活。例如，Michener（1946）观察到一个无刺蜂蜂巢在14个月内先后被 *Tetragonisca angustula* 无刺蜂的3个不同蜂群所占据。如果是人工饲养无刺蜂蜂群或定期进行检查的蜂群，我们可以对这种无刺蜂蜂群连续的存活历史更有信心。

研究者对无刺蜂蜂群的平均预期寿命估测需要频繁且不间断的监测。Slaa（2006）就进行了这样的研究，来定量4年内哥斯达黎加无刺蜂蜂种的年度蜂群死亡率。她估算出不同蜂种的年度生存概率高达96%。基于此，她进一步推断出蜂群平均寿命约为23年。这比居住在相同栖息地的非洲化蜜蜂的蜂群寿命要长得多（Slaa，2006）。Eltz等（2002）以及Kajobe和Roubik（2006）发现乌干达和马来西亚地区的无刺蜂蜂种的年度蜂群存活概率为85%～88%，表明平均蜂群寿命为4～7年。无刺蜂长时间的蜂群寿命结合其低分蜂性（第4章）提示，无刺蜂蜂群遵循着高蜂群存活率和低蜂群复制率的生活史策略（Slaa，2006）。

图 1.14　贝朗普雷图的圣保罗大学"Bloco 7"墙上的一个 *Trigona hyalinata* 无刺蜂蜂巢，在撰写本文时该蜂巢已经有 18 年的历史了［建立于 2002 年，摄影：C. Grüter（2016）］

注：这个蜂巢内无刺蜂的任何改变都极有可能会被每天从这里经过的众多昆虫学家注意到。

1.8　无刺蜂对于人类的重要性

人类开始饲养无刺蜂并利用无刺蜂蜂产品已有上千年的历史（Crane，1992, 1999; Jones，2013; Źrałka et al., 2014, 2018; Quezada-Euán，2018; Quezada-Euán and Alves，2020）。特别是玛雅人曾将无刺蜂文化纳入了他们的社会、经济及宗教活动中（Cortopassi-Laurino et al., 2006; Rosales，2013; Quezada-Euán，2018; Źrałka et al., 2018; Quezada-Euán and Alves，2020）。例如，玛雅人曾采用无刺蜂蜂蜜和巢质同墨西哥阿兹特克人（Nogueira-Neto，1951; Quezada-Euán，2018）进行贸易交换，并且在 16 世纪还将无刺蜂蜂蜜和巢质作为向西班牙统治者缴税的一种方式（Jones，2013; Quezada-Euán，2018）。在许多土著部落的传统和文化中，无刺蜂也具有很重要的地位，例如巴西的 Kayapó 族（Camargo and Posey，1990; Quezada-Euán et al., 2018）、巴拉圭的 Aché 族（有时他们也被称为"蜂蜜文明"）（Nogueira-Neto，1951）、乌干达 Abayanda 俾格米族（Byarugaba，2004）、马达加斯加的农业社会（Randrianandrasana and Berenbaum，2015）、喜马拉雅的土著社会（Bhatta et al., 2020）以及澳大利亚的 Yolngu 原住民社会（Fijn and Baynes-Rock，2018）[11]。

---

[11] 欧洲文献中关于无刺蜂以及它们如何被人类利用的报告最早刊登在 1557 年出版的 *Warhaftig Historia* 中。德国雇佣兵 Hans Staden 在该书中提到他被巴西图比南巴部落囚禁了好几个月，并描述了他观察到的该部落人是如何定期采集无刺蜂蜂蜜供人食用的（Engels，2009, 2013）。

无刺蜂养殖为热带地区很多族群的人们带来了持续收入（Cortopassi-Laurino et al., 2006; Kwapong et al., 2010; Alves，2013; Tornyie and Kwapong, 2015; Quezada-Euán et al., 2018）。人们养殖了数十种不同的无刺蜂来生产蜂蜜或巢质（Crane，1992; Cortopassi-Laurino et al., 2006）。其中最常用的无刺蜂蜂种包括澳洲的 *Tetragonula carbonaria* 无刺蜂（Heard，2016; Chapman et al., 2018）和新热带区的 *Melipona beecheii*、*M. eburnea*、*M. quadrifasciata*、*M. scutellaris* 和 *Tetragonisca angustula* 无刺蜂（Jaffé et al., 2015; Quezada-Euán, 2018; Quezada-Euán and Alves，2020）。近年来人们也采用无刺蜂来促进农作物授粉（第 9 章）（Heard，1999; Slaa et al., 2006; Giannini et al., 2015）。

1.8.1 无刺蜂养殖

中美洲玛雅人自哥伦布时代以来就开始进行大规模的集约化无刺蜂养殖（图 1.15）（Źrałka et al., 2014, 2018; Quezada-Euán，2018）[12]。图 1.15A 所示的土制蜂箱可追溯到玛雅文化的原始阶段（公元前 100 年至公元 300 年），这也是已知的最古老的玛雅蜂箱（Źrałka et al., 2014, 2018）。该文物和其他文物表明中美洲在过去的 1 700 ～ 2 000 年已经在不断地开展以 *Melipona beecheii* 为主要饲养品种的无刺蜂养殖（Crane，1992; Źrałka et al., 2014, 2018; Quezada-Euán，2018; Quezada-Euán and Alves，2020）。

人工饲养无刺蜂的主要蜂产品包括蜂蜜、树脂和巢质（Nogueira-Neto, 1997; Cortopassi-Laurino et al., 2006; Kwapong et al., 2010; Jaffé et al., 2015; Heard，2016; Quezada-Euán，2018; Quezada-Euán et al., 2018），但也有很多人饲养无刺蜂仅仅是为了娱乐和休闲（Carvalho et al., 2018）。人们从野外收集蜂群并将其转移到木箱、原木或黏土罐中（图 1.16）。在过去的几十年中人类开发了各种类型的蜂箱，其中最著名的是巴西的 Paulo Nogueira-Neto 箱（Nogueira-Neto，1997）。经验丰富的养蜂人可以通过分割已有蜂群来建立新的蜂群，这也是古代玛雅人已经成功开展的一种做法（Quezada-Euán，2018; Źrałka et al., 2018）。

---

[12] 15 世纪 50 年代在尤卡坦担任政府官员的托马斯·洛佩斯·梅德尔（Tomás López Medel）在他的《De los tres elementos》一书中提到了玛雅的养蜂人和蜂蜜生产情况，"我不相信印度的任何地方及其西班牙会像尤卡坦那样生产大量的蜂蜜和蜂蜡，因此这些省份的印第安人和当地人……不满足于从山上和普通地方得来的蜂蜜和蜂蜡，他们还在他们的房屋和其他地方安置了许多蜂巢"（源自 Źrałka et al., 2018）。

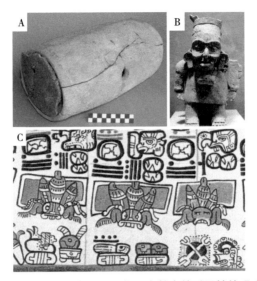

图 1.15　显示无刺蜂在玛雅文化和宗教中的重要性的玛雅文物

注：A 为危地马拉 Nakum 遗址中发现的前古典时期的黏土蜂巢（Źrałka et al., 2014, 2018）。目前还不清楚这个蜂巢是用来饲养无刺蜂的还是作为蜂巢的一个展示品。B 为戴着无刺蜂蜂巢项链的神像（源自 Źrałka et al., 2014）。C 为《马德里抄本》(*Madrid Codex*) 第 105 页中 "Mulzencab" 或 "Ah-Muzen-Cab" 神图。《马德里抄本》(*Madrid Codex* 或 Tro-Cortesianus codex) 是现存仅有的 3 种玛雅书籍之一，其年代可追溯至公元 900—1 500 年。

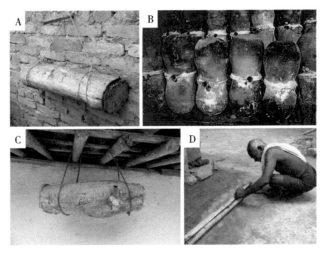

图 1.16　不同类型的原木和黏土无刺蜂蜂巢

注：A 为来自中国云南省西双版纳的 *Tetragonula* 属无刺蜂蜂群的原木蜂巢（摄影：周冰峰）。B 为来自墨西哥的 *Scaptotrigona pectoralis* 无刺蜂的泥罐巢（Vit et al., 2004）。C 为巴西东北部伯南布哥州 Sítio Xixá 的 *Scaptotrigona* sp. 无刺蜂的原木蜂巢（Carvalho et al., 2018）。D 为印度南部卡尼人用来饲养无刺蜂的竹木（Kumar et al., 2012）。

在美洲，特别是在巴西和墨西哥，人们已经普遍在人造蜂箱中饲养无刺蜂［Jaffé et al., 2015; 参见 Quezada-Euán（2018）对墨西哥无刺蜂养殖的一个全面描述］，但是在非洲却不那么普遍，传统上人们更喜欢饲养西方蜜蜂，而且该地区的人们为了从野生无刺蜂巢中收获蜂蜜和巢质，常常会破坏无刺蜂蜂群（Cortopassi-Laurino et al., 2006）。近年来，安哥拉和坦桑尼亚等其他非洲国家也开始尝试无刺蜂养殖，并且也对无刺蜂饲养管理有了越来越多的关注［Cortopassi-Laurino et al., 2006; 参见 Kwapong 等（2010）对加纳无刺蜂养殖的介绍］。澳洲的无刺蜂养殖也越来越盛行，在那里无刺蜂蜂群被用于生产蜂蜜、授粉农作物（如澳洲坚果）或保护生态环境（Heard, 1999, 2016）。而亚洲热带地区却很少人工饲养无刺蜂蜂群，在那里人们可以从本地的 *Apis cerana* 中华蜜蜂和 *A. dorsata* 大蜜蜂中获得大量蜂蜜（Cortopassi-Laurino et al., 2006; Kumar et al., 2012）。与此同时，印度南部的卡尼族人已经会利用竹筒饲养 *Tetragonula iridipennis* 无刺蜂（图 1.16D）（Kumar et al., 2012），而且如中国东南部区域等亚洲其他区域也在不断发展无刺蜂养殖（Pan et al., 2020）。

1.8.2　无刺蜂蜂产品的药用价值

人类收集无刺蜂产品的主要是因为无刺蜂蜂蜜和巢质极具药用价值（Cortopassi-Laurino et al., 2006; Kumar et al., 2012; Choudhari et al., 2012; Chanchao, 2013; Rodríguez-Malaver, 2013; Zamora et al., 2013; Carvalho et al., 2018; Quezada-Euán et al., 2018; Bhatta et al., 2020）。无刺蜂蜂蜜和巢质本身及其所含某些化合物具有许多有益功能（Vit et al., 2004; Rosales, 2013; Rao et al., 2016; Amin et al., 2018）。印度尼泊尔族群曾使用 *Lepidotrigona arcifera* 无刺蜂的蜂蜜来治疗各种健康问题，包括腹泻和哮喘等（Biswa et al., 2017）。乌干达 Abayanda 俾格米部落曾使用无刺蜂蜂蜜来治疗便秘（Byarugaba, 2004）。阿根廷北部地区的族群曾使用 *Plebeia* 属无刺蜂蜂蜜治疗与消化系统、呼吸系统和生殖系统有关的多种疾病（Flores et al., 2018）。在玻利维亚、危地马拉、墨西哥和委内瑞拉，人们常用 *Tetragonisca angustula* 无刺蜂蜂蜜来治疗许多轻症中的眼疾（Vit et al., 2004; Dardón et al., 2013; Ferrufino and Vit, 2013）。当前需要研究者通过进一步试验来评估无刺蜂蜂产品对不同疾病治疗方法的有效性。

1.8.3　无刺蜂的精神和宗教重要性

在某些土著社会中人们认为无刺蜂是神圣的，在这些土著社会的宇宙观和神话传说中，蜂经常扮演了重要角色（Carvalho et al., 2018; Quezada-Euán, 2018; Quezada-Euán et al., 2018）。特别是 *Melipona beecheii* 无刺蜂（玛雅人称其为 Xunan kab 或"淑女蜂"）在玛雅文化中具有悠久的文化和宗教历史，这种无刺蜂的蜂蜜也被玛雅人认为是神圣的（Cortopassi-Laurino et al., 2006; Quzada-Euán，2018; Źrałka et al., 2018）。在《马德里抄本》（*Madrid Codex*）中多次将玛雅文化中蜂蜜和蜂的众神之一"Muzen Cab"或"Ah-Muzen-cab"描绘为一只从正面看起来像是处于防御位置的蜂（图1.15 C）。巴西的 Kayapó 土著部落会在宗教仪式中使用无刺蜂的巢质和蜂蜜（Cortopassi-Laurino et al., 2006; Quezada-Euán et al., 2018），而阿根廷西北部社群中人们会使用蜂蜜为安第斯社会文化中的"地球之母"和生育女神——Pachamama 献祭（Flores et al., 2018）。然而，在马达加斯加的某些地区食用无刺蜂蜂蜜被视为禁忌，美洲一些社群则将花粉视为"污物"摒弃（Randrianandrasana and Berenbaum，2015; Quezada-Euán et al., 2018）。

1.8.4　无刺蜂蜂产品的其他用途

无刺蜂的巢质和蜂蜡无论是天然状态还是经过加工后，都是具有多用途的建巢材料（Stearman et al., 2008），曾被人类用于日常生活或充当食物。玛雅人曾通过一种名为"脱蜡铸造"的工艺来使用巢质为其金饰和小雕像制作模具（Schwarz，1945; Crane，1992; Jones，2013）。澳大利亚的原住民艺术家曾使用无刺蜂巢质来保护自己的岩画不受雨水侵害（Schwarz，1945）。无刺蜂的巢质和树脂也曾被人类在武器（例如箭和吹箭筒等）制造过程中用作黏合剂或密封材料，玻利维亚原住民 Yuquí 部落（Stearman et al., 2008）、厄瓜多尔 Kitchwa 部落等制作独木舟时也曾用到无刺蜂的巢质和树脂［更多例子参见 Schwarz（1945）和 Quezada-Euán 等（2018）］。

在安第斯山脉地区，无刺蜂蜂蜜被人们作为壮阳剂添加到酒精饮料以及对安第斯地区社群具有重要文化意义的吉开酒（*chicha*）中（Flores et al., 2018）。人们有时还用无刺蜂花粉作为食品和酒精饮料的补充剂（Flores et al., 2018）。另外，在墨西哥（Ramos-Elorduy，2002）、巴西、厄瓜多尔、巴拉圭和澳大利亚等地（Fijn and Baynes-Rock, 2018; Quezada-Euán et al., 2018），人们还会食用无刺蜂的幼虫和蜂蛹。

1.9　人类活动对无刺蜂的新挑战

像许多其他昆虫一样，无刺蜂也面临着许多新挑战，例如农药、新病原体、气候变化等（Potts et al., 2010, 2016; Ramírez et al., 2013; Lima et al., 2016; Tsvetkov et al., 2017; Giannini et al., 2017, 2020; Sánchez-Bayo and Wyckhuys, 2019; Guimarães-Cestaro et al., 2020）。当今无刺蜂经受的最严峻的挑战是环境的迅速变化，引起这种变化的原因主要是自然和半自然土地在迅速转变为城市或集约化耕地（Winfree et al., 2009; Potts et al., 2010, 2016）。仅在2019年，地球由于森林砍伐就损失了38 000km^{2}的原始热带森林，其面积约等于荷兰的国土面积[13]。这导致很多经过人类改造过的自然环境不能再提供足够的食物或筑巢空间来维持无刺蜂的生存（Brown and Albrecht, 2001; Cairns et al., 2005; Brosi et al., 2008; Ramírez et al., 2013; Kaluza et al., 2018）。无刺蜂更易遭受森林砍伐之害，因为大多数无刺蜂蜂种生活在树洞中（第3章），并且同期大量开花型的树木是许多无刺蜂的主要食物来源（第8章）。鉴于无刺蜂可以为天然植物和栽培植物提供授粉服务，因此亟须深入研究森林砍伐是如何影响无刺蜂种群的。一项评估森林砍伐对无刺蜂多样性影响的研究发现，随着巴西当地森林砍伐程度的增加，巴西西部的一个栖息地的无刺蜂蜂种数量有所减少（图1.17）（Brown and Oliveira, 2014）。但是同时值得注意的是，某些无刺蜂蜂种表现出了对高度改变的环境令人惊讶的适应能力（第3章）。

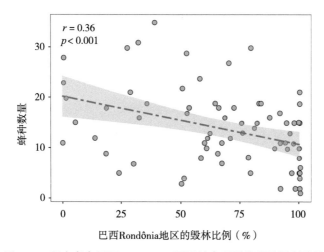

图1.17　研究者在巴西 Rondônia 不同地点采样发现的无刺蜂物种数量（研究者是在无刺蜂正在采集资源时对其进行采样的）

注：x轴显示取样地点 500 m 范围内的毁林面积的百分比。灰色区域显示 95% 的置信区间［根据 Brown and Oliveira（2014）修改］。

---

[13] 数据来源：https://www.globalforestwatch.org/.

1.9.1 杀虫剂

有越来越多的证据表明，农药导致了温带地区的蜂数量减少（Henry et al., 2012; Gill et al., 2012; Goulson et al., 2015; Siviter et al., 2018 a）。例如，新烟碱类化合物（例如吡虫啉、噻虫嗪）和苯并吡唑类药物（例如氟虫腈）是神经毒素，会损害无刺蜂中枢神经系统的功能（Goulson et al., 2015; Samuelson et al., 2016; Siviter et al., 2018 b）。低剂量农药可能不会立即杀死蜂个体和蜂群，但它们可能通过损害蜂的学习和记忆能力、导航能力和繁殖力而产生所谓的亚致死作用（Henry et al., 2012; Gill et al., 2012; Balbuena et al., 2015; Samuelson et al., 2016; Siviter et al., 2018 a）。考虑到农药对蜜蜂和熊蜂的不利影响以及巴西等国家农药的使用量增加（Santos et al., 2018），人们关注到农药对无刺蜂的不利影响也就有迹可循，并且越来越多的研究评估了不同无刺蜂蜂种对热带地区常用的农药和其他农用化学品的反应（Lima et al., 2016）。令人担忧的是，研究者评估了不同农药的毒性（例如，以导致50%死亡率的剂量测定值，LD_{50}），研究结果表明，氟虫腈、吡虫啉或噻虫嗪等农药对无刺蜂的毒性甚至要比对蜜蜂的毒性大得多（Valdovinos-Núñez et al., 2009; Jacob et al., 2013; Sarto et al., 2014; Costa et al., 2015; Moreira et al., 2018; Jacob et al., 2019 a, 2019 b）。研究报道发现，农药对无刺蜂的作用是多种多样的（Lima et al., 2016）。例如含有农药的幼虫食物会增加无刺蜂幼虫死亡率，损害其大脑发育，降低成蜂活动能力，并导致蜂王巢室中的无刺蜂幼虫发育成工蜂（Tomé et al., 2012; Barbosa et al., 2015; Santos et al., 2016; Seide et al., 2018）。摄入了亚致死剂量农药的成年无刺蜂工蜂活动能力下降，交哺行为和触角感触行为反应迟缓，这也表明农药可能会损害无刺蜂蜂群的社会性行为（Boff et al., 2018; Araujo et al., 2019; Jacob et al., 2019 a, 2019 b）。迄今为止，大多数研究都是在实验室条件下进行的，目前我们对实验室条件下所用农药剂量与成年无刺蜂和幼虫在自然条件下所接触到的农药剂量是否一致还不明确。但是Gómez-Escobar等（2018）根据农药说明书在杧果园中反复施用农药（多杀霉素）测试了其对无刺蜂的影响。他们的数据表明，放置在该果园中的 *Scaptotrigona mexicana* 蜂群群势低于对照果园中的蜂群。研究结果也同样发现农药对无刺蜂的危害似乎比对蜜蜂的危害更大（Gómez-Escobar et al., 2018）。我们需要更多的研究来评估在田间条件下广泛使用农药对无刺蜂的影响，这也将帮助人们更好地认识农药使用对无刺蜂的种群水平的潜在影响。

参考文献

Almeida-Muradian LB（2013）*Tetragonisca angustula* Pot-Honey Compared to *Apis mellifera* Honey from Brazil. In: Vit P, Pedro SRM, Roubik DW（eds）Pot-Honey: A legacy of stingless bees. Springer, New York, NY, pp 375–382

Alves RMO（2013）Production and Marketing of Pot-Honey. In: Vit P, Pedro SRM, Roubik DW（eds）Pot-Honey: A legacy of stingless bees. Springer, New York, NY, pp 541–556

Amdam GV, Aase ALT, Seehuus SC, Fondrk MK, Norberg K, Hartfelder K（2005）Social reversal of immunosenescence in honey bee workers. Experimental Gerontology 40:939–947

Amdam GV, Simões ZL, Hagen A, Norberg K, Schrøder K, Mikkelsen Ø, Kirkwood TB, Omholt SW（2004）Hormonal control of the yolk precursor vitellogenin regulates immune function and longevity in honeybees. Experimental Gerontology 39:767–773

Amin FAZ, Sabri S, Mohammad SM, Ismail M, Chan KW, Ismail N, Norhaizan ME, Zawawi N（2018）Therapeutic Properties of Stingless Bee Honey in Comparison with European Bee Honey. Advances in Pharmacological Sciences 2018:6179596

Araujo RS, Lopes MP, Barbosa WF, Gonçalves WG, Fernandes KM, Martins GF, Tavares MG（2019）Spinosad-mediated effects on survival, overall group activity and the midgut of workers of *Partamona helleri*（Hymenoptera: Apidae）. Ecotoxicology and Environmental Safety 175:148–154

Ávila S, Beux MR, Ribani RH, Zambiazi RC（2018）Stingless bee honey: Quality parameters, bioactive compounds, health-promotion properties and modification detection strategies. Trends in Food Science & Technology 81:37–50

Balbuena MS, Tison L, Hahn M-L, Greggers U, Menzel R, Farina WM（2015）Effects of sublethal doses of glyphosate on honeybee navigation. Journal of Experimental Biology 218:2799–2805

Baptistella AR, Souza CCM, Santana WC, Soares AEE（2012）Techniques for the In Vitro Production of Queens in Stingless Bees（Apidae, Meliponini）. Sociobiology 59:297–310

Barbosa FM, Alves RMO, Souza BA, Carvalho CAL（2013）Nest architecture of the stingless bee *Geotrigona subterranea* (Friese, 1901) (Hymenoptera: Apidae: Meliponini). Biota Neotropica 13:147–152

Barbosa W, Tomé HVV, Cupertino Bernardes R, Lima M, Smagghe G, Guedes RN (2015) Biopesticide-induced behavioral and morphological alterations in the stingless bee *Melipona quadrifasciata*. Environmental Toxicology and Chemistry 34:2149–2158

Bassindale R (1955) The biology of the Stingless Bee *Trigona* (*Hypotrigona*) *gribodoi* Magretti (Meliponidae). Proceedings of the Zoological Society of London 125:49–62

Baudier KM, Ostwald MM, Grüter C, Segers FHID, Roubik DW, Pavlic TP, Pratt SC, Fewell JH (2019) Changing of the guard: mixed specialization and flexibility in nest defense (*Tetragonisca angustula*). Behavioral Ecology 30:1041–1049

Bhatta C, Gonzalez VH, Smith D (2020) Traditional uses and relative cultural importance of *Tetragonula iridipennis* (Smith) (Hymenoptera: Apidae: Meliponini) in Nepal. Journal of Melittology, in press

Beckers R, Goss S, Deneubourg JL, Pasteels JM (1989) Colony size, communication and ant foraging strategy. Psyche 96:239–256

Bego LR (1983) On some aspects of bionomics in *Melipona bicolor bicolor* Lepeletier (Hymenoptera, Apidae, Meliponinae). Revista Brasileira de Entomologia 27:211–224

Biesmeijer JC, Tóth E (1998) Individual foraging, activity level and longevity in the stingless bee *Melipona beecheii* in Costa Rica (Hymenoptera, Apidae, Meliponinae). Insectes Sociaux 45:427–443

Bijlsma L, Bruijn LLM de, Martens EP, Sommeijer MJ (2006) Water content of stingless bee honeys (Apidae, Meliponini): interspecific variation and comparison with honey of *Apis mellifera*. Apidologie 37:480–486

Biluca FC, Bernal J, Valverde S, Ares AM, Gonzaga LV, Costa ACO, Fett R (2019) Determination of Free Amino Acids in Stingless Bee (Meliponinae) Honey. Food Analytical Methods 12:902–907

Biluca FC, Braghini F, Gonzaga LV, Costa ACO, Fett R (2016) Physicochemical profiles, minerals and bioactive compounds of stingless bee honey (Meliponinae). Journal of Food Composition and Analysis 50:61–69

Biswa R, Sarkar A, Khewa S (Subba) (2017) Ethnomedicinal uses of honey of stingless bee by *Nepali* community of Darjeeling foothills of West Bengal, India. Indian Journal of Traditional Knowledge 16:648–653

Bitondi MMG, Simões ZLP (1996) The relationship between level of pollen in the diet, vitellogenin and juvenile hormone titres in Africanized *Apis mellifera* workers. Journal of Apicultural Research 35:27–36

Boff S, Friedel A, Mussury RM, Lenis PR, Raizer J (2018) Changes in social behavior are induced by pesticide ingestion in a Neotropical stingless bee. Ecotoxicology and Environmental Safety 164:548–553

Boongird S, Michener CD (2010) Pollen and propolis collecting by male stingless bees (Hymenoptera: Apidae). Journal of the Kansas Entomological Society 83:47–50

Bourke AFG (1999) Colony size, social complexity and reproductive conflict in social insects. Journal of Evolutionary Biology 12:245–257

Brosi BJ, Daily GC, Shih TM, Oviedo F, Durán G (2008) The effects of forest fragmentation on bee communities in tropical countryside: Bee communities and tropical forest fragmentation. Journal of Applied Ecology 45:773–783

Brown JC, Albrecht C (2001) The effect of tropical deforestation on stingless bees of the genus *Melipona* (Insecta: Hymenoptera: Apidae: Meliponini) in central Rondonia, Brazil. Journal of Biogeography 28:623–634

Brown JC, Oliveira ML (2014) The impact of agricultural colonization and deforestation on stingless bee (Apidae: Meliponini) composition and richness in Rondônia, Brazil. Apidologie 45:172–188

Bustamante NCR (2006) Divisão de Trabalho em Três Espécies de Abelhas do Gênero *Melipona* (Hymenoptera, Apidae) na Amazônia, Brasileira. PhD Thesis, Universidade Federal do Amazonas, Brazil

Byarugaba D (2004) Stingless bees (Hymenoptera: Apidae) of Bwindi impenetrable forest, Uganda and Abayanda indigenous knowledge. International Journal of Tropical Insect Science 24:117–121

Cairns CE, Villanueva-Gutiérrez R, Koptur S, Bray DB (2005) Bee populations, forest disturbance, and africanization in Mexico. Biotropica 37:686–692

Camargo CA (1974) Notas sobre a morfologia e biologia de *Plebeia* (*Schwarziana*) *quadripunctata quadripunctata* (Hymenoptera, Apidae, Meliponinae). Studia

Entomologica 17:433–470

Camargo JMR, Pedro SRM (2008) Revision of the species of *Melipona* of the *fuliginosa* group (Hymenoptera, Apoidea, Apidae, Meliponini). Revista Brasileira de Entomologia 52:411–427

Camargo JMF, Pedro SRM (2007) Notes on the bionomy of *Trichotrigona extranea* Camargo & Moure (Hymenoptera, Apidae, Meliponini). Revista Brasileira de Entomologia 51:72–81

Camargo JMF, Posey DA (1990) O conhecimento dos Kayapó sobre as abelhas sociais sem ferrão (Meliponinae, Apidae, Hymenoptera): notas adicionais. Boletim do Museu Paraense Emílio Goeldi, Zoologia 6:17–42

Camargo JMF, Roubik DW (1991) Systematics and bionomics of the apoid obligate necrophages: the *Trigona hypogea* group (Hymenoptera: Apidae; Meliponinae). Biological Journal of the Linnean Society 44:13–39

Camargo JMF, Pedro SRM (2013) Meliponini Lepeletier, 1836. In: Moure JS, Urban D, Melo GAR (eds) Catalogue of Bees (Hymenoptera, Apoidea) in the Neotropical Region - online version. Available at http://www.moure.cria.org.br/catalogue

Cardoso R (2010) Divisao etária de trabalho em operárias de *Frieseomelitta varia* (Hymenoptera, Apidae, Apini). BSc Thesis, University of São Paulo, Ribeirão Preto, Brazil

Carvalho WJ de, Fujimura PT, Bonetti AM, Goulart LR, Cloonan K, Silva NM da, Araújo ECB, Ueira-Vieira C, Leal WS (2017) Characterization of antennal sensilla, larvae morphology and olfactory genes of *Melipona scutellaris* stingless bee. PLoS ONE 12:e0174857

Carvalho RMA, Martins CF, Nóbrega Alves RR, Alves ÂGC (2018) Do emotions influence the motivations and preferences of keepers of stingless bees? Journal of Ethnobiology and Ethnomedicine 14:47

Çelemli ÖG (2013) Chemical Properties of Propolis Collected by Stingless Bees. In: Vit P, Pedro SRM, Roubik DW (eds) Pot-Honey: A legacy of stingless bees. Springer, New York, NY, pp 525–537

Chalcoff VR, Gleiser G, Ezcurra C, Aizen MA (2017) Pollinator type and secondarily climate are related to nectar sugar composition across the angiosperms. Evolutionary Ecology 31:585–602

Chanchao C (2013) Bioactivity of Honey and Propolis of *Tetragonula laeviceps* in Thailand. In: Vit P, Pedro SRM, Roubik DW (eds) Pot-Honey: A legacy of stingless bees. Springer, New York, NY, pp 495–505

Chapman NC, Byatt M, Cocenza RDS, Nguyen LM, Heard TA, Latty T, Oldroyd BP (2018) Anthropogenic hive movements are changing the genetic structure of a stingless bee (*Tetragonula carbonaria*) population along the east coast of Australia. Conservation Genetics 19:619–627

Chinh TX, Sommeijer MJ, Boot WJ, Michener CD (2005) Nest and colony characteristics of three stingless bee species in Vietnam with the first description of the nest of *Lisotrigona carpenteri* (Hymenoptera: Apidae: Meliponini). Journal of the Kansas Entomological Society 78:363–372

Choudhari MK, Punekar SA, Ranade RV, Paknikar KM (2012) Antimicrobial activity of stingless bee (*Trigona sp.*) propolis used in the folk medicine of Western Maharashtra, India. Journal of Ethnopharmacology 141:363–367

Cortopassi-Laurino M, Imperatriz-Fonseca VL, Roubik DW, Dollin A, Heard T, Aguilar I, Venturieri GC, Eardley C, Nogueira-Neto P (2006) Global meliponiculture: challenges and opportunities. Apidologie 37:275–292

Cortopassi-Laurino M, Nogueira-Neto P (2003) Notas sobre a bionomia de *Tetragonisca weyrauchi* Schwarz, 1943 (Apidae, Meliponini). Acta Amazonica 33:643–650

Costa LM da, Grella TC, Barbosa RA, Malaspina O, Nocelli RCF (2015) Determination of acute lethal doses (LD50 and LC50) of imidacloprid for the native bee *Melipona scutellaris* Latreille, 1811 (Hymenoptera: Apidae). Sociobiology 62:578–582

Couvillon MJ (2012) The dance legacy of Karl von Frisch. Insectes Sociaux 59:297–306

Crane E (1992) The Past and Present Status of Beekeeping with Stingless Bees. Bee World 73:29–42

Crane E (1999) The World History of Beekeeping and Honey Hunting. Routledge, New York

Cruz-Landim C da (2000) Ovarian development in Meliponine bees (Hymenoptera: Apidae): the effect of queen presence and food on worker ovary development and egg production. Genetics and Molecular Biology 23:83–88

da Silva DLN, Zucchi R, Kerr WE (1972) Biological and behavioural aspects of the reproduction in some species of *Melipona* (Hymenoptera, Apidae, Meliponinae). Animal Behaviour 20:123–132

Darchen R (1977) L'Acclimatation des Trigones Africanes (Apidae, Trigonini) en France. Apidologie 8:147–154

Darchen R (1969) Sur la biologie de *Trigona* (*Apotrigona*) *nebulata* komiensis Cock. I. Biologia Gabonica 5:151–183

Dardón MJ, Maldonado-Aguilera C, Enríquez E (2013) The Pot-Honey of Guatemalan Bees. In: Vit P, Pedro SRM, Roubik DW (eds) Pot-Honey: A legacy of stingless bees. Springer, New York, NY, pp 395–408

Deliza R, Vit P (2013) Sensory Evaluation of Stingless Bee Pot-Honey. In: Vit P, Pedro SRM, Roubik DW (eds) Pot-Honey: A legacy of stingless bees. Springer, New York, NY, pp 349–361

Dornhaus A, Powell S, Bengston S (2012) Group size and its effects on collective organization. Annual Review of Entomology 57:123–141

dos Santos CG, Blochtein B, Megiolaro FL, Imperatriz-Fonseca VL (2010) Age polyethism in *Plebeia emerina* (Friese) (Hymenoptera: Apidae) colonies related to propolis handling. Neotropical Entomology 39:691–696

Drumond PM, Zucchi R, Yamane S, Sakagami SF (1998) Oviposition behavior of the stingless bees: XX. *Plebeia* (*Plebeia*) *julianii* which forms very small brood batches (Hymenoptera: Apidae, Meliponinae). Entomological Science 1:195–205

Duarte R, Souza J, Soares AEE (2016) Nest Architecture of *Tetragona clavipes* (Fabricius) (Hymenoptera, Apidae, Meliponini). Sociobiology 63:813–818

Eardley CD (2004) Taxonomic revision of the African stingless bees (Apoidea: Apidae: Apinae: Meliponini). African Plant Protection 10:63–96

Eltz T, Brühl CA, Van Der Kaars S, Linsenmair EK (2002) Determinants of stingless bee nest density in lowland dipterocarp forests of Sabah, Malaysia. Oecologia 131:27–34

Engel MS, Rozen JG, Sepúlveda-Cano PA, Smith CS, Thomas JC, Ospina Torres R, González VH (2019) Nest architecture, immature stages, and ethnoentomology of a new species of *Trigonisca* from northern Colombia (Hymenoptera, Apidae).

American Museum Novitates 3942:1–33

Engels W (2009) The first record of Brazilian stingless bees published 450 years ago by Hans Staden. Genetics and Molecular Research 8:738–743

Engels W (2013) Staden's First Report in 1557 on the Collection of Stingless Bee Honey by Indians in Brazil. In: Vit P, Pedro SRM, Roubik DW (eds) Pot-Honey: A legacy of stingless bees. Springer, New York, NY, pp 241–246

Exley EM (1980) New Species and Records of Quasihesma Exley (hymenoptera: Apoidea: Euryglossinae). Australian Journal of Entomology 19:161–170

Ferrufino U, Vit P (2013) Pot-Honey of Six Meliponines from Amboró National Park, Bolivia. In: Vit P, Pedro SRM, Roubik DW (eds) Pot-Honey: A legacy of stingless bees. Springer, New York, NY, pp 409–416

Fijn N, Baynes-Rock M (2018) A Social Ecology of Stingless Bees. Human Ecology 46:207–216

Fletcher MT, Hungerford NL, Webber D, Carpinelli de Jesus M, Zhang J, Stone ISJ, Blanchfield JT, Zawawi N (2020) Stingless bee honey, a novel source of trehalulose: a biologically active disaccharide with health benefits. Scientific Reports 10:12128

Flores FF, Hilgert NI, Lupo LC (2018) Melliferous insects and the uses assigned to their products in the northern Yungas of Salta, Argentina. Journal of Ethnobiology and Ethnomedicine 14:27

Fuenmayor CA, Díaz-Moreno AC, Zuluaga-Domínguez CM, Quicazán MC (2013) Honey of Colombian Stingless Bees: Nutritional Characteristics and Physicochemical Quality Indicators. In: Vit P, Pedro SRM, Roubik DW (eds) Pot-Honey: A legacy of stingless bees. Springer, New York, NY, pp 383–394

Giannini TC, Boff S, Cordeiro GD, Cartolano EA, Veiga AK, Imperatriz-Fonseca VL, Saraiva AM (2015) Crop pollinators in Brazil: a review of reported interactions. Apidologie 46:209–223

Giannini TC, Costa WF, Borges RC, Miranda L, da Costa CPW, Saraiva AM, Imperatriz Fonseca VL (2020) Climate change in the Eastern Amazon: crop-pollinator and occurrence-restricted bees are potentially more affected. Regional Environmental Change 20:9

Giannini TC, Maia-Silva C, Acosta AL, Jaffé R, Carvalho AT, Martins CF, Zanella FCV, Carvalho CAL, Hrncir M, Saraiva AM, Siqueira JO, Imperatriz-Fonseca VL

(2017) Protecting a managed bee pollinator against climate change: strategies for an area with extreme climatic conditions and socioeconomic vulnerability. Apidologie 48:784–794

Gill RJ, Ramos-Rodriguez O, Raine NE (2012) Combined pesticide exposure severely affects individual- and colony-level traits in bees. Nature 491:105–108

Gomes RLC, Menezes C, Contrera FAL (2015) Worker longevity in an Amazonian *Melipona* (Apidae, Meliponini) species: effects of season and age at foraging onset. Apidologie 46:133–143

Gómez-Escobar E, Liedo P, Montoya P, Méndez-Villarreal A, Guzmán M, Vandame R, Sánchez D (2018) Effect of GF-120 (Spinosad) Aerial Sprays on Colonies of the Stingless Bee *Scaptotrigona mexicana* (Hymenoptera: Apidae) and the Honey Bee (Hymenoptera: Apidae). Journal of Economic Entomology 111:1711–1715

Goudie F, Oldroyd BP (2014) Thelytoky in the honey bee. Apidologie 45:306–326

Goulson D, Nicholls E, Botias C, Rotheray EL (2015) Bee declines driven by combined stress from parasites, pesticides, and lack of flowers. Science 347:1255957–1255957

Grosso AF, Bego LR (2002) Labor division, average life span, survival curve, and nest architecture of *Tetragonisca angustula angustula* (Hymenoptera, Apinae, Meliponini). Sociobiology 40:615–637

Grüter C (2018) Repeated switches from cooperative to selfish worker oviposition during stingless bee evolution. Journal of Evolutionary Biology 31:1843–1851

Grüter C, Kärcher M, Ratnieks FLW (2011) The natural history of nest defence in a stingless bee, *Tetragonisca angustula* (Latreille) (Hymenoptera: Apidae), with two distinct types of entrance guards. Neotropical Entomology 40:55–61

Grüter C, Segers FHID, Menezes C, Vollet-Neto A, Falcon T, von Zuben LG, Bitondi MMG, Nascimento FS, Almeida EAB (2017) Repeated evolution of soldier sub-castes suggests parasitism drives social complexity in stingless bees. Nature Communications 8:4

Grüter C, von Zuben LG, Segers F, Cunningham JP (2016) Warfare in stingless bees. Insectes Sociaux 63:223–236

Guimarães-Cestaro L, Martins MF, Martínez LC, Alves MLTMF, Guidugli-Lazzarini KR, Nocelli RCF, Malaspina O, Serrão JE, Teixeira ÉW (2020) Occurrence of virus, microsporidia, and pesticide residues in three species of stingless bees(Apidae: Meliponini)in the field. Science of Nature 107:16

Halcroft M, Haigh AM, Spooner-Hart R (2013 a)Ontogenic time and worker longevity in the Australian stingless bee, *Austroplebeia australis*. Insectes Sociaux 60:259–264

Halcroft MT, Haigh AM, Holmes SP, Spooner-Hart RN (2013 b)The thermal environment of nests of the Australian stingless bee, *Austroplebeia australis*. Insectes Sociaux 60:497–506

Hammel B, Vollet-Neto A, Menezes C, Nascimento FS, Engels W, Grüter C (2016) Soldiers in a stingless bee: work rate and task repertoire suggest guards are an elite force. The American Naturalist 187:120–129

Hammond RL, Keller L (2004)Conflict over male parentage in social insects. PLoS Biology 2:e248

Heard T (2016)The Australian Native Bee Book: keeping stingless bee hives for pets, pollination and sugarbag honey. Sugarbag Bees, Brisbane, Australia

Heard TA (1999)The role of stingless bees in crop pollination. Annual Review of Entomology 44:183–206

Hebling NJ, Kerr WE, Kerr FS (1964)Divisão de trabalho entre operárias de *Trigona* (*Scaptotrigona*)*xanthotricha* Moure. Papéis Avulsos de Zoologia, São Paulo 16:115–127

Henry M, Beguin M, Requier F, Rollin O, Odoux J-F, Aupinel P, Aptel J, Tchamitchian S, Decourtye A (2012)A common pesticide decreases foraging success and survival in honey bees. Science 336:348–350

Hofstede FE, Sommeijer MJ (2006)Effect of food availability on individual foraging specialisation in the stingless bee *Plebeia tobagoensis* (Hymenoptera, Meliponini). Apidologie 37:387

Hölldobler B (1976)Tournaments and slavery in a desert ant. Science 192:912–914

Hrncir M (2009)Mobilizing the Foraging Force Mechanical Signals in Stingless Bee Recruitment. In: Jarau S, Hrncir M (eds)Food Exploitation by Social Insects: Ecological, Behavioral, and Theoretical Approaches. CRC Press, Taylor & Francis Group, Boca Raton, Florida

Hrncir M, Jarau S, Barth FG (2016)Stingless bees (Meliponini): senses and behavior. Journal of Comparative Physiology A 202:597–601

Hrncir M, Maia-Silva C (2013) On the diversity of forging-related traits in stingless bees. In: Vit P, Pedro SRM, Roubik DW (eds) Pot-Honey: A legacy of stingless bees. Springer, New York, pp 201–215

Hrncir M, Maia-Silva C, da Silva Teixeira-Souza VH, Imperatriz-Fonseca VL (2019) Stingless bees and their adaptations to extreme environments. Journal of Comparative Physiology A 205:415–426

I'Anson Price R, Grüter C (2015) Why, when and where did honey bee dance communication evolve? Frontiers in Ecology and Evolution 3:1–7

Imperatriz-Fonseca VL (1978) Studies on *Paratrigona subnuda* (Moure) (Hymenoptera, Apidae, Meliponinae) III. Queen supersedure. Boletim de Zoologia, Universidade de São Paulo 3:153–162

Imperatriz-Fonseca VL, Zucchi R (1995) Virgin queens in stingless bee (Apidae, Meliponinae) colonies: a review. Apidologie 26:231–244

Inoue T, Sakagami SF, Salmah S, Yamane S (1984) The process of colony multiplication in the Sumatran stingless bee *Trigona* (*Tetragonula*) *laeviceps*. Biotropica 16:100–111

Inoue T, Salmah S, Sakagami SF (1996) Individual variations in worker polyethism of the Sumatran stingless bee, *Trigona* (*Tetragonula*) *minangkabau* (Apidae, Meliponinae). Japanese Journal of Entomology 64:641–668

Jacob CR de O, Zanardi OZ, Malaquias JB, Souza Silva CA, Yamamoto PT (2019a) The impact of four widely used neonicotinoid insecticides on *Tetragonisca angustula* (Latreille) (Hymenoptera: Apidae). Chemosphere 224:65–70

Jacob CRO, Malaquias JB, Zanardi OZ, Silva CAS, Jacob JFO, Yamamoto PT (2019b) Oral acute toxicity and impact of neonicotinoids on *Apis mellifera* L. and *Scaptotrigona postica* Latreille (Hymenoptera: Apidae). Ecotoxicology 28:744–753

Jacob CRO, Soares HM, Carvalho SM, Nocelli RCF, Malaspina O (2013) Acute Toxicity of Fipronil to the Stingless Bee *Scaptotrigona postica* Latreille. Bulletin of Environmental Contamination and Toxicology 90:69–72

Jaffé R, Pope N, Carvalho AT, Maia UM, Blochtein B, Carvalho CAL de, Carvalho-Zilse GA, Freitas BM, Menezes C, Ribeiro M de F, Venturieri GC, Imperatriz-Fonseca VL (2015) Bees for Development: Brazilian Survey Reveals How to Optimize Stingless Beekeeping. PLoS ONE 10:e0121157

Jarau S (2009) Chemical Communication during Food Exploitation in Stingless Bees. In: Jarau S, Hrncir M (eds) Food Exploitation by Social Insects: Ecological, Behavioral, and Theoretical Approaches. CRC University Press, Boca Raton, Florida

Jones R (2013) Stingless Bees: A Historical Perspective. In: Vit P, Pedro SRM, Roubik DW (eds) Pot-Honey: A legacy of stingless bees. Springer, New York, NY, pp 219–227

Kajobe R (2007) Pollen foraging by *Apis mellifera* and stingless bees *Meliponula bocandei* and *Meliponula nebulat*a in Bwindi Impenetrable National Park, Uganda. African Journal of Ecology 45:265–274

Kajobe R, Roubik DW (2006) Honey-making bee colony abundance and predation by apes and humans in a Uganda forest reserve. Biotropica 38:210–218

Kaluza BF, Wallace HM, Heard TA, Minden V, Klein A, Leonhardt SD (2018) Social bees are fitter in more biodiverse environments. Scientific Reports 8:1–10

Karsai I, Wenzel JW (1998) Productivity, individual-level and colony-level flexibility, and organization of work as consequences of colony size. Proceedings of the National Academy of Sciences of the United States of America 95:8665–8669

Kerr WE, de Lello E (1962) Sting glands in stingless bees: a vestigial character (Hymenoptera: Apidae). Journal of the New York Entomological Society 70:190–214

Kerr WE, Sakagami SF, Zucchi R, Portugal-Araújo V, Camargo JMF (1967) Observações sobre a arquitetura dos ninhos e comportamento de algumas espécies de abelhas sem ferrão das vizinhanças de Manaus, Amazonas (Hymenoptera, Apoidea). Conselho Nacional de Pesquisas Rio de Janeiro, pp 255–309

Kerr WE, Zucchi R, Nakadaira JT, Butolo JE (1962) Reproduction in the social bees (Hymenoptera: Apidae). Journal of the New York Entomological Society 70:265–276

Koch H (2010) Combining morphology and DNA barcoding resolves the taxonomy of western Malagasy *Liotrigona* Moure, 1961 (Hymenoptera: Apidae: Meliponini). African Invertebrates 51:413–421

Kumar MS, Singh AJAR, Alagumuthu G (2012) Traditional beekeeping of stingless bee (*Trigona* sp.) by Kani tribes of Western Ghats, Tamil Nadu, India. Indian Journal of Traditional Knowledge 11:342–345

Kwapong P, Aidoo K, Combey R, Karikari A（2010）Stingless bees: Importance, management and utilisation. Unimax MacMillan LTD, Ghana

Leão KL（2019）Desenvolvimento colonial em abelhas nativas sem ferrão Amazônicas（Apidae: Meliponini）: Tamanho populacional, Nutrição e Alocação fenotípica. Universidade Federal do Pará & Embrapa Amazônia Oriental, Brazil

Leonhardt SD（2017）Chemical ecology of stingless bees. Journal of Chemical Ecology 43:385–402

Lichtenberg EM, Imperatriz-Fonseca VL, Nieh JC（2010）Behavioral suites mediate group-level foraging dynamics in communities of tropical stingless bees. Insectes Sociaux 57:105–113

Lima M a. P, Martins GF, Oliveira EE, Guedes RNC（2016）Agrochemical-induced stress in stingless bees: peculiarities, underlying basis, and challenges. Journal of Comparative Physiology A 202:733–747

Lindauer M, Kerr WE（1960）Communication between the workers of stingless bees. Bee World 41:29–71

Lindauer M, Kerr WE（1958）Die gegenseitige Verständigung bei den stachellosen Bienen. Journal of Comparative Physiology A 41:405–434

Lopes BSC, Campbell AJ, Contrera FAL（2020）Queen loss changes behavior and increases longevity in a stingless bee. Behavioral Ecology and Sociobiology 74:35

Luna-Lucena D, Rabico F, Simões ZLP（2019）Reproductive capacity and castes in eusocial stingless bees（Hymenoptera: Apidae）. Current Opinion in Insect Science 31:20–28

Maia-Silva C, Hrncir M, Silva CI, Imperatriz-Fonseca VL（2015）Survival strategies of stingless bees（*Melipona subnitida*）in an unpredictable environment, the Brazilian tropical dry forest. Apidologie 46:631643

Massaro FC, Brooks PR, Wallace HM, Russell FD（2011）Cerumen of Australian stingless bees（*Tetragonula carbonaria*）: gas chromatography-mass spectrometry fingerprints and potential anti-inflammatory properties. Naturwissenschaften 98:329–337

Mateus S, Ferreira-Caliman MJ, Menezes C, Grüter C（2019）Beyond temporal-polyethism: division of labor in the eusocial bee *Melipona marginata*. Insectes Sociaux 66:317–328

Melo GAR (2020) Stingless Bees (Meliponini). In: Starr CK (ed) Encyclopedia of Social Insects. Springer International Publishing, Cham, pp 1–18

Menezes C, Vollet-Neto A, Contrera FAFL, Venturieri GC, Imperatriz-Fonseca VL (2013) The role of useful microorganisms to stingless bees and stingless beekeeping. In: Vit P, Pedro SRM, Roubik DW (eds) Pot-Honey. Springer, pp 153–171

Michener CD (2007) The bees of the world, 2nd edn. The Johns Hopkins University Press, Baltimore

Michener CD (1974) The Social Behavior of the Bees. Harvard University Press, Cambridge, Massachusetts

Michener CD (1946) Notes on the habits of some Panamanian stingless bees (Hymenoptera, Apidae). Journal of the New York Entomological Society 54:179–197

Month-Juris E, Ravaiano SV, Lopes DM, Salomão TMF, Martins GF (2020) Morphological assessment of the sensilla of the antennal flagellum in different castes of the stingless bee *Tetragonisca fiebrigi*. Journal of Zoology 310:110–125

Moreira DR, Gigliolli AAS, Falco JRP, Julio AHF, Volnistem EA, Chagas F, Toledo VAA, Ruvolo-Takasusuki MCC (2018) Toxicity and effects of the neonicotinoid thiamethoxam on *Scaptotrigona bipunctata* lepeletier, 1836 (Hymenoptera: Apidae). Environmental Toxicology 33:463–475

Müller F (1874) The habits of various insects. Nature 10:102–103

Namu FN, Wittmann D (2016) An African stingless bee *Plebeina hildebrandti* Friese nest size and design (Apidae, Meliponini). African Journal of Ecology 55:111–114

Nieh JC (2004) Recruitment communication in stingless bees (Hymenoptera, Apidae, Meliponini). Apidologie 35:159–182

Nieh JC, Kruizinga K, Barreto LS, Contrera FAL, Imperatriz-Fonseca VL (2005) Effect of group size on the aggression strategy of an extirpating stingless bee, *Trigona spinipes*. Insectes Sociaux 52:147–154

Njoya MTM, Wittmann D (2013) Tasks partitioning among workers of *Meliponula* (*Meliplebeia*) *becarrii* (Meliponini) in Cameroon. International Journal of Environmental Sciences 3:1796–1805

Njoya MTM, Seino RA, Wittmann D, Kenneth T (2018) Nest Architecture and Colony Characteristics of *Meliponula bocandei* (Hymenoptera, Apidae, Meliponini) in

Cameroon. International Journal of Research in Agricultural Sciences 5:274–279

Nogueira DS, Júnior JEDS, Oliveira FFD, Oliveira MLD (2019) Review of *Scaura* Schwarz, 1938 (Hymenoptera: Apidae: Meliponini). Zootaxa 4712:451–496

Nogueira-Neto P (1970) Behavior problems related to the pillages made by some parasitic stingless bees (Meliponinae, Apidae). In: Aronson LR (ed) Development and evolution of behavior: Essays in memory of TC Schneirla. W. H. Freeman, San Francisco, California, pp 416–434

Nogueira-Neto P (1997) Vida e Criação de Abelhas Indígenas Sem Ferrão. Editora Nogueirapis, São Paulo

Nogueira-Neto P (1951) Stingless Bees and their Study. Bee World 32:73–76

Nordin A, Sainik NQAV, Chowdhury SR, Saim AB, Idrus RBH (2018) Physicochemical properties of stingless bee honey from around the globe: A comprehensive review. Journal of Food Composition and Analysis 73:91–102

Nweze JA, Okafor JI, Nweze EI, Nweze JE (2017) Evaluation of physicochemical and antioxidant properties of two stingless bee honeys: a comparison with *Apis mellifera* honey from Nsukka, Nigeria. BMC Research Notes 10:566

Oddo LP, Heard TA, Rodríguez-Malaver A, Pérez RA, Fernández-Muiño M, Sancho MT, Sesta G, Lusco L, Vit P (2008) Composition and Antioxidant Activity of *Trigona carbonaria* Honey from Australia. Journal of Medicinal Food 11:789–794

Oster GF, Wilson EO (1978) Caste and Ecology in the Social Insects. Princeton University Press, Princeton

Pan P, Wang S, Zhong Y, Xu H, Wang Z (2020) New record of the stingless bee *Tetragonula gressitti* (Sakagami, 1978) in Southwest China (Hymenoptera: Apidae: Meliponini). J Apic Res (in press)

Pedro SRM, Cordeiro GD (2015) A new species of the stingless bee *Trichotrigona* (Hymenoptera: Apidae, Meliponini). Zootaxa 3956:389–402

Portugal-Araojo DE, Kerr WE (1959) A case of sibling species among social bees. Revista Brasileira de Biologia 19:223–228

Potts SG, Biesmeijer JC, Kremen C, Neumann P, Schweiger O, Kunin WE (2010) Global pollinator declines: trends, impacts and drivers. Trends in Ecology & Evolution 25:345–353

Potts SG, Imperatriz-Fonseca V, Ngo HT, Aizen MA, Biesmeijer JC, Breeze TD, Dicks

LV, Garibaldi LA, Hill R, Settele J (2016) Safeguarding pollinators and their values to human well-being. Nature 540:220–229

Quezada-Euán JJG (2018) Stingless Bees of Mexico: The Biology, Management and Conservation of an Ancient Heritage. Springer, Cham

Quezada-Euán JJG, Alves DA (2020) Meliponiculture. In: Starr C (ed) Encyclopedia of Social Insects. Springer International Publishing, Cham, pp 1–6

Quezada-Euán JJG, González-Acereto JA (2002) Notes on the nest habits and host range of cleptobiotic *Lestrimelitta niitkib* (Ayala 1999) (Hymenoptera: Meliponini) from the Yucatan Peninsula, Mexico. Acta Zoológica Mexicana 86:245–249

Quezada-Euán JJG, Nates-Parra G, Maués MM, Roubik DW, Imperatriz-Fonseca VL (2018) The economic and cultural values of stingless bees (Hymenoptera: Meliponini) among ethnic groups of tropical America. Sociobiology 65:534–557

Quintal RB, Roubik DW (2013) Melipona Bees in the Scientific World: Western Cultural Views. In: Vit P, Pedro SRM, Roubik DW (eds) Pot-Honey: A legacy of stingless bees. Springer, New York, NY, pp 247–259

Ramírez VM, Calvillo LM, Kevan PG (2013) Effects of Human Disturbance and Habitat Fragmentation on Stingless Bees. In: Vit P, Pedro SRM, Roubik DW (eds) Pot-Honey: A legacy of stingless bees. Springer, New York, NY, pp 269–282

Ramos-Elorduy J (2002) Edible insects of chiapas, Mexico. Ecology of Food and Nutrition 41:271–299

Randrianandrasana M, Berenbaum MR (2015) Edible Non-Crustacean Arthropods in Rural Communities of Madagascar. etbi 35:354–383

Rao PV, Krishnan KT, Salleh N, Gan SH (2016) Biological and therapeutic effects of honey produced by honey bees and stingless bees: a comparative review. Revista Brasileira de Farmacognosia 26:657–664

Rasmussen C, Cameron S (2010) Global stingless bee phylogeny supports ancient divergence, vicariance, and long distance dispersal. Biological Journal of the Linnean Society 99:206–232

Rasmussen C, Thomas JC, Engel MS (2017) A New Genus of Eastern Hemisphere Stingless Bees (Hymenoptera: Apidae), with a Key to the Supraspecific Groups of Indomalayan and Australasian Meliponini. American Museum Novitates 3888:1–33

Ribeiro MF, Wenseleers T, Santos Filho PS, Alves DA (2006) Miniature queens in stingless bees: basic facts and evolutionary hypotheses. Apidologie 37:191–206

Rodríguez-Malaver AJ (2013) Antioxidant Activity of Pot-Honey. In: Vit P, Pedro SRM, Roubik DW (eds) Pot-Honey: A legacy of stingless bees. Springer, New York, NY, pp 475–480

Roldão-Sbordoni YS, Nascimento FS, Mateus S (2018) Estimating colonies of *Plebeia droryana* (Friese, 1900) (Hymenoptera, Apidae, Meliponini): adults, brood and nest structure. Sociobiology 65:280–284

Rosales GRO (2013) Medicinal Uses of *Melipona beecheii* Honey, by the Ancient Maya. In: Vit P, Pedro SRM, Roubik D (eds) Pot-Honey: A legacy of stingless bees. Springer, New York, NY, pp 229–240

Roubik DW (1989) Ecology and Natural History of Tropical Bees. Cambridge University Press, New York

Roubik DW (2006) Stingless bee nesting biology. Apidologie 37:124–143

Roubik DW (2018) 100 Species of Meliponines (Apidae: Meliponini) in a Parcel of Western Amazonian Forest at Yasuní Biosphere Reserve, Ecuador. In: Vit P, Pedro SRM, Roubik DW (eds) Pot-Pollen in Stingless Bee Melittology. Springer International Publishing, Cham, pp 189–206

Roubik DW (1982) Seasonality in colony food storage, brood production and adult survivorship: studies of *Melipona* in tropical forest (Hymenoptera: Apidae). Journal of the Kansas Entomological Society 55:789–800

Roubik DW (1983) Nest and colony characteristics of stingless bees from Panama (Hymenoptera: Apidae). Journal of the Kansas Entomological Society 56:327–355

Roubik DW (1979) Nest and colony characteristics of stingless bees from French Guiana (Hymenoptera: Apidae). Journal of the Kansas Entomological Society 52:443–470

Roubik DW, Patiño JEM (2009) *Trigona corvina*: an ecological study based on unusual nest structure and pollen analysis. Psyche 2009:268756

Sakagami S, Inoue T, Yamane S, Salmah S (1983) Nest architecture and colony composition of the Sumatran stingless bee *Trigona* (*Tetragonula*) *laeviceps*. Kontyu 51:100–111

Sakagami SF (1971) Ethosoziologischer Vergleich zwischen Honigbienen und stachellosen Bienen. Zeitschrift für Tierpsychologie 28:337–350

Sakagami SF (1982) Stingless bees. In: Hermann HR (ed) Social Insects III. Academic Press, New York, pp 361–423

Sakagami SF, Camilo C, Zucchi R (1973) Oviposition behavior of a Brazilian stingless bee, *Plebeia* (*Friesella*) *schrottkyi*, with some remarks on the behavioral evolution in stingless bees. Journal of the Faculty of Science Hokkaido University Series VI, Zoology 19:163–189

Samuelson EEW, Chen-Wishart ZP, Gill RJ, Leadbeater E (2016) Effect of acute pesticide exposure on bee spatial working memory using an analogue of the radial-arm maze. Scientific Reports 6:1–11

Sánchez-Bayo F, Wyckhuys KAG (2019) Worldwide decline of the entomofauna: A review of its drivers. Biological Conservation 232:8–27

Santos CF (2012) Biologia reprodutiva de rainhas e machos de *Tetragonisca angustula* (Hymenoptera: Meliponini). PhD Thesis, Universidade de São Paulo, Brazil

Santos CF, Nunes-Silva P, Halinski R, Blochtein B (2014) Diapause in Stingless Bees (Hymenoptera: Apidae). Sociobiology 61:369–377

Santos DCJ (2013) Divisão de trabalho e sua relação com a dinâmica dos hidrocarbonetos cuticulares em *Melipona scutellaris* (Hymenoptera, Apidae, Meliponini). University of São Paulo, Ribeirão Preto, Brazil, MSc Thesis

Santos CF, Acosta AL, Dorneles AL, Santos PDS, Blochtein B (2016) Queens become workers: pesticides alter caste differentiation in bees. Scientific Reports 6:31605

Santos CF, Otesbelgue A, Blochtein B (2018) The dilemma of agricultural pollination in Brazil: Beekeeping growth and insecticide use. PLoS ONE 13:e0200286

Sarto MCL, Oliveira EE, Guedes RNC, Campos LAO (2014) Differential insecticide susceptibility of the Neotropical stingless bee *Melipona quadrifasciata* and the honey bee *Apis mellifera*. Apidologie 45:626–636

Schwarz HF (1948) Stingless Bees (Meliponidae) of the Western Hemisphere. Bulletin of the American Museum of Natural History 90:1–546

Schwarz HF (1939) The Indo-Malayan species of *Trigona*. Bulletin of the American Museum of Natural History 76:83–141

Schwarz HF (1945) The wax of stingless bees (Meliponidae) and the uses to which it has been put. Journal of the New York Entomological Society 53:137–144

Se KW, Ibrahim RKR, Wahab RA, Ghoshal SK (2018) Accurate evaluation of sugar contents in stingless bee (*Heterotrigona itama*) honey using a swift scheme. Journal of Food Composition and Analysis 66:46–54

Seehuus S-C, Norberg K, Gimsa U, Krekling T, Amdam GV (2006) Reproductive protein protects functionally sterile honey bee workers from oxidative stress. Proceedings of the National Academy of Sciences of the United States of America 103:962–967

Segers FHID, Menezes C, Vollet-Neto A, Lambert D, Grüter C (2015) Soldier production in a stingless bee depends on rearing location and nurse behaviour. Behavioral Ecology and Sociobiology 69:613–623

Seide VE, Bernardes RC, Pereira EJG, Lima MAP (2018) Glyphosate is lethal and Cry toxins alter the development of the stingless bee *Melipona quadrifasciata*. Environmental Pollution 243:1854–1860

Shackleton K, Al Toufailia H, Balfour NJ, Nascimento FS, Alves DA, Ratnieks FLW (2015) Appetite for self-destruction: suicidal biting as a nest defense strategy in *Trigona* stingless bees. Behavioral Ecology and Sociobiology 69:273–281

Simões D, Bego LR (1979) Estudo da regulação social em *Nannotrigona* (*Scaptotrigona*) *postica* Latreille, em duas colónias (normal e com rainhas virgens), com especial referência ao polietismo etário (Hym., Apidae, Meliponinae). Boletim de Zoologia, Universidade de São Paulo 4:89–98

Siviter H, Brown MJF, Leadbeater E (2018a) Sulfoxaflor exposure reduces bumblebee reproductive success. Nature 561:109–112

Siviter H, Koricheva J, Brown MJF, Leadbeater E (2018b) Quantifying the impact of pesticides on learning and memory in bees. Journal of Applied Ecology 55:2812–2821

Slaa EJ (2006) Population dynamics of a stingless bee community in the seasonal dry lowlands of Costa Rica. Insectes Sociaux 53:70–79

Slaa EJ, Chaves LAS, Malagodi-Braga KS, Hofstede FE (2006) Stingless bees in applied pollination: practice and perspectives. Apidologie 37:293–315

Smedal B, Brynem M, Kreibich CD, Amdam GV (2009) Brood pheromone suppresses physiology of extreme longevity in honeybees (*Apis mellifera*). Journal of Experimental Biology 212:3795–3801

Sommeijer MJ (1984) Distribution of labour among workers of *Melipona favosa* F.:

age-polyethism and worker oviposition. Insectes Sociaux 31:171–184

Sommeijer MJ, de Bruijn LL, Meeuwsen F, Slaa EJ (2003) Reproductive behaviour of stingless bees: nest departures of non-accepted gynes and nuptial flights in *Melipona favosa* (Hymenoptera: Apidae, Meliponini). Entomologische Berichten-Nederlandsche Entomologische Vereenigung 63:7–13

Sommeijer MJ, Houtekamer JL, Bos W (1984) Cell construction and egg-laying in *Trigona nigra paupera* with a note on the adaptive significance of oviposition behaviour of stingless bees. Insectes Sociaux 31:199–217

Souza B, Roubik DW, Barth O, Heard T, Enriquez E, Carvalho C, Villas-Boas J, Marchini L, Locatelli J, Persano-Oddo L (2006) Composition of stingless bee honey: Setting quality standards. Interciencia 31:867

Starr CK, Sakagami SF (1987) An extraordinary concentration of stingless bee colonies in the Philippines, with notes on nest structure (Hymenoptera: Apidae: *Trigona spp.*). Insectes Sociaux 34:96–107

Stearman AM, Stierlin E, Sigman ME, Roubik DW, Dorrien D (2008) Stradivarius in the Jungle: Traditional Knowledge and the Use of "Black Beeswax" Among the Yuquí of the Bolivian Amazon. Human Ecology 36:149–159

Tavares MG, Lopes DM, Campos LAO (2017) An overview of cytogenetics of the tribe Meliponini (Hymenoptera: Apidae). Genetica 145:241–258

Terada Y, Garofalo CA, Sakagami SF (1975) Age-survival curves for workers of two eusocial bees (*Apis mellifera* and *Plebeia droryana*) in a subtropical climate, with notes on worker polyethism in *P. droryana*. Journal of Apicultural Research 14:161–170

Tomé HVV, Martins GF, Lima MAP, Campos LAO, Guedes RNC (2012) Imidacloprid-Induced Impairment of Mushroom Bodies and Behavior of the Native Stingless Bee *Melipona quadrifasciata anthidioides*. PLoS ONE 7:e38406

Tornyie F, Kwapong PK (2015) Nesting ecology of stingless bees and potential threats to their survival within selected landscapes in the northern Volta region of Ghana. African Journal of Ecology 53:398–405

Tóth E, Queller DC, Dollin A, Strassmann JE (2004) Conflict over male parentage in stingless bees. Insectes Sociaux 51:1–11

Travenzoli NM, Cardoso DC, Werneck HA, Fernandes-Salomão TM, Tavares MG,

Lopes DM (2019) The evolution of haploid chromosome numbers in Meliponini. PLoS ONE 14:e0224463

Tsvetkov N, Samson-Robert O, Sood K, Patel HS, Malena DA, Gajiwala PH, Maciukiewicz P, Fournier V, Zayed A (2017) Chronic exposure to neonicotinoids reduces honey bee health near corn crops. Science 356:1395–1397

Tuksitha L, Chen YLS, Chen YL, Wong KY, Peng CC (2018) Antioxidant and antibacterial capacity of stingless bee honey from Borneo (Sarawak). Journal of Asia-Pacific Entomology 21:563–570

Valdovinos-Núñez GR, Quezada-Euán JJG, Ancona-Xiu P, Moo-Valle H, Carmona A, Sánchez ER (2009) Comparative Toxicity of Pesticides to Stingless Bees (Hymenoptera: Apidae: Meliponini). Journal of Economical Entomology 102:1737–1742

van Benthem FDJ, Imperatriz-Fonseca VL, Velthuis HHW (1995) Biology of the stingless bee *Plebeia remota* (Holmberg): observations and evolutionary implications. Insectes Sociaux 42:71–87

van Veen JW, Arce Arce HG, Sommeijer MJ (2004) Production of queens and drones in *Melipona beecheii* (Meliponini) in relation to colony development and resource availability. Proceedings of the Netherlands Entomological Society 15:35–39

van Veen JW, Sommeijer MJ, Meeuwsen F (1997) Behaviour of drones in *Melipona* (Apidae, Meliponinae). Insectes Sociaux 44:435–447

Velthuis HHW (1976) Environmental, genetic and endocrine influences in stingless bee caste determination. In: Lüscher M (ed) Phase and Caste Determination in Social Insects: Endocrine Aspects. Pergamon Press, Washington D.C., pp 35–53

Vit P (2013) *Melipona favosa* Pot-Honey from Venezuela. In: Vit P, Pedro SRM, Roubik DW (eds) Pot-Honey: A legacy of stingless bees. Springer, New York, NY, pp 363–373

Vit P, Medina M, Enríquez E (2004) Quality standards for medicinal uses of Meliponinae honey in Guatemala, Mexico and Venezuela. Bee World 85:2–5

Vit P, Pedro SR, Roubik D (2013) Pot-honey: a legacy of stingless bees. Springer, New York

von Frisch K (1967) The dance language and orientation of bees. Harvard University Press, Cambridge von Ihering H（1886）Der Stachel der Meliponen. Entomologische Nachrichten 12:177–191

von Ihering H（1903）Biologie der stachellosen Honigbienen Brasiliens. Zoologische Jahrbücher Abteilung für Systematik Ökologie und Geographie der Tiere 19:179–287

Wille A（1983）Biology of the stingless bees. Annual Review of Entomology 28:41–64

Wille A（1966）Notes on two species of ground nesting stingless bees（*Trigona mirandula* and *T. buchwaldi*）from the pacific rain forest of Costa Rica. Revista de Biologia Tropical 14:251–277

Wille A（1969）A new species of stingless bee *Trigona*（*Plebeia*）from Costa Rica, with descriptions of its general behavior and cluster-type nest. Revista de Biologia Tropical 15:299–313

Wille A, Michener CD（1973）The nest architecture of stingless bees with special reference to those of Costa Rica. Revista de Biologia Tropical 21:9–278

Winfree R, Aguilar R, Vázquez DP, LeBuhn G, Aizen MA（2009）A meta-analysis of bees' responses to anthropogenic disturbance. Ecology 90:2068–2076

Witter S, Blochtein B, Andrade F, Wolff LF, Imperatriz-Fonseca VL（2007）Meliponicultura no Rio Grande do Sul: contribuição sobre a biologia e conservação de *Plebeia nigriceps*（Friese 1901）（Apidae, Meliponini）. Bioscience Journal, Uberlândia 23:134–140

Wittwer B, Elgar MA（2018）Cryptic castes, social context and colony defence in a social bee, *Tetragonula carbonaria*. Ethology 124:617–622

Zamora G, Arias ML, Aguilar I, Umaña E（2013）Costa Rican Pot-Honey: Its Medicinal Use and Antibacterial Effect. In: Vit P, Pedro SRM, Roubik DW（eds）Pot-Honey: A legacy of stingless bees. Springer, New York, NY, pp 507–512

Żrałka J, Helmke C, Sotelo L, Koszkul W（2018）The discovery of a beehive and identification of apiaries among the ancient Maya. Latin American Antiquity 29:514–531

Żrałka J, Koszkul W, Radnicka K, Sotelo Santos LE, Hermes B（2014）Excavations in Nakum structure 99: new data on Protoclassic rituals and Precolumbian Maya beekeeping. Estudios de Cultura Maya 44:85–117

Żrałka J, Helmke C, Sotelo L, Koszkul W (2018) The discovery of a beehive and identification of apiaries among the ancient Maya. Lat Am Antiq 29:514–531

2 无刺蜂的进化和物种多样性

无刺蜂（Meliponini 麦蜂族）属于具花粉筐蜂，是一个单系群体，其特征是后足上有特殊的花粉携带结构（"花粉篮"或花粉筐）（第 1 章；图 1.1）（Michener，2007）。凹面花粉筐的演变使得具花粉筐蜂能够以有效的方式运输大量的花粉甚至是树脂（Martins et al., 2014）[1]。具花粉筐蜂还包括了其他 3 个类群：高度真社会性的蜜蜂（Apini 蜜蜂族）、原始社会性的熊蜂（Bombini 熊蜂族）和大多数独居的兰蜂（Euglossini 兰蜂族）（Grimaldi and Engel，2005；Michener，2007）。这里的真社会性（有时也称为高级社会性和超级有机体）一词通常用于指代具有形态上完全不同的蜂王和工蜂级型的物种（Michener，1974；Michener，2007；Danforth et al., 2013；Boomsma and Gawne，2018）。个别研究者也认为，常年以蜂群形式生活（Michener，2007）或成蜂之间存在广泛食物交换（Michener，1974）才是高度真社会性蜂群的定性特征。而在原始社会性蜂群中，蜂王和工蜂虽然在形态上看起来很相似，却能通过二者体型大小来进行区分（Michener，1974，2007；Danforth et al., 2013）[2]。人们可以通过许多形态学特征将无刺蜂与其他具花粉筐蜂区分开来，如无刺蜂的前翅翅脉退化，有刚毛簇，后足基跗节基部外角没有小突起（用以将花粉压入花粉篮内），后翅中存在翅轭，且螫针退化（图 1.1，图 1.3）（Michener，2007）。

---

[1] 有研究者猜测花粉筐之所以进化出来，与蜂类为建设蜂巢而加强收集和应用树脂有关（Melo，2020）。

[2] 研究者通过更详尽的观察研究发现二者在行为和生理上也存在很多其他不同。研究者对原始社会性这一术语的最新的详尽探讨参见 Boomsma 和 Gawne（2018）。

研究者对于 4 类具花粉筐蜂类群之间的进化关系一直争论不休。而其中有两个关键且相互关联的问题：①无刺蜂是与蜜蜂进化关系更近还是与熊蜂进化关系更近；②无刺蜂是独立于蜜蜂进化出了高度真社会性，还是无刺蜂的高度真社会性源自无刺蜂和蜜蜂的共同祖先中所具有的高度真社会性。在《物种起源》（*Darwin*，1859，第 7 章）一书中，达尔文是最早推测出具花粉筐蜂中 3 个真社会性类群间的进化关系的研究者。基于 *Melipona beecheii* 无刺蜂蜂巢具有巢脾结构，他认为无刺蜂是介于熊蜂和蜜蜂之间的蜂类，但与熊蜂的进化关系更近。在过去的几十年里有大量研究探索了这一假设，并指出了不同具花粉筐蜂类群之间的进化关系。而这些研究面临的一个关键挑战是这些具花粉筐蜂的表型数据和分子数据不同，往往得到的结论也不同（Danforth et al., 2013; Almeida and Porto, 2014; Porto et al., 2016）。主要依据表型（形态学/行为学）数据的相关研究表明，Meliponini 麦蜂族和 Apini 蜜蜂族是姊妹群，并且它们的共同祖先曾发生了一次高度真社会性进化（图 2.1 A）（Schultz et al., 2001; Noll, 2002; Winston and Michener, 1977），而分子研究则主要表明 Meliponini 麦蜂族和 Bombini 熊蜂族是姊妹群，并且 Apini 蜜蜂族和 Meliponini 麦蜂族中具有两个独立的高度真社会性起源（图 2.1 B，图 2.1 C）（Lockhart and Cameron, 2001; Cameron and Mardulyn, 2001; Rasmussen and Cameron, 2007; Kawakita et al., 2008; Cardinal and Danforth, 2011, 2013; Martins et al., 2014; Martins and Melo, 2016; Romiguier et al., 2016; Bossert et al., 2017, 2019）。Payne（2014）所进行的一项综合分析中，结合了形态学、行为学和分子数据，来解答具花粉筐蜂之间的系统发育关系，其研究结果更支持上述的第二种情况。研究者最近的分析也进一步支持了第二种情况，即具花粉筐蜂的高度真社会性的双重起源，以及无刺蜂和熊蜂之间是姊妹群关系。而且根据目前相关研究最支持的情况（图 2.1 C），在蜜蜂族、熊蜂族和麦蜂族的共同祖先中，已经发生了一次社会性进化（Romiguier et al., 2015; Bossert et al., 2017, 2019）。这表明，当今蜜蜂和无刺蜂的祖先是相互独立进化出的高度社会性生活方式，即不同形态雌蜂级型分化、多年生蜂群生活方式和广泛的食物共享（Melo, 2020）。两个独立的高度真社会性的不同起源将有助于解释为什么蜜蜂和无刺蜂在蜂群繁殖（第 4 章）、养育幼虫（第 5 章）、蜂群防御（第 7 章）或采集交流（第 10 章）（表 1.2）等方面上有如此巨大的差异。

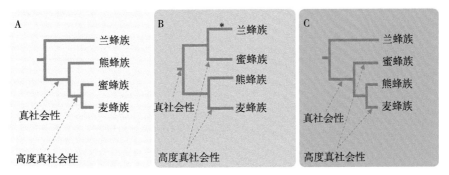

图 2.1　现存 4 种具花粉筐蜂族群的 3 种可能存在的系统发育假设

注：图 A 为 Michener（1944）和 Noll（2002）提出的；图 B 中的红色星标是指 Euglossini 兰蜂族已经丧失了真社会性（Cardinal and Danforth，2011）。

2.1　当今无刺蜂的多样性和分布

2.1.1　当今有多少种无刺蜂？

目前，人们已经记述到的无刺蜂有约 58 个属，其中包括约 550 个种（表 2.1 列出了 552 种；图 1.2，图 2.2），这使得麦蜂族无刺蜂成为具花粉筐蜂类群中最大和最具多样化的类群。相比之下，蜜蜂约有 11 个种，熊蜂约有 250 个种，兰蜂有 200～250 个种（Michener，2007; Danforth et al., 2013; Ascher and Pickering，2018）。此外还存在许多未被记述过的无刺蜂蜂种（Eardley，2004; Michener，2007; Rasmussen and Camargo，2008; Freitas et al., 2009; Rasmussen and Cameron，2010; Pedro，2014; Hurtado–Burillo et al., 2017; Ndungu et al., 2017; Roubik，2018）。例如，微小型的新热带植物区 *Plebeia minima* 无刺蜂就是一个可能由几个蜂种组成的无刺蜂种群（Drummond et al., 2000）。Roubik（2018）在对厄瓜多尔亚马孙地区高度多样化的 Yasuní 国家公园的相对小型地块进行调研时发现，该区域仅仅 *Trigonisca* 属就有 16 种未被记述过的蜂种。他在 Yasuní 发现的 100 种无刺蜂蜂种中总共有 43 种是科学界新发现的无刺蜂蜂种（Roubik，2018）。Pedro（2014）在调查巴西无刺蜂时发现，仅在巴西圣保罗大学昆虫标本馆中就存在 244 个已被研究者记录过的蜂种和 89 个尚未被报道过的蜂种，需要研究者进行仔细的分类学修订。因此，仅新热带区就可能有超过 500 种无刺蜂。

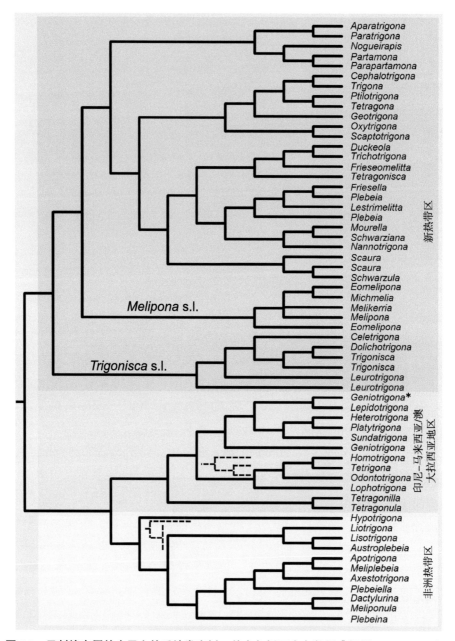

图 2.2 无刺蜂在属的水平上的系统发育树，其中包括了大多数属[根据 Rasmussen 和 Cameron（2010）修改]

注：虚线显示了备选方案；s.l.=sensu lato（广义的）。基于该系统发育树，Rasmussen 和 Cameron（2010）提议将 *Schwarzula* 属同义归入 *Scaura* 属，*Dolichotrigona* 属同义归入 *Trigonisca* 属。*Geniotrigona incisa*（与 *Lepidotrigona* 亲缘关系密切的 *Geniotrigona**）后来被归入 *Wallacetrigona* 新属中，名为 *W. incisa*（Rasmussen et al., 2017）（表 2.1）。需要注意，*Leurotrigona* 属和 *Plebeia* 属不是单系的。

表2.1展示了无刺蜂相关属列表，该表中的属名称主要依据Moure（1961）的观点并且由Rasmussen和Cameron（2007, 2010）的更新。也有其他人提出了不同的（"集总"）分类系统来避免蜂属过多（Wille and Michener, 1973; Wille, 1979; Michener, 1990, 2007）。笔者此处遵循Sakagami（1982）的推论，即Moure的分类群代表了在行为、生态和遗传差异方面表现显著的自然类群（Rasmussen and Cameron, 2010; Roubik, 2018）。Sakagami（1982）还指出，这两种分类系统通常不会相互矛盾，因此能显示出实质性的对应关系（Rasmussen, 2008; Rasmussen et al., 2017提供了不同属名的比较）[3]。

2.1.2 哪里可以找到无刺蜂？

当今的无刺蜂生活在非洲、亚洲、澳大利亚和美洲的热带及亚热带地区（图2.3）。新热带区的无刺蜂最为丰富且物种繁多（约426种），北至古巴和墨西哥（锡那罗亚州），南达阿根廷（图2.3）都有无刺蜂出现。分布在阿根廷最南端的无刺蜂蜂种（*Plebeia* spp.）可以在圣路易斯、圣达菲以及包括布宜诺斯艾利斯市在内的布宜诺斯艾利斯省内找到（Roig-Alsina et al., 2013; Mazzeo and Torretta, 2015; Roig-Alsina and Alvarez, 2017）。在非洲（约有36种无刺蜂，其中包括了马达加斯加特有的7种无刺蜂[4]，无刺蜂蜂种在赤道地区多样化最高（Eardley, 2004; Eardley and Kwapong, 2013; Faber Anguilet et al., 2015）。北方的撒哈拉沙漠是一道天然屏障。而南方的无刺蜂分布可远至南非和马达加斯加南部。大多数非洲无刺蜂蜂种生活在热带森林和热带草原中。有研究者曾报道了2个存在于沙漠地区的无刺蜂蜂种（*Hypotrigona penna*，*Liotrigona* sp.）（Eardley and Kwapong, 2013）。在亚洲和澳大利亚（约90种），无刺蜂的分布范围从西部的印度（可能还有巴基斯坦东南部）（Rasmussen, 2013）延伸到东部的所罗门群岛，从北部的尼泊尔、中国（云南、海南和台湾）延伸到南部的澳大利亚（Rasmussen, 2008, 2013; Pan et al., 2020）。*Tetragonula carbonaria*无刺蜂分

---

[3] 需要注意的是很多无刺蜂蜂种的学名已经发生改变，有些甚至不止一次改名。本书中使用的无刺蜂学名与引用文献中的无刺蜂学名的差异已经通过咨询Camargo和Pedro（2013）、Rasmussen和Cameron（2007, 2010）、Rasmussen等（2017）和Eardly（2004）得以解决。

[4] 非洲无刺蜂蜂种数量因参考资料不同而有很大差异。例如，Kerr和Maule（1964）提到32种；Eardley（2004）提到有19种；Anguilet等（2015）提到有21种；Kajobe和Roubik（2018）提到有24种。Eardley和Kwapong（2013）强调非洲无刺蜂的多样性可能比目前所知的要高。基于这些参考文献和其他参考资料（Michener, 2007; Rasmussen, 2008），笔者在此已经将36种无刺蜂算作非洲无刺蜂蜂种（第2.4.2节），但需要研究者对非洲属进行分类修订（Eardley and Kwapong, 2013）。

布在澳大利亚（新南威尔士州）包括了温带气候地区的最南段（Halcroft et al., 2013）。系统发育分析证实，在麦蜂族无刺蜂进化史的早期阶段，麦蜂族主要进化成了 3 个不同的类群：非洲热带谱系、印尼 – 马来西亚 / 澳大拉西亚谱系和新热带谱系（图 2.2）（Rasmussen and Cameron，2010）。

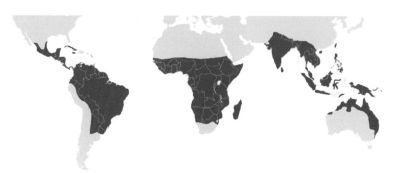

图 2.3　当代无刺蜂的分布位置（蓝色区域）

注：Kerr and Maule（1964）；Sakagami（1982）；Camargo et al.（1988）；Pauly et al.（2001）；Camargo and Pedro（2003）；Eardley（2004）；Rasmussen（2008）；Rasmussen（2013）；Vit et al.（2013）；Dollin et al.（2015）（更多细节参见正文）。

虽然人们发现的无刺蜂通常是栖息在低地雨林中，特别是亚马孙流域栖息的无刺蜂多样性最高（Biesmeijer and Slaa，2006; Melo，2020），但是无刺蜂也可以在中国台湾海拔高达 2 500 m 的地方栖息繁衍（*Lepidotrigona hoozana*），甚至可以在哥伦比亚、厄瓜多尔和秘鲁安第斯山脉海拔 3 400 m 的地方（*Parapartamona* spp.）以及苏门答腊岛、婆罗洲或新几内亚的海拔 2 500 m 的地方栖息繁衍（Roubik，1989; Gonzalez and Smith-Pardo，2003; Sung et al., 2011）。无刺蜂蜂群为了在这些海拔高度上得以生存，需要能够在一定程度上保持巢温，它们可能会在一年中最冷的时期停止繁育幼虫（第 3 章，第 5 章）。

2.2　无刺蜂的起源

无刺蜂的化石记录状况好于大多数其他蜂，目前已知有 12 种已灭绝的无刺蜂（Engel and Michener，2013）。人们在琥珀（树脂化石）和硬树脂（幼琥珀）中发现了无刺蜂化石，这些琥珀中的无刺蜂通常保存完好（图 2.4）。存在这种相对完好化石记录的原因可能是因为无刺蜂本身会收集大量树木树脂用于筑巢、防御和觅食，因此最终被困在树脂中。尽管无刺蜂的化石记录相对完好，但我们对无刺蜂的进化历史或它们如何在世界各地这么多不同的地方栖息的相关信息仍然知之甚少。

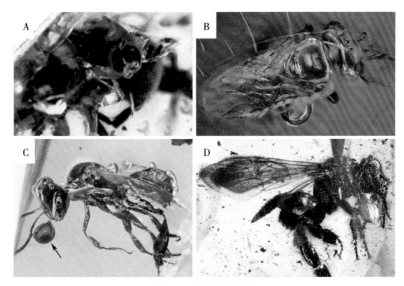

图 2.4 无刺蜂化石

注：A 为 *Cretotrigona prisca* 工蜂，它是目前已知的最古老的无刺蜂化石，可追溯到距今 7 000 万—6 500 万年前（Michener and Grimaldi，1988；Engel，2000 a）。B 为发现于波罗的海的琥珀样本中的一只始新世 *Exebotrigona velteni* 无刺蜂（Engel and Michener，2013，已由 Michael S. Engel©，于 2020 年批准）。C 为最常见的 *Proplebeia dominicana* 无刺蜂化石，这只无刺蜂头部还带着一个兰花（*Globosites apicola*）的花粉块（箭头）（Poinar，2016，www.schweizerbart.de/journals/njgpa）。D 为波罗的海琥珀中的始新世中期 *Melissites trigona*（Melikertini）蜂（Engel，2011）。

现存最古老的化石无刺蜂是 *Cretotrigona prisca* 无刺蜂，它是一只小型工蜂（体长约 5mm），于北美新泽西琥珀中发现了它（图 2.4A）（Michener and Grimaldi，1988 a，1988 b）。据推测这只无刺蜂生活在白垩纪晚期（马斯特里赫特期），距今 7 000 万—6 500 万年前（Engel，2000 a）。该 *Cretotrigona prisca* 无刺蜂也是已确定的蜜蜂科最古老的化石和真社会性蜂最古老的化石（Engel，2000 a）。人们通过其小型的后躯（腹部）（图 1.8）将其认定为工蜂，因为工蜂的卵巢是退化的。这一特征表明 *C. prisca* 是一种高度真社会性蜂（Michener and Grimaldi，1988 a，1988 b；Engel，2000 a）。在始新世中期（距今约 4 500 万年前）的波罗的海琥珀中，人们发现了 3 种无刺蜂化石（*Exebotrigona velteni*[5]，*Liotrigonopsis rozeni* 和 *Kelneriapis eozenica*）（Engel，

---

[5] 研究者最初认为这个蜂种化石是源自中国抚顺的琥珀（Engel and Michener，2013），但是研究者已经重新分析并且认定这个化石是来自卢台特期的波罗的海的琥珀（距今 4 800 万—4 100 万年前）（M.S. Engel，2020）。

2001 a; Engel and Michener，2013）。*Problebeia dominicana* 化石是最常见的无刺蜂化石，常见于多米尼加和墨西哥琥珀中，其历史可以追溯到中新世早期（距今 2 000 万—1 500 万年前）（Camargo et al., 2000）。*Nogueirapis silacea* 无刺蜂化石也是来自中新世早期墨西哥琥珀，该蜂种已经灭绝，但其所属蜂属中有 4 个种仍尚存于世（表 2.1）。

表 2.1　无刺蜂的属列表及属内的种的数量（参考文献参见第 2.4 节的种属列表）

区域	属	种的数量
新热带区	*Aparatrigona* Moure, 1951	2
	Camargoia Moure, 1989	3
	Celetrigona Moure, 1950	4
	Cephalotrigona Schwarz, 1940	5
	Duckeola Moure, 1944	2
	Friesella Moure, 1946	1
	Frieseomelitta Ihering, 1912	16
	Geotrigona Moure, 1943	22
	Lestrimelitta Friese, 1903	23
	Leurotrigona Moure, 1950	4
	Melipona Illiger, 1806	74
	Meliwillea Roubik, Lobo and Camargo, 1997	1
	Mourella Schwarz, 1946	1
	Nannotrigona Cockerell, 1922	10
	Nogueirapis Moure, 1953	4
	Oxytrigona Cockerell, 1917	11
	Parapartamona Schwarz, 1948	7
	Paratrigona Schwarz, 1938	32
	Paratrigonoides Camargo and Roubik, 2005	1
	Partamona Schwarz, 1938	32
	Plebeia Schwarz, 1938	42
	Plectoplebeia Melo, 2016	1
	Ptilotrigona Moure, 1951	3
	Scaptotrigona Moure, 1942	23
	Scaura Schwarz, 1938	8
	Scaura (formerly *Schwarzula* Moure, 1946)	2
	Schwarziana Moure, 1943	4
	Tetragona Lepeletier and Audinet-Serville, 1828	13
	Tetragonisca Moure, 1946	4
	Trichotrigona Camargo and Moure, 1983	2
	Trigona Jurine, 1807	32
	Trigonisca (formerly *Dolichotrigona* Moure, 1950)	10
	Trigonisca Moure, 1950	27

续表

区域	属	种的数量
	31 个属	426
非洲热带区	*Apotrigona* Moure, 1961	1
	Axestotrigona Moure, 1961	7
	Cleptotrigona Moure, 1961	1
	Dactylurina Cockerell, 1934	2
	Hypotrigona Cockerell, 1934	5
	Liotrigona Moure, 1961	12
	Meliplebeia Moure, 1961	4
	Meliponula Cockerell, 1934	1
	Plebeiella Moure, 1961	2
	Plebeina Moure, 1961	1
	10 个属	36
印尼 – 马来半岛 / 澳大拉西亚区	*Austroplebeia* Moure, 1961	5
	Geniotrigona Moure, 1961	2
	Heterotrigona Schwarz, 1939	3
	Homotrigona Moure, 1961	4
	Lepidotrigona Schwarz, 1939	13
	Lisotrigona Moure, 1961	6
	Lophotrigona Moure, 1961	1
	Odontotrigona Moure, 1961	1
	Papuatrigona Michener and Sakagami, 1990	1
	Pariotrigona Moure, 1961	1
	Platytrigona Moure, 1961	5
	Sahulotrigona Engel and Rasmussen, 2017	2
	Sundatrigona Inoue and Sakagami, 1993	2
	Tetragonilla Moure, 1961	4
	Tetragonula Moure, 1961	34
	Tetrigona Moure, 1961	5
	Wallacetrigona Engel and Rasmussen, 2017	1
	17 个属	90
总数	58 个属	552

那么，基于这些化石年龄以及我们目前对当今无刺蜂系统发育关系的理解，我们能否知道无刺蜂作为一个类群的存世时间到底是多久呢？*Cretotrigona prisca* 蜂化石看似是一种真社会性蜂，它与一些现存的无刺蜂

有着惊人的相似之处。因此，无刺蜂出现的时间点最晚也应该是早于 6 500 万—7 000 万年前。但是无刺蜂出现的时间点又一定是晚于 1.2 亿年前的，因为 1.2 亿年前这个时间点也是开花植物多样化和蜂作为一个类群出现的时间（Cardinal and Danforth，2013;Danforth et al., 2013）。最近两项探索蜂进化的研究表明，具花粉筐蜂出现的时间点为距今 8 700 万—8 400 万年前（Cardinal and Danforth，2013;Martins et al., 2014）。这表明无刺蜂是于距今 8 700 万—7 000 万年前的白垩纪晚期进化而来（Rasmussen and Cameron,2010; Cardinal and Danforth,2011）。

2.3　无刺蜂的生物地理学

当今的无刺蜂大多数栖息在热带陆地，但它们最初是源自哪里？它们是从哪些线路扩散到世界不同地区的？这些都仍然是引人注目的未解之谜。例如，无刺蜂是如何定居到瓜德罗普岛、马达加斯加、澳大利亚、所罗门群岛等这些孤立大陆的？过去 1 亿年中错综复杂的地质历史、大陆板块的运动以及气候和海平面的急剧变化导致了复杂的隔离分化史（即分裂祖先种群并导致物种形成的物理和生物屏障）及生物扩散事件（Camargo and Pedro,1992; Rasmussen and Cameron,2010）。而对无刺蜂来说，即使是狭窄的水屏障也是一个巨大的困难。无刺蜂分蜂是通过将筑巢材料从母巢运往新巢来渐进式地建立新蜂群，这种分蜂模式意味着无刺蜂的分蜂距离很短，通常小于 300 m（第 4 章）。这种情况使得无刺蜂分蜂群不太可能跨水域扩张到很远的地方（Kerr and Maule,1964）[6]。无刺蜂这一生物学特征可以通过以下观察结果例证。Francisco 等（2016，2017）研究了巴西海岸线附近岛屿上常见的新热带无刺蜂 *Tetragonisca angustula* 和熊蜂 *Bombus morio* 的种群遗传结构。这些岛屿中许多岛屿与大陆相距不到 5 km。尽管这些岛屿与大陆相对较近，但这些岛屿中的一些岛屿和领地上却完全没有 *T. angustula* 无刺蜂存在，而其他岛屿领地中存在的 *T. angustula* 无刺蜂则非常有可能是由于人类引进而来的（Francisco et al., 2017）。无刺蜂蜂群遗传结构意味着无刺蜂蜂王（以及分蜂群）无法穿过几百米以上的水屏障。不同的是，这些岛屿中的大多数岛屿上都存在熊蜂，熊蜂种群的遗传结构结合人们观察到的在水面上飞行的熊蜂蜂王表明，几千

---

[6] 蜜蜂分蜂群移居的距离要比无刺蜂远得多（Seeley，2010），而据发现熊蜂群是单蜂王，这种单蜂王也能轻松跨越水域。

米宽的水屏障并不能构成 *Bombus morio* 熊蜂迁移的物理障碍（Francisco et al., 2016）。

2.3.1 无刺蜂生物地理分布的相关设想

研究者目前已经提出了几种设想来解释目前无刺蜂的泛热带分布。这些设想大致分为两类（图 2.5）：一类是认为无刺蜂起源于热带非洲（Wille，1979; Michener，2007; Rasmussen and Cameron，2010），另一类则认为热带美洲才是无刺蜂的起源地（Kerr and Maule，1964; Michener，1990; Kajobe and Roubik，2018; Melo，2020）。这两种观点都试图解释新热带无刺蜂和非洲热带无刺蜂蜂种间的遗传差异（图 2.2）（Rasmussen and Cameron，2010; Bossert et al., 2019）。在地球平均温度较高的古近纪时期，无刺蜂的扩张史涉及范围相当大。

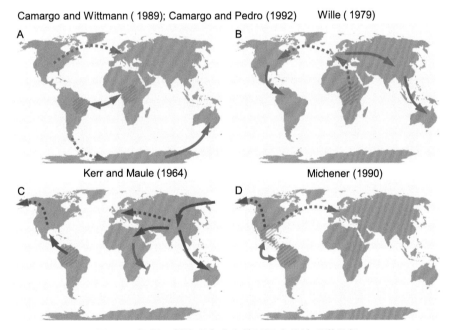

图 2.5　解释无刺蜂现今分布的不同生物地理学设想

注：图中带条纹的区域表示设想中假设的无刺蜂起源地。带有实线箭头表示无刺蜂迁移到目前仍然存在无刺蜂的地方。带虚线箭头表示无刺蜂迁移到目前不存在无刺蜂的地方。相关描述见正文。

Camargo 和 Wittmann（1989）提出了非洲起源说与美洲起源说以外的其他设想（图 2.5 A）（Michener，1979; Camargo and Pedro，1992）。他们认为，无刺蜂起源于距今 13 000 万—10 000 万年前的冈瓦纳大陆，当时非洲和南美洲仍然是相连的。这种设想认为后来迁移到澳洲的无刺蜂可能是通过南极洲迁移过去的，或者是通过北部路线到了欧亚大陆然后再迁移到东南亚和大洋洲的。有证据表明确实存在从南美洲、南极洲到新西兰、澳大利亚的生物扩散事件，这些生物扩散事件是通过南桑威奇群岛和南乔治亚岛实现的，而南桑威奇群岛和南乔治亚岛在白垩纪晚期一些时期中曾形成了一个连续的陆桥（Morley，2003）。在提出这个假设时，人们仍然认为 *Cretotrigona prisca* 蜂化石是更古老的无刺蜂（Michener and Grimaldi，1988 b）。而鉴于 Engel（2000 a，2000 b）重新评估了 *C. prisca* 化石的年龄，而且研究者在无刺蜂和具花粉筐蜂系统发育关系方面取得了一些最新进展（Rasmussen and Cameron，2007，2010；Cardinal and Danforth，2013；Martins et al.，2014），这种冈瓦纳大陆起源说似乎不太可能。相反，更有可能的是，不断增长的南大西洋将非洲和南美洲分开之后（非洲和南美洲之间的直接陆地连接消失于 1 亿年前左右，Morley，2003），自然界才进化出了具花粉筐蜂。

非洲起源论

非洲起源论是 Wille（1979）（图 2.5 B）以及 Rasmussen 和 Cameron（2010）提出的。Wille（1979）认为非洲无刺蜂蜂种比新热带无刺蜂蜂种更"原始"[但 Michener（1990）反对 Wille 研究中对这种特征的阐述]。他认为无刺蜂在始新世期间首先从非洲迁移到欧洲，然后可能是通过格陵兰岛迁移到北美和亚洲（Wille，1979）。这就可以解释为什么在始新世中期（约 4 500 万年前）的欧洲波罗的海琥珀中发现了无刺蜂化石。然而，这些事件发生的时间与新泽西晚白垩纪琥珀中 *Cretotrigona prisca* 无刺蜂化石的存在相矛盾。

Rasmussen 和 Cameron（2010）也提出无刺蜂的起源地在非洲，并认为非洲和南美洲之间的岛链可能促使了无刺蜂种群间保持着联系，直到约 7 000 万年前新大陆和旧大陆物种发生分离后，这种联系才得以终止。非洲和南美洲之间的岛链可以解释在白垩纪—第三纪边界之前，南大西洋的植物为什么会同时出现在南大西洋两岸（Rasmussen and Cameron，2010）。然而，植物能够通过风、水和鸟类媒介穿越这些水屏障（Morley，2003），而如上所述，无刺蜂蜂群不太可能像植物一样轻松地穿越南大西洋。Kerr 和 Maule（1964）提出了另一种无刺蜂扩散模式，他们认为无刺蜂扩散的方式可能是通过浮岛或漂浮的天然木筏。在亚马孙河的河面上，特别是在大雨之后，可以看到带有大量动物的浮岛和漂浮的树木，他们的浮岛扩散论的灵感就是受这一现象的启发而来。他们认为，无刺蜂蜂群也能够以同样的方式越过宽阔的水屏障。浮岛运送的无刺蜂蜂群可存活 20～40 d（Kerr and Maule，1964）。然而，由于许多加勒比岛屿上都没有无刺蜂存在，因此 Kerr 和 Maule（1964）也不太相信他们自己的猜想。Michener（1979）也反对这种猜想，他指出尽管中美洲和南美洲的无刺蜂种类繁多，但在安的列斯群岛的一些岛屿上只存在两种无刺蜂。这两种无刺蜂中的其中一种，*Melipona beecheii* 无刺蜂可能是由人类带到古巴和牙买加的，因为自玛雅时代以来养蜂人就盛行养 *Melipona beecheii* 无刺蜂（第 1 章）（Michener，1979; Camargo et al.，1988; Genaro，2008）。而另一种 *Melipona variegatipes* 无刺蜂则只存在于小安的列斯群岛中的 3 个岛屿上（瓜德罗普岛、多梅尼卡岛和蒙特塞拉特岛）（Camargo et al.，1988; Camargo and Pedro，2013）。如果安的列斯群岛的蜂群是通过漂流而来的，那么安的列斯群岛上就应该会有来自中美洲和南美洲的其他更常见的无刺蜂蜂种栖息。因此，无论是 Kerr 和 Maule（1964）还是 Michener（1979）认为无刺蜂蜂群是采用漂流的方式穿越大型水屏障进行生物扩散的说法不太让人信服。*M. variegatipes* 无刺蜂是如何以及从哪里到达小安的列斯群岛仍然是一个谜（该蜂种是从美洲和南美洲的亲缘蜂中分裂成群的，例如距今 3 000 万—2 000 万年前 *M. beecheii*、*M. fasciculata* 和 *M. grandis* 就是这种情况，参见 Rasmussen 和 Cameron，2010）。大安的列斯群岛（伊斯帕尼奥拉岛、古巴、牙买加、波多黎各等）上并没有出现 *M. variegatipes* 无刺蜂，而且 *M. variegatipes* 无刺蜂与南美洲的 *Melipona* 蜂种形态相似，这一事实表明，小安的列斯群岛等一些较小岛屿上的无刺蜂来自南美洲无刺蜂的移居定殖，并随后出现了定殖蜂群的局部灭绝（Camargo et al.，1988）。因此，无刺蜂仍然有可能是从南美洲建造陆桥或通过水面扩散而来的（Iturralde-Vinent and MacPhee，1999; Hedges，2006;Ali，2012）。

无刺蜂从非洲扩散到南美洲是采用浮岛和木筏作为扩散方式的说法并不是无稽之谈，有一些原因可以用来解释其可信度。系统发育证据表明，在始新世中期，灵长类动物就从非洲穿越大西洋来到了南美洲（Antoine et al.，2012; Bond et al.，2015），而研究者认为用浮岛漂流解释该事件是合理的（Houle，1998; Bond et al.，2015）。Houle（1998）认为 5 000 万年前在合适的风力条件下，一个浮岛大约需要一周时间就能越过大西洋水屏障。而在 7 000 万年前，这一过程需要的时间甚至更少，而且当时的古风向似乎有助于南大西洋的西向跨越（Houle，1998）。一些带有蜂蜜和花粉储备的无刺蜂蜂群在这一过程中得以在水上生存。一株含有许多相同蜂种的多个巢穴的漂浮树就可能会导致无刺蜂从非洲到南美洲的成功迁徙（Michener，1979）。然而，目前缺乏明确的系统发育证据证明无刺蜂横跨南大西洋的扩散事件（如灵长类动物上所出现的情况），在这种情况下浮岛假说仍然有些不能令人信服，因为系统发育证据才是可以用来证实关于各种令人费解的扩散事件的猜想的有力证据。此外，目前缺少支持无刺蜂非洲起源论的化石证据，且非洲无刺蜂的多样性低，这都使人们继续对无刺蜂非洲起源论存疑。

美洲起源论

Kerr 和 Maule（1964）提出了无刺蜂新热带起源论（图 2.5 C），因为新热带是生物数量最多和多样性最丰富的生物家园，这一设想似乎是合理的（Rasmussen and Cameron，2010；表 2.1）。此外，与 Wille（1979）的设想相反，Kerr 和 Maule（1964）认为一些新热带蜂种更为典型且具有更多的"原始"特征（即这些特征被认为在表型上与祖先表现的特征相似），笔者认为这一现象支持了他的猜想，即无刺蜂起源于南美洲。而后，在始新世期间无刺蜂首先扩展到北美，随后通过白令海峡扩散到亚洲。在渐新世期间，无刺蜂能够通过亚洲到达非洲、澳大利亚和亚洲的其他岛屿（Kerr and Maule，1964）。与大多数其他设想一样，这一假设的合理性在很大程度上依赖于在过去 8 000 万年中的某个时间点连接现在遥远大陆的陆桥的存在。

根据 Michener（1990）的系统发育分析（图 2.5 D），他也猜测无刺蜂是在新热带区进化的，当时北美大部分地区仍存在热带气候。这个假设还认为无刺蜂就是在现在的北美洲上进化出来的。无刺蜂从北美洲而后通过白令海峡迁移到亚洲，然后通过格陵兰岛迁移到了欧洲。Michener（1990）没有提供有关这些不同迁徙的潜在时间线的详细信息，而 Rasmussen 和 Cameron（2010）则认为美洲起源可能性不大，因为在过去 1 亿年的大部分时间里，南美洲和北美洲都是分开的。Kerr 和 Maule（1964）及 Michener（1990）的假设里都需要南美洲和北美洲之间相连。

尽管存在这些障碍，美洲作为无刺蜂出生地仍然是有可能的，这主要是因为我们有实物证据表明，无刺蜂（*Cretotrigona prisca*）在其进化的早期，距今 7 000 万—6 500 万年前就曾出现在北美洲（Kajobe and Roubik，2018）。而上文中提到，研究者认为美洲起源论不可信的部分原因是南美洲与北美洲在过去 1 亿年的漫长时期内是相互分离的。然而在大陆分离的早期，陆地连接的细节往往不为人所知（Morley，2003），而加勒比地区的构造也很复杂且存在争议（Pindell et al.，1988；Morley，2003；Ali，2012）。有证据表明，美洲（今天的墨西哥尤卡坦和哥伦比亚）岛之间曾存在洲际陆桥，例如在白垩纪晚期和古新世早期（距今 8 000 万—6 500 万年前）阿维斯海脊和大安的列斯群岛就曾形成了美洲岛之间的洲际陆桥（Iturralde-Vinent and MacPhee，1999; Morley，2003）。这可能使得无刺蜂向两个方向迁移，并使无刺蜂在进化出来之后不久就出现在整个热带美洲（Michener，1990）。大型陆地脊椎动物的化石为这种扩散途径的设想提供了证据，这些化石包括了坎帕尼亚时期（距今 8 300 万—7 200 万年前）的黑龙和角龙化石，以及蛇和蜥蜴等小型脊椎动物的化石（Morley，2003）。此外，系统发育分析表明具花粉筐蜂的近亲 *Centridini* 蜂在距今 8 500 万—7 500 万年前也是使用这种南美洲北美洲连接的方式迁移的（Martins and Melo，2016）。动物的交换扩散在白垩纪—古生代边界时进一步增加（Morley，2003）。随后，在中始新世（距今 4 900 万—3 900 万年前）期间南北美洲的连接发生断裂分离，这意味着北美洲和南美洲的无刺蜂种群也暂时分离开来。

具花粉筐蜂和 *Centridini* 蜂之间的近亲关系（Danforth et al.，2013; Martins et al.，2014）支持了无刺蜂起源于新热带地区的观点（Melo，2020）。研究者认为 *Centridini* 蜂是在南美洲热带地区进化而来的，并在白垩纪晚期约 8 600 万年从具花粉筐蜂中分化出去（Martins and Melo，2016）。因此，人们认为具花粉筐蜂也是在美洲进化和分化出来的（Martins et al.，2014; Martins and Melo，2016）。这使得无刺蜂的具花粉筐蜂祖先出现的时间在美洲无刺蜂出现前不久。

起源自北美洲的无刺蜂可能还有两条其他迁移路线：连接东亚和北美洲的 Beringia 路线和通过格陵兰岛连接北美洲和欧洲的 Thulean 路线（Brikiatis，2014）。在白垩纪晚期（约 6 500 万年前）、古新世晚期（约 5 800 万年前）和始新世早期的温暖时期，东亚和北美洲之间存在着跨越白令陆桥的陆地连接（Morley，2003; Brikiatis，2014）。而通往欧亚大陆的图里安（Thulean）廊道则由于其高纬度，直到晚古新世（距今 6 000 万—5 400 万年前）和早始新世才在南格陵兰岛上建立了北美洲和欧洲的陆地连接（Morley，2003）。研究者认为这种相对短暂的陆地连接发生在晚古新世/早始新世的气候适宜期（约 5 500 万年前）（Zachos et al.，2001; Brikiatis，2014），这种陆地连接为依赖温暖气候的动物的扩散提供了一条路线（Morley，2003）。研究者认为其他蜂种是在距今 6 900 万—4 700 万年前的时间窗口中通过图里安廊道进行了扩散（Praz and Packer，2014）。那么，无刺蜂也有可能通过这种陆地连接从亚洲迁移到欧洲。在古新世（距今6 600万—5 600万年前）期间，动物可以利用欧洲和亚洲之间的路线进行相互迁移扩散，直到从始新世到渐新世（距今5 000万—3 500万年前）的气候变化，这条路线才不被动物使用（Rasmussen and Cameron，2010）。Meliponini族无刺蜂也可能是通过亚洲到达非洲大陆。始新世中期气候最佳的时候（约4 000万年前）可能是无刺蜂进行扩散的一个时间段（Bohaty et al.，2009）。有证据表明在这个时期里很多灵长类、啮齿类和其他亚洲陆地哺乳动物群均从亚洲扩散到了非洲（Chaimanee et al.，2012）。

北方无刺蜂的灭绝与"大断裂"事件

无刺蜂化石显示在始新世到渐新世（距今5 000万—3 500万年前）期间时，北半球也有无刺蜂存在，在这期间北半球的气候发生了缓慢但持续的变化，最初天气以温暖潮湿为主，且北半球主要是潮湿的森林栖息地，而后天气逐渐变冷且栖息地也逐渐变为更为干燥开放的草原（Zachose et al.，2001），这种气候变化可能促使了古北界的无刺蜂蜂种的灭绝（Grimaldi and Engel，2005; Rasmussen and Cameron，2010）。人们认为这种全球降温也造成了其他具花粉筐蜂物种的大量灭绝（Engel，2001 a,2001 b; Grimaldi and Engel，2005）。这些受影响的种群之一就有现已灭绝的Melikertini蜂族群，它是已知的Meliponini族无刺蜂亲缘关系最近的族群。这个蜂族群是由居住在欧洲的几个形态各异的社会性蜂属组成。与无刺蜂的不同之处在于，它们有完整的翅脉和发达的螫刺（图2.4）（Engel，2001 a）。据显示，在始新世—新世过渡时期Melikertine族蜂就已经灭绝了，这个时期与被称为"grande coupure"（大断裂）的灭绝事件发生的时期相吻合（Engel，2001 a）。

无刺蜂不像Melikertini族蜂那样惨遭灭绝，但是在这一全球变冷时期内无刺蜂的分布也发生了巨大的变化。我们尚不清楚无刺蜂为什么不能像熊蜂或蜜蜂那样在较冷的栖息地生存，但研究者推测这可能是因为在较冷条件下，无刺蜂的温度调节能力不足以保持哺育幼虫的最适宜的条件（第3章），也可能是因为在热带和亚热带地区之外，缺少无刺蜂筑巢的适宜巢址以及蜂巢材料（Ortiz-Mora et al.，1995）。鉴于无刺蜂的筑巢习性极具多样性，后一种解释似乎不太可能（第3章）。无刺蜂可能早在距今2 000万—1 500万年前就通过中美洲重新移居到北美洲南部地区（Rasmussen and Cameron，2010）。人们在多米尼加和墨西哥的琥珀中发现了与现存的*Plebeia*属无刺蜂相似的*Proplebeia dominicana*无刺蜂（Moure and Camargo，1978）（图2.4），这可能就是由于无刺蜂重新移居至北美洲的结果。

马达加斯加无刺蜂的迁移定居

无刺蜂是如何迁移到马达加斯加的呢？这仍然是一个有趣的问题。目前人们记述了 7 个马达加斯加地区特有的无刺蜂蜂种，它们都属于 *Liotrigona* 属，研究者在非洲大陆也发现了这个属的蜂种（Pauly et al., 2001; Koch, 2010; Pauly and Fabre Anguilet, 2013）。目前还不知道这些无刺蜂究竟是由出现在马达加斯加岛上的一个单一祖先蜂种扩散开来的，还是由来自非洲大陆的不同 *Liotrigona* 属无刺蜂蜂种蜂群的多次迁移造成的（Koch, 2010）。无刺蜂迁移至马达加斯加这一事实实在令人费解，因为马达加斯加和非洲大陆这两块陆地早在 1.6 亿年前就已经分开了（Yoder and Nowak, 2006）。因此，不能用隔离分化来解释马达加斯加无刺蜂的存在（Fuller et al., 2005）。此外，在过去的 1.2 亿年里，马达加斯加和非洲大陆之间是始终保持着约 400 km 的距离（Yoder and Nowak, 2006）。McCall（1997）认为在距今 4 500 万—2 600 万年前可能有一座陆桥将非洲和马达加斯加连接起来，但这也受到了其他人的质疑，原因之一是非洲和马达加斯加之间的海渠十分深（Rogers et al., 2000; Yoder and Nowak, 2006）。有研究者曾提出过一个漂流假说来解释马达加斯加岛上出现的 *allodopine* 蜜蜂（Fuller et al., 2005），他认为蜜蜂可能借助枯萎的茎和树枝中或搭乘了被冲入莫桑比克海峡的植被筏实现了跨海峡旅行。然而，由于 *allodapines* 蜜蜂采用了一种非常不同的扩散方式，这种扩散方式使漂流成为这种蜂族群的更为合理的选择（Fuller et al., 2005）。目前，人们认为漂流筏假说是一些脊椎动物族群迁移到马达加斯加岛的最合理解释（Yoder and Nowak, 2006）。最近研究者对西方蜜蜂亚种的线粒体 DNA 分析表明，距今 3 万—2 万年前科摩罗群岛可能是让蜜蜂实现从非洲大陆穿越到马达加斯加的垫脚石（Techer et al., 2017）。那么在早期水位较低的时期，无刺蜂会不会也使用了这条路线呢？因此，需要将马达加斯加无刺蜂与非洲大陆的 *Liotrigona* 蜂种进行详细的系统发育比较，将对确定马达加斯加无刺蜂迁移发生的时间有很大的指导意义。

澳大利亚无刺蜂的迁移定殖

澳大利亚有两个属的无刺蜂（*Austroplebeia* 属与 *Tetragonula* 属），它们似乎来自不同的迁移潮。*Austroplebeia* 属无刺蜂与非洲热带无刺蜂蜂种的亲缘关系较之与 *Tetragonula* 属的亲缘关系更近（图 2.2），这表明 *Austroplebeia* 属无刺蜂是从非洲经亚洲迁移到澳大利亚的（Rasmussen and Cameron，2007）。*Braunsapis* 属蜜蜂（Apidae: Xylocopinae）也有类似的从非洲扩散到澳大拉西亚大陆的情况（Fuller et al., 2005），它起源于早中新世时的热带非洲，于大约1 700万年前扩散到亚洲，在晚中新世时迁移到了澳大利亚（Fuller et al., 2005）。

在澳大利亚发现的 7 种 *Tetragonula* 属无刺蜂（(Dollin et al., 1997; Franck et al., 2004）似乎是来自从印尼－马来西亚地区到澳大利亚的 3 个不同的扩散事件（Franck et al., 2004）。研究者认为一些 *Tetragonula* 属无刺蜂是在更近的时候迁移到澳大利亚的，当时更新世时期陆桥周期性通过约克角半岛和新几内亚连接澳大利亚，最后一个陆桥大约在 10 000 年前消失了（Dollin et al., 1997; Franck et al., 2004）。由于马来群岛复杂的地质历史，很难知道无刺蜂这些不同的迁徙是如何以及何时发生的（Michaux，2010; Van Welzen et al., 2011）。而且该地区海平面有很大的波动，这些波动对暴露的陆地表面的面积和位置也有相当大的影响。

2.4 无刺蜂的种属清单

2.4.1 新热带区无刺蜂蜂种

***Aparatrigona* 属无刺蜂**（Moure，1951）

该属无刺蜂体型小（工蜂体长 4～5 mm）且颜色相对较深，有醒目的黄色斑纹，类似 *Paratrigona* 属［Michener（2007）认为这是 *Paratrigona* 种，但 Camargo 和 Moure（1994）将其提升为属］。从北部的哥斯达黎加到南部的巴西都有该蜂种出现。有 2 个种（Camargo and Pedro，2013）。

- *A. impunctata*（Ducke, 1916）
- *A. isopterophila*（Schwarz, 1934）

***Camargoia* 属无刺蜂**（Moure，1989）

该属无刺蜂中等体型（体长 7～8 mm），面部的下半部分偏黄色，与 *Tetragona* 属和 *Ptilotrigona* 属亲缘关系较近（Camargo，1996; Michener，2007）。主要出现在巴西和法属圭亚那的北部地区，为地面筑巢型无刺蜂。有 3 个种（Camargo and Pedro，2013）。

- *C. camargoi*（Moure, 1989）
- *C. nordestina*（Camargo, 1996）
- *C. pilicornis*（Ducke, 1910）

Celetrigona 属无刺蜂（Moure，1950）

该属无刺蜂为小型蜂（体长 3～5 mm），主要发现于亚马孙雨林地区，与 *Leurotrigona-Trigonisca* 群亲缘关系密切（Camargo and Pedro，2009; Rasmussen and Cameron，2010）。有 4 个种（Camargo and Pedro，2013）。

- *C. euclydiana*（Camargo and Pedro, 2009）
- *C. hirsuticornis*（Camargo and Pedro, 2009）
- *C. longicornis*（Friese, 1903）
- *C. manauara*（Camargo and Pedro, 2009）

Cephalotrigona 属无刺蜂（Schwarz，1940）

该属无刺蜂相对体型较大（体长 6～10 mm），是大型 *Trigona* 属的姊妹群，从墨西哥到阿根廷均有分布（Schwarz，1948; Michener，2007; Rasmussen and Cameron，2010; Camargo and Pedro，2013）。值得注意的是该属无刺蜂头部较大且其蜂巢内有较大的食物罐。该属无刺蜂主要在树洞里筑巢，但也会在地里筑巢（Schwarz，1948; Michener，2007）。有 5 个种。

- *C. capitata*（Smith, 1854）
- *C. eburneiventer*（Schwarz, 1948）
- *C. femorata*（Smith, 1854）
- *C. oaxacana*（Ayala, 1999）
- *C. zexmeniae*（Cockerell, 1912）

Duckeola 属无刺蜂（Moure，1944）

该属无刺蜂体型相对较大（体长 8～9 mm）且很强壮，主要分布在亚马孙河流域（Michener，2007; Camargo and Pedro，2013）。该属无刺蜂与 *Trichotrigona*、*Frieseomelitta* 和 *Tetragonisca* 属无刺蜂亲缘关系密切（Rasmussen and Cameron，2010）。有 2 个种（Camargo and Pedro，2013）。

- *D. ghilianii*（Spinola, 1853）
- *D. pavani*（Moure, 1963）

Friesella 属无刺蜂（Moure，1946）

该属无刺蜂体型非常小（体长约 3 mm）且很胆小，被发现于巴西南部（Camargo and Pedro，2013）。该属无刺蜂与 *Plebeia* 属亲缘关系密切（Rasmussen and Cameron，2010）。有 1 个种。

- *F. schrottkyi*（Friese, 1900）

Frieseomelitta 属无刺蜂（Ihering，1912）

该属无刺蜂体型为小到中型且身形细长（体长 4～6.5 mm），从北部的墨西哥到南部的巴西和玻利维亚均有分布（Michener，2007; Camargo and Pedro，2013）。该属无刺蜂会建造簇状育虫巢室以及拉长的储物罐（第 3 章）。该属无刺蜂的有些蜂种蜂群中有颜色特殊的守卫蜂级型（Grüter et al., 2017）。该属与 *Trichotrigona*、*Duckeola* 与 *Tetragonisca* 属无刺蜂亲缘关系密切（Rasmussen and Cameron，2010）。该属有 16 个种。

- *F. dispar*（Moure, 1950）
- *F. doederleini*（Friese, 1900）
- *F. flavicornis*（Fabricius, 1798）
- *F. francoi*（Moure, 1946）
- *F. freiremaiai*（Moure, 1963）
- *F. languida*（Moure, 1990）
- *F. lehmanni*（Friese, 1901）
- *F. longipes*（Smith, 1854）
- *F. meadewaldoi*（Cockerell, 1915）
- *F. nigra*（Cresson, 1878）
- *F. paranigra*（Schwarz, 1940）
- *F. paupera*（Provancher, 1888）
- *F. portoi*（Friese, 1900）
- *F. silvestrii*（Friese, 1902）
- *F. trichocerata*（Moure, 1990）
- *F. varia*（Lepeletier, 1836）

Geotrigona 属无刺蜂（Moure，1943）

该属无刺蜂体型中等且较为强壮（体长 5～6.5 mm），从北部的墨西哥到南部的阿根廷北部均有分布（Michener，2007; Camargo and Pedro，2013）。该属无刺蜂与 *Tetragona* 属亲缘关系密切（Rasmussen and Cameron，2010）。其蜂巢通常建在地里的洞穴中（第 3 章）。该属有 22 个种。

- *G. acapulconis*（Strand, 1919）
- *G. aequinoctialis*（Ducke, 1925）
- *G. argentina*（Camargo and Moure, 1996）

- *G. chiriquiensis*（Schwarz, 1951）
- *G. fulvatra*（Camargo and Moure, 1996）
- *G. fulvohirta*（Friese, 1900）
- *G. fumipennis*（Camargo and Moure, 1996）
- *G. joearroyoi*（Gonzalez and Engel, 2012）
- *G. kaba*（Gonzalez and Sepúlveda, 2007）
- *G. kraussi*（Schwarz, 1951）
- *G. kwyrakai*（Camargo and Moure, 1996）
- *G. leucogastra*（Cockerell, 1914）
- *G. lutzi*（Camargo and Moure, 1996）
- *G. mattogrossensis*（Ducke, 1925）
- *G. mombuca*（Smith, 1863）
- *G. subfulva*（Camargo and Moure, 1996）
- *G. subgrisea*（Cockerell, 1920）
- *G. subnigra*（Schwarz, 1940）
- *G. subterranea*（Friese, 1901）
- *G. tellurica*（Camargo and Moure, 1996）
- *G. terricola*（Camargo and Moure, 1996）
- *G. xanthopoda*（Camargo and Moure, 1996）

Lestrimelitta 属无刺蜂（Friese，1903）

该属无刺蜂体型为小到中型（体长 4～7 mm），颜色呈黑色且有光泽，强壮，从北部的墨西哥到南部的阿根廷北部均有分布（Marchi and Melo, 2006; Michener, 2007; Gonzalez and Griswold, 2012; Camargo and Pedro, 2013）。该属无刺蜂营完全盗食寄生型生活方式（第 7 章）。该属无刺蜂与 *Plebeia* 和 *Friesella* 属亲缘关系密切（Rasmussen and Cameron，2010）。该属有 23 个种。

- *L. catira*（Gonzalez and Griswold, 2012）
- *L. chacoana*（Roig Alsina, 2010）
- *L. chamelensis*（Ayala, 1999）
- *L. ciliata*（Marchi and Melo, 2006）
- *L. danuncia*（Oliveira and Marchi, 2005）
- *L. ehrhardti*（Friese, 1931）

- *L. glaberrima*（Oliveira and Marchi, 2005）
- *L. glabrata*（Camargo and Moure, 1989）
- *L. guyanensis*（Roubik, 1980）
- *L. huilensis*（Gonzalez and Griswold, 2012）
- *L. limao*（Smith, 1863）
- *L. maracaia*（Marchi and Melo, 2006）
- *L. monodonta*（Camargo and Moure, 1989）
- *L. mourei*（Oliveira and Marchi, 2005）
- *L. nana*（Melo, 2003）
- *L. niitkib*（Ayala, 1999）
- *L. opita*（Gonzalez and Griswold, 2012）
- *L. rufa*（Friese, 1903）
- *L. rufipes*（Friese, 1903）
- *L. similis*（Marchi and Melo, 2006）
- *L. spinosa*（Marchi and Melo, 2006）
- *L. sulina*（Marchi and Melo, 2006）
- *L. tropica*（Marchi and Melo, 2006）

***Leurotrigona* 属无刺蜂（Friese，1903）**

该属无刺蜂体型微小（体长 2～3 mm），与 *Celetrigona* 与 *Trigonisca* 属的其他微小型蜂亲缘关系密切（Camargo and Pedro, 2005; Rasmussen and Cameron, 2010）。从北部的哥伦比亚和法属圭亚那到南部的巴拉圭均有分布（Camargo and Pedro, 2013）。根据 Rasmussen 和 Cameron（2010），该属属于并系群（图2.2）。Engel 等（2019）给 *L. crispula* 和 *L. pusilla* 提出的新亚属名称为 *Exochotrigona*。该属有 4 个种。

- *L. crispula*（Pedro and Camargo, 2009）
- *L. gracilis*（Pedro and Camargo, 2009）
- *L. muelleri*（Friese, 1900）
- *L. pusilla*（Moure and Camargo, in Moure et al., 1988）

***Melipona* 属无刺蜂（Illiger，1806）**

该属无刺蜂体型大且强壮（体长 8～15 mm），通常生活在蜂群规模相对较小的蜂群中（第 1 章; Michener, 2007）。该属中体型最大的蜂种——

Melipona fuliginosa 种及 *M. titania* 种（与 *M. fallax* 一起组成 *M. fuliginosa* 种群）的体型大小与大蜜蜂 *Apis dorsata* 相当。分布范围从北部的墨西哥到南部的阿根廷。该属无刺蜂的一个独特的特征是蜂王幼蜂是在工蜂巢室等大的巢室中繁育而成的。该属的蜂王决定系统可能有遗传因素（第5章）。*Melipona* 属约有74个种，是最大的无刺蜂属。一些作者根据形态特征将该属进一步划分为4个亚属（Camargo and Pedro，2013; Michener，2007）。分子数据显示其中3个亚属是单系的（*Michmelia*、*Melikerria* 和 *Melipona*），而1个亚属（*Eomelipona*）是多系的（Ramírez et al., 2010; Rasmussen and Cameron，2010）。

***Melipona*（*Eomelipona*）亚属**

- *M.amazonica*（Schulz, 1905）
- *M.asilvai*（Moure, 1971）
- *M.bicolor*（Lepeletier, 1836）
- *M.bradleyi*（Schwarz, 1932）
- *M.carrikeri*（Cockerell, 1919）
- *M.concinnula**（Cockerell, 1919）
- *M.illustris*（Schwarz, 1932）
- *M.marginata*（Lepeletier, 1836）
- *M.micheneri*（Schwarz, 1951）
- *M.obscurior*（Moure, 1971）
- *M.ogilviei*（Schwarz, 1932）
- *M.puncticollis*（Friese, 1902）
- *M.schwarzi*（Moure, 1963）
- *M.torrida*（Friese, 1916）
- *M.tumupasae*（Schwarz, 1932）

* 据知该名称仅源自委内瑞拉的一个样本，且根据 Schwarz（1932, pp. 419）以及 Camargo 和 Pedro（2013）的研究，它可能是 *M. ogilviei* 的本种同物异名。

***Melipona*（*Melikerria*）亚属**

- *M. ambigua*（Roubik and Camargo, 2012）
- *M. beecheii*（Bennett, 1831）

- *M. compressipes*（Fabricius, 1804）
- *M. fasciculata*（Smith, 1854）
- *M. grandis*（Guérin, 1844）
- *M. insularis*（Roubik and Camargo, 2012）
- *M. interrupta*（Latreille, 1811）
- *M. quinquefasciata*（Lepeletier, 1836）
- *M. salti*（Schwarz, 1932）
- *M. triplaridis*（Cockerell, 1925）

***Melipona*（*Melipona*）亚属**

- *M. baeri*（Vachal, 1904）
- *M. favosa*（Fabricius, 1798）
- *M. fuscata*（Lepeletier, 1836）
- *M. lunulata*（Friese, 1900）
- *M. lupitae*（Ayala, 1999）
- *M. mandacaia*（Smith, 1863）
- *M. orbignyi*（Guérin, 1844）
- *M. peruviana*（Friese, 1900）
- *M. phenax*（Cockerell, 1919）
- *M. quadrifasciata*（Lepeletier, 1836）
- *M. subnitida*（Ducke, 1910）
- *M. variegatipes*（Gribodo, 1893）
- *M. yucatanica*（Camargo, Moure and Roubik, 1988）

***Melipona*（*Michmelia*）亚属**

- *M. apiformis*（Buysson, in Du Buysson and Marshall, 1892）
- *M. belizeae*（Schwarz, 1932）
- *M. boliviana*（Schwarz, 1932）
- *M. brachychaeta*（Moure, 1950）
- *M. captiosa*（Moure, 1962）
- *M. capixaba*（Moure and Camargo, 1994）
- *M. colimana*（Ayala, 1999）
- *M. costaricensis*（Cockerell, 1919）

- *M. cramptoni*（Cockerell, 1920）
- *M. crinita*（Moure and Kerr, 1950）
- *M. dubia*（Moure and Kerr, 1950）
- *M. eburnean*（Friese, 1900）
- *M. fallax*（Camargo and Pedro, 2008）
- *M. fasciata*（Latreille, 1811）
- *M. flavolineata*（Friese, 1900）
- *M. fuliginosa*（Lepeletier, 1836）
- *M. fulva*（Lepeletier, 1836）
- *M. fuscopilosa*（Moure and Kerr, 1950）
- *M. illota*（Cockerell, 1919）
- *M. indecisa*（Cockerell, 1919）
- *M. lateralis*（Erichson, 1848）
- *M. melanoventer*（Schwarz, 1932）
- *M. mimetica*（Cockerell, 1914）
- *M. mondury*（Smith, 1863）
- *M. nebulosa*（Camargo, 1988）
- *M. nigrescens*（Friese, 1900）
- *M. nitidifrons*（Benoist, 1933）
- *M. panamica*（Cockerell, 1912）
- *M. paraensis*（Ducke, 1916）
- *M. rufescens*（Friese, 1900）
- *M. rufiventris*（Lepeletier, 1836）
- *M. scutellaris*（Latreille, 1811）
- *M. seminigra*（Friese, 1903）
- *M. solani*（Cockerell, 1912）
- *M. titania*（Gribodo, 1893）
- *M. trinitatis*（Cockerell, 1919）

Meliwillea 属无刺蜂（Roubik, Lobo and Camargo, 1997）

该属无刺蜂体型为小到中型（体长 5～6 mm），只有1个种，与 *Scaptotrigona* 属亲缘关系密切，发现于中美洲的山区（Roubik et al., 1997）。

- *M. bivea*（Roubik, Lobo and Camargo, 1997）

Mourella 属无刺蜂（Schwarz，1946）

该属无刺蜂体型为小到中型（体长 5～6 mm），颜色呈深色，只有 1 个种，与 *Schwarziana* 属亲缘关系密切（Rasmussen and Cameron，2010），发现于南美洲中部（Camargo and Pedro，2013）。

- *M. caerulea*（Friese, 1900）

Nannotrigona 属无刺蜂（Cockerell，1922）

该属无刺蜂体型小（体长 3.5～5 mm）且胆小，颜色呈深色，在新热带区非常普遍。该属无刺蜂在城市化和受干扰地区也能顺利生存（Rasmussen and Gonzalez，2017）。该属无刺蜂与 *Mourella* 与 *Schwarziana* 属无刺蜂亲缘关系密切（Rasmussen and Cameron，2010）。该属有 10 个种（Rasmussen and Gonzalez，2017）。

- *N. camargoi*（Rasmussen and Gonzalez, 2017）
- *N. chapadana*（Schwarz, 1938）
- *N. dutrae*（Friese, 1901）
- *N. melanocera*（Schwarz, 1938）
- *N. mellaria*（Smith, 1862）
- *N. perilampoides*（Cresson, 1878）
- *N. punctata*（Smith, 1854）
- *N. schultzei*（Friese, 1901）
- *N. testaceicornis*（Lepeletier, 1836）
- *N. tristella*（Cockerell, 1922）

Nogueirapis 属无刺蜂（Moure，1953）

该属无刺蜂体型小（体长 3.5～5.5 mm）且有明显黄色花纹，与 *Partamona* 属亲缘关系密切。该属有 4 个种，这些蜂种可能都是地里筑巢型。分布范围从墨西哥到南美洲北部（Michener，2007; Rasmussen and Cameron，2010; Ayala and Engel，2014）。

- *N. butteli*（Friese, 1900）
- *N. costaricana*（Ayala and Engel, 2014）
- *N. minor*（Moure and Camargo, 1982）
- *N. mirandula*（Cockerell, 1917）

Oxytrigona 属无刺蜂（Cockerell，1917）

该属无刺蜂中等体型（体长 5～6 mm），工蜂头部呈明显的红色，工蜂还有大的上颚腺用于分泌腐蚀性的防御物质（第 7 章）。其亲缘关系最近的属是 *Scaptotrigona* 属（Rasmussen and Cameron，2010）。该属包含 11 个种，广泛分布在新热带区（Pedro and Camargo，2013）。

- *O. chocoana*（Gonzalez and Roubik, 2008）
- *O. daemoniaca*（Camargo, 1984）
- *O. flaveola*（Friese, 1900）
- *O. huaoranii*（Gonzalez and Roubik, 2008）
- *O. ignis*（Camargo, 1984）
- *O. isthmina*（Gonzalez and Roubik, 2008）
- *O. mediorufa*（Cockerell, 1913）
- *O. mellicolor*（Packard, 1869）
- *O. mulfordi*（Schwarz, 1948）
- *O. obscura*（Friese, 1900）
- *O. tataira*（Smith, 1863）

Parapartamona 属无刺蜂（Schwarz，1948）

该属无刺蜂体型中等，与 *Partamona* 属亲缘关系密切（Rasmussen and Cameron，2010）。目前只知道该属存在于海拔 3 400 m 的哥伦比亚、厄瓜多尔和秘鲁的安第斯山脉。当前研究者已经鉴别出该属的 7 个种，但是该属蜂种的鉴定仍然存在争议（Gonzalez and Smith-Pardo，2003; Camargo and Pedro，2013）。

- *P. brevipilosa*（Schwarz, 1948）
- *P. caliensis*（Schwarz, 1948）
- *P. fumata*（Moure, 1995）
- *P. imberbis*（Moure, 1995）
- *P. tungurahuana*（Schwarz, 1948）
- *P. vittigera*（Moure, 1995）
- *P. zonata*（Smith, 1854）

Paratrigona 属无刺蜂（Moure，1951）

该属无刺蜂体型相对较小（体长 4～6 mm），胸部和面部有明显的黄色花纹(Camargo and Moure, 1994; Michener, 2007; Gonzalez and Griswold, 2011)。该属包括的蜂种十分丰富(32个种),分布范围从墨西哥至阿根廷。与 *Aparatrigona* 属亲缘关系密切(Rasmussen and Cameron, 2010)。

- *P. anduzei*(Schwarz, 1943)
- *P. catabolonota*(Camargo and Moure, 1994)
- *P. compsa*(Camargo and Moure, 1994)
- *P. crassicornis*(Camargo and Moure, 1994)
- *P. eutaeniata*(Camargo and Moure, 1994)
- *P. euxanthospila*(Camargo and Moure, 1994)
- *P. femoralis*(Camargo and Moure, 1994)
- *P. glabella*(Camargo and Moure, 1994)
- *P. guatemalensis*(Schwarz, 1938)
- *P. guigliae*(Moure, 1960)
- *P. haeckeli*(Friese, 1900)
- *P. incerta*(Camargo and Moure, 1994)
- *P. lineata*(Lepeletier, 1836)
- *P. lineatifrons*(Schwarz, 1938)
- *P. lophocoryphe*(Moure, 1963)
- *P. lundelli*(Schwarz, 1938)
- *P. melanaspis*(Camargo and Moure, 1994)
- *P. myrmecophila*(Moure, 1989)
- *P. nuda*(Schwarz, 1943)
- *P. onorei*(Camargo and Moure, 1994)
- *P. opaca*(Cockerell, 1917)
- *P. ornaticeps*(Schwarz, 1938)
- *P. pacifica*(Schwarz, 1943)
- *P. pannosa*(Moure, 1989)
- *P. peltata*(Spinola, 1853)
- *P. permixta*(Camargo and Moure, 1994)
- *P. prosopiformis*(Gribodo, 1893)
- *P. rinconi*(Camargo and Moure, 1994)
- *P. scapisetosa*(Gonzalez and Griswold, 2011)

- *P. subnuda*（Moure, 1947）
- *P. uwa*（Gonzalez and Vélez, 2007）
- *P. wasbaueri*（Gonzalez and Griswold, 2011）

***Paratrigonoides* 属无刺蜂（Camargo and Roubik，2005）**

该属形态上与 *Aparatrigona* 和 *Paratrigona* 属非常相似，有人认为它们是姊妹群（Michener，2007）。据知该属有 1 个种，存在于哥伦比亚（Camargo and Pedro，2013）。

- *P. mayri*（Camargo and Roubik, 2005）

***Partamona* 属无刺蜂（Schwarz，1938）**

该属无刺蜂体型为小到中型（体长 4.5～7 mm），该属无刺蜂常会建造精细的蜂巢入口结构（例如"蛤蟆嘴"，第 3 章）（Camargo and Pedro，2003）。该属无刺蜂与 *Parapartamona* 属亲缘关系密切（Rasmussen and Cameron，2010）。该属包括的蜂种十分丰富（32 种），从墨西哥到阿根廷均发现了该属无刺蜂存在（Camargo and Pedro，2013）。

- *P. aequatoriana*（Camargo, 1980）
- *P. ailyae*（Camargo, 1980）
- *P. auripennis*（Pedro and Camargo, 2003）
- *P. batesi*（Pedro and Camargo, 2003）
- *P. bilineata*（Say, 1837）
- *P. chapadicola*（Pedro and Camargo, 2003）
- *P. combinata*（Pedro and Camargo, 2003）
- *P. criptica*（Pedro and Camargo, 2003）
- *P. cupira*（Smith, 1863）
- *P. epiphytophila*（Pedro and Camargo, 2003）
- *P. ferreirai*（Pedro and Camargo, 2003）
- *P. grandipennis*（Schwarz, 1951）
- *P. gregaria*（Pedro and Camargo, 2003）
- *P. helleri*（Friese, 1900）
- *P. littoralis*（Pedro and Camargo, 2003）
- *P. mourei*（Camargo, 1980）
- *P. mulata*（Moure, in Camargo, 1980）

- *P. musarum*(Cockerell, 1917)
- *P. nhambiquara*(Pedro and Camargo, 2003)
- *P. nigrior*(Cockerell, 1925)
- *P. orizabaensis*(Strand, 1919)
- *P. pearsoni*(Schwarz, 1938)
- *P. peckolti*(Friese, 1901)
- *P. rustica*(Pedro and Camargo, 2003)
- *P. seridoensis*(Pedro and Camargo, 2003)
- *P. sooretamae*(Pedro and Camargo, 2003)
- *P. subtilis*(Pedro and Camargo, 2003)
- *P. testacea*(Klug, 1807)
- *P. vicina*(Camargo, 1980)
- *P. vitae*(Pedro and Camargo, 2003)
- *P. xanthogastra*(Pedro and Camargo, 1997)
- *P. yungarum*(Pedro and Camargo, 2003)

***Plebeia* 属无刺蜂（Schwarz，1938）**

该属无刺蜂为小型蜂（体长 3～6 mm），是一个物种丰富（约 42 个种）的多源群（Rasmussen and Cameron，2010; Alvarez et al., 2016），分布范围从墨西哥到阿根廷（Camargo and Pedro，2013）。

- *P. alvarengai*(Moure, 1994)
- *P. catamarcensis*(Holmberg, 1903)
- *P. cora*(Ayala, 1999)
- *P. droryana*(Friese, 1900)
- *P. emerina*(Friese, 1900)
- *P. emerinoides**(Silvestri, 1902)
- *P. flavocincta*(Cockerell, 1912)
- *P. franki*(Friese, 1900)
- *P. fraterna*(Laroca and Rodriguez-Parilli, 2009)
- *P. frontalis*(Friese, 1911)
- *P. fulvopilosa*(Ayala, 1999)
- *P. goeldiana*(Friese, 1900)
- *P. grapiuna*(Melo and Costa, 2009)

- *P. guazurary*（Alvarez, Rasmussen and Abrahamovich, 2016）
- *P. jatiformis*（Cockerell, 1912）
- *P. julianii*（Moure, 1962）
- *P. kerri*（Moure, 1950）
- *P. llorentei*（Ayala, 1999）
- *P. lucii*（Moure, 2004）
- *P. malaris*（Moure, 1962）
- *P. manantlensis*（Ayala, 1999）
- *P. margaritae*（Moure, 1962）
- *P. melanica*（Ayala, 1999）
- *P. meridionalis*（Ducke, 1916）
- *P. mexica*（Ayala, 1999）
- *P. minima*（Gribodo, 1893）
- *P. molesta*（Puls, in Strobel, 1868）
- *P. mosquito*（Smith, 1863）
- *P. moureana*（Ayala, 1999）
- *P. nigriceps*（Friese, 1901）
- *P. parkeri*（Ayala, 1999）
- *P. peruvicola*（Moure, 1994）
- *P. phrynostoma*（Moure, 2004）
- *P. poecilochroa*（Moure and Camargo, 1993）
- *P. pugnax*（Moure, in litt.）
- *P. pulchra*（Ayala, 1999）
- *P. remota*（Holmberg, 1903）
- *P. saiqui*（Friese, 1900）
- *P. tica*（Wille, 1969）
- *P. tobagoensis*（Melo, 2003）
- *P. variicolor*（Ducke, 1916）
- *P. wittmanni*（Moure and Camargo, 1989）

* Camargo 和 Pedro（2013）认为 *P. emerinoides* 是 *P. nigriceps* 的次定同物异名，但 Roig-Alsina 和 Alvarez（2017）认为这是一个合理的蜂种。

Plectoplebeia 属无刺蜂（Melo，2016）

该属无刺蜂体型为小到中型（体长 5.5～6 mm），有 1 个种（以前是 *Plebeia intermedia*）（Melo，2016）。据知该属存在于阿根廷、玻利维亚和秘鲁（Flores et al., 2015; Alvarez et al., 2016; Melo，2016）。

- *P. nigrifacies*（Friese, 1900）

Ptilotrigona 属无刺蜂（Moure，1951）

该属无刺蜂体型相对较大（体长 8～10 mm），其亲缘关系最接近的属是 *Tetragona* 属（Rasmussen and Cameron，2010）。该属无刺蜂蜂群内几乎不储存蜂蜜（Camargo and Pedro，2004）。分布范围从北部的哥斯达黎加到南部的巴西北部地区，穴居，有 3 个种（Camargo and Pedro，2013）。

- *P. lurida*（Smith, 1854）
- *P. occidentalis*（Schulz, 1904）
- *P. pereneae*（Schwarz, 1943）

Scaptotrigona 属无刺蜂（Moure，1942）

该属无刺蜂体型为小到中型且很强壮（体长 5～7 mm），与 *Oxytrigona* 属亲缘关系密切（Rasmussen and Cameron，2010）。该属大约有 22 个穴居的蜂种（Michener, 2007; Camargo and Pedro, 2013; Hurtado-Burillo et al., 2017）。分布范围从墨西哥到阿根廷。注意这里被视为 *Scaptotrigona depilis* 种的（Moure, 1942）是指在圣保罗地区常被称为 *Scaptotrigona* aff. depilis 的蜂种。为了更好地评估 *Scaptotrigona depilis* 无刺蜂覆盖范围亟须进行分类学修订。

- *S. affabra*（Moure, 1989）
- *S. barrocoloradensis*（Schwarz, 1951）
- *S. bipunctata*（Lepeletier, 1836）
- *S. depilis*（Moure, 1942）
- *S. emersoni*（Schwarz, 1938）
- *S. fulvicutis*（Moure, 1964）
- *S. hellwegeri*（Friese, 1900）
- *S. jujuyensis*（Schrottky, 1911）
- *S. limae*（Brèthes, 1920）
- *S. luteipennis*（Friese, 1902）

- *S. mexicana*（Guérin, 1844）
- *S. ochrotricha*（Buysson, in Du Buysson and Marshall, 1892）
- *S. panamensis*（Cockerell, 1913）
- *S. pectoralis*（Dalla Torre, 1896）
- *S. polysticta*（Moure, 1950）
- *S. postica*（Latreille, 1807）
- *S. subobscuripennis*（Schwarz, 1951）
- *S. tricolorata*（Camargo, 1988）
- *S. tubiba*（Smith, 1863）
- *S. turusiri*（Janvier, 1955）
- *S. wheeleri*（Cockerell, 1913）
- *S. xanthotricha*（Moure, 1950）

Scaura 属无刺蜂（Schwarz，1938）

该属无刺蜂为小型蜂（体长 3～6 mm）（Schwarz, 1948），目前包含 10 个种（Rasmussen and Cameron，2010; Camargo and Pedro，2013; Nogueira et al., 2019）。分布范围从墨西哥到巴西南部。以前该属无刺蜂有两个蜂种属于 *Schwarzula* 属，但是 Rasmussen 和 Cameron（2010）建议将这两个属归属于 *Scaura* 属形成该属下的一个单系属（Michener，2007）。此外，Nogueira 等（2019）建议将 *S. tenuis* 与 *S. latitarsis* 同物异名，但根据 Rasmussen 和 Cameron（2010）的分子系统发育分析这将产生一个多系物种，因此在这里笔者没有遵循 Nogueira 等（2019）的建议。

- *S. amazonica*（Nogueira, Oliveira and Oliveira, 2019）
- *S. argyrea*（Cockerell, 1912）
- *S. aspera*（Nogueira and Oliveira, 2019）
- *S. atlantica*（Melo, 2004）
- *S. cearensis*（Nogueira, Santos Júnior and Oliveira, 2019）
- *S. latitarsis*（Friese, 1900）
- *S. longula*（Lepeletier, 1836）
- *S. tenuis*（Ducke, 1916）
- *S.*（formerly *Schwarzula*）*coccidophila*（Camargo and Pedro, 2002）
- *S.*（formerly *Schwarzula*）*timida*（Silvestri, 1902）

***Schwarziana* 属无刺蜂**（Moure，1943）

该属无刺蜂体型中等（体长 6～7.5 mm），只存在于巴西东部、巴拉圭和阿根廷北部（Camargo and Pedro，2013; Melo，2015）。与 *Mourella* 属亲缘关系密切（Rasmussen and Cameron，2010）。该属无刺蜂蜂群主要在地里筑巢，且有微小型蜂王（第 3 章，第 5 章）。有 4 个种。

- *S. bocainensis*（Melo，2015）
- *S. chapadensis*（Melo，2015）
- *S. mourei*（Melo，2003）
- *S. quadripunctata*（Lepeletier，1836）

***Tetragona* 属无刺蜂**（Lepeletier and Audinet-Serville，1828）

该属无刺蜂体型中等（体长 5～8 mm），分布范围从墨西哥到阿根廷（Camargo and Pedro，2013; Roig-Alsina et al.，2013）。与 *Ptilotrigona* 和 *Camargoia* 属亲缘关系密切（Michener，2007; Rasmussen and Cameron，2010）。目前有 13 个种。

- *T. beebei*（Schwarz，1938）
- *T. clavipes*（Fabricius，1804）
- *T. dissecta*（Moure，2000）
- *T. dorsalis*（Smith，1854）
- *T. essequiboensis*（Schwarz，1940）
- *T. goettei*（Friese，1900）
- *T. handlirschii*（Friese，1900）
- *T. kaieteurensis*（Schwarz，1938）
- *T. mayarum*（Cockerell，1912）
- *T. perangulata*（Cockerell，1917）
- *T. quadrangula*（Lepeletier，1836）
- *T. truncata*（Moure，1971）
- *T. ziegleri*（Friese，1900）

***Tetragonisca* 属无刺蜂**（Moure，1946）

该属无刺蜂为小型淡黄色蜂（体长 4～5 mm），分布范围从墨西哥到阿根廷（Camargo and Pedro，2013）。亲缘关系最近的属是 *Duckeola*、*Frieseomelitta* 和 *Trichotrigona*（Rasmussen and Cameron，2010）。其中的 *T. angustula* 和 *T. fiebrigi*

无刺蜂分化出了一种兵蜂级型（Grüter et al., 2012, 2017; Segers et al., 2015）。该属无刺蜂包含了穴居型、地巢型和暴露建巢型的蜂种（第3章）。目前研究者描述该属无刺蜂中有4个种，但是Francisco等（2014）认为 *T. angustula* 和 *T. fiebrigi* 是正处于物种分化过程中的亚种。

- *T. angustula*（Latreille, 1811）
- *T. buchwaldi*（Friese, 1925）
- *T. fiebrigi*（Schwarz, 1938）
- *T. weyrauchi*（Schwarz, 1943）

Trichotrigona 属无刺蜂（Camargo and Moure，1983）

该属无刺蜂为小型蜂（体长 4～5 mm），身体上有黑色和黄色的条纹，十分罕见，到目前为止只在亚马孙地区发现过该属无刺蜂。目前研究者描述该属无刺蜂有2个种，据观察这2种蜂似乎不储存食物，研究者猜测它们是通过盗食寄生与 *Frieseomelitta* 属无刺蜂生活在一起的（Camargo and Moure, 1983; Pedro and Cordeiro, 2015）。

- *T. camargoiana*（Pedro and Cordeiro, 2015）
- *T. extranea*（Camargo and Moure, 1983）

Trigona 属无刺蜂（Jurine，1807）

该属无刺蜂是高度多样化的族群，工蜂体型有大有小（体长 5.5～11 mm），分布范围从墨西哥到阿根廷（Camargo and Pedro, 2013）。该属无刺蜂蜂群规模通常较大（第1章；表1.3）；该属中的一些蜂种在地面上筑巢，有些则是在白蚁、蚂蚁巢穴中筑巢（第3章）。目前研究者描述该属包含32个种。

- *T. albipennis*（Almeida, 1995）
- *T. amalthea*（Olivier, 1789）
- *T. amazonensis*（Ducke, 1916）
- *T. branneri*（Cockerell, 1912）
- *T. braueri*（Friese, 1900）
- *T. chanchamayoensis*（Schwarz, 1948）
- *T. cilipes*（Fabricius, 1804）
- *T. corvina*（Cockerell, 1913）
- *T. crassipes*（Fabricius, 1793）

- *T. dallatorreana*（Friese, 1900）
- *T. dimidiata*（Smith, 1854）
- *T. ferricauda*（Cockerell, 1917）
- *T. fulviventris*（Guérin, 1844）
- *T. fuscipennis*（Friese, 1900）
- *T. guianae*（Cockerell, 1910）
- *T. hyalinata*（Lepeletier, 1836）
- *T. hypogea*（Silvestri, 1902）
- *T. lacteipennis*（Friese, 1900）
- *T. muzoensis*（Schwarz, 1948）
- *T. necrophaga*（Camargo and Roubik, 1991）
- *T. nigerrima*（Cresson, 1878）
- *T. pallens*（Fabricius, 1798）
- *T. pampana*（Strand, 1910）
- *T. pellucida*（Cockerell, 1912）
- *T. permodica*（Almeida, 1995）
- *T. recursa*（Smith, 1863）
- *T. sesquipedalis*（Almeida, 1984）
- *T. silvestriana*（Vachal, 1908）
- *T. spinipes*（Fabricius, 1793）
- *T. truculenta*（Almeida, 1984）
- *T. venezuelana*（Schwarz, 1948）
- *T. williana*（Friese, 1900）

Trigonisca 属无刺蜂（Moure，1950）

该属无刺蜂是微型蜂（体长 2～4 mm），与 *Celetrigona* 属和 *Leurotrigona* 属中的一些微型蜂亲缘关系密切（Camargo and Pedro，2005; Michener，2007; Rasmussen and Cameron，2010; Engel et al., 2019）。研究者前期将 10 个种归入了 *Dolichotrigona* 属，但 Rasmussen 和 Cameron（2010）提出将它们都归入 *Trigonisca* 属以创建一个单系属（图 2.2）。人们常常认为该属无刺蜂是一种扰民蜂，因为该属无刺蜂喜欢飞入人类的眼睛、耳朵和鼻子里去收集人类的分泌物（Camargo and Pedro，2005）。研究者目前记述该属包含 37 个种，分布范围从墨西哥到阿根廷（Camargo and Pedro，2013; Roig-Alsina et al., 2013;

Alvarez and Lucia，2018; Engel et al., 2019）。
- *T. atomaria*（Cockerell, 1917）
- *T. azteca*（Ayala, 1999）
- *T. bidentata*（Albuquerque and Camargo, 2007）
- *T. browni*（Camargo and Pedro, 2005）
- *T. buyssoni*（Friese, 1902）
- *T. ceophloei*（Schwarz, 1938）
- *T. chachapoya*（Camargo and Pedro, 2005）
- *T. clavicornis*（Camargo and Pedro, 2005）
- *T. discolor*（Wille, 1965）
- *T. dobzhanskyi*（Moure, 1950）
- *T. duckei*（Friese, 1900）
- *T. extrema*（Albuquerque and Camargo, 2007）
- *T. flavicans*（Moure, 1950）
- *T. fraissei*（Friese, 1901）
- *T. graeffei*（Friese, 1900）
- *T. hirticornis*（Albuquerque and Camargo, 2007）
- *T. intermedia*（Moure, 1990）
- *T. longitarsis*（Ducke, 1916）
- *T. martinezi*（Brèthes, 1920）
- *T. maya*（Ayala, 1999）
- *T. mepecheu*（Engel and Gonzalez, 2019）
- *T. mendersoni*（Camargo and Pedro, 2005）
- *T. meridionalis*（Albuquerque and Camargo, 2007）
- *T. mixteca*（Ayala, 1999）
- *T. moratoi*（Camargo and Pedro, 2005）
- *T. nataliae*（Moure, 1950）
- *T. pediculana*（Fabricius, 1804）
- *T. pipioli*（Ayala, 1999）
- *T. rondoni*（Camargo and Pedro, 2005）
- *T. roubiki*（Albuquerque and Camargo, 2007）
- *T. sachamiski*（Alvarez and Lucia, 2018）

- *T. schulthessi*（Friese, 1900）
- *T. tavaresi*（Camargo and Pedro, 2005）
- *T. townsendi*（Cockerell, 1911）
- *T. unidentata*（Albuquerque and Camargo, 2007）
- *T. variegatifrons*（Albuquerque and Camargo, 2007）
- *T. vitrifrons*（Albuquerque and Camargo, 2007）

2.4.2 非洲热带区无刺蜂蜂种

Apotrigona 属无刺蜂（Moure，1950）

该属无刺蜂为小到中型蜂（体长 5～7 mm）。目前该属有 1 个种，人们用多个名字称呼这个种（例如 *Meliponula nebulata* 或 *Apotrigona infuscata*）（Eardley, 2004; Rasmussen and Cameron, 2007）。Michener（2007）认为该属是 *Plebeiella* 的异名同属，但是分子数据显示这将造成这个属成为一个非单系群（Rasmussen and Cameron, 2007）。塞拉利昂、科特迪瓦、尼日利亚、加蓬、刚果和乌干达均报告了该属无刺蜂的存在（Faber Anguilet et al., 2015; Ascher and Pickering, 2018）。

- *A. nebulata*（Smith, 1854）

Axestotrigona 属无刺蜂（Moure，1961）

该属无刺蜂为中等体型（体长 5.5～7.5 mm）（Eardley, 2004; Rasmussen and Cameron, 2007）。该属中有几个蜂种形态不定，有时有不同的颜色形态（Eardley, 2004; Rasmussen and Cameron, 2007; Michener, 2007; Pauly et al., 2013; Ndungu et al., 2017）。目前该属包含约 6 个种（Eardley, 2004; Pauly et al., 2013; Ndungu et al., 2017）。据报告该属分布范围从冈比亚到肯尼亚，南到安哥拉和南非（Michener, 2007; Asher and Pickering, 2018）。

- *A. cameroonensis*（Friese, 1900）
- *A. erythra*（Schletterer, 1891）
- *A. ferruginea**（Lepeletier, 1841）
- *A. simpsoni*（Moure, 1961）
- *A. richardsi*（Darchen, 1981）
- *A. togoensis*（Stadelmann, 1895）

*该蜂种有两种形态：红棕色蜂和黑色蜂，很可能代表了两个蜂种（Ndungu et al., 2017）。

Cleptotrigona 属无刺蜂（Moure，1961）

该属无刺蜂体型小（体长 3.5 ~ 4 mm）而强壮，颜色呈黑色，专抢 *Hypotrigona* 属等其他无刺蜂的蜂巢（第 7 章）（Eardley，2004）。目前该属无刺蜂只有 1 个已认证的蜂种，但是 Moure（1961）报告了两个。该属无刺蜂可能是 *Liotrigona* 属的姊妹群（Michener，2007）。分布范围从利比里亚到坦桑尼亚，南到安哥拉和南非（Michener，2007; Ascher and Pickering，2018）。

- *C. cubiceps*（Friese，1912）

Dactylurina 属无刺蜂（Cockerell，1934）

该属无刺蜂体型纤细，中等大小（体长 5 ~ 7 mm），该属的特殊之处在于它们建造垂直型巢脾，且蜂巢一般为暴露型（Eardley，2004）。该属有 2 个种，与 *Meliponula* 属亲缘关系密切（Rasmussen and Cameron，2007）。该属广泛分布在热带非洲，分布范围从肯尼亚和坦桑尼亚到刚果、安哥拉和利比里亚（Michener，2007; Ascher and Pickering，2018）。

- *D. schmidti*（Stadelmann，1895）
- *D. staudingeri*（Gribodo，1893）

Hypotrigona 属无刺蜂（Cockerell，1934）

该属无刺蜂为微小型蜂（体长 2 ~ 4 mm），很难用形态学特征来区分（Eardley，2004; Michener 2007）。该属无刺蜂在非洲热带地区广泛分布且数量众多，分布范围从东部的索马里到西部的塞内加尔，从北部的苏丹到南部的南非（Michener，2007; Ascher and Pickering，2018）。目前该属无刺蜂大约有 5 个种（Eardley，2004; Michener，2007）。

- *H. araujoi*（Michener，1959）
- *H. braunsi*（Kohl，1894）
- *H. gribodoi*（Magretti，1884）
- *H. squamuligera**（Benoist，1937）
- *H. ruspolii*（Magretti，1898）

* Pauly 等（2013）认为 *Hypotrigona penna*（Eardley，2004）是 *H. squamuligera* 蜂的次定同物异名。

Liotrigona 属无刺蜂（Moure，1961）

该属无刺蜂为微型蜂（体长 2 ~ 4 mm），与另一个微型蜂属 *Hypotrigona* 属亲缘关系密切（Rasmussen and Cameron，2007）。该属有 12 个种，是物种最丰富的非洲属（Koch，2010; Pauly et al., 2001, 2013; Faber Anguilet et al., 2015）。该属广泛分布在撒哈拉以南的非洲地区，也是马达加斯加地区发现的唯一无刺蜂属（有7个地区特有种）（Pauly et al., 2001; Michener，2007; Koch，2010）。

- *L. baleensis*（Pauly and Hora, 2013）
- *L. betsimisaraka*（Pauli, 2001）
- *L. bitika*（Brooks and Michener, 1988）
- *L. bottegoi*（Magretti, 1895）
- *L. bouyssoui**（Vachal, 1903）
- *L. chromensis*（Pauly, 2001）
- *L. gabonensis*（Pauly and Fabra Anguilet, 2013）
- *L. kinzelbachi*（Koch, 2010）
- *L. madecassa*（Saussure, 1890）
- *L. mahafalya*（Brooks and Michener, 1988）
- *L. nilssoni*（Michener, 1990）
- *L. voeltzkovi*（Friese, 1900）

* Eardley（2004）认为这是 *Hypotrigona* 属的，但是 Pauly 等（2013）认为这是 *Liotrigona parvula* 的一个先定同物异名（Darchen, 1971）。

Meliplebeia 属无刺蜂（Moure，1961）

该属无刺蜂体型小到中型（体长 4～7 mm）（Eardley，2004; Michener，2007）。分布范围从赤道非洲到南非（Michener，2007; Ascher and Pickering，2018）。目前该属有4个种。

- *M. beccarii*（Gribodo, 1879）
- *M. gambiana*（Moure, 1961）
- *M. ogouensis*（Vachal, 1903）
- *M. roubiki*（Eardley, 2004）

Meliponula 属无刺蜂（Cockerell，1934）

该属无刺蜂是蜂群规模最大的非洲无刺蜂（体长6.5～8 mm），身体部分呈黄色，因此容易辨认（Eardley，2004; Michener，2007）。该属有1个种，分布范围从西部的塞内加尔到东部的肯尼亚以及南部的博茨瓦纳和纳米比亚（Eardley，2004; Ascher and Pickering，2018）。

- *M. bocandei*（Spinola, 1853）

Plebeiella 属无刺蜂（Moure，1961）

该属无刺蜂有 2 个小体型（体长 4～5.5 mm）的地面筑巢型蜂种（Eardley，2004; Estienne et al., 2017）。分布范围从多哥到肯尼亚和南部的纳米比亚和赞比亚（Eardley，2004; Ascher and Pickering，2018）。

- *P. griswoldorum*（Eardley, 2004）
- *P. lendliana*（Friese, 1900）

Plebeina 属无刺蜂（Moure，1961）

该属无刺蜂为小型蜂（体长 3～5 mm），生活在陆生白蚁巢穴中（Eardley，2004）。分布范围从塞内加尔到肯尼亚，南至南非（Eardley，2004; Ascher and Pickering，2018）。有人曾用几个不同的名字来称呼该属无刺蜂的一个易变种或几个蜂种（Michener，2007; Kajobe and Roubik，2018）。研究者认为 *Plebeina hildebrandti*（Friese, 1900）和 *P. denoiti*（Vachal, 1903）是 *P. armata* 的次定同物异名（Kajobe and Roubik，2018）。

- *P. armata*（**Magretti**, 1895）

2.4.3 印尼 – 马来半岛生物地理区域和澳大拉西亚区的无刺蜂蜂种 [7]

Austroplebeia 属无刺蜂（Moure，1961）

该属无刺蜂体型小而健壮（体长 3～4 mm），在其盾片上有明显的黄色花纹，在面部通常也有黄色花纹（Michener，2007; Dollin et al., 2015）。它们与 *Lisotrigona* 属以及非洲的 *Liotrigona* 属亲缘关系最密切，其中与非洲的 *Liotrigona* 属近亲关系十分有趣。其最有可能的生物扩散方式是从非洲经印度洋地区扩散到澳大利亚（Rasmussen and Cameron, 2010）。在澳大利亚北部和东部地区以及新几内亚均发现了该属无刺蜂（Dollin et al., 2015）。该属的特殊之处在于会建造由螺旋巢室构成的簇状巢脾（第 3 章），该属约有 5 个种（Rasmussen，2008; Rasmussen et al., 2017）。

- *A. australis*（Friese, 1898）
- *A. cassia*（Cockerell, 1910）
- *A. cincta*（Mocsáry in Friese, 1898）
- *A. essingtoni*（Cockerell, 1905）

---

[7] 参考 Rasmussen 等（2017）的文献附录，见印尼 – 马来半岛和大洋洲区生物地理区域和澳大拉西亚区域无刺蜂种属的次定同物异名列表。

- *A. magna*（Dollin, Dollin, and Rasmussen, 2015）

Geniotrigona 属无刺蜂（Moure，1961）

该属无刺蜂体型健壮，为中型蜂（体长 5～7 mm），呈黑色和深棕色。分布范围从缅甸和老挝到苏门答腊和加里曼丹（印度尼西亚）（Rasmussen，2008; Rasmussen et al., 2017; Ascher and Pickering，2018）。该属有 2 个种。

- *G. lacteifasciata*（Cameron, 1902）
- *G. thoracica*（Smith, 1857）

Heterotrigona 属无刺蜂（Schwarz，1939）

该属无刺蜂体型为相对小到中型（体长 4～7 mm），分布范围为泰国到印度尼西亚和菲律宾（Engel and Rasmussen, 2017; Rasmussen et al., 2017; Ascher and Pickering，2018）。有 3 个种。

- *H. bakeri*（Cockerell, 1919）
- *H. erythrogastra*（Cameron, 1902）
- *H. itama*（Cockerell, 1918）

Homotrigona 属无刺蜂（Moure，1961）

该属无刺蜂体型为中到大型且体格强壮（体长 7～8 mm），分布范围为泰国和越南到印度尼西亚（Rasmussen, 2008; Rasmussen et al., 2017; Ascher and Pickering, 2018）。该属有 4 个种。

- *H. aliceae*（Cockerell, 1929）
- *H. anamitica*（Friese, 1909）
- *H. fimbriata*（Smith, 1857）
- *H. lutea*（Bingham, 1897）

Lepidotrigona 属无刺蜂（Schwarz，1939）

该属无刺蜂为小到中型（体长 4～5.5 mm），分布范围从印度到中国台湾和菲律宾，南至爪哇（Michener，2007）。该属有 13 个种（Rasmussen, 2008; Rasmussen et al., 2017; Attasopa et al., 2018）。

- *L. arciferal*（Cockerell, 1929）
- *L. doipaensis*（Schwarz, 1939）
- *L. flavibasis*（Cockerell, 1929）
- *L. hoozana*（Strand, 1913）
- *L. javanica*（Gribodoo, 1891）

- *L. latebalteata*（Cameron, 1902）
- *L. latipes*（Friese, 1900）
- *L. nitidiventris*（Smith, 1857）
- *L. palavanica*（Cockerell, 1915）
- *L. satun*（Attasopa and Bänziger, 2018）
- *L. terminata*（Smith, 1878）
- *L. trochanterica*（Cockerell, 1920）
- *L. ventralis*（Smith, 1857）

***Lisotrigona* 属无刺蜂（Moure，1961）**

该属无刺蜂为微型到小体型蜂（体长 2.5～4 mm），分布范围从印度和斯里兰卡到马来西亚（Rasmussen，2008）。该属约有 6 个种（Engel，2000 b; Rasmussen，2008; Viraktamath and Jose，2017）。

- *L. cacciae*（Nurse, 1907）
- *L. carpenteri*（Engel, 2000）
- *L. chandrai*（Viraktamath and Sajan Jose, 2017）
- *L. furva*（Engel, 2000）
- *L. mohandasi*（Jobiraj and Narendran, 2004）
- *L. revanai*（Viraktamath and Sajan Jose, 2017）

***Lophotrigona* 属无刺蜂（Moure，1961）**

该属无刺蜂体型为中到大型且体格强壮（体长 7～8 mm），在食物源处属于优势群，分布范围从缅甸和泰国到印度尼西亚（Rasmussen，2008）。该属有 1 个种。

- *L. canifrons*（Smith, 1857）

***Odontotrigona* 属无刺蜂（Moure，1961）**

该属无刺蜂体型中等（体长约 6 mm），黑色蜂，主要发现于马来西亚（Rasmussen，2008; Ascher and Pickering，2018），该属有 1 个种。

- *O. haematoptera*（Cockerell, 1919）

***Papuatrigona* 属无刺蜂（Michener and Sakagami，1991）**

该属无刺蜂有 1 个种，为小型至中型无刺蜂（体长 4.5～5 mm），于新几内亚和印度尼西亚发现了该属无刺蜂（Michener，2007; Rasmussen，2008）。

- *P. genalis*（Friese, 1908）

Pariotrigona 属无刺蜂（Moure，1961）

该属无刺蜂为微型蜂（体长 2.5～3 mm）。分布范围从北部的泰国到南部的婆罗洲和苏门答腊（Michener，2007）。据记述，该属有 1 个种，与 *Hypotrigona* 属亲缘关系密切（Michener，2007; Rasmussen and Cameron, 2010; Rasmussen et al., 2017）。

- *P. pendleburyi*（Schwarz, 1939）

Platytrigona 属无刺蜂（Moure，1961）

该属无刺蜂体型中等，发现于文莱、印度尼西亚、马来西亚和巴布亚新几内亚（Rasmussen, 2008; Rasmussen et al., 2017; Ascher and Pickering, 2018）。据记述，该属有 5 个种。

- *P. flaviventris*（Friese, 1908）
- *P. hobbyi*（Schwarz, 1937）
- *P. keyensis*（Friese, 1901）
- *P. lamingtonia*（Cockerell, 1929）
- *P. planifrons*（Smith, 1865）

Sahulotrigona 属无刺蜂（Engel and Rasmussen，2017）

该属无刺蜂为小到中型（体长 5～6 mm），于新几内亚发现，该属有 2 个种。

- *S. atricornis*（Smith, 1865）
- *S. paradisaea*（Engel and Rasmussen, 2017）

Sundatrigona 属无刺蜂（Inoue and Sakagami，1993）

该属无刺蜂体型小（体长 3～4mm），黑色且有光泽。在印度尼西亚和马来西亚发现了该属的 2 个种（Schwarz, 1937; Rasmussen, 2008）。

- *S. lieftincki*（Sakagami and Inoue, 1987）
- *S. moorei*（Schwarz, 1937）

Tetragonilla 属无刺蜂（Moure，1961）

该属无刺蜂体型中等（5～6 mm），与 *Tetragonula* 属亲缘关系密切。分布范围从缅甸和泰国到马来西亚和印度尼西亚。该属蜂的深色翅膀有明显的白色尖端，类似于 *Tetrigona* 属和一些新热带 *Frieseomelitta* 属无刺蜂，有 4 个种。

- *T. atripes*（Smith, 1857）

- *T. collina*（Smith, 1857）
- *T. fuscibasis*（Cockerell, 1920）
- *T. rufibasalis*（Cockerell, 1918）

***Tetragonula* 属无刺蜂（Moure，1961）**

该属无刺蜂为小到中型（体长 3～7 mm），广泛分布在亚洲热带和澳大利亚。该属物种丰富且数量庞大，大约有 34 个种（Rasmussen, 2008; Rasmussen et al., 2017; Shanas and Faseeh, 2019）。3 个种已经被分配到 *Flavotetragonula* 亚属（Shanas and Faseeh，2019）。该属蜂种表现出极大的生物多样性，例如它们的筑巢行为（第 3 章）。

- *T. bengalensis*（Cameron, 1897）
- *T. biroi*（Friese, 1898）
- *T. carbonaria*（Smith, 1854）
- *T. clypearis*（Friese, 1909）
- *T. dapitanensis*（Cockerell, 1925）
- *T. davenporti*（Franck, 2004）
- *T. drescheri*（Schwarz, 1939）
- *T. fuscobalteata*（Cameron, 1908）
- *T. geissleri*（Cockerell, 1918）
- *T. hirashimai*（Sakagami, 1978）
- *T. hockingsi*（Cockerell, 1929）
- *T. iridipennis*（Smith, 1854）
- *T. laeviceps*（Smith, 1857）
- *T. malaipanae*（Engel, Michener, Boontop, 2017）
- *T. melanocephala*（Gribodo, 1893）
- *T. melina*（Gribodo, 1893）
- *T. mellipes*（Friese, 1898）
- *T. Minangkabau*（Sakagami and Inoue, 1985）
- *T. minor*（Sakagami, 1978）
- *T. pagdeni*（Schwarz, 1939）
- *T. pagdeniformis*（Sakagami, 1978）
- *T. penangensis*（Cockerell, 1919）
- *T. perlucipinnae*（Faseeh and Shanas, 2019）

- *T. reepeni*（Friese, 1918）
- *T. ruficornis*（Smith, 1870）
- *T. sapiens*（Cockerell, 1911）
- *T. sarawakensis*（Schwarz, 1937）
- *T. sirindhornae*（Michener and Boongird, 2004）
- *T. testaceitarsis*（Cameron, 1901）
- *T. travancorica*（Shanas and Faseeh, 2019）
- *T. zucchii*（Sakagami, 1978）

***Tetragonula*（*Flavotetragonula*）**

- *T. calophyllae*（Shanas and Faseeh, 2019）
- *T. gressitti*（Sakagami, 1978）
- *T. praeterita*（Walker, 1860）

***Tetrigona* 属无刺蜂（Moure，1961）**

该属无刺蜂体型中到大型（体长 6～8 mm）。分布范围在北部的缅甸、老挝和越南和南部的印度尼西亚、东帝汶（Rasmussen, 2008）。该属有 5 个种。

- *T. apicalis*（Smith, 1857）
- *T. binghami*（Schwarz, 1937）
- *T. melanoleuca*（Cockerell, 1929）
- *T. peninsularis*（Cockerell, 1927）
- *T. vidua*（Lepeletier, 1836）

***Wallacetrigona* 属无刺蜂（Engel and Rasmussen，2017）**

该属无刺蜂体型中等（体长 5～7 mm）。目前已知在印度尼西亚的华莱士线以东存在该属的 1 个种（Rasmussen et al., 2017）。

- *W. incisa*（Sakagami and Inoue, 1989）

参考文献

Ali JR（2012）Colonizing the Caribbean: is the GAARlandia land-bridge hypothesis gaining a foothold? Journal of Biogeography 39:431–433

Almeida EA, Porto DS（2014）Investigating Eusociality in Bees while Trusting the Uncertainty. Sociobiology 61:355–368

Alvarez LJ, Lucia M (2018) Una especie nueva de *Trigonisca* y nuevos registros de abejas sin aguijón para la Argentina (Hymenoptera: Apidae)/ A new species of *Trigonisca* and new records of stingless bees for Argentina (Hymenoptera: Apidae). Caldasia 40:232–245

Alvarez LJ, Rasmussen C, Abrahamovich AH (2016) Nueva especie de *Plebeia* Schwarz, clave para las especies argentinas de *Plebeia* y comentarios sobre *Plectoplebeia* en la Argentina (Hymenoptera: Meliponini). Revista del Museo Argentino de Ciencias Naturales 18:65–74

Antoine P-O, Marivaux L, Croft DA, Billet G, Ganerød M, Jaramillo C, Martin T, Orliac MJ, Tejada J, Altamirano AJ (2012) Middle Eocene rodents from Peruvian Amazonia reveal the pattern and timing of caviomorph origins and biogeography. Proceedings of the Royal Society of London B: Biological Sciences 279:1319–1326

Ascher J, Pickering J (2018) Discover Life bee species guide and world checklist (Hymenoptera: Apoidea: Anthophila). http://www.discoverlife.org/mp/20q?guide=Apoidea_species

Attasopa K, Bänziger H, Disayathanoowat T, Packer L (2018) A new species of *Lepidotrigona* (Hymenoptera: Apidae) from Thailand with the description of males of *L. flavibasis* and *L. doipaensis* and comments on asymmetrical genitalia in bees. Zootaxa 4442:63–82

Ayala R, Engel MS (2014) A new stingless bee species of the genus *Nogueirapis* from Costa Rica (Hymenoptera: Apidae). Journal of Melittology 37:1–9

Biesmeijer JC, Slaa EJ (2006) The structure of eusocial bee assemblages in Brazil. Apidologie 37:240–258

Bohaty SM, Zachos JC, Florindo F, Delaney ML (2009) Coupled greenhouse warming and deep-sea acidification in the middle Eocene. Paleoceanography 24:1–16

Bond M, Tejedor MF, Campbell Jr KE, Chornogubsky L, Novo N, Goin F (2015) Eocene primates of South America and the African origins of New World monkeys. Nature 520:538–541

Boomsma JJ, Gawne R (2018) Superorganismality and caste differentiation as points of no return: how the major evolutionary transitions were lost in translation. Biological Reviews 93:28–54

Bossert S, Murray EA, Almeida EAB, Brady SG, Blaimer BB, Danforth BN (2019)

Combining transcriptomes and ultraconserved elements to illuminate the phylogeny of Apidae. Molecular Phylogenetics and Evolution 130:121–131

Bossert S, Murray EA, Blaimer BB, Danforth BN (2017) The impact of GC bias on phylogenetic accuracy using targeted enrichment phylogenomic data. Molecular Phylogenetics and Evolution 111:149–157

Brikiatis L (2014) The De Geer, Thulean and Beringia routes: key concepts for understanding early Cenozoic biogeography. Journal of Biogeography 41:1036–1054

Camargo JMF de, Pedro SRM (2009) Neotropical Meliponini: the genus *Celetrigona* Moure (Hymenoptera: Apidae, Apinae). Zootaxa 2155:37–54

Camargo JMF (1996) Meliponini neotropicais: o gênero *Camargoia* Moure, 1989 (Apinae, Apidae, Hymenoptera). Arquivos de Zoologia (São Paulo) 33:71–92

Camargo JMF, Moure JS (1994) Meliponinae neotropicais: os gêneros *Paratrigona* Schwarz, 1938 e *Aparatrigona* Moure, 1951 (Hymenoptera, Apidae). Arquivos de Zoologia 32:33–109

Camargo JMF, Pedro SRM (2003) Neotropical Meliponini: the genus *Partamona* Schwarz, 1939 (Hymenoptera, Apidae, Apinae)-bionomy and biogeography. Revista Brasileira de Entomologia 47:311–372

Camargo JMF, Grimaldi DG, Pedro SRM (2000) The Extinct Fauna of Stingless Bees (Hymenoptera: Apidae: Meliponini) in Dominican Amber: Two New Species and Redescription of the Male of *Proplebeia dominicana* (Wille and Chandler). American Museum Novitates 2000:1–24

Camargo JMF, Moure JS, Roubik DW (1988) *Melipona yucatanica*, new species (Hymenoptera: Apidae: Meliponinae); stingless bee dispersal across the Caribbean arc and post-eocene vicariance. Pan-Pacific Entomologist 64:147–157

Camargo JMF, Moure JSF (1983) *Trichotrigona*, un novo genero de Meliponinae (Hymenoptera, Apidae) do Rio Negro, Amazonas Brasil. Acta Amazonica 13:421–429

Camargo JMF, Pedro SRM (1992) Systematics, phylogeny and biogeography of the Meliponinae (Hymenoptera, Apidae): a mini-review. Apidologie 23:509–522

Camargo JMF, Pedro SRM (2005) Meliponini Neotropicais: o gênero *Dolichotrigona* Moure (Hymenoptera, Apidae, Apinae). Revista Brasileira de Entomologia 49:69–92

Camargo JMF, Wittmann D (1989) Nest architecture and distribution of the primitive stingless bee, *Mourella caerulea* (hymenoptera, apidae, meliponinae): Evidence for the origin of *Plebeia* (s. lat.) on the gondwana continent. Studies on Neotropical Fauna and Environment 24:213–229

Cameron SA, Mardulyn P (2001) Multiple Molecular Data Sets Suggest Independent Origins of Highly Eusocial Behavior in Bees (Hymenoptera:Apinae). Systematic Biology 50:194–214

Cardinal S, Danforth BN (2011) The Antiquity and Evolutionary History of Social Behavior in Bees. PLoS ONE 6:e21086

Cardinal S, Danforth BN (2013) Bees diversified in the age of eudicots. Proceedings of the Royal Society of London Series B-Biological Sciences 280:20122686

Chaimanee Y, Chavasseau O, Beard KC, Kyaw AA, Soe AN, Sein C, Lazzari V, Marivaux L, Marandat B, Swe M (2012) Late Middle Eocene primate from Myanmar and the initial anthropoid colonization of Africa. Proceedings of the National Academy of Sciences of the United States of America 109:10293–10297

Danforth BN, Cardinal S, Praz C, Almeida EAB, Michez D (2013) The impact of molecular data on our understanding of bee phylogeny and evolution. Annual Review of Entomology 58:57–78

Darwin C (1859) On the origin of species by means of natural selection. John Murray, London

Dollin AE, Dollin LJ, Rasmussen C (2015) Australian and New Guinean Stingless Bees of the Genus *Austroplebeia* Moure (Hymenoptera: Apidae) - a revision. Zootaxa 4047:1–73

Dollin AE, Dollin LJ, Sakagami SF (1997) Australian stingless bees of the genus *Trigona* (Hymenoptera: Apidae). Invertebrate Systematics 11:861–896

Eardley CD, Kwapong P (2013) Taxonomy as a Tool for Conservation of African Stingless Bees and Their Honey. In: Vit P, Pedro SRM, Roubik D (eds) Pot-Honey: A legacy of stingless bees. Springer New York, New York, NY, pp 261–268

Eardley CD (2004) Taxonomic revision of the African stingless bees (Apoidea: Apidae: Apinae: Meliponini). African Plant Protection 10:63–96

Engel MS (2001 a) A monograph of the Baltic amber bees and evolution of the Apoidea (Hymenoptera). Bulletin of the American Museum of Natural History 259:1–192

Engel MS (2001 b) Monophyly and extensive extinction of advanced eusocial bees: insights from an unexpected Eocene diversity. Proceedings of the National Academy of Sciences of the United States of America 98:1661–1664

Engel MS (2000 a) A new interpretation of the oldest fossil bee (Hymenoptera: Apidae). American Museum Novitates 3296:1–11

Engel MS (2000 b) A review of the Indo-Malayan meliponine genus *Lisotrigona*, with two new species (Hymenoptera: Apidae). Oriental Insects 34:229–237

Engel MS, Michener CD (2013) A minute stingless bee in Eocene Fushan amber from northeastern China (Hymenoptera: Apidae). Journal of Melittology 14:1–10

Engel MS, Rasmussen C (2017) A new subgenus of *Heterotrigona* from New Guinea (Hymenoptera: Apidae). Journal of Melittology 73:1–16

Engel MS, Rozen JG, Sepúlveda-Cano PA, Smith CS, Thomas JC, Ospina Torres R, González VH (2019) Nest architecture, immature stages, and ethnoentomology of a new species of *Trigonisca* from northern Colombia (Hymenoptera, Apidae). American Museum Novitates 3942:1–33

Estienne V, Mundry R, Kühl HS, Boesch C (2017) Exploitation of underground bee nests by three sympatric consumers in Loango National Park, Gabon. Biotropica 49:101–109

Faber Anguilet E, Nguyen BK, Bengone Ndong T, Haubruge E, Francis F (2015) Meliponini and Apini in Africa (Apidae: Apinae): a review on the challenges and stakes bound to their diversity and their distribution. Biotechnologie, Agronomie, Société et Environnement, Biotechnology, Agronomy, Society and Environment 19:1–10

Flores FF, Lupo LC, Hilgert NI (2015) Recursos tróficos utilizados por *Plebeia intermedia* (Apidae, Meliponini) en la localidad de Baritú, Salta, Argentina: Caracterización botánica de sus mieles. Boletín de la Sociedad Argentina de Botánica 50:515–529

Francisco FO, Santiago LR, Brito RM, Oldroyd BP, Arias MC (2014) Hybridization and asymmetric introgression between *Tetragonisca angustula* and *Tetragonisca fiebrigi*. Apidologie 45:1–9

Francisco FO, Santiago LR, Mizusawa YM, Oldroyd BP, Arias MC (2016) Genetic structure of island and mainland populations of a Neotropical bumble bee species. Journal of Insect Conservation 20:383–394

Francisco FO, Santiago LR, Mizusawa YM, Oldroyd BP, Arias MC (2017) Population structuring of the ubiquitous stingless bee *Tetragonisca angustula* in southern Brazil as revealed by microsatellite and mitochondrial markers. Insect Science 24:877–890

Franck P, Cameron E, Good G, Rasplus J-Y, Oldroyd BP（2004）Nest architecture and genetic differentiation in a species complex of Australian stingless bees. Molecular Ecology 13:2317–2331

Freitas BM, Imperatriz-Fonseca VL, Medina LM, Kleinert A de MP, Galetto L, Nates-Parra G, Quezada-Euán JJG（2009）Diversity, threats and conservation of native bees in the Neotropics. Apidologie 40:332–346

Fuller S, Schwarz M, Tierney S（2005）Phylogenetics of the allodapine bee genus *Braunsapis*: historical biogeography and long-range dispersal over water. Journal of Biogeography 32:2135–2144

Genaro JA（2008）Origins, composition and distribution of the bees of Cuba（Hymenoptera: Apoidea: Anthophila）. Insecta Mundi 0052:1–16

Gonzalez V, Griswold T（2011）Two new species of *Paratrigona* Schwarz and the male of *Paratrigona ornaticeps*（Schwarz）（Hymenoptera, Apidae）. ZooKeys 120:9–25

Gonzalez V, Griswold TT（2012）New species and previously unknown males of Neotropical cleptobiotic stingless bees（Hymenoptera, Apidae, *Lestrimelitta*）. Caldasia 34:227–245

Gonzalez VH, Smith-Pardo A（2003）New Distribution Records and Taxonomic Comments on *Parapartamona*（Hymenoptera: Apidae: Meliponini）. Journal of the Kansas Entomological Society 76:655–657

Grimaldi D, Engel MS（2005）Evolution of the Insects. Cambridge University Press, Cambridge

Grüter C, Menezes C, Imperatriz-Fonseca VL, Ratnieks FLW（2012）A morphologically specialized soldier caste improves colony defence in a Neotropical eusocial bee. Proceedings of the National Academy of Sciences of the United States of America 109:1182–1186

Grüter C, Segers FHID, Menezes C, Vollet-Neto A, Falcon T, von Zuben LG, Bitondi MMG, Nascimento FS, Almeida EAB（2017）Repeated evolution of soldier sub-castes suggests parasitism drives social complexity in stingless bees. Nature Communications 8:4

Halcroft M, Spooner-Hart R, Dollin LA（2013）Australian stingless bees. In: Vit P,

Pedro SRM, Roubik DW (eds) Pot-Honey. Springer, New York, pp 35–72

Hedges SB (2006) Paleogeography of the Antilles and origin of West Indian terrestrial vertebrates. Annals of the Missouri Botanical Garden 93:231–244

Houle A (1998) Floating islands: a mode of long-distance dispersal for small and medium-sized terrestrial vertebrates. Diversity and Distributions 4:201–216

Hurtado-Burillo M, May-Itzá WDJ, Quezada-Eúan JJG, Rúa PDL, Ruiz C (2017) Multilocus species delimitation in Mesoamerican *Scaptotrigona* stingless bees (Apidae: Meliponini) supports the existence of cryptic species. Systematic Entomology 42:171–181

Iturralde-Vinent M, MacPhee RDE (1999) Paleogeography of the Caribbean region: implications for Cenozoic biogeography. Bulletin of the American Museum of Natural History 238:1–95

Kajobe R, Roubik DW (2018) Nesting Ecology of Stingless Bees in Africa. In: Vit P, Pedro SRM, Roubik DW (eds) Pot-Pollen in Stingless Bee Melittology. Springer International Publishing, Cham, pp 229–240

Kawakita A, Ascher JS, Sota T, Kato M, Roubik DW (2008) Phylogenetic analysis of the corbiculate bee tribes based on 12 nuclear protein-coding genes (Hymenoptera: Apoidea: Apidae). Apidologie 39:163–175

Kerr WE, Maule V (1964) Geographic distribution of stingless bees and its implications (Hymenoptera: Apidae). Journal of the New York Entomological Society 72:2–18

Koch H (2010) Combining morphology and DNA barcoding resolves the taxonomy of western Malagasy *Liotrigona* Moure, 1961 (Hymenoptera: Apidae: Meliponini). African Invertebrates 51:413–421

Lockhart PJ, Cameron SA (2001) Trees for bees. Trends in Ecology & Evolution 16:84–88

Marchi P, Melo GAR (2006) Revisão taxonômica das espécies brasileiras de abelhas do gênero *Lestrimelitta* Friese (Hymenoptera, Apidae, Meliponina). Revista Brasileira de Entomologia 50:6–30

Martins AC, Melo GA, Renner SS (2014) The corbiculate bees arose from New World oil-collecting bees: Implications for the origin of pollen baskets. Molecular Phylogenetics and Evolution 80:88–94

Martins AC, Melo GAR (2016) The New World oil-collecting bees *Centris* and *Epicharis* (Hymenoptera, Apidae): molecular phylogeny and biogeographic history. Zoologica Scripta 45:22–33

Mazzeo NM, Torretta JP (2015) Wild bees (Hymenoptera: Apoidea) in an urban botanical garden in Buenos Aires, Argentina. Studies on Neotropical Fauna and Environment 50:182–193

McCall RA (1997) Implications of recent geological investigations of the Mozambique Channel for the mammalian colonization of Madagascar. Proceedings of the Royal Society of London Series B: Biological Sciences 264:663–665

Melo GA (2015) New species of the stingless bee genus *Schwarziana* (Hymenoptera, Apidae). Revista Brasileira de Entomologia 59:290–293

Melo GAR (2016) *Plectoplebeia*, a new Neotropical genus of stingless bees (Hymenoptera: Apidae). Zoologia 33:e20150153

Melo GAR (2020) Stingless Bees (Meliponini). In: Starr CK (ed) Encyclopedia of Social Insects. Springer International Publishing, Cham, pp 1–18 Michaux B (2010) Biogeology of Wallacea: geotectonic models, areas of endemism, and natural biogeographical units. Biological Journal of the Linnean Society 101:193–212

Michener CD (1944) Comparative external morphology, phylogeny, and a classification of the bees (Hymenoptera). Bulletin of the American Museum of Natural History 82:151–326

Michener CD (1974) The Social Behavior of the Bees. Harvard University Press, Cambridge

Michener CD (1979) Biogeography of the Bees. Annals of the Missouri Botanical Garden 66:277–347

Michener CD (1990) Classification of the Apidae (Hymenoptera). Appendix: *Trigona genalis* Friese, a hitherto unplaced New Guinea species. University of Kansas Science Bulletin 54:75–164

Michener CD (2007) The bees of the world, 2nd edn. The Johns Hopkins University Press, Baltimore

Michener CD, Grimaldi DA (1988a) The oldest fossil bee: Apoid history, evolutionary stasis, and antiquity of social behavior. Proceedings of the National Academy of Sciences of the United States of America 85:6424–6426

Michener CD, Grimaldi DA (1988 b) A *Trigona* from late Cretaceous amber of New Jersey (Hymenoptera, Apidae, Meliponinae). American Museum Novitates 2917:1–10

Morley RJ (2003) Interplate dispersal paths for megathermal angiosperms. Perspectives in Plant Ecology, Evolution and Systematics 6:5–20

Moure JS (1961) A preliminary supraspecific classification of the Old World meliponine bees. Studia Entomologica 4:181–242

Moure JS, Camargo JMF (1978) A Fossil Stingless Bee from Copal (Hymenoptera: Apidae). Journal of the Kansas Entomological Society 51:560–566

Ndungu NN, Kiatoko N, Ciosi M, Salifu D, Nyansera D, Masiga D, Raina SK (2017) Identification of stingless bees (Hymenoptera: Apidae) in Kenya using morphometrics and DNA barcoding. Journal of Apicultural Research 56:341–353

Nogueira DS, Júnior JEDS, Oliveira FFD, Oliveira MLD (2019) Review of *Scaura* Schwarz, 1938 (Hymenoptera: Apidae: Meliponini). Zootaxa 4712:451–496

Noll FB (2002) Behavioral Phylogeny of corbiculate Apidae (Hymenoptera; Apinae), with special reference to social behavior. Cladistics 18:137–153

Ortiz-Mora RA, Veen JWV, Corrales G, Sommeijer MJ, Rica C (1995) Influence of altitude on the distribution of stingless bees (Hymenoptera Apidae: Meliponinae). Apiacta 4:1–3

Pan P, Wang S, Zhong Y, Xu H, Wang Z (2020) New record of the stingless bee *Tetragonula gressitti* (Sakagami, 1978) in Southwest China (Hymenoptera: Apidae: Meliponini). J Apic Res (in press)

Pauly A, Brooks RW, Nilsson LA, Pesenko YA, Eardley CD, Terzo M, Griswold T, Schwarz M, Patiny S, Munzinger J, Barbier Y (2001) Hymenoptera Apoidea de Madagascar et des îles voisines. Annales du Musée royal de l'Afrique centrale (Sciences Zoologiques), Tervuren (Belgique) 286:1–406

Pauly A, Fabre Anguilet E (2013) Description de *Liotrigona gabonensis* sp. nov., et quelques corrections à la synonymie des espèces africaines de mélipones (Hymenoptera : Apoidea : Apinae : Meliponini). Belgian Journal of Entomology 15:1–15

Pedro SRM, Cordeiro GD (2015) A new species of the stingless bee *Trichotrigona* (Hymenoptera: Apidae, Meliponini). Zootaxa 3956:389–402

Pedro SRM (2014) The Stingless Bee Fauna In Brazil (Hymenoptera: Apidae). Sociobiology 61:348–354

Pindell JL, Cande SC, Pitman WC, Rowley DB, Dewey JF, Labrecque J, Haxby W (1988) A plate-kinematic framework for models of Caribbean evolution. Tectonophysics 155:121–138

Porto DS, Vilhelmsen L, Almeida EAB (2016) Comparative morphology of the mandibles and head structures of corbiculate bees (Hymenoptera: Apidae: Apini). Systematic Entomology 41:339–368

Praz CJ, Packer L (2014) Phylogenetic position of the bee genera *Ancyla* and *Tarsalia* (Hymenoptera: Apidae): A remarkable base compositional bias and an early Paleogene geodispersal from North America to the Old World. Molecular Phylogenetics and Evolution 81:258–270

Ramírez SR, Nieh JC, Quental TB, Roubik DW, Imperatriz-Fonseca VL, Pierce NE (2010) A molecular phylogeny of the stingless bee genus *Melipona* (Hymenoptera: Apidae). Molecular Phylogenetics and Evolution 56:519–525

Rasmussen C (2008) Catalog of the Indo-Malayan/Australasian stingless bees (Hymenoptera: Apidae: Meliponini). Zootaxa 1935:1–80

Rasmussen C (2013) Stingless bees (Hymenoptera: Apidae: Meliponini) of the Indian subcontinent: Diversity, taxonomy and current status of knowledge. Zootaxa 3647:401–428

Rasmussen C, Camargo JMF (2008) A molecular phylogeny and the evolution of nest architecture and behavior in *Trigona s.s.* (Hymenoptera: Apidae: Meliponini). Apidologie 39:102–118

Rasmussen C, Cameron SA (2007) A molecular phylogeny of the Old World stingless bees (Hymenoptera: Apidae: Meliponini) and the non-monophyly of the large genus *Trigona*. Systematic Entomology 32:26–39

Rasmussen C, Cameron SA (2010) Global stingless bee phylogeny supports ancient divergence, vicariance, and long distance dispersal. Biological Journal of the Linnean Society 99:206–232

Rasmussen C, Gonzalez VH (2017) The neotropical stingless bee genus *Nannotrigona* Cockerell (Hymenoptera: Apidae: Meliponini): An illustrated key, notes on the types, and designation of lectotypes. Zootaxa 4299:191–220

Rasmussen C, Thomas JC, Engel MS (2017) A New Genus of Eastern Hemisphere Stingless Bees (Hymenoptera: Apidae), with a Key to the Supraspecific Groups of Indomalayan and Australasian Meliponini. American Museum Novitates 2017:1–33

Rogers RR, Hartman JH, Krause DW (2000) Stratigraphic Analysis of Upper Cretaceous Rocks in the Mahajanga Basin, Northwestern Madagascar: Implications for Ancient and Modern Faunas. The Journal of Geology 108:275–301

Roig-Alsina A, Alvarez LJ (2017) Southern distributional limits of Meliponini bees (Hymenoptera, Apidae) in the Neotropics: taxonomic notes and distribution of *Plebeia droryana* and *P. emerinoides* in Argentina. Zootaxa 4244: 261–268

Roig-Alsina A, Vossler FG, Gennari GP (2013) Stingless Bees in Argentina. In: Vit P, Pedro SRM, Roubik D (eds) Pot-Honey: A legacy of stingless bees. Springer New York, New York, NY, pp 125–134

Romiguier J, Cameron SA, Woodard SH, Fischman BJ, Keller L, Praz CJ (2016) Phylogenomics Controlling for Base Compositional Bias Reveals a Single Origin of Eusociality in Corbiculate Bees. Molecular Biology & Evolution 33:670–678

Roubik DW, Arturo J, Segura L, Camargo JMFD (1997) New stingless bee genus endemic to Central American cloudforests: phylogenetic and biogeographic implications (Hymenoptera: Apidae: Meliponini). Systematic Entomology 22:67–80

Roubik DW (1989) Ecology and Natural History of Tropical Bees. Cambridge University Press, New York

Roubik DW (2018) 100 Species of Meliponines (Apidae: Meliponini) in a Parcel of Western Amazonian Forest at Yasuní Biosphere Reserve, Ecuador. In: Vit P, Pedro SRM, Roubik DW (eds) Pot-Pollen in Stingless Bee Melittology. Springer International Publishing, Cham, pp 189–206

Sakagami SF (1982) Stingless bees. In: Hermann HR (ed) Social Insects Ⅲ. Academic Press, New York, pp 361–423

Schultz TR, Engel MS, Aschier JS (2001) Evidence for the origin of eusociality in the corbiculate bees (Hymenoptera: Apidae). Journal of the Kansas Entomological Society 74:10–16

Schwarz HF (1937) Results of the Oxford University Sarawak (Borneo) expedition:

Bornean stingless bees of the genus *Trigona*. Bulletin of the American Museum of Natural History, 73: 281–328

Schwarz HF (1948) Stingless Bees (Meliponidae) of the Western Hemisphere. Bulletin of the American Museum of Natural History 90:1–546

Seeley TD (2010) Honeybee Democracy. Princeton University Press, Princeton

Segers FHID, Menezes C, Vollet-Neto A, Lambert D, Grüter C (2015) Soldier production in a stingless bee depends on rearing location and nurse behaviour. Behavioral Ecology and Sociobiology 69:613–623

Shanas S, Faseeh P (2019) A new subgenus and three new species of stingless bees (Hymenoptera: Apidae: Apinae: Meliponini) from India. Entomon 44:33–48

Sung I-H, Yamane S, Lu S-S, Ho K-K (2011) Climatological influences on the flight activity of stingless bees (*Lepidotrigona hoozana*) and honeybees (*Apis cerana*) in Taiwan (Hymenoptera, Apidae). Sociobiology 58:835–850

Techer MA, Clémencet J, Simiand C, Preeaduth S, Azali HA, Reynaud B, Hélène D (2017) Large-scale mitochondrial DNA analysis of native honey bee *Apis mellifera* populations reveals a new African subgroup private to the South West Indian Ocean islands. BMC Genetics 18:53

Van Welzen PC, Parnell JAN, Slik JWF (2011) Wallace's Line and plant distributions: two or three phytogeographical areas and where to group Java? Biological Journal of the Linnean Society 103:531–545

Viraktamath S, Jose KS (2017) Two new species of *Lisotrigona* Moure (Hymenoptera: Apidae: Meliponini) from India with notes on nest structure. The Bioscan 12:21–28

Vit P, Pedro SRM, Roubik DW (2013) Pot-honey: a legacy of stingless bees. Springer, New York

Wille A (1979) Phylogeny and relationships among the genera and subgenera of the stingless bees (Meliponinae) of the world. Revista de Biologia Tropical 27:241–277

Wille A, Michener CD (1973) The nest architecture of stingless bees with special reference to those of Costa Rica. Revista de Biologia Tropical 21:9–278

Winston ML, Michener CD (1977) Dual origin of highly social behavior among bees. Proceedings of the National Academy of Sciences of the United States of America 74:1135–1137

Yoder AD, Nowak MD (2006) Has Vicariance or Dispersal Been the Predominant Biogeographic Force in Madagascar? Only Time Will Tell. Annual Review of Ecology, Evolution, and Systematics 37:405–431

Zachos J, Pagani M, Sloan L, Thomas E, Billups K (2001) Trends, rhythms, and aberrations in global climate 65 Ma to present. Science 292:686–693

3 筑巢生物学

蜂巢是蜜蜂度过大部分时间的地方。工蜂只有在生命的最后阶段，才会开始定期离开巢穴，去寻找资源或守卫巢穴的入口（第6章，第8章）。在完全黑暗的巢穴中，蜂会频繁地触碰同伴、分享食物以及交换许多关于各种蜂群需求的化学和触觉信息。蜂巢也是收集资源的储存地，更是蜂繁衍后代和躲避天敌的地方。考虑蜂巢这些重要作用，显然蜂巢的位置、建筑特征以及它所蕴藏的材料和资源对蜂群至关重要。例如蜂巢的位置和结构会影响天敌的捕食强度（Fowler，1979；Slaa，2006），也会影响幼虫对气候条件的感应（Jones and Oldroyd，2006）。与蜜蜂相比，找到一个合适的筑巢位置并建立所需的结构对于无刺蜂来说可能更为重要，因为无刺蜂蜂群无法将不合适的巢换成一个适宜生存的巢（第4章）。

无刺蜂的筑巢特征表现出极大的多样性，似乎并不遵循进化关系。近缘物种在筑巢的关键方面可能存在差异，育虫巢室的排列（蜂巢或蜂群）或巢穴类型的使用（例如，新热带区的 *Tetragonisca* 属包括一个在地下筑巢的物种，两个在洞穴中筑巢的物种，以及一个建造暴露巢穴的物种），在一些物种中，例如在育虫巢穴（包壳）周围是否有防护层等因物种不同而有所区别。

3.1 筑巢地点

无刺蜂作为一个群体，已经适应了范围广泛的筑巢地点。它们可以生活在蚂蚁和白蚁的地上和地下巢穴中，也可以生活在树木、树枝、岩石或人类建筑中的孔洞里，大部分则生活在自建的暴露树上的蜂巢里（图3.1，表3.1）（Schwarz，1948；Wille and Michener，1973；Michener，1974；Roubik，2006；

Roubik，1983）。

图 3.1 无刺蜂不同的筑巢方式

注：A 为 *Tetragonisca angustula* 无刺蜂在金属管（直径约 3 cm）中筑巢。B 为在房屋墙壁上的 *Lestrimelitta limao* 无刺蜂巢（摄影：Lucas von Zuben）。C 为 *Trigona spinipes* 无刺蜂在树上的裸露蜂巢。D 为 *Trigona hyalinata* 无刺蜂附着在墙上的巢。E 为 *Partamona helleri* 无刺蜂的泥巢，在墙壁的凹陷处有其典型的"蛤蟆嘴"入口。F 为生活在一个电箱中的 *Frieseomelitta varia* 无刺蜂蜂群，该蜂种会在巢穴周围的表面盖一层黑树脂，该图中这层黑树脂是在建筑物的墙壁上，这种行为的功能尚不清楚（照片 A，C-F 是笔者在巴西 Ribeirão Preto 拍摄的）。

表 3.1　中美洲 145 个无刺蜂蜂种的筑巢习性

筑巢习性（专一或普遍）	物种百分比（%）
地面	11.7%（17）
树洞[a]	65.5%（95）
白蚁巢穴[b]	9%（13）
蚂蚁巢穴	2.1%（3）
暴露或部分暴露	11.7%（17）

注：数据来源于表 2（Wille and Michener，1973）。
a. 它们也经常在人工洞穴中筑巢。
b. 地上、地下均有发现。

3.1.1　地上筑巢型无刺蜂

大多数无刺蜂会在树干或树枝中预先存在的空穴中筑巢（图 3.2），这些空穴可以通过相对较小的入口孔进入（图 3.3）。无刺蜂选定的树木通常很大（植株胸径 > 50 cm）并且是活树（Roubik，1983; Eltz et al., 2003; Samejima et al., 2004; Siqueira et al., 2012; Silva and Ramalho，2014; Arena et al., 2018），并且无刺蜂体型越大越喜欢在较大的树上筑巢（Hubbell and Johnson，1977）[1]。然而，有些蜂种的蜂群常在枯树上筑巢，例如新热带区的 *Frieseomelitta* 属、*Trichotrigona extranea* 无刺蜂，以及亚洲的 *Tetragonula melanocephala*、*Tetragonilla rufibasalis* 无刺蜂（Camargo and Moure，1983; Samejima et al., 2004）。如果巢在树干中，它们通常相对靠近地面（<5 m 高，Kerr et al., 1967; Eltz et al., 2003; Cortopassi-Laurino et al., 2009; Silva and Ramalho，2014）。然而一些蜂种，特别是 *Trigona* 和 *Oxytrigona* 属无刺蜂则常在 10～25 m 高处筑巢（Roubik，1983）。如 *Melipona nigra* 等其他蜂种则是在树底根洞内或根间筑巢（Camargo，1970; Michener，1974; Eltz et al., 2003; Rasmussen and Camargo，2008）。筑巢高度可能会影响蜂群所承受的捕食压力和蜂群内的微环境。例如研究者发现，对于 *Tetragonisca angustula* 无刺蜂来说，在树越高的地方筑

---

[1] Hubbell 和 Johnson（1977）提出树木大小与大体型蜂种存在之间的联系可能不仅可以解释没有大树的地区也没有大体型蜂种存在的现象，也可以明确大体型蜂种重新定殖于次生林栖息地的能力。因此，次生林的蜂种的重新定殖顺序与蜂体型大小相关。这反过来也可能会影响森林管理（Samejima et al., 2004）。

巢的蜂群就越不易死亡，这可能是由于高处通常遭受蚂蚁的攻击更少（Slaa，2006）。

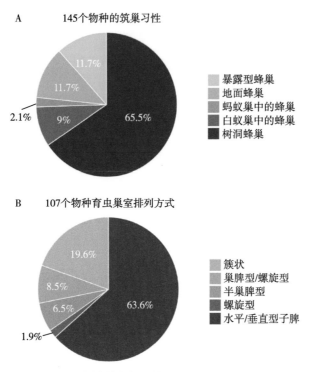

图 3.2　无刺蜂的筑巢习性及育虫巢室的排列方式

注：A 为中美洲 145 个无刺蜂蜂种的筑巢习性，数据来自表 3.1。B 为表 1.3 所列 107 个物种的育虫巢室的排列类型分布。最常见的类型是子脾型，有的是垂直型，有的则是水平型。"半巢脾型"包含一些可变蜂种（表 1.3）。

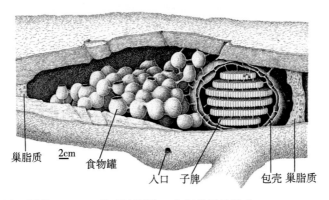

图 3.3　*Melipona grandis* 无刺蜂在一个树枝里的巢（Camargo，1970）

注：不同功能部分的描述见图中文字。

大多数无刺蜂蜂种似乎并不会选择特定树种，而是随机利用合适的筑巢地点（Hubbell and Johnson，1977；Fowler，1979；Eltz et al.，2003；Siqueira et al.，2012；Tornyie and Kwapong，2015；Silva and Ramalho，2014）。研究者对巴西 716 株有无刺蜂栖息的树木调查发现，无刺蜂用于筑巢的树木分属 56 个不同的科（Cortopassi-Laurino et al.，2009；Eltz et al.，2003；Siqueira et al.，2012；Silva and Ramalho，2014）。然而有些无刺蜂蜂种似乎表现出树种偏好性。例如研究者在巴西"Cerrado"（热带草原）研究了 *Melipona quadrifasciata* 无刺蜂蜂群的 48 个栖息地，其中 46 个是同一个树种 *Caryocar brasiliense*，然而事实上这一地区还有许多其他潜在的筑巢地点可用（Antonini and Martins，2003；Martins et al.，2004）。同样，Kerr［1984，被 Roubik（1989）引用］报道在巴西马托格罗索州的 200 种可用树种中只有 12 种树有蜂群筑巢[2]。

蜂群对洞穴大小的要求是具有物种特异性的。在巴拿马对 30 种无刺蜂蜂群进行研究时发现，无刺蜂巢穴大小范围最小低至 0.1 L（例如 *Plebeia minima* 无刺蜂），最大高达 30 L 以上（例如 *Cephalotrigona zexmeniae* 无刺蜂）（Roubik，1983）。甚至有研究发现亚洲 *Lophotrigona canifrons* 无刺蜂在超过 300 L 的巨大洞穴中筑巢（Inoue et al.，1993）。如果洞穴太大，蜂群会使用片状（或板状）巢脂质（Michener，1974）来限制居住空间，这种巢脂质是由巢质、树脂、泥浆、粪便或植物材料构成的坚硬保护层，能够包围住巢腔（Ihering，1903；Schwarz，1948；Michener，1974；Wille，1983）。

许多洞巢型无刺蜂可以利用人类建筑，在屋檐下以及墙壁、电箱里甚至金属管中的中空空间里筑巢（图 3.1）。特别是小体型的新热带区 *Tetragonisca angustula* 无刺蜂，就像 Schwarz（1948）所说的这种无刺蜂具有"将人类原本打算用于其他用途的空洞做成它们的巢穴的神奇行为"。目前人们已经在信箱、金属罐、缝纫机和水管阀门中都发现过这种无刺蜂的存在（Schwarz，1948）。图 3.1 A 展示了在直径不超过 3 cm 的长金属管中生活的无刺蜂，并已经在那里生活了数年之久。*T. angustula* 无刺蜂这种利用各种天然和人造空穴的非凡能力可能是其成为许多新热带区最常见无刺蜂的原因之一（Freitas，2001；Slaa，2006；Velez-Ruiz et al.，2013）。

---

[2] Roubik（1989）没有说明 Kerr 研究的蜂群数量，因此很难确定这份报告是否确实表明了无刺蜂对某些树木有强烈偏好。

有时，蜂群会利用其他动物废弃的洞穴，如鸟类或者甲虫和蛾幼虫（Schwarz，1948；Darchen，1969；Michener，1974；Camargo and Pedro，2002）。*Tetragonisca angustula* 无刺蜂偶尔会使用棕灶鸟废弃的泥巢，而 *Paratrigona pacifica* 无刺蜂则好像会定期重复使用这些鸟巢（Schwarz，1948；Camargo and Moure，1994）。

3.1.2 地下筑巢型无刺蜂

相当多的无刺蜂蜂种会利用地下的洞穴，这些洞穴通常是废弃的蚂蚁、白蚁或啮齿动物的巢穴，这其中就包括了美洲的 *Camargoia*、*Geotrigona*、*Melipona*、*Mourella*、*Nogueirapis*、*Paratrigona*、*Partamona*、*Schwarziana* 和 *Tetragonisa* 属无刺蜂，或非洲的 *Meliplebeia*、*Plebeiella* 和 *Plebeina* 属无刺蜂（da Silva et al., 1972; Wille，1966; Michener，1974; Fowler，1979; Camargo，1996; Camargo and Pedro，2003; Eardley，2004; Njoya，2009; Vossler et al., 2010; Barbosa et al., 2013; Njoya et al., 2017; Galaschi-Teixeira et al., 2018）。例如有研究者认为 *Geotrigona mombuca*、*Schwarziana quadripunctata* 和 *Paratrigona lineata* 无刺蜂主要使用废弃的切叶蚁巢穴（Schwarz，1948; Michener，1974; Wille，1983），而 *Tetragonisca buchwaldi* 无刺蜂则是使用小型哺乳动物制造的洞穴（Wille，1966）。因此这似乎表明无刺蜂不会自己挖掘洞穴筑巢，但它们会扩大现有的洞穴（Michener，1974；曾看到 *Scaptotrigona bipunctata* 无刺蜂在房屋底部移除大量泥土来为它们的巢穴腾出空间）。无刺蜂偶尔也会利用由雨水冲刷形成的洞穴或岩石中的天然裂缝（Wille，1966; Bänziger et al., 2011; Barbosa et al., 2013）。

无刺蜂的巢穴大多数建在地下 1 m 以内：*Nogueirapis mirandula* 无刺蜂蜂群通常在地下 25～30 cm 深度筑巢，*Tetragonisca buchwaldi* 无刺蜂的巢穴则在 15～35 cm 的深度（Wille，1966）。Estienne 等（2017）研究了非洲 *Plebeiella lenliana* 无刺蜂的 73 个巢穴，发现巢腔的平均深度为（49.8 cm± 19.8 cm）（范围为 14～117 cm）。而 *Paratrigona lineata*、*Geotrigona subterranean*、*G. argentina* 无刺蜂则经常在地下 1 m 甚至更深处筑巢（Wille，1983; Vossler et al., 2010; Barbosa et al., 2013），据发现，*Geotrigona subterranean* 和 *Schwarziana quadripunctata* 无刺蜂巢穴建在地下约 3 m 处（Schwarz，1948）。在越深的地方筑巢并不一定能减少由哺乳动物捕食者带来的危害，至少对于深度小于 1 m 的巢穴来说是这样，反而非垂直巢穴入口更难被捕食者发现（Estienne et al., 2017）（第 7 章）。

无刺蜂许多地下巢穴的一个显著特征是会从巢穴底部伸出垂直突起管，这个结构可能起到排水管的作用，以去除巢内多余的蒸汽和水分（图3.4）（Portugal-Araujo，1963；Wille，1966；Camargo，1970；Wille and Michener，1973；Njoya，2009；Njoya et al.，2017）。无刺蜂巢穴有时有多个"排水管"，而 *Geotrigona subterranea*、*G. mombuca* 和 *Tetragonisca buchwaldi* 无刺蜂的巢穴则似乎没有排水管结构，但它们的巢穴底部有开口，这一结构可能具有相同的用途（Wille，1966；Barbosa et al.，2013）。排水通道本身很可能是由无刺蜂自己创造的。一些树上筑巢的诸如 *Trigona* 或 *Tetragona* 属无刺蜂也有这种排水管结构（Roubik，2006）。地面筑巢型无刺蜂也有其他方法来防止水进入巢穴，例如，*Schwarziana* 和 *Geotrigona* 属无刺蜂蜂群会建造一个带沟的包壳（图3.3及下文）来防止水进入蜂巢内部（Roubik，1989）。

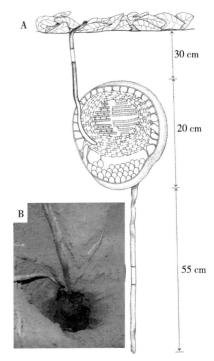

图 3.4 无刺蜂的地下巢穴

注：A 为 *Nogueirapis mirandula* 无刺蜂的地下巢穴，有垂直排水管（Wille，1966）。这个巢穴是在哥斯达黎加的雨林中研究绘制的。B 为巴西 *Geotrigona subterranean* 无刺蜂的地下巢穴（Barbosa et al.，2013）。

3.1.3 在活蚂蚁或白蚁巢穴中筑巢

很多种无刺蜂生活在地下或地上的蚂蚁（例如 *Azteca*、*Camponotus*、*Crematogaster* 属）或白蚁（特别是 *Nasutitermes* 属，但也包括 *Constrictotermes*、*Macrotermes*、*Microcerotermes*、*Odontotermes* 和 *Pseudocanthotermes* 属）巢穴中分隔出的部分（例如 Wille and Michener, Roubik, 2006; Michael Hrncir, pers. commun）。Wille 和 Michener（1973）发现他们研究的所有无刺蜂种中有 9% 可以和白蚁一起筑巢（表 3.1），其中大多数甚至是完全依赖白蚁筑巢。来自美洲的 *Aparatrigona*、*Partamona*、*Scaura*、*Trigona* 属无刺蜂（Schwarz，1948；Roubik，1983；Wille，1983；Camargo and Pedro，2003；Rasmussen，2004；Rasmussen and

Camargo，2008）和来自非洲的 *Apotrigona*、*Axestotrigona*、*Plebeina* 属（Namu，2008; Darchen, 1969; Namu and Wittmann, 2016）就是这种情况。图 3.5 展示了不同 *Partamona* 属无刺蜂在活跃或废弃的白蚁巢穴中筑的巢穴。诸如 *Partamona peckolti* 和 *Trigona amalthea* 等某些无刺蜂种的蜂群在某些地区生活在白蚁巢中，但在其他地区则不然（Schwarz, 1948; Camargo and Pedro, 2003）。尽管许多无刺蜂蜂群可能会利用已经存在的空穴，但研究者猜测 *Aparatrigona isopterophila*、*Apotrigona nebulata* 和 *Scaura* 属无刺蜂会挖掘通道进入白蚁巢穴并移走相应部位的白蚁（Darchen, 1969; Michener, 1974; Roubik, 1983）。

Darchen（1972）认为白蚁中的兵蚁会攻击入侵蜂来保卫它们的巢穴，他提出蜂群的存在可能是被入侵蚁群非自愿接受的，而不是协同共生（Schwarz, 1948）。只需要小洞穴的无刺蜂蜂种，如 *Scaura latitarsis* 不太可能给白蚁带来巨大的危害，但有时 *Partamona* 属无刺蜂会在白蚁巢穴中占据多个巢穴（图 3.5），在这种情况下可能会怀疑白蚁群落是否因为无刺蜂而失去宝贵的生活空间。*Partamona* 属无刺蜂入住对白蚁的一个潜在益处是 *Partamona* 无刺蜂蜂群极具攻击性，可以帮助白蚁抵御它们共同的天敌

图 3.5 在白蚁巢穴中筑巢的 *Partamona* 属蜂群
（**手绘图源自 Camargo and Pedro，2003**）

注：A 为巴西帕拉州的在一个废弃的白蚁巢中的 *P. auripennis* 无刺蜂。B 为在巴西亚马孙州的一个 *Nasutitermes acangussu* 白蚁活巢中的 *P. batesi* 无刺蜂巢的聚合巢。图中可以看到 11 个不同的入口。C 是在一个附着在一个房子的屋顶的白蚁活巢中的一个 *Partamona gregaria* 无刺蜂巢（两个较低的入口）和一个 *Partamona vicina* 无刺蜂巢（较高的入口）。

（Roubik，1989）。

在一些 *Partamona* 属无刺蜂的巢穴入口附近可以发现一些功能不明确的类似前庭的结构，该前庭是由泥浆和树脂制成的复杂结构与工蜂建造的空储罐构成的（图 3.6）。研究者目前推测这些结构的功能一方面可以使守卫蜂得以休息，另一方面可以拖延诸如盗蜂（第 7 章）等攻击者，或者可以阻止诸如食蚁兽等脊椎动物捕食者（Camargo and Pedro, 2003; Roubik, 2006）的入侵。有时也可以在无刺蜂巢穴中发现一些更复杂的"假巢"结构，这些结构可能也是为了分散攻击者的注意力（Camargo and Pedro，2003）。

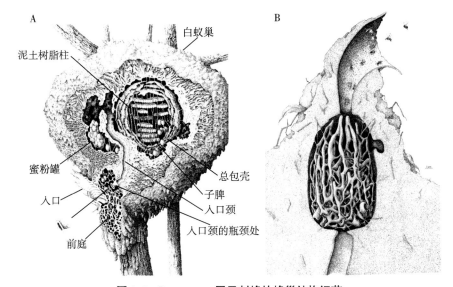

图 3.6 *Partamona* 属无刺蜂的蜂巢结构细节

注：A 为 *Partamona pearsoni* 无刺蜂蜂巢的纵向切割面，该巢位于巴西亚马孙州的一个 *Nasutitermes* cf. *peruanus* 白蚁活巢中。在左下角是通往子巢连接通道的瓶颈处细节图。比例尺为 5.0 cm（手绘图源自 Camargo and Pedro, 2003）。B 为与 *P. testacea* 无刺蜂巢穴入口相连的前庭（手绘图源自 Camargo, 1970）。

考虑到蚂蚁会对无刺蜂构成重大威胁（第 7 章）（Schwarz，1948；Slaa, 2006），无刺蜂和蚂蚁之间的关系更加令人费解。甚至有些无刺蜂蜂种专门与蚂蚁一起筑巢（表 3.1）（Wille and Michener，1973；Sakagami et al., 1989）。特别是 *Camponotus* 属蚂蚁的巢穴是无刺蜂最喜欢的筑巢地点（Schwarz, 1948; Kerr et al., 1967）。无刺蜂和 *Camponotus* 属蚂蚁一起筑巢的案例就包括与一个 *Camponotus* sp. 蚁共享树洞的一个 *Trigona braueri* 无刺蜂蜂群（Ihering, 1903），再如与 *C. sericeiventris* 蚂蚁生活在一起的是 *Scaptotrigona bipunctata*

无刺蜂（Schwarz，1948）。据报道 *Trigona chanchamoyoensis* 无刺蜂常与 *C. senex* 蚂蚁住在一起（Camargo and Posey，1990），*Patrigona peltata* 无刺蜂也常栖息在 *Camponotus* 属蚂蚁的地表巢穴中（Wille and Michener，1973）。另外，研究者发现在秘鲁热带雨林栖息地中 *Partamona testacea* 无刺蜂的筑巢位置与子弹蚁（*Paraponera clavata*）或切叶蚁（*Atta sp.*）巢穴高度关联（Bordoni et al.，2020）。其他无刺蜂与蚂蚁同住的相关报道还提到，一个 *Paratrigona opaca* 无刺蜂蜂群与一个攻击性的 *Dolichoderus bispinosus* 蚂蚁群共同居住在一个前期是 *Azteca* 蚂蚁巢的巢穴中（Schwarz，1948; Kerr et al.，1967; Camargo and Moure，1994）。*Trigona cilipes* 无刺蜂喜欢与 *Azteca* 或 *Crematogaster* 属蚁群共同筑巢（Kerr et al.，1967），但研究者也曾在纸胡蜂 *Epipona tatua*（Rasmussen，2004）巢穴中发现过该蜂[3]。目前尚不清楚共用一个巢穴的无刺蜂和蚂蚁之间通常存在多少联系，但 Roubik（1989）观察到 *Nannotrigona mellaria* 无刺蜂与它们的蚂蚁邻居生活空间十分紧密，因此二者间的频繁联系是不可避免的。他猜测无刺蜂可能会使用化学伪装来避免受到攻击。生活在 *Camponotus senex* 蚁巢中的 *Paratrigona opaca* 无刺蜂就是双方冲突的典型例子：位于入口管内的守卫蜂会积极捍卫它们的巢穴入口并杀死靠得太近的工蚁（Kerr et al.，1967）。还有研究者观测到一个蜂巢的入口管上覆盖了大量黏性树脂，这可能是为了防止蚂蚁爬到蜂巢入口管上并进入蜂巢（第 7 章）。*Partamona testacea* 无刺蜂和子弹蚁反而对彼此表现低水平攻击，Bordoni 等（2020）猜测可能是因为双方阵营的成员对对方的气味彼此熟悉，以保持和平共处。

  对于无刺蜂和其他群居昆虫之间的关联筑巢是如何启动和维持的，我们仍然知之甚少。Sakagami 等（1989）描述了亚洲 *Sundatrigona moorei* 无刺蜂如何与 *Crematogaster* 蚁一起筑巢：无刺蜂首先在宿主巢穴附近的植物上堆积树脂，接着在蚁巢的表面建造入口管。随后无刺蜂将入口延伸到蚂蚁窝内，并通过建造树脂墙来创造出一个无蚂蚁区域。无刺蜂通过不断更换巢穴壁来缓慢增加洞穴尺寸（Sakagami et al.，1989）。快速建造树脂墙以创建一个最初的小保护区似乎是无刺蜂在蚂蚁或白蚁巢穴内建立新巢穴的关键步骤（Kerr et al.，1967）。无刺蜂可能会从与蚂蚁和白蚁的关联筑巢中受益，因为这些宿主可以保护无刺蜂以抵御天敌。此外，无刺蜂在另一个物种的巢穴内筑巢有

---

[3] 研究者后来认为该蜂是 *Trigona lacteipennis*（Rasmussen and Camargo，2008）。

助于其巢穴内的气候控制和内环境稳定（见下文）。

3.1.4 暴露型巢穴

许多无刺蜂生活在暴露的或部分暴露的巢穴中（表 3.1）。非洲的 *Dactylurina* 属无刺蜂会建造悬挂于大树枝下的蜂巢来保护蜂群免受恶劣天气的影响（Darchen and Pain，1966; Michener，1974）。美洲几种 *Trigona* 属（例如 *T. corvina*、*T. spinipes*、*T. nigerrima*）和 *Tetragonisca weyrauchi* 无刺蜂会构建完全暴露的巢穴（图 3.1）（Michener，1974; Roubik，2006; Rasmussen and Camargo，2008）。为了弥补蜂巢缺乏墙壁保护，无刺蜂工蜂们会用硬质巢脂质（也称为巢脂质层板）包裹住蜂群，这些巢脂质是由巢质、植物材料、泥土或粪便（脊椎动物和蜂的粪便）组成的（Wille，1983; Wille and Michener，1973）。*Trigona corvina* 无刺蜂的蜂巢中复合巢脂质层可以达到 24 cm 厚（Michener，1974），蜂巢重量可以达到 140 kg（Roubik and Patiño，2009）。最大的无刺蜂蜂巢可能是由 *T. amazonensis* 无刺蜂建造的：它们的暴露型蜂巢可高达 6 m，宽至 1 m，重达数百千克（Roubik，1989）。这种无刺蜂蜂巢的最外层很薄而且通常质地较脆。研究者还在暴露型的 *Trigona* 属蜂巢（例如 *T. amalthea*、*T. corvina*、*T. dallatorreana*、*T. nigerrima* 或 *T. spinipes*）和部分暴露的 *Partamona* 属无刺蜂蜂巢（*P. cupira* 和 *P. testacea*）中偶然发现另一种大型结构——盾片（Ihering，1903; Schwarz，1948; Nogueira-Neto，1962; Wille and Michener，1973; Sakagami，1982; Wille，1983; Rasmussen and Camargo，2008，图 16）[4]。Schwarz（1948）将这种盾片描述为碗状的混合有蜡和树脂的陶土材料。Wille（1983）将这种盾片描述为蜂巢一侧的巢脂质沉积物，而 Roubik 和 Patiño（2009）则将包围着整个蜂群的全部平面称为盾片，并且他们发现 *T. corvina* 无刺蜂蜂巢的盾片主要由无刺蜂粪便中的花粉外壁组成（无刺蜂工蜂在巢内排便，这与蜜蜂不同，见第 6 章）。这与 Nogueira-Neto（1962）的观察结果一致，即盾片由无刺蜂粪便、死蜂、巢质、树脂和孵化茧的残骸组成。目前尚不清楚盾片是否具有防御功能（Roubik，2006），是否是蜂巢的支撑结构（Nogueira-Neto，1962），还是仅仅是蜂巢里堆积的废物（Rasmussen and Camargo，2008）。

---

[4] von Ihering（1940）（Nogueira-Neto，1962）描述了在巴西东北部人们使用蜂巢盾片捕鱼的场景。人们会将这种盾片研磨成粉末，煮熟并放入水中的篮子里。这种粉末可能因为含有较高含量的乙酰胆碱，因此会毒死鱼类。

3.1.5 筑巢习性的种内差异

一些蜂种在筑巢习性方面可分为多个类别，这些蜂种的筑巢习性要么在种群内表现出差异（墨西哥的 *Trigona fulviventris* 无刺蜂就是既有树穴型又有地穴型）（Buchwald and Breed，2005），要么在不同地区表现出不同的筑巢行为。例如，虽然 Schwarz（1948）提出 *Plebeia minima* 无刺蜂偏爱树洞和空的甲虫洞穴，但 Roubik（1983）发现 *Plebeia minima* 无刺蜂部分蜂巢暴露在棕榈树外，并且在棕榈树侧被厚厚树脂墙包围。*Trigona cilipes* 无刺蜂在南美洲会在空心树的白蚁群旁筑巢，但在中美洲则完全不会和白蚁关联筑巢（Roubik，1983）。另一个有趣的例子是有人在巴西一些地区的地下发现了肉食性 *T. hypogea* 无刺蜂蜂巢（Schwarz，1948），但在巴西其他地区以及巴拿马地区 *T. hypogea* 无刺蜂蜂巢则只在树洞中被发现（更多筑巢习惯变化的例子的相关讨论参见 Roubik，1983；Gonzalez et al., 2018）。然而，基于这些描述，我们尚不清楚这些筑巢习惯的差异报道是否表明无刺蜂筑巢习惯具有种内行为可塑性，还是说不同的作者可能描述就是不同的蜂种。例如，*Plebeia minima* 无刺蜂可能代表一个物种集（第 2 章），黑色和红棕色变形的 *Axestotrigona ferruginea* 无刺蜂种群之间存在筑巢习惯的差异，这进一步表明这些差异代表两个不同的物种（Ndungu et al., 2017）。

3.2 无刺蜂蜂巢巢体建筑结构、建筑材料及无刺蜂的建筑行为

无刺蜂蜂巢主要由一种叫作巢质的物质构成，巢质是由无刺蜂生产的蜡与树脂和其他黏性植物物质混合而成（巢质也可能含有工蜂分泌的唾液，Massaro et al., 2011）。巢质中添加树脂至少有两个优点：一是混入树脂的蜡较之纯蜡形成了更为坚固的筑巢材料；二是无刺蜂收集的树脂已被证明具有抗菌特性，可以减少真菌和细菌的生长（Messer，1985；Velikova et al., 2000；Chapuisat et al., 2007；Muli et al., 2008；Simone-Finstrom and Spivak，2010；Choudhari et al., 2012；Massaro et al., 2014；Torres et al., 2018；Lavinas et al., 2019）。这种特性对栖息在温暖潮湿环境中的无刺蜂尤为重要。

无刺蜂蜂蜡的成分比蜂蜜蜂蜡更简单，并且熔点较低（Blomquist et al., 1985；Milborrow et al., 1987；Koedam et al., 2002）。无刺蜂蜂蜡包含碳氢化合物的百分比更高[例如，研究者报道，*Trigonisca atomaria* 和 *T.buyssoni* 无刺蜂蜂蜡的碳氢化合物含量为 60%～70%，Blomquist et al., 1985；

Milborrow 等（1987）报道 *Austroplebeia australis* 无刺蜂蜂蜡的碳氢化合物含量高达 90%，而 Winston 等（1987）报道 *Apis mellifera* 蜜蜂蜂蜡的碳氢化合物含量仅为约 14%]。第二重要的化合物是酯类［Blomquist 等（1985）报道无刺蜂巢质中酯类含量从 6%（*A. australis* 无刺蜂）到 25%（*T. atomaria* 和 *T.buyssoni* 无刺蜂）不等；Milborrow 等（1987）报道发现 *Apis mellifera* 蜜蜂的巢蜡酯类含量约达 35%（Winston，1987）]。但是，无刺蜂蜂巢巢质中树脂含量会随蜂种和蜂巢的不同功能部分而发生变化，有些巢质中树脂的含量可能超过 40%（Schwarz，1945；Roubik，2006）。例如，在树茎和树枝上筑巢的无刺蜂体型越小（例如 *Trigonisca* 或 *Plebeia* 属无刺蜂），则其巢质中使用的树脂也就越少，而使用的蜡就相对更多（Roubik，2006）。

巢质中用的蜡质主要来源于蜡堆或直接来自泌蜡的工蜂（图 3.7 A），这些蜡与一些树脂堆（图 3.7 B）中的树脂相结合，而树脂堆在整个巢穴中均有分布，但通常靠近蜂巢入口。在某些蜂种（例如 *Melipona* 或 *Partamona* 属无刺蜂）中，泥浆和粪便也是重要的筑巢材料，经常会被单独使用，也会与蜡、树脂或植物材料混合使用［研究者观察到一些 *Melipona* 属无刺蜂甚至会添加少量石头和植物种子（Garcia et al., 1992; Roubik, 2006）][5]。这些情况导致无刺蜂蜂巢的巢脂质成分各不相同[6]。由于树脂是蜂巢的重要组成部分（以及用于防御，第 7 章），无刺蜂会大量收集树脂，尤其是那些建造暴露蜂巢需要树脂作为保护层的蜂种（Michener，1974）。

如前所述，无刺蜂常用巢脂质包裹蜂巢来保护巢穴（Schwarz，1948; Michener，1974）。巢脂质有时可分为厚巢脂质板（图 3.1）和衬里巢脂质，厚巢脂质板主要用于从自然洞穴中分隔并分隔出蜂巢，衬里巢脂质则是蜂巢巢壁上薄层结构（Wille and Michener，1973）。对于地下筑巢型的无刺蜂蜂种

---

[5] 由于巢质是由不同的材料组成，因此巢质的颜色也多种多样，从几乎是纯白色的（如 *Plebeia mosquito* 无刺蜂）到黑色（如 *Melipona fasciculata* 无刺蜂）。土著人已经会将无刺蜂的巢质用于各种用途，包括制作蜡烛（第 1 章）（Schwarz，1945）。

[6] "蜂胶"是蜜蜂研究人员和养蜂人使用的一个术语，它既指代蜜蜂采集的树脂（Winston，1987），也指代加入了蜜蜂分泌物的树脂（Simone-Finstrom and Spivak，2010），蜜蜂将其作为筑巢和修复材料。无刺蜂研究人员偶尔会使用"蜂胶"（或"地胶"）作为巢质的同义词（Massaro et al., 2011）或等同于"巢脂质"（Greco et al., 2010）。由于这种模糊性和无刺蜂使用材料的多样性，笔者在文中就没有对无刺蜂使用"蜂胶"这一术语，而是根据物质组成用树脂（或类似树脂的物质）、巢质或巢脂质等替代。

来说，巢脂质层可能对蜂巢防水十分重要（Wille，1966）。无刺蜂蜂群如果需要空间来扩大种群规模，可以通过重新建造巢脂质壁来增大洞穴（Michener，1974）。

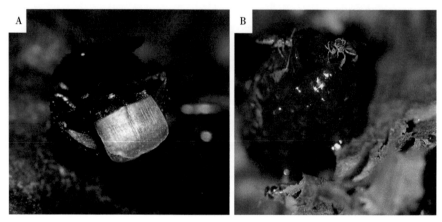

图 3.7　无刺蜂分泌的蜡片及树脂堆

注：A 为从 *Scaptotrigona depilis* 无刺蜂工蜂的腹部背侧分泌的蜡片（摄影：Cristiano Menezes）。B 为在 *Tetragonisca angustula* 无刺蜂蜂巢入口附近的树脂堆上的工蜂。在整个蜂巢中可以找到几个树脂堆（摄影：C. Grüter）。

3.2.1　入口

无刺蜂蜂巢的入口有大有小，小的蜂巢入口小而不显眼，仅有一只蜂的头大小（图 3.8）（例如 *Frieseomelitta* 属[7]和一些 *Melipona* 属无刺蜂蜂巢入口），大的入口管则大到 50 多厘米长且制作精良（例如 *Lestrimelitta*、*Geniotrigona*、*Scaptotrigona* 属无刺蜂蜂巢入口，Wille and Michener，1973）。Camargo 和 Pedro（2003）曾报道了一个与蚂蚁一起筑巢的 *Trigona cilipes* 无刺蜂的蜂巢，有 1.2 m 长的入口管。生活在泰国石灰岩洞穴中的 *Lepidotrigona terminata* 无刺蜂蜂巢入口管甚至能长达 1.5 m 以上（Bänziger et al.，2011）。长的入口管会由于其重量较重而定期脱落（Roubik，2006; pers. obs.）。通常人们可以根据蜂巢入口的特征来判别生活在该巢穴中的蜂种（Roubik，2006; Ndungu et al.，2019）。

---

[7] 一些 *Frieseomelitta* 属无刺蜂，例如 *F. varia*，有一个习惯，它们会在入口周围覆盖一层薄薄的黑树脂（图 3.1 F）。这个区域可以延伸到 1 m² 以上，这可能是利用视觉或嗅觉信号来定位，或者可能在蜂群防御中发挥作用，例如使蚂蚁或蜘蛛不敢接近。

3 筑巢生物学

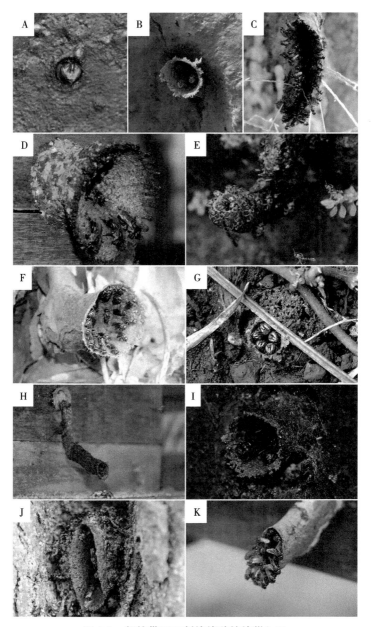

图 3.8 新热带区无刺蜂蜂种的蜂巢入口

注：A 为 *Frieseomelitta languida* 无刺蜂蜂巢入口。B 为 *Friesella schrottki* 无刺蜂蜂巢入口。C 为 *Lestrimelitta limao* 无刺蜂蜂巢入口。D 为 *Melipona seminigra* 无刺蜂蜂巢入口。E 为 *Tetragonisca angustula* 无刺蜂蜂巢入口。F 为 *Scaptotrigona bipunctata* 无刺蜂蜂巢入口。G 为地里的 *Paratrigona lineata* 无刺蜂蜂巢入口。H 为 *Leurotrigona muelleri* 无刺蜂蜂巢入口，其入口下方有废弃物堆。I 为 *Scaura longula* 无刺蜂蜂巢入口。J 为 *Tetragona clavipes* 无刺蜂蜂巢入口。K 为 *Scaptotrigona depilis* 无刺蜂蜂巢入口（所有照片由笔者在巴西拍摄）。

无刺蜂蜂巢的入口孔的相对大小在不同蜂种之间差异很大（Biesmeijer et al., 2007; Couvillon et al., 2008）。蜂巢入口是归巢无刺蜂的重要视觉标志（Camargo and Pedro，2003），而且入口通常是无刺蜂在新巢址筑巢时建造的第一个蜂巢结构（Wille and Michener，1973）。蜂巢出入口越大越有利于无刺蜂自由出入，但蜂群可能也就需要更多的守卫蜂来防御（Biesmeijer et al., 2007; Couvillon et al., 2008）。研究者在一些 *Partamona* 属无刺蜂蜂巢中发现了一种奇特的出入口结构：*P. helleri* 工蜂建造了一个大的外部泥制入口，这个入口通向一个小的相邻入口。采集蜂高速冲入这种外部入口，并从外部入口的天花板反弹到内部的小入口里（图 3.1）（Couvillon et al., 2008; Shackleton et al., 2019）。由于这种特殊外观的蜂巢入口，当地人称一些 *Partamona* 属无刺蜂为"蛤蟆嘴"或"蟾蜍嘴"（图 3.1）。

无刺蜂蜂巢入口管的质地有的硬而脆，有的质软且可变形。入口管通常只有靠近开口的部分是柔软且有弹性的，这有助于工蜂在夜间关闭蜂巢入口（第 7 章）（Wille and Michener，1973; Roubik，1989; Grüter et al., 2011）。无刺蜂可以在入口管上打孔并在上面涂上树脂小液滴，这可能会阻止蚂蚁或蜘蛛进入蜂巢（Wille and Michener，1973; Roubik，2006），因此有时能在覆盖有巢质或树脂的蜂巢入口管上看到蚂蚁尸体（*Tetragonisca angustula* 无刺蜂的蜂巢入口，图 7.2）。虽然大多数无刺蜂蜂种的蜂巢入口都是由巢质制成的，但许多 *Melipona* 属无刺蜂则是使用泥土建造蜂巢入口（Schwarz，1948）。*M.fuliginosa* 无刺蜂还会使用混合树脂的植物种子来建造蜂巢入口（Roubik，1989）。

无刺蜂蜂巢的入口管通常会向巢内继续延伸直至通向食物储藏区附近（Michener，1974; Roubik，1983）。这种内部入口管往往比外部入口管厚。有时，蜂巢会有两个以上的外部入口，特别是一些 *Partamona* 属无刺蜂的蜂巢（Schwarz，1948）。*Plebeia jatiformis* 和 *P. droryana* 无刺蜂蜂巢通常有 2 个甚至 3 个入口孔（Roubik，1983）。*P. droryana* 无刺蜂蜂巢的两个入口孔的大小不同，小入口的位置会比大入口高出几厘米。这种较小入口的功能尚不清楚，因为通常采集蜂并不使用这个小入口，而是被守卫蜂占据。这种较小入口可能为守卫蜂提供额外空间，或在蜂巢空气流通中发挥作用（Paulo Nogueira Neto）。无刺蜂建造的最细入口可能是发现于泰国石灰岩中 *Pariotrigona pendleburyi* 无刺蜂蜂巢。它们的蜂巢入口管是由几十个小管连接成珊瑚状的团块（图 3.9）（Bänziger et al., 2011）。一只守卫蜂可值守多个小管。

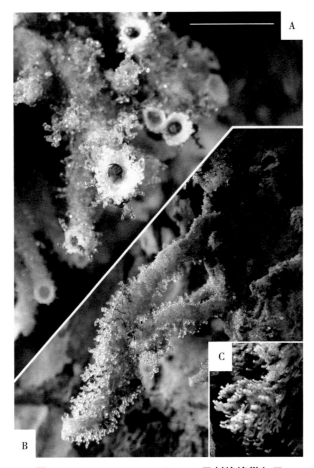

图 3.9 *Pariotrigona pendleburyi* 无刺蜂蜂巢入口

注：它是由许多小管组成的（A、B），这些小管形成了珊瑚状的入口丛（C）（源自 Bänziger et al., 2011）。一只守卫蜂可守护多个小管。

3.2.2 产卵育虫区

总包壳

多种无刺蜂，特别是造巢脾型蜂巢的无刺蜂蜂种（见下文），蜂巢的产卵区是被包壳包围住的，包壳是一种保护性结构，由几层柔软的巢质组成，将产卵区与蜂巢的其余部分分隔开（图 3.3）（Schwarz, 1948; Wille, 1983; Michener, 2007）[8]。对于一些无刺蜂蜂种来说，如澳大利亚的 *Austroplebeia*

---

[8] 无刺蜂蜂群通常只有一个产卵腔，但是研究者已经观察到有些蜂巢有几个产卵腔（例如，*Plebeia frontalis* 无刺蜂蜂巢，Wille and Michener, 1973，图 16）

australis 和 *A.cincta* 无刺蜂、美国的一些 *Plebeia* 属无刺蜂，产卵区包壳是否存在、不完整或缺失等情况存在种内差异（Michener，1961; Kerr et al., 1967; Drumond et al., 1995; van Benthem et al., 1995）。Kerr 等（1967）将巴西亚马孙地区某些 *Melipona* 属无刺蜂蜂巢内没有包壳的原因归结于该地的持续高温。无刺蜂孵卵需要恒定的环境，因而促使了无刺蜂建造包壳，这一观点也可以解释建造巢脾型蜂巢的 *Scaura latitarsis*、*Trigona cilipes* 和 *Plebeia wittmanni* 无刺蜂为什么不在蜂巢中建造包壳结构：因为 *Scaura latitarsis*、*Trigona cilipes* 无刺蜂的蜂巢建在活白蚁和蚂蚁巢中（Kerr et al., 1967; Camargo，1970）；而 *Plebeia wittmanni* 无刺蜂生活在花岗岩块的裂缝中（Wittmann，1989）。这些栖息地有可能创造更恒定的温度，从而不需要建造包壳结构。包壳的巢质层数也取决于季节。亚洲 *Lepidotrigona ventralis* 无刺蜂蜂巢在冬季的包壳的层数约为夏季的 4 倍（Chinh et al., 2005）。*Austroplebeia australis* 和 *Plebeia remota* 无刺蜂蜂群则只在一年中最冷的月份建造包壳结构（van Benthem et al., 1995；*A. australis* 无刺蜂在夏季会建造一层非常薄的包壳，Halcroft et al., 2013）。包壳的强度进一步取决于巢中可用蜡的量（Kolmes and Sommeijer，1992）。无刺蜂蜂巢内的包壳不是封闭的，其内有允许工蜂进出产卵区的通道（Schwarz，1948）。包壳的另一个功能是能够让寄生的苍蝇或蜂远离蜂卵（Roubik，2006）。

育虫巢室及其排列

育虫巢室由柔软的巢质构成，每个巢室可以用来孵育一只无刺蜂。无刺蜂蜂巢中的育虫巢室数量变异度很大，例如亚洲 *Lisotrigona carpenteri* 无刺蜂蜂巢中仅有几个育虫巢室，而一些美国的 *Trigona* 属无刺蜂蜂巢中有高达 80 000 个以上的育虫巢室（Michener，1946）。无刺蜂的雄蜂和工蜂是在大小相同的育虫巢室中孵育的。无刺蜂蜂王的巢室则通常比工蜂巢室大一些（Engels and Imperatriz-Fonseca，1990）。也有研究者在这些蜂王巢室中罕见地发现巨大的雄蜂（Nogueira-Neto，1997; Alves et al., 2009）。*Melipona* 和 *Trichotrigona* 属无刺蜂的蜂王是在工蜂巢室一样大小的巢室中孵育的，而某些无刺蜂蜂种会在工蜂巢室一样大小的巢室中孵育侏儒蜂王，同时在较大的蜂王巢室中孵育大蜂王（第 5 章）（Camargo，1974; Wenseleers et al., 2005; Ribeiro et al., 2006; Camargo and Pedro，2007）。*Tetragonisca angustula* 无刺蜂蜂巢子脾中央的育虫巢室体积较大，是用来饲养大体型的守卫兵蜂的（第 6 章，第 7 章）（Segers et al., 2015）。无刺蜂育虫巢室的排列方式主要有 3 种类型：

水平巢脾型、垂直巢脾型和簇状巢室。但是也存在一些变体和中间形式（图 3.2）。

(a) 水平型子脾

最常见的排列形式就是水平型子脾（表 1.3；图 3.3，图 3.10）(Wille, 1983)。每层子脾是由单层的向上开口的单巢室组成。固体的巢质细柱将水平子脾层固定在适当的位置（图 3.11）。无刺蜂蜂巢内子脾的层数随物种、蜂群和季节等因素发生很大变异，可以从少于 5 层到高达 40 层不等（Schwarz, 1948; Wille and Michener, 1973; Barbosa et al., 2013）。群势越强的蜂群往往建造的子脾越大，即直径更大的子脾层（van Benthem et al., 1995; Alves et al., 2009）。"火蜂" *Oxytrigona tataira*、*Tetragona clavipes* 以及澳大利亚"糖袋"蜂 *Tetragonula carbonaria* 的产卵区子脾会相互连接形成螺旋状（表 1.3；图 3.10）(Michener, 1974; Souza et al., 2007; Duarte et al., 2016)。这种螺旋型子脾不一定是一个蜂种特有的特征，也会存在于一个蜂种的某些蜂群中，而在该蜂种的其他蜂群中则并不存在（Wille, 1983; Drumond et al., 1995），有时同一个蜂群也只是偶尔会建造螺旋状子脾（例如 *Melipona*、*Plebeia*、*Plebeina*、*Nannotrigona*、*Trigona* 和 *Tetragona* 属无刺蜂）(Kerr, 1948; Roubik, 2006)。甚至在诸如 *Oxytrigona mellicolor* 无刺蜂的一个蜂群可以同时找到螺旋状子脾和圆盘状子脾（Schwarz, 1948）。这表明在圆盘状和螺旋状之间切换只需要对筑巢行为进行微小的调整（Oldroyd and Pratt, 2015; Cardoso et al., 2020），但尚不明确无刺蜂为什么有时会构建圆盘状子脾而有时会构建螺旋状子脾。

无刺蜂育虫巢室是氩饲的，即在蜂王产卵之前，一些工蜂会将育虫食物反刍到育虫巢室直到充满巢室的 2/3（图 5.1 A，图 5.1 B）。蜂王产卵后巢室会立即被关闭并保持密封，直到新出房蜂准备出房（第 5 章）。一旦发育中的幼虫在育虫巢室内结茧，工蜂们就会开始移走这些巢室上的大部分巢质以便在其他地方重复利用（Kerr, 1948; Michener, 1974）。因此，无刺蜂与蜜蜂不同，无刺蜂的每个育虫巢房都仅使用一次。有时，无刺蜂会在幼虫出房后完好无损地保留该巢室的巢质底部，从而形成一个坚固的巢质平台，并在此之上建造新的巢室。这种巢质平台被研究者称为 *trochoblast*（Ihering, 1903; van Benthem et al., 1995），它可以用新材料加固或由旧巢室地板的残余物组成。

图 3.10　子脾类型

注：A 为 *Tetragonisca fiebrigi* 无刺蜂的水平型子脾。工蜂通过在一层旧子脾中心的支柱上建立一个新巢室来开始新巢脾层的建造（摄影：C. Grüter）。B 为 *Duckeola ghiliani* 无刺蜂建设的排列更不规则的子巢脾（"半巢脾型"），没有包壳的保护（摄影：C. Grüter）。C 为 *Nannotrigona testaceicornis* 无刺蜂的子脾呈螺旋状排布（摄影：C. Menezes）。D 为 *Leurotrigona muelleri* 无刺蜂的簇型育虫巢室（摄影：C. Menezes）。E 为 *Scaura longula* 无刺蜂建立的由单层巢室构成的垂直型子脾（Cortopassi-Laurino and Nogueira-Neto, 2018）。F 为非洲 *Dactylurina staudingeri* 无刺蜂的垂直型子脾，由双层巢室构成（图 3.11）。箭头指的是分隔巢脾的巢质柱（Njoya, 2009）。

无刺蜂工蜂通过在现有的子脾中央顶部制作一根细蜡柱开始在该子脾层顶部建造新的子脾层，然后在这层子脾上建造单巢室（图 3.11 A）。一层子脾上建造的第一个巢室是圆柱形的，但是一旦在其周围建造了 6 个巢室以后，它就会被限制成为典型的六边形。无刺蜂与蜜蜂不同，其新巢室通常是按顺序依次建造的（Sakagami，1982）。然而有些无刺蜂蜂种会在子脾的不同位置同时建造多个巢室（Sakagami and Zucchi，1974）。每个巢室由几个工蜂建造而成。在蜂王产卵后，工蜂会最后给巢室构建一个柔软的巢质圈用以密封巢室（Cepeda，2006）。无刺蜂工蜂通常还未完成一层巢脾就开始建造新一层巢脾（图 3.10，图 3.11），建一层巢脾需要 3～10 d 的时间（*Paratrigona opaca*, Schwarz, 1948; *Tetragonisca angustula*, Segers et al., 2015）。这种建筑策略的结果就是中心的幼虫总是比外围的幼虫虫龄更大，因此出房时间更早（图 3.11）。

当没有空间可以向上建造蜂巢时，工蜂们会开始利用下方旧蜂巢中的新出房蜂出房后留下的空间，在育虫腔的底部建造新的巢脾层（图 3.11）。*Melipona bicolor* 无刺蜂（可能还有 *Paratrigona subnuda* 无刺蜂）是个例外，因为它们的新巢脾层总是向上构建的。Nogueira-Neto（1997）推测可能是因为它们的蜂巢重量不断增加会导致其慢慢向下沉，从而不断为下一个要建在上面的巢脾层创造新的空间。非洲的 *Plebeiella lentliana* 和 *Plebeina armata* 无刺蜂（= *P. hildebrandti*）则是造脾作巢蜂种中例外中的例外（Portugal-Araujo,1963）。它们建造巢脾是绕着巢脾中轴不断在新的育虫巢室前沿构建更新的巢室，这样就伴随着出现了出房蜂巢室的后沿，导致整个巢脾看起来像一个缓慢移动的螺旋盘（图 3.11 B，图 3.11 C）。因此它们不是在旧巢室顶部构建新的巢脾层，而是不断地重建巢脾。

那么无刺蜂工蜂是如何知道下一步在何处建造巢室，才能实现特定蜂种巢穴的设计图纸呢？我们对此知之甚少。其他社会昆虫似乎并不简单地遵循着与生俱来的巢穴设计，而是将看似简单的行为规则与现有结构本身提供的信息相结合，这一过程可称之为交互建设（Camazine et al., 2001; Oldroyd and Pratt, 2015; Khuong et al., 2016; Perna and Theraulaz, 2017; Cardoso et al., 2020）。例如，Michener（1961）观察了 *Tetragonula carbonaria* 无刺蜂的筑巢行为，并提到携带微量巢质的无刺蜂是以看似随机的方式四处游荡，直到它们碰巧遇到一个凸出点，然后它就将巢质放到这个点上。凸出点凸起越高，无刺蜂在上面堆放大量建筑材料的可能性就越大。这种情况进而导致了蜂巢中柱子的建造形成。他猜测其他筑巢活动可能同样受到现有结构的引导和刺激（Michener，1974）。

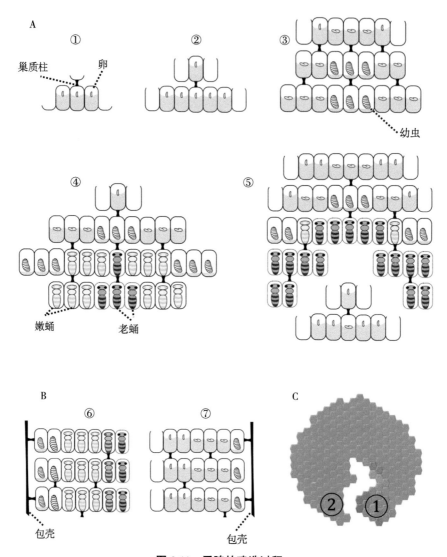

图 3.11　子脾的建造过程

注：A 为工蜂在一个已经存在的旧巢脾顶部建造一个单巢室而后开始新巢脾层的建造。新巢脾的巢室随后紧挨着这个第一巢室进行建造（①~④）。如果没有足够的空间往上建，工蜂就会腾出出房蜂留下的旧巢脾所在空间用于建造新巢脾⑤（根据 Sakagami，1982 年修改）。B、C 为一些非洲无刺蜂蜂种的蜂巢是以螺旋状建造的。新巢室会顺着巢脾的半径缘建造（C①，绿色细胞），而日龄较大的幼虫则沿着螺旋巢脾的滞后缘出房（C②）。

（b）簇状巢室

第二种最常见的育虫巢室排列类型是巢室簇，这种簇通常被薄巢质连接在一起。人们在许多亲缘关系较远的无刺蜂蜂种中（例如，美洲的 *Trigonisca*、

Frieseomelitta、*Leurotrigona*、Australian *Austroplebeia* 以及非洲的 *Hypotrigona*）发现了成簇排列的育虫巢室（图 3.10）。建造簇状巢室的无刺蜂蜂种往往体型较小，并且这些蜂种中大多数的蜂巢子区没有包壳（*Austroplebeia australis* 和 *A. cincta* 无刺蜂是例外）（Michener，1961；Camargo，1970；Wille and Michener，1973；Roubik，2006）。这些蜂种通常在不规则的小管腔中筑巢（Wille and Michener，1973；Michener，1974），因此它们难以在子区周围建立包壳。而且这种育虫巢室排列方式使有些无刺蜂蜂群得以利用造脾蜂无法使用的不规则腔。

不同的蜂种建造的簇状巢也有差异，有些是接近球形的（例如，*Austroplebeia*、*Leurotrigona* 属无刺蜂），有些则是垂直伸长的（例如，*Frieseomelitta* 属无刺蜂）。而且一些蜂种的簇状巢的巢室间有些是相互接触的（例如，*Austroplebeia* 属无刺蜂），而其他蜂种的簇状巢的巢室间则分得很开（例如，*Frieseomelitta* 属无刺蜂）（Wille and Michener，1973）。此外，在 *Austroplebeia* 和 *Trigonisca* 属无刺蜂中单个巢室可以向不同方向开口，但其他蜂种例如 *Frieseomelitta* 属无刺蜂的巢室则只能向上开口。簇状巢中新巢室是建造在簇状的外围或新出房蜂出房后形成的中空中心空间中（Michener，1961），巢室的构建可以向四面八方推进，但通常是从中心向外围推进。

（c）半脾型

在诸如美洲的 *Duckeola ghiliani*、*Friesella schrottkyi*、*Plebeia juliani* 以及澳洲的 *Tetragonula hockingsi* 无刺蜂蜂巢中育虫巢室的排列更不规则（*D. ghiliani*、*F. schrottkyi* 和 *P. juliani* 无刺蜂子区也没有包壳）（图 3.10）（Kerr et al.，1967；Drumond et al.，1998；Sakagami et al.，1973；Njoya，2009；Brito et al.，2012；Oldroyd and Pratt，2015；Pedro and Cordeiro，2015）。育虫巢室也是成群相连，但是是以不规则的方式排列的，以至于它们看起来介于脾状和簇状之间（"半脾型"）。

对于亚洲 *Tetragonula iridipennis* 无刺蜂，来自不同类群的蜂群的巢室排列方式也各不相同，有像 *T. hockingsi* 无刺蜂（这两种无刺蜂形态上也不同，表明它们可能代表了亲缘关系较近的种）的巢室一样呈簇状排列的，也有呈不规则半脾型的（Francoy et al.，2016）。在 *Friesella schrottkyi* 无刺蜂的相同蜂群中可以发现巢室排列的规则整齐程度上也有高有低（Sakagami et al.，1973）。这些观察结果表明，和螺旋巢脾的案例类似，半脾型到脾型或簇型之间的过渡可能也只需要对建造行为进行微小的调整。有人提出，这些行为差异可能与在产卵和食物供应过程中（第 5 章）蜂王和工蜂间相互作用有关，但这需要进一步研究（Oldroyd and Pratt，2015）。

(d) 垂直型巢脾

非洲的 *Dactylurina* 属无刺蜂和美洲的 *Scaura longula* 属无刺蜂是唯一能构建垂直型巢脾的无刺蜂。*Dactylurina* 属无刺蜂的垂直巢脾层是由两层水平朝向开口巢室构成，*Scaura* 属无刺蜂的垂直巢脾层则是由一层巢室组成（图 3.10，图 3.12）（Darchen and Pain，1966; Nogueira-Neto，1997; Eardley，2004; Njoya et al., 2016）。*Dactylurina* 属无刺蜂的巢脾巢室是从顶部向底部构建的（Wille and Michener，1973），并且子脾区被包壳包围。子脾层由巢质柱固定在一起（图 3.10）（Njoya et al., 2016）。有趣的是，*Dactylurina* 属无刺蜂的子脾区分为上半部分和下半部分，两部分由一个巢质片隔开（图 3.12）（Darchen and Pain，1966; Michener，1974; Njoya et al., 2016）。每部分包含 6～10 个双层的子脾层。这种上下部分分离的作用尚不清楚，但这种分隔结构可以提高子脾垂直排列的结构稳定性。

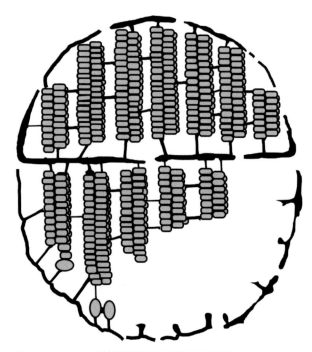

图 3.12 *Dactylurina staudingeri* 无刺蜂的垂直型巢脾（根据 Darchen and Pain，1966 修改）

注：巢室呈两列，巢脾层间以及巢脾与包围子脾的包壳间通过巢质柱相互连接。图中靠下的位置显示了 3 个较大的蜂王巢室。

3.2.3 食物储存区

无刺蜂与蜜蜂和熊蜂相比，它们的食物储存罐与育虫巢室在位置和形状上有明显不同（图 3.3，图 3.10）。无刺蜂的食物储存罐由软巢质制成，根据所处的季节不同蜂群中食物储存罐数量可能从几个到数百个不等（Schwarz，1948）。无刺蜂蜂巢内储存的蜂蜜量在种间和种内差异很大，从几克到几千克（通常 <5 kg）不等（第 1 章）（Roubik，1983）。*Trichotrigona* 属无刺蜂蜂群则看似不会在蜂巢储存任何食物，它们很可能从其他蜂群窃取食物（Camargo and Pedro，2007; Pedro and Cordeiro，2015）。

在那些会建造包壳无刺蜂蜂种的蜂巢内，食物罐通常在包壳之外（图 3.3，图 3.4），但也有一些例外（Camargo，1970）。食品罐能够直接附着在蜂巢的巢脂质内壁，或者通过巢柱与之相连。无刺蜂的食物储存罐比育虫巢室体积大，一些体型较大的诸如 *Melipona* 或 *Cephalotrigona* 属无刺蜂蜂巢中储存罐可以达到一个鸡蛋大小。在某些蜂种中，蜜罐和花粉罐是分开的，而在其他蜂种中它们是混在一起的。通常蜜罐和花粉罐的形状相同，但美洲的 *Frieseomelitta* 属无刺蜂蜂巢中二者的形状和大小都不同。蜜罐更圆更小，而花粉罐更大更细长（图 3.13）（Kerr et al., 1967）。无刺蜂蜂巢中的储存罐的大小在同一蜂群的蜂巢中也可能有很大差异（Michener，1974）。盗食寄生型的 *Lestrimelitta* 和 *Cleptotrigona* 属无刺蜂则将从其他蜂群中掠夺来的花粉、蜂蜜和育虫食物混在一起储存在同种储存罐中（Michener，1974；Roubik，2006）。

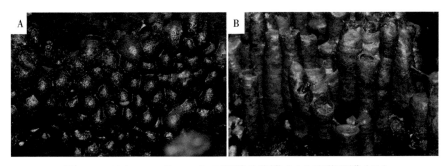

图 3.13 *Frieseomelitta varia* 无刺蜂的蜜罐（A）和花粉罐（B）

注：与许多其他蜂种不同的是，*Frieseomelitta varia* 无刺蜂的这两种食物罐在大小和形状上都不一样（摄影：Cristiano Menezes）。

3.2.4 监禁室

包括 *Plebeia* spp.、*Friesomelitta varia*、*F. silvestrii*、*Friesella schrottki*、*Schwarziana quadripunctata* 和 *Tetragonisca angustula* 在内的多个无刺蜂蜂种都会构建蜂王"监禁室"（第 4 章）（图 3.14）（Juliani，1962; Sakagami，1982; Roubik，2006; Engels and Imperatriz-Fonseca，1990）。在某些情况下，这种监禁室是重新使用的食物罐（Juliani，1962）。无刺蜂蜂群中能够构建多个监禁室，每个监禁室里都有一个处女蜂王。当巢中仍有在产卵的蜂王时，这些监禁室可以保护处女蜂王免于被工蜂杀死。这些监禁室通常不是完全密封的，而是有开口允许工蜂进入，或监禁的处女蜂王通过开口与工蜂交哺来接收工蜂送来的食物（Drumond et al.，1995）。*Plebeia subnuda* 和 *P. droryana* 无刺蜂的处女蜂王偶尔会离开监禁室游走（Imperatriz-Fonseca，1977; Imperatriz-Fonseca and Zucchi，1995），这表明处女蜂王是自愿留在监禁室里以避免受到工蜂或产卵蜂王的攻击。有研究偶然观察到处女蜂王会自己建造蜂王巢室，这一观察结果进一步支持了处女蜂王会自我禁闭的观点（Sakagami，1982; Imperatriz-Fonseca and Zucchi，1995）。处女蜂王可以在监禁室内待上数月（Drumond et al.，1995）。在某些情况下，例如 *Frieseomelitta varia* 无刺蜂中，已经观察到哺育蜂日龄的工蜂会看守监禁室（Sakagami，1982）。*Leurotrigona muelleri* 和 *Celetrigona longicornis* 无刺蜂护卫工蜂的防护屏障甚至完全取代了监禁室壁（Terada，1974，引自 Imperatriz-Fonseca，1977; Sakagami，1982）。

图 3.14 *Plebeia droryana* 蜂的处女蜂王的蜂王"监禁室"
（摄影：Cristiano Menezes）

3.3 无刺蜂对蜂巢内部气候条件控制

保持巢穴内相对恒定的气候条件对于许多社会性昆虫的群体功能维持很重要（Jones and Oldroyd，2006）。例如幼虫孵育就需要温度保持在一定范围内：在西方蜜蜂中幼虫孵育就需要 33～36 ℃的恒定温度以实现最佳发育条件。不管是被动措施（例如巢址选择）还是主动措施（飞行肌颤抖、通风、蒸发冷却）都可以使蜜蜂获得非常稳定的环境条件（Heinrich，1993; Bujok et al., 2002; Tautz et al., 2003; Jones and Oldroyd，2006）。无刺蜂的幼虫孵育也需要恒定的环境条件。例如，*Scaptotrigona depilis* 无刺蜂的孵育幼虫的温度不在 26～34℃ 温度范围之内，幼虫的死亡率会急剧增加（Vollet-Neto et al., 2015）。如果温度超过 35.5° C，非洲的 *Plebeina armata* 无刺蜂幼虫就会死亡（Moritz and Crewe，1988），温度 ≥ 34° C 则会导致美洲的 *Melipona interrupta* 无刺蜂幼虫死亡率大幅增加（Becker et al., 2018）。如果生活在 21° C 以下的 *Melipona marginata* 无刺蜂的雄蜂会产生多倍体精子（Kerr，1972）。相反，分布在澳洲的暖温带地区的 *Austroplebeia australis* 无刺蜂即使是在寒冷季节，蜂群也会继续孵育幼虫，并且幼虫能够在 0℃ 以下的温度下存活（Halcroft et al., 2013）。然而目前尚不清楚非最优孵育温度是否会产生不易察觉到的亚致死的后果，例如，蜜蜂中就发现非最优温度会对工蜂行为表现产生不利影响（Tautz et al., 2003; Jones et al., 2005）。

人们通常认为无刺蜂缺乏有效的气温控制能力，并且有人提出这就是为什么无刺蜂大部分仅存在于热带和亚热带地区的原因（Wille，1983; Engels et al., 1995）。然而，即使在热带和亚热带地区，温度也有很大差异，并且由于位置、海拔高度、季节或时段的不同，温度会在 0° C 以下到 40° C 以上范围内变化（Darchen，1973; Halcroft et al., 2013; Sung et al., 2011）。因此，温度调节对于无刺蜂来说可能也很重要。截至目前，相关研究检测过的蜂种在维持气温稳态的能力方面似乎差异很大。例如，非洲的 *Apotrigona nebulata* 无刺蜂（Darchen，1969）几乎完全没有温度调节能力，而美洲的 *Trigona spinipes* 无刺蜂维持蜂巢内恒温的能力与西方蜜蜂不相上下（Zucchi and Sakagami，1972）。Zucchi 和 Sakagami（1972）测定了 *T. spinipes* 无刺蜂蜂巢的各个部分在 24 h 内的温度变化，他们发现虽然蜂巢外围的温度在 15.5～28 ℃波动，但产卵育虫区的温度却高得多，温度在 34.1～36.0 ℃。其他具有良好温度调节能力的无刺蜂蜂种还包括美洲的 *Melipona fuliginosa*、*M.*

rufiventris、*M. Seminigra*、*Partamona cupira*、*Scaptotrigona depilis* 、*S. postica* 无刺蜂以及亚洲的 *Lepidotrigona ventralis* 无刺蜂，而 *Austroplebeia essingtoni*、*Friesomelitta varia*、*Leurotrigoneps muellerevella* 和 *Tetragonula laeviceps* 无刺蜂蜂种的蜂巢温度则或多或少与环境温度相近（Zucchi and Sakagami，1972; Wille and Orozco，1975; Wille，1976; Roubik and Peralta，1983; Engels et al.，1995; Sung et al., 2008; Sakagami et al., 1983; Ayton et al., 2016）。靠后的这些无刺蜂蜂种蜂巢都是簇状型巢室并且没有包壳，这就解释了为什么它们的蜂巢内部温度波动更大。而 *Austroplebeia essingtoni* 无刺蜂确实展现出将蜂巢的相对湿度保持在 65% 左右的能力，这意味着在炎热潮湿的夏季这种蜂的巢内湿度可以保持在低于其周围环境湿度的水平，从而可以减少真菌的生长（Ayton et al., 2016）。研究者认为这种情况可能与该蜂种蜂巢内的废物堆吸收了水分，工蜂从蜂群中移除这些废弃物的同时可以去除蜂巢内多余的水分有关。

3.3.1 主动气候控制

很多无刺蜂蜂种是通过蜂振翅进而产生气流来实现蜂巢降温的（Nogueira-Neto，1948; Wille，1983）。研究者认为工蜂会停留在巢穴的不同部位来促使空气通过巢穴入口离开蜂巢，这些工蜂通常几只排成一排（Sakagami and Oniki，1963; Michener，1974; Roubik，2006）。工蜂经常将头部朝向入口，而不是腹部朝向入口（Sakagami and Oniki，1963; Sakagami and Zucchi，1967）。Moritz 和 Crewe（1988）在研究两种非洲无刺蜂种——*Hypotrigona gribodoi* 和 *Plebeina armata* 无刺蜂时发现无刺蜂在 1～7 h 内采用振翅置换整个蜂巢的空气。*Scaptotrigona depilis*（Vollet-Neto et al., 2015）和 *Amazonian Melipona crinita* 无刺蜂蜂种蜂群中振翅蜂的数量与外界环境温度呈正相关，在内部温度超过 37°C 以上工蜂就会在蜂巢入口处开始振翅通风（Cortopassi-Laurino and Nogueira-Neto，2003）。

Darchen（1972）研究了非洲科特迪瓦大草原上的麦蜂族无刺蜂，他观察到，在一天中最热的时间段工蜂们会在蜂巢的外层挖出许多小孔，这可能是为了促进空气流通。建造暴露型蜂巢的 *Trigona spinipes* 无刺蜂的工蜂也会采用这种降温策略（Roubik，2006）。无刺蜂另一种避免过热的方法是工蜂从产卵区撤离，研究者在 *Scaptotrigona postica* 无刺蜂应对高温时观察到这一现象（Engels et al., 1995）。Nogueira-Neto（1948）发现在较冷的冬天无刺蜂也会做振翅扇风行为，他认为这更有可能是用来调节蜂巢湿度的，而不是为了降温。他提出通风对于减少蜂巢中霉菌的生长十分重要。此外，空气交换对于蜂巢氧气补充也很重要（Nogueira-Neto，1948; Moritz and Crewe，1988）。

众所周知，蜜蜂在环境温度很高时，会使用蒸发降温作为温度调节的一种方法（Lindauer，1954）。为此蜜蜂会收集水分并在整个巢的表面涂上少量的水，它们甚至会采用一种称为"舌抽打"的行为来更快地蒸发它们自己吐出的水（Winston，1987）。但是，目前仍然缺少确凿证据来证明无刺蜂收集水是为了蒸发降温，Macías-Macías 等（2011）发现生活在墨西哥高原地区的 *Melipona colimana* 无刺蜂蜂群的工蜂会吐水并扇动翅膀来应对高温。因此这种无刺蜂在面临高温时确实会利用水分蒸发来给蜂群降温。研究发现当孵卵区的温度达到临界水平时，*Scaptotrigona depilis* 无刺蜂的工蜂就会增加集水量，这一现象也进一步证实无刺蜂会使用水来降温（Vollet-Neto et al., 2015）。该作者还发现，在一天中最热的时候，*Scaptotrigona depilis* 无刺蜂蜂箱中的温度比空对照蜂箱中的温度低约 1 ℃，而在较冷时间段 *Scaptotrigona depilis* 无刺蜂蜂巢的温度则高于对照空蜂箱的温度。这些发现进一步表明，某些无刺蜂蜂种可能真的会使用蒸发降温，而研究者观察到无刺蜂工蜂会因巢内高温而吐水，也观察到高温时产卵区有水滴出现，这些现象则进一步证实蒸发降温确实存在于无刺蜂中。但是，有些无刺蜂蜂种则并不会用水来给巢穴降温（Roubik and Peralta，1983; Engels et al., 1995; Roubik，2006）。

如前所述，居住在某些地区的无刺蜂经常会面临较低的环境温度，而提高蜂巢内温度的行为策略将改善幼虫的生存环境（Engels et al., 1995; Halcroft et al., 2013）。蜜蜂会通过振动翅肌来产生热量（Winston，1987; Heinrich，1993），并且有证据表明一些无刺蜂也会使用相同的升温策略（Roldão-Sbordoni et al., 2019）。当墨西哥高地蜂种 *Melipona colmana* 无刺蜂暴露在低温下时，蜂群中的工蜂会食用糖浆并提高它们胸部温度（Macías-Macías et al., 2011），而与之亲缘关系较近的低地蜂种 *Melipona beecheii* 无刺蜂则因面临的极端温度变化较小而并没有表现出这种行为（Macías-Macías et al., 2011）。科学家在 *Scaptotrigona postica* 无刺蜂上观察到了一种有趣的行为，这种行为可能在温度调节中起到了作用。即当 *Scaptotrigona postica* 无刺蜂蜂群暴露在低温时，蜂群中的工蜂就会盖住育虫巢室，可能是为了通过飞行肌来加热育虫巢室（Engels et al., 1995），蜜蜂中也观察到相似的现象（Bujok et al., 2002）。

3.3.2 被动气候控制

对于许多无刺蜂蜂种来说，保持最佳巢穴温度最基础的方式是选择一个合适的巢穴并建造与之相适应的筑巢结构（Jones and Oldroyd，2006）。木头、蜡质、巢质、泥土或石头的绝缘性保证了蜂群不易过热。此外，巢穴和蜂穴的物理特性（例如位置、大小和形状）也将决定环境气流进入蜂巢并在蜂巢内循环的程度（Roubik，2006）。产卵区热量的保持或损失也可能取决于育虫巢室的排列方式，因为在造脾筑巢型蜂种中集中建造的层状巢脾的巢室可能比某些簇状巢中松散、孤立的巢室更能保持热量（Oldroyd and Pratt，2015）。

3.4 巢穴结构的演变

无刺蜂蜂巢结构的不同似乎与其系统发育无关，因此很难确定特定的蜂巢特征是祖传的还是衍生的。一些研究者提出无刺蜂簇状巢室进化上代表了一种祖先特征状态，与熊蜂和一些兰蜂的育虫巢室排列相似（Michener，1964; Camargo，1970; Wille and Michener，1973）。这意味着水平型巢脾是从簇状巢室进化而来的。然而，无刺蜂种间系统发育关系表明，从巢脾型到簇型的转变是反复出现的。Wille 和 Michener（1973）认为在澳洲 *Austroplebeia* 属或一些新热带区 *Trigonisca* 属无刺蜂建造的簇状巢中发现的向各个方向开口的相对球形的巢室排列方式代表了祖先状态，而 *Frieseomelitta* spp. 无刺蜂构建的簇状巢中椭圆形且向上开口的巢室排列方式则是衍生的。事实上，*Frieseomelitta* 属和 *Trichotrigona* 属（图 2.2）的系统发育地位与新热带区蜂种（主要是造脾筑巢型无刺蜂属）的亲缘关系最近，这种现象表明簇状巢衍生自子脾型巢。

Michener（1961）则认为在 *Austroplebeia* 属无刺蜂蜂巢中发现的简单单层包壳代表了祖先状态。多层层叠的包壳则是后期出现的用以改善产卵育虫区的气候控制。原始无刺蜂蜂巢好像在巢腔中会用一层巢质做内衬（Wille and Michener，1973）。在新热带区大属 *Trigona* 属无刺蜂中经常发现它们建造暴露型巢穴时，常将植物或粪便材料混入巢质和巢脂质，这些有可能是衍生特征（Michener，1964; Sakagami，1982; Rasmussen and Camargo，2008）。无刺蜂建造暴露型巢穴的行为使得蜂群不再依赖于寻找合适的洞穴（第1章）（Rasmussen and Camargo，2008）。然而，随着我们对无刺蜂系统发育的理解不断加深，研究者需要在此基础上进行更科学的分析来检验关于不同筑巢习性性状的进化起源和转变的假设［类似分析参见 Rasmussen 和 Camargo（2008）对 *Trigona* 属无刺蜂的分析］。

3.5 无刺蜂群落及其与其他生物体的联系

3.5.1 蜂群群落与蜂群密度

无刺蜂蜂群的自然密度变化很大，每平方千米蜂群数可以从几个到 1 600 个蜂群不等（Fowler，1979; Oliveira et al., 1995; Breed et al., 1999; Eltz et al., 2002; Batista et al., 2003; Samejima et al., 2004; Kajobe and Roubik，2006; Slaa, 2006; Siqueira et al., 2012; Tornyie and Kwapong，2015; Bobadoye et al., 2017）。人们对于影响无刺蜂的筑巢密度和物种多样性的因素知之甚少，但可能有 3 个因素发挥了关键作用，即①食物可利用度（Hubbell and Johnson，1977; Eltz et al., 2002; Arena et al., 2018）；② 巢址可利用度（Michener，1974; Eltz et al., 2002; Inoue et al., 1993）；③ 人类对环境改造度（Kajobe and Roubik，2006; Slaa，2006; Pioker-Hara et al., 2014; Cairns et al., 2005; Brosi，2009; Freitas et al., 2009）。这 3 个因素在某种程度上相互关联，并且与栖息地质量的其他关键因素相关，例如植物物种多样性和森林覆盖率。不同的研究在对塑造无刺蜂群落的不同因素的重要性方面形成了鲜明的对比。这突出表明了巢址丰富度、食物可利用度和人类影响这 3 个因素的重要性会因栖息地和无刺蜂群落的不同而不同。

（a）食物源可利用度

Hubbell 和 Johnson（1977）以及 Eltz 等（2002）发现食物源可利用度是中美洲和东南亚栖息地的无刺蜂蜂巢密度的最佳预测指标，而食巢天敌或可用于筑巢的巢址数量对他们研究的无刺蜂来说似乎不那么重要。哥伦比亚安第斯森林的无刺蜂蜂种丰富度和丰度也与食物可利用度呈正相关（Gutiérrez-Chacón et al., 2018）。植物多样性以及稳定可利用的食物也是澳大利亚 *Tetragonula carbonaria* 无刺蜂的种群生长和种群数量的最佳预测指标（Kaluza et al., 2018）。然而也有其他研究发现食物源与蜂巢密度之间本质上没有密切的相关性（Kajobe and Roubik, 2006; Brosi, 2009）。

当食物源比较集中时，蜂巢分布也会相对集中（Michener，1946），而食物源分布得比较均匀分散时，可能会导致蜂巢分布更均匀分散（Hubbell and Johnson，1977; Eltz et al., 2002）。此外，如果蜂群之间的食物竞争激烈，那么不难猜想到蜂巢的分布将更加均匀，表现为蜂巢间距基本一致（第 8 章）。Hubbell 和 Johnson（1977）也发现具有群聚性和好斗性的蜂种间分布更为均匀，而非群聚性蜂种的分布则更随机[9]。一些无刺蜂蜂种（例如 *Trigona fulviventris*）则似乎是通过主动阻止同类在其蜂巢附近筑巢来直接影响蜂巢分布的（Hubbell and Johnson，1977）。

（b）巢址可利用度

在某些情况下，蜂巢的密度和分布取决于适宜巢址的可利用度（Michener，1974；Eltz et al.，2002）。例如有研究实验性地增加了亚洲的 *Tetragonula minangkabau* 无刺蜂的适宜筑巢点，结果其蜂群密度增加了 250%（Inoue et al.，1993）。蚂蚁似乎是筑巢地点的重要竞争者（Nogueira-Neto，1954；Inoue et al.，1993）。哥斯达黎加地区的无刺蜂的分蜂频率与可用空巢地点数呈正相关，并且随着森林被砍伐所致的筑巢地点减少，也导致了蜂巢密度的降低（Slaa，2006）。Lichtenberg 等（2017）也同样发现在森林被高度砍伐的地区，越不依赖树木筑巢的蜂种越能更好地生存。

大量报道发现在许多聚集性筑巢点出现了极高密度的蜂巢分布（Starr and Sakagami，1987；Roubik，1989；Eltz et al.，2003；Leonhardt et al.，2010）。甚至有些研究发现在一株树中聚集了二三十个蜂群（且通常属于多个蜂种）（Schwarz，1948；Roubik，1989；Roubik，1983；Camargo and Pedro，2002）。Starr 和 Sakagami（1987）甚至在菲律宾的一个竹农舍中发现了 *Tetragonula* 属下两个蜂种的 84 个蜂群。有研究者在榕树根部发现了几个到几十个亚洲 *Tetragonilla collina* 无刺蜂的蜂巢（Roubik，1989），也有人发现美洲的 *Partamona cupira* 无刺蜂可以在悬崖边形成上百个蜂巢的聚集体（Roubik，1983）。对于 *Scaptotrigona luteipennis*、*S. postica* 和非洲的 *Axestotrigona ferruginea* 无刺蜂蜂种，研究者偶尔发现几个蜂群共享一个连续的空穴，这些空穴由巢脂质板分隔开来（Schwarz，1948；Roubik，1983）。此类聚集性蜂巢或多或少是由潜在巢址聚集而产生的，而不是由于食物供应充足而产生的。

（c）人类影响

越来越多的证据表明，人类活动会对无刺蜂的丰富度和多样性产生负面影响（第 1 章）。人类干扰的后果则是因无刺蜂蜂种而异的。有些无刺蜂在受干扰的、森林砍伐的或城市的环境中表现相对较好，例如那些可以利

---

[9] 研究者将 *Tetragonisca angustula* 无刺蜂蜂群从蜂巢密度低的地区移到蜂巢密度高的地区，发现迁移后的蜂群产生的工蜂体型更小且生物量下降，该发现凸显了食物源竞争的影响（Segers et al.，2016）。

用人类建筑中小空穴的蜂种或竞争优势物种（Michener，1946; Batista et al., 2003; Jaffé et al., 2016; Lichtenberg et al., 2017; Kiatoko et al., 2018; Cely-Santos and Philpott，2019），而其他蜂种则需要无人干扰的天然森林（Brown and Albrecht，2001; Batista et al., 2003; Pioker-Hara et al., 2014; Brosi，2009; Brown and de Oliveira，2014）。有趣的是，Lichtenberg 等（2017）发现食性越广的蜂种受森林被砍伐的影响越大，这可能是因为森林被砍伐地区的资源更少，而且这类蜂主要采用单蜂觅食的策略，这使它们在竞争上不如成群觅食的蜂种（第 8 章）。

尽管研究者观察到一些无刺蜂蜂种可以在受干扰的地区生长良好［例如新热带区的 *Tetragonisca angustula* 和 *Trigona spinipes* 无刺蜂以及肯尼亚的 *Axestotrigona ferruginea*（红棕色变形）和 *Hypotrigona gribodoi* 无刺蜂］（Giannini et al., 2015; Jaffe et al., 2016; Kiatoko et al., 2017），总体上无刺蜂的蜂巢密度和物种多样性还是受到了森林砍伐以及自然栖息地向农业和城市景观转变的负面影响（图 1.15）（Klein et al., 2002; Cairns et al., 2005; Slaa，2006; Brosi et al., 2007, 2008; Brosi，2009; Freitas et al., 2009; Venturieri，2009; Ramírez et al., 2013; Pioker-Hara et al., 2014; Kennedy et al., 2013; Brown and Oliveria，2014; Kiatoko et al., 2017; Arena et al., 2018; Smith and Mayfield，2018）。例如，Cairns 等（2005）比较了墨西哥地区不同地点的无刺蜂的丰度和多样性，他们发现无刺蜂的多样性与人为干扰的程度呈负相关。在巴西的两个不同无刺蜂生物群落也发现了类似的情况，即在塞拉多（热带雨林）（Pioker-Hara et al., 2014）和亚马孙盆地（Brown and Albrecht，2001）中 Melipona 族无刺蜂蜂种的丰度和多样性均会随受干扰程度的加深而减少。具有生物多样性的森林对于无刺蜂族群来说尤其重要，因为它们提供了更多样化的筑巢选择和更稳定的食物来源（Brosi et al., 2007; Kaluza et al., 2017, 2018; Smith and Mayfield，2018）[10]。这也就解释了为什么哥斯达黎加和哥伦比亚的森林覆盖率减少和植物物种丰富度降低会与无刺蜂丰度和物种丰富度呈负相关（Brosi et al., 2008; Brosi，2009; Gutiérrez-Chacón et al., 2018）。同理，澳洲植物生物多样性较低和森林片较小的地区，*Tetragonula carbonaria* 无刺蜂蜂群生长缓慢且蜂巢密度也较低（Kaluza et al., 2018; Smith and Mayfield，

---

[10] 另一种解释可能是，被砍伐地区的无刺蜂受到疾病、寄生虫和捕食者的影响更大（Brosi et al., 2007）。

2018）。在距离哥斯达黎加森林地区越远的地方，无刺蜂的多样性和丰度也越低（而蜜蜂表现出相反的趋势）（Brosi et al., 2007），从这些研究中我们可以得到相反的结果。Brown 和 Albrecht（2001）的研究表明，中等程度的森林砍伐就足以对麦蜂族无刺蜂多样性产生负面影响；Brosi（2009）、Kaluza（2018）和 Arena（2018）等的数据表明，只要提供很小的森林块和天然植被区域就能对无刺蜂产生积极影响。

3.5.2　无刺蜂蜂群间的紧密联系

混合蜂群是由两种无刺蜂组成的蜂群，已在自然界中发现了混合蜂群，并且混合蜂群可人工创造。在这些情况下，一种无刺蜂可接受另一蜂种无刺蜂的幼虫和新出房工蜂［Michener（1974）提到一个蜂巢中既有 *Scaptotrigona postica* 无刺蜂工蜂，又有 *Melipona marginata* 无刺蜂工蜂］。然而，混合蜂群在自然界中可能非常罕见（Roubik, 1981）。在这个 *Melipona fuliginosa* 和 *M. fasciata* 共享蜂巢的案例里，前者建造了这个蜂巢，而后者则添加了自己的蜂巢结构，例如储物罐。这两个蜂种的巢脾层位于巢内的不同位置（Roubik, 1981）。*M. fasciata* 工蜂数量较少，在蜂巢入口处观察到它们将食物转移给 *M. fuliginosa* 工蜂，这可能是为了获得进巢许可。Menezes 等（2009 b）报道了一个自发形成的混合蜂群，这个混合蜂群主要由 *Scaptotrigona depilis* 无刺蜂蜂群组成的，而后被 *Nannotrigona testaceicornis* 无刺蜂入侵。在研究期间，该混合蜂群中始终没有发现 *N. testaceicornis* 蜂王，因此该混合蜂群可能是由 *N. testaceicornis* 无刺蜂工蜂误入 *S. depilis* 无刺蜂蜂群所形成的（Menezes et al., 2009 b）。

亚马孙地区的 *Trichotrigona* 和 *Frieseomelitta* 属无刺蜂蜂群之间似乎存在着一种微妙的联系，但我们对其知之甚少。*Trichotrigona extranea* 和 *T. camargoiana* 无刺蜂蜂群看起来没有储存任何食物，但它们似乎总是在 *Frieseomelitta* 无刺蜂蜂群附近筑巢（Camargo and Pedro, 2007; Pedro and Cordeiro, 2015），这是一个属水平亲缘关系较近的蜂种（Rasmussen and Cameron, 2010）。这些无刺蜂蜂群间可能会共享同一个空穴，仅由巢脂质板隔开（Camargo and Pedro, 2007）。鉴于这种蜂巢的排布方式，而且几乎很少观察到这些无刺蜂去访花，因此研究者有理由怀疑 *Trichotrigona* 无刺蜂在使用 *Frieseomelitta* 无刺蜂蜂群储存的食物，因此 *Trichotrigona* 无刺蜂属于 *Frieseomelitta* 无刺蜂的专性偷窃寄生蜂（Camargo and Pedro, 2007; Pedro and Cordeiro, 2015）。

3.5.3 寄食昆虫

鉴于我们对大多数无刺蜂蜂种的了解有限，因此对生活在无刺蜂巢内的生物知之甚少也就不足为奇了。然而，寄食昆虫似乎是普遍存在的。据 Salt（1929）的粗略统计，在无刺蜂巢中发现了来自 10 个目、19 个科、30 个属和 37 个种的节肢动物，其中甲虫尤为常见。其他研究者报道在 *Geotrigona*、*Trigona* 和 *Melipona* 等新热带属无刺蜂蜂种中，一些节肢动物类昆虫居在蜂巢的废物堆，包括弹尾目、耳螟、木虱、多足类、假蝎子和螨类（Roubik, 2006; Barbosa et al., 2013）。它们主要以分解的物质和真菌为食（Roubik, 2006; Barbosa et al., 2013）。在 *Melipona* 属无刺蜂蜂巢中，甲虫以其内的花粉、粪便和生长在巢穴较潮湿部分的真菌为食（Roubik and Wheeler, 1982; Roubik, 2006）。这些昆虫对于无刺蜂的作用可能是互惠共生，因为它们减少了蜂巢中的废弃物和真菌的生长。

Portugal-Araujo（1963）提到两个非洲蜂种的天然地下蜂巢中有一种小甲虫的幼虫，可能是小蜂巢甲虫（SHB）*Aethina tumida*[11]，这种小蜂巢甲虫在过去几年中已成为蜜蜂中的一种新兴害虫（Neumann and Elzen, 2004）。在筑造暴露型蜂巢的蜂种 *Dactylurina staudingeri* 无刺蜂的蜂巢中也发现了这种小蜂巢甲虫 [Mutsaers, 2006, 被 Halcraft 等（2011）引用]。这种甲虫对健康的无刺蜂群几乎没有或不会造成损害，但可以通过消耗储存的食物来破坏虚弱的蜂群（Portugal-Araujo, 1963）。Greco 等（2010）报道发现了澳洲的 *Tetragonula carbonaria* 无刺蜂蜂巢中有这种入侵甲虫。*Haptoncus luteolus* 则是另一种寄食于几种亚洲无刺蜂蜂巢的甲虫。这种甲虫以花粉为食，与小蜂巢甲虫一样，也有可能会破坏和摧毁重度感染的蜂群（Krishnan et al., 2015）。*Tetragonula minangkabau* 无刺蜂蜂群则偶尔会被 *Procoryphaeus wallacei* 甲虫的幼虫破坏（Inoue et al., 1993）。*Cleidostethus meliponae* 甲虫则好像是完全寄生在非洲无刺蜂的蜂巢中（Schwarz, 1948）。*Amphicrossus* 属的甲虫幼虫主要生活在非洲的 *Apotrigona nebulata* 无刺蜂蜂巢中，在那里它们主要采食花粉和蜂蜜（Darchen, 1969）。这种寄生虫也可能会对虚弱的蜂群产生破坏性影响。某些种类的甲虫可以通过附着在工蜂的后腿上搭"便车"而传播开来（Roubik and Wheeler, 1982）。Leiodidae 科的甲虫甚至可以在无刺蜂的泥浆收

---

[11] Portugal-Araujo 认为这种甲虫属于 *Aecthina* 属，但他文中提到了这种幼虫造成的损害，与属名为小蜂巢甲虫（*Aethina*）类似，但是笔者没有找到一种名为 *Aecthina* 的甲虫属，因此怀疑他文中其实说的是小蜂巢甲虫。

集点更换宿主蜂种（Roubik and Wheeler，1982），而其他甲虫则在蜂群间食物转移（主要是盗蜂）期间传播（Roubik，1989）。

Scaura（以前称为 *Schwarzula*）*coccidophila* 无刺蜂蜂群会饲养和照料 *Cryptostigma* 属的软鳞昆虫，偶尔甚至多达 200 只（Camargo and Pedro，2002）。饲养这些寄生生物可能会大有裨益，因为蜂群可以从它们那里获得蜜露和蜂蜡。Camargo 和 Pedro（2002）观察发现 *S. coccidophila* 无刺蜂蜂群通过这些供蜡昆虫能够在蜂巢内堆存大量蜡，而这种情况在无刺蜂蜂巢中是不常见的。这些无刺蜂的另一个罕见特征是它们使用不添加树脂的纯蜡作为筑巢材料（Camargo and Pedro，2002）。反之，介壳虫也可能会从这种伙伴关系中受益，因为这种模式增强了其抵御天敌的能力。

Michiliid 苍蝇能够食用无刺蜂蜂群粪便，它们大量出现在 *Melipona capixaba* 无刺蜂的蜂巢中（Melo，1996）。*Phorid* 苍蝇（*Phoridae* 的几个属）经常在无刺蜂蜂巢内的垃圾堆、打开的花粉罐和受损的幼虫上产卵，如果工蜂不设法清除废物或修复损伤，这种情况可能会成为一个隐患，例如在弱蜂群里出现这种情况就会成为大麻烦（第 7.1 节）（Sakagami and Zucchi，1967；Nogueira-Neto，1997），*Phorid* 蝇幼虫会在几天内杀死无刺蜂蜂群（Nogueira-Neto，1997）。由于这些潜在的危害，研究者观察到无刺蜂工蜂会清除各种各样的苍蝇或甲虫的幼虫（Schwarz，1948；Halcroft et al.，2011）。无刺蜂清除烦人的昆虫的另一种方法就是用蜡和树脂覆盖它们（"木乃伊化"），例如研究者观察到，栖息在 *Tetragonula carbonaria* 无刺蜂蜂巢中的苍蝇幼虫和栖息在 *T. carbonaria* 或 *Tetragonisca angustula* 无刺蜂蜂巢中的甲虫就被这样清除了（Schwarz，1948; Greco et al.，2010）。

螨虫是西方蜜蜂的主要健康威胁（最著名的是狄斯瓦螨 *Varroa destructor*），而在无刺蜂蜂巢中也发现了很多种螨虫（特别是来自 Acaridae 科、Blattisociidae 科、Laelapidae 科和 Pyemotidae 科）（Schwarz，1948; Delfinado-Baker et al.，1983; Fain and Heard，1987; Delfinado-Baker and Baker，1988; Baker et al.，1984; Menezes et al.，2009 a; Vijayakumar and Jayaraj，2013; Radhakrishnan and Ramaraju，2017; Da-Costa et al.，2019）。它们带来的风险尚不明确，并且可能因情况而异。一些螨类被认为是有益的，因为这些螨虫能吃掉线虫或有害真菌并抑制有害生物的生长（Roubik，1989）。例如，有证据表明在 *Scaptotrigona* 属无刺蜂蜂巢的育虫巢室内发现的 *Neotydeolus therapeutikos* 螨类能大大降低幼虫死亡率（Roubik，1989）。还有一些研

究在无刺蜂蜂巢内的食物罐附近或育虫巢室内发现了螨虫，在那里它们附着在幼虫或蛹上（与寄生于蜜蜂的 *Varroa destructor* 螨相似）（Salt，1929；Vijayakumar and Jayaraj，2013；Roubik，1989；Menezes et al.，2009 a）。研究者在成年工蜂的身体上也发现过螨虫（Schwarz，1948；Radhakrishnan and Ramaraju，2017）。Schwarz（1948）就提到过一个 *Partamona testacea* 无刺蜂蜂群，这个蜂群中约 25% 的工蜂都被螨虫侵扰。由于这些螨虫多发现于工蜂的后胫骨或其附近（在 *Trigona corvina* 无刺蜂中也观察到这种现象），他怀疑这些螨虫可能会吃工蜂身上的花粉粒。研究者已经在 *Tetragonula iridipennis* 蜂王身上发现了 Pyemotes 族螨，并怀疑这种螨虫会杀死严重感染的蜂王（Vijayakumar and Jayaraj，2013）。在新热带区，*Pyemotes tritici* 螨则可能是无刺蜂的重大威胁，据报道它会杀死一些 *Melipona* 属、*Tetragonisca angustula* 和 *Frieseomelitta varia* 无刺蜂蜂群（Kerr et al.，1996；Nogueira-Neto，1997；Menezes et al.，2009 a）。当来自邻近蜂群的工蜂从被感染的巢穴中偷取食物或由于蜂农操作时，这些螨虫很可能传播到其他蜂群。

3.5.4 微生物

无刺蜂蜂巢是许多微生物的家园（Rosa et al.，2003；Promnuan et al.，2009；Menezes et al.，2013；Morais et al.，2013；Cambronero-Heinrichs et al.，2019）。无刺蜂与微生物（主要是细菌、酵母菌和霉菌）之间的功能关系仍知之甚少（Menezes et al.，2013）。一些物种，例如无刺蜂，会在巢穴中长期存放食物，微生物的互利共生行为将有助于防止食物变质，这种互利共生是非常有益的。事实上，微生物似乎在无刺蜂蜂蜜成熟过程中发挥了重要作用，也参与了巢内储存的花粉（"罐粉"）、蜂蜜（"罐蜜"）或育虫食物（Vit et al.，2013，2018）的生化修饰。例如，研究者已经在 *Melipona* 属和 *Trigona* 属无刺蜂蜂巢的食物存储处及幼虫食物中鉴定出芽孢杆菌属的细菌，并怀疑这些细菌通过分泌酶参与蜂群的食物代谢转化和发酵（第 1 章）（Gilliam et al.，1985，1990；Morais et al.，2013）。这些酶在蜜蜂食用花粉前能够软化花粉外壁，它们还能够降低花粉对有害微生物的易感性（Villegas-Plazas et al.，2018）。有趣的是，与新鲜收集的花粉相比，*Scaptotrigona depilis* 无刺蜂工蜂更喜欢旧食物罐里的发酵花粉（Vollet-Neto et al.，2017）。研究者在 *Melipona scutellaris* 无刺蜂的幼虫食物中发现了多黏类芽孢杆菌（*Paenibacillus polymyxa*），它似乎可以保护幼虫抵御食虫的真菌（Menegatti et al.，2018）。用抗生素处理蜂群杀死所有细菌会导致蜂群的死亡，这表明细菌对蜂群存活起着至关重要的作用（Machado，1971；

Gilliam et al., 1990）。

  Ptilotrigona 属无刺蜂（*P. lurida* 和 *P. pereneae*）则依赖酵母（*Candida* sp.）来促进储存花粉的干燥和长期保存（Camargo and Pedro,2004），因此研究者猜测其他各种酵母也会通过将酶分泌在无刺蜂蜂巢内来转化和保存食物（Rosa et al., 2003; Menezes et al., 2013）。例如，在 *Tetragonisca angustula* 和 *Frieseomelitta varia* 无刺蜂蜂巢中都可以发现 *Starmerella meliponinorum* 酵母，研究者推测这种酵母可以改善蜂巢内储存食物的营养价值［Rosa et al.,2003; Morais et al. ,2013; 参见 Villegas–Plazas 等（2018）的综述］。然而，并非所有从无刺蜂蜂巢的食物罐中分离出来的酵母都对无刺蜂有益，有些可能会导致蜂蜜或花粉变质（Rosa et al., 2003）。

  最近，Menezes 等（2015）和 Paludo 等（2018）发现一种有趣的现象，即一种无刺蜂和一种微生物之间存在固定的一种关联。*Scaptotrigona depilis* 无刺蜂只有在有 *Zygosaccharomyces* 属的真菌存在的情况下才能存活（第 5 章）。这种真菌在育虫巢室内生长，而幼虫则以这种真菌为食（图 5.1 C）。在没有这种真菌的情况下幼虫大多数都会死亡。然而这些真菌的主要功能并不是为无刺蜂幼虫提供营养的，而是为其提供了必需的类固醇前体（Paludo et al., 2018）。在其他蜂种中也发现了这种真菌，并且似乎这种真菌与无刺蜂相关性更强（Menezes et al., 2013）。

参考文献

Alves D, Imperatriz-Fonseca V, Santos-Filho P（2009）Production of workers, queens and males in *Plebeia remota* colonies（Hymenoptera, Apidae, Meliponini）, a stingless bee with reproductive diapause. Genet Mol Res 8（2）:672–683

Antonini Y, Martins RP（2003）The value of a tree species（*Caryocar brasiliense*）for a stingless bee *Melipona quadrifasciata quadrifasciata*. Journal of Insect Conservation 7（3）:167–174

Arena MVN, Martines MR, da Silva TN, Destéfani FC, Mascotti JCS, Silva-Zacarin ECM, Toppa RH（2018）Multiple-scale approach for evaluating the occupation of stingless bees in Atlantic forest patches. Forest Ecology and Management 430:509–516

Ayton S, Tomlinson S, Phillips RD, Dixon KW, Withers PC（2016）Phenophysiological variation of a bee that regulates hive humidity, but not hive temperature. Journal of Experimental Biology 219（10）:1552–1562

Baker E, Flechtmann C, Delfinado-Baker M (1984) Acari domum meliponinarum brasiliensium habitantes. VI. New species of Bisternalis Hunter (Laelapidae: Acari). International Journal of Acarology 10 (3):181–189

Bänziger H, Pumikong S, Srimuang K (2011) The remarkable nest entrance of tear drinking *Pariotrigona klossi* and other stingless bees nesting in limestone cavities (Hymenoptera: Apidae). Journal of the Kansas Entomological Society 84:22–35

Barbosa FM, Alves RMdO, Souza BdA, Carvalho CALd (2013) Nest architecture of the stingless bee *Geotrigona subterranea* (Friese, 1901) (Hymenoptera: Apidae: Meliponini). Biota Neotropica 13 (1):147–152

Bassindale R (1955) The biology of the Stingless Bee *Trigona* (*Hypotrigona*) *gribodoi* Magretti (Meliponidae). Proceedings of the Zoological Society of London 125 (1):49–62

Batista MA, Ramalho M, Soares AE (2003) Nesting sites and abundance of Meliponini (Hymenoptera: Apidae) in heterogeneous habitats of the Atlantic rain forest, Bahia, Brazil. Lundiana 4 (1):19–23

Becker T, Pequeno PACL, Carvalho-Zilse GA (2018) Impact of environmental temperatures on mortality, sex and caste ratios in *Melipona interrupta* Latreille (Hymenoptera, Apidae). Science of Nature 105:55

Biesmeijer JC, Slaa EJ, Koedam D (2007) How stingless bees solve traffic problems. Entomologische Berichten-Nederlandsche Entomologische Vereeniging 67 (1/2):7–13

Blomquist GJ, Roubik DW, Buchmann SL (1985) Wax chemistry of two stingless bees of the *Trigonisca* group (Apidae: Meliponinae). Comparative Biochemistry and Physiology Part B: Comparative Biochemistry 82 (1):137–142

Bobadoye BO, Ndegwa PN, Irungu L, Ayuka F, Kajobe R (2017) Floral resources sustaining African meliponine bee species (Hymenoptera: Meliponini) in a fragile habitat of Kenya. Journal of Biology and Life Science 8:42–58

Bordoni A, Mocilnik G, Forni G, Bercigli M, Giove CDV, Luchetti A, Turillazzi S, Dapporto L, Marconi M (2020) Two aggressive neighbours living peacefully: the nesting association between a stingless bee and the bullet ant. Insectes Sociaux 67:103–112

Breed MD, McGlynn TP, Sanctuary MD, Stocker EM, Cruz R (1999) Distribution and abundance of colonies of selected meliponine species in a Costa Rican tropical wet forest. Journal of Tropical Ecology 15（6）:765–777

Brito RM, Schaerf TM, Myerscough MR, Heard TA, Oldroyd BP（2012）Brood comb construction by the stingless bees *Tetragonula hockingsi* and *Tetragonula carbonaria*. Swarm Intelligence 6（2）:151–176

Brosi BJ（2009）The complex responses of social stingless bees（Apidae: Meliponini）to tropical deforestation. Forest Ecology and Management 258（9）:1830–1837

Brosi BJ, Daily GC, Ehrlich PR（2007）Bee community shifts with landscape context in a tropical countryside. Ecological Applications 17（2）:418–430

Brosi BJ, Daily GC, Shih TM, Oviedo F, Durán G（2008）The effects of forest fragmentation on bee communities in tropical countryside. Journal of Applied Ecology 45（3）:773–783

Brown JC, Albrecht C（2001）The effect of tropical deforestation on stingless bees of the genus *Melipona*（Insecta: Hymenoptera: Apidae: Meliponini）in central Rondonia, Brazil. Journal of Biogeography 28（5）:623–634

Brown JC, de Oliveira ML（2014）The impact of agricultural colonization and deforestation on stingless bee（Apidae: Meliponini）composition and richness in Rondônia, Brazil. Apidologie 45:172–188

Brütsch T, Jaffuel G, Vallat A, Turlings TC, Chapuisat M（2017）Wood ants produce a potent antimicrobial agent by applying formic acid on tree-collected resin. Ecology and Evolution 7（7）:2249–2254

Buchwald R, Breed MD（2005）Nestmate recognition cues in a stingless bee, *Trigona fulviventris*. Animal Behaviour 70:1331–1337

Bujok B, Kleinhenz M, Fuchs S, Tautz J（2002）Hot spots in the bee hive. Naturwissenschaften 89:299–301

Cairns CE, Villanueva-Gutiérrez R, Koptur S, Bray DB（2005）Bee populations, forest disturbance, and africanization in Mexico. Biotropica 37（4）:686–692

Camargo J, Moure J（1983）*Trichotrigona*, un novo genero de Meliponinae（Hymenoptera, Apidae）do Rio Negro, Amazonas Brasil. Acta Amazonica 13（2）:421–429

Camargo JD（1970）Ninhos e biologia de algumas espécies de Meliponídeos（Hymenoptera: Apidae）da região de Porto Velho, Território de Rondônia, Brasil.

Revista de Biologia Tropical 16（2）:207–239

Camargo Jd（1974）Notas sobre a morfologia e biologia de *Plebeia*（*Schwarziana*）*quadripunctata quadripunctata*（Hymenoptera, Apidae, Meliponinae）. Studia Entomologica 17:433–470

Camargo JM（1996）Meliponini neotropicais: o gênero *Camargoia* Moure, 1989（Apinae, Apidae, Hymenoptera）. Arquivos de Zoologia（São Paulo）33（2-3）:71–92

Camargo JM, Moure JS（1994）Meliponinae neotropicais: os gêneros *Paratrigona* Schwarz, 1938 e *Aparatrigona* Moure, 1951（Hymenoptera, Apidae）. Arquivos de Zoologia 32（2）:33–109

Camargo JM, Pedro SR（2002）Mutualistic association between a tiny Amazonian stingless bee and a wax-producing scale insect. Biotropica 34（3）:446–451

Camargo JM, Pedro SR（2003）Neotropical Meliponini: the genus *Partamona* Schwarz, 1939（Hymenoptera, Apidae, Apinae）-bionomy and biogeography. Revista Brasileira de Entomologia 47（3）:311–372

Camargo JM, Pedro SR（2007）Notes on the bionomy of *Trichotrigona extranea* Camargo & Moure（Hymenoptera, Apidae, Meliponini）. Revista Brasileira de Entomologia 51（1）:72–81

Camargo JM, Posey DA（1990）O conhecimento dos Kayapó sobre as abelhas sociais sem ferrão（Meliponinae, Apidae, Hymenoptera）: notas adicionais. Boletim do Museu Paraense Emílio Goeldi, Zoologia 6:17–42

Camargo JM, Roubik DW（1991）Systematics and bionomics of the apoid obligate necrophages: the *Trigona hypogea* group（Hymenoptera: Apidae; Meliponinae）. Biological Journal of the Linnean Society 44（1）:13–39

Camargo JMF, Pedro SRM（2004）Meliponini neotropicais: o gênero *Ptilotrigona* Moure（Hymenoptera, Apidae, Apinae）. Revista Brasileira de Entomologia 48:353–377

Camazine S, Deneubourg J-L, Franks NR, Sneyd J, Theraulaz G, Bonabeau E（2001）Self-Organization in Biological Systems. Princeton studies in complexity. Princeton University Press, Princeton

Cambronero-Heinrichs JC, Matarrita-Carranza B, Murillo-Cruz C, Araya-Valverde E, Chavarría M, Pinto-Tomás AA（2019）Phylogenetic analyses of antibiotic-producing *Streptomyces* sp. isolates obtained from the stingless-bee *Tetragonisca angustula*

(Apidae: Meliponini). Microbiol Read Engl 165:292–301

Cardoso SSS, Cartwright JHE, Checa AG, Escribano B, Osuna-Mascaró AJ, Sainz-Díaz CI (2020) The bee *Tetragonula* builds its comb like a crystal. Journal of The Royal Society Interface 17:20200187

Cely-Santos M, Philpott SM (2019) Local and landscape habitat influences on bee diversity in agricultural landscapes in Anolaima, Colombia. J Insect Conserv 23:133–146

Cepeda OI (2006) Division of labor during brood production in stingless bees with special reference to individual participation. Apidologie 37 (2):175

Chapuisat M, Oppliger A, Magliano P, Christe P (2007) Wood ants use resin to protect themselves against pathogens. Proceedings of the Royal Society of London B: Biological Sciences 274 (1621):2013–2017

Chinh T, Sommeijer M, Boot W, Michener C (2005) Nest and colony characteristics of three stingless bee species in Vietnam with the first description of the nest of *Lisotrigona carpenteri* (Hymenoptera: Apidae: Meliponini). Journal of the Kansas Entomological Society 78 (4):363–372

Choudhari MK, Punekar SA, Ranade RV, Paknikar KM (2012) Antimicrobial activity of stingless bee (*Trigona* sp.) propolis used in the folk medicine of Western Maharashtra, India. Journal of Ethnopharmacology 141:363–367

Christe P, Oppliger A, Bancala F, Castella G, Chapuisat M (2003) Evidence for collective medication in ants. Ecology Letters 6 (1):19–22

Cortopassi-Laurino M, Alves DA, Imperatriz-Fonseca VL (2009) Arvores neotropicais, recursos importantes para a nidivicacao de abelhas sem ferrao (Apidae, meliponini). Mensagem doce 100

Cortopassi-Laurino M, Nogueira-Neto P (2003) Notas sobre a bionomia de *Tetragonisca weyrauchi* Schwarz, 1943 (Apidae, Meliponini). Acta Amazonica 33:643–650

Cortopassi-Laurino M, Nogueira-Neto P (2018) Stingless Bees from Brazil. University of São Paulo Press, São Paulo

Couvillon MJ, Wenseleers T, Imperatriz-Fonseca VL, Nogueira-Neto P, Ratnieks FLW (2008) Comparative study in stingless bees (Meliponini) demonstrates that nest entrance size predicts traffic and defensivity. Journal of Evolutionary Biology 21:194–201

Da-Costa T, Rodighero LF, Silva GLD, Ferla NJ, Blochtein B (2019) Two new species of Tydeidae (Acari: Prostigmata) associated with stingless bees. Zootaxa 4652:101–112

da Silva DLN, Zucchi R, Kerr WE (1972) Biological and behavioural aspects of the reproduction in some species of *Melipona* (Hymenoptera, Apidae, Meliponinae). Animal Behaviour 20 (1):123–132

Darchen R (1969) Sur la biologie de *Trigona* (*Apotrigona*) *nebulata* komiensis Cock. I. Biologia Gabonica 5:151–183

Darchen R (1972) Ecologie de quelques trigones (*Trigona* sp.) de la savane de Lamto (Cote D'Ivoire). Apidologie 3 (4):341–367

Darchen R (1973) La thermoregulation et l'écologie de quelques espèces d'abeilles sociales d'Afrique (Apidae, Trigonini et *Apis mellifica* Var. *adansonii*). Apidologie 4:341–370

Darchen R, Pain J (1966) Le nid de *Trigona* (*Dactylurina*) *staudingeri* Gribodoi (Hymenoptera, Apidae). Biologia Gabonica 2:25–35

Delfinado-Baker M, Baker E (1988) New mites (Acari: Laelapidae) from the nests of stingless bees (Apidae: Meliponinae) from Asia. International Journal of Acarology 14 (3):127–136

Drumond P, Bego L, Melo G (1995) Nest architecture of the stingless bee *Plebeia poecilochroa* Moure and Camargo, 1993 and related considerations (Hymenoptera, Apidae, Meliponinae). Iheringia, Ser Zool, Porto Alegre 79:39–49

Drumond PM, Zucchi R, Yamane S, Sakagami SF (1998) Oviposition behavior of the stingless bees: XX. *Plebeia* (*Plebeia*) *julianii* which forms very small brood batches (Hymenoptera: Apidae, Meliponinae). Entomological Science 1 (2):195–205

Duarte R, Souza J, Soares AEE (2016) Nest Architecture of *Tetragona clavipes* (Fabricius) (Hymenoptera, Apidae, Meliponini). Sociobiology 63 (2):813–818

Eardley C (2004) Taxonomic revision of the African stingless bees (Apoidea: Apidae: Apinae: Meliponini). African plant protection 10 (2):63–96

Eltz T, Brühl CA, Imiyabir Z, Linsenmair KE (2003) Nesting and nest trees of stingless bees (Apidae: Meliponini) in lowland dipterocarp forests in Sabah, Malaysia, with implications for forest management. Forest Ecology and Management 172 (2):301–313

Eltz T, Brühl CA, Van Der Kaars S, Linsenmair EK (2002) Determinants of stingless bee nest density in lowland dipterocarp forests of Sabah, Malaysia. Oecologia 131

(1):27–34

Engels W, Imperatriz-Fonseca VL (1990) Caste development, reproductive strategies, and control of fertility in honey bees and stingless bees. In: Engels W (ed) Social Insects. Springer, Heidelberg, pp 167–230

Engels W, Rosenkranz P, Engels E (1995) Thermoregulation in the nest of the Neotropical stingless bee *Scaptotrigona postica* and a hypothesis on the evolution of temperature homeostasis in highly eusocial bees. Studies on Neotropical Fauna and Environment 30 (4):193–205

Estienne V, Mundry R, Kühl HS, Boesch C (2016) Exploitation of underground bee nests by three sympatric consumers in Loango National Park, Gabon. Biotropica 49:101–109

Fain A, Heard T (1987) Description and life cycle of *Cerophagus trigona* spec. nov. (Acari, Acaridae), associated with the stingless bee *Trigona carbonaria* Smith in Australia. Bulletin de L'Institut Royal des Sciences Natruelles de Belgique 57:197–202

Fowler H (1979) Responses by a stingless bee to a subtropical environment. Revista de Biologia Tropical 27:111–118

Francoy T, Bonatti V, Viraktamath S, Rajankar B (2016) Wing morphometrics indicates the existence of two distinct phenotypic clusters within population of *Tetragonula iridipennis* (Apidae: Meliponini) from India. Insectes Sociaux 63 (1):109–115

Freitas GS (2001) Levantamento de ninhos de meliponíneos (Hymenoptera, Apidae) em área urbana: Campus da USP, Ribeirao Preto-SP. Master Thesis, University of Sao Paulo

Freitas BM, Imperatriz-Fonseca VL, Medina LM, Kleinert A de MP, Galetto L, Nates-Parra G, Quezada-Euán JJG (2009) Diversity, threats and conservation of native bees in the Neotropics. Apidologie 40:332–346

Galaschi-Teixeira JS, Falcon T, Ferreira-Caliman MJ, Witter S, Francoy TM (2018) Morphological, chemical, and molecular analyses differentiate populations of the subterranean nesting stingless bee *Mourella caerulea* (Apidae: Meliponini). Apidologie 49:367–377

Garcia MV, de Olivera ML, de Olivera Campos LA (1992) Use of Seeds of *Coussapoa asperifolia magnifolia* (Cecropiaceae) by Stingless Bees in the Central Amazonian

Forest (Hymenoptera: Apidae: Meliponinae). Entomologia Generalis 17:255–258

Giannini TC, Boff S, Cordeiro GD, Cartolano EA, Veiga AK, Imperatriz-Fonseca VL, Saraiva AM (2015) Crop pollinators in Brazil: a review of reported interactions. Apidologie 46:209–223

Gilliam M, Buchmann SL, Lorenz BJ, Roubik DW (1985) Microbiology of the Larval Provisions of the Stingless Bee, *Trigona hypogea*, an Obligate Necrophage. Biotropica 17:28–31

Gilliam M, Roubik D, Lorenz B (1990) Microorganisms associated with pollen, honey, and brood provisions in the nest of a stingless bee, *Melipona fasciata*. Apidologie 21:89–97

Gonzalez VH, Amith JD, Stein TJ (2018) Nesting ecology and the cultural importance of stingless bees to speakers of Yoloxóchitl Mixtec, an endangered language in Guerrero, Mexico. Apidologie 49:625–636

Greco MK, Hoffmann D, Dollin A, Duncan M, Spooner-Hart R, Neumann P (2010) The alternative Pharaoh approach: stingless bees mummify beetle parasites alive. Naturwissenschaften 97 (3):319–323

Grüter C, Kärcher M, Ratnieks FLW (2011) The natural history of nest defence in a stingless bee, *Tetragonisca angustula* (Latreille) (Hymenoptera: Apidae), with two distinct types of entrance guards. Neotropical Entomology 40:55–61

Gutiérrez-Chacón C, Dormann CF, Klein A-M (2018) Forest-edge associated bees benefit from the proportion of tropical forest regardless of its edge length. Biological Conservation 220:149–160

Halcroft M, Haigh A, Holmes S, Spooner-Hart R (2013) The thermal environment of nests of the Australian stingless bee, *Austroplebeia australis*. Insectes Sociaux 60 (4):497–506

Halcroft M, Spooner-Hart R, Neumann P (2011) Behavioral defense strategies of the stingless bee, *Austroplebeia australis*, against the small hive beetle, *Aethina tumida*. Insectes Sociaux 58 (2):245–253

Heinrich B (1993) The hot-blooded insects: mechanisms and evolution of thermoregulation. Harvard University Press, Cambridge

Hubbell SP, Johnson LK (1977) Competition and nest spacing in a tropical stingless bee community. Ecology 58:949–963

Ihering Hv（1903）Biologie der stachellosen Honigbienen Brasiliens. Zoologische Jahrbuecher Abteilung fuer Systematik Oekologie und Geographie der Tiere 19:179–287

Imperatriz-Fonseca V, Zucchi R（1995）Virgin queens in stingless bee（Apidae, Meliponinae）colonies: a review. Apidologie 26（3）:231–244

Imperatriz-Fonseca VL（1977）Studies on *Paratrigona subnuda*（Moure）Hymenoptera, Apidae, Meliponinae-II. Behaviour of the virgin queen. Boletim de Zoologia 2（2）:169–182

Inoue T, Nakamura K, Salmah S, Abbas I（1993）Population dynamics of animals in unpredictably-changing tropical environments. Journal of Biosciences 18（4）:425–455

Jaffé R, Castilla A, Pope N, Imperatriz-Fonseca VL, Metzger JP, Arias MC, Jha S（2016）Landscape genetics of a tropical rescue pollinator. Conservation Genetics 17:267–278

Jones JC, Helliwell P, Beekman M, Maleszka R, Oldroyd BP（2005）The effects of rearing temperature on developmental stability and learning and memory in the honey bee, *Apis mellifera*. Journal of Comparative Physiology A-Neuroethology Sensory Neural and Behavioral Physiology 191:1121–1129

Jones JC, Oldroyd BP（2006）Nest thermoregulation in social insects. Advances in Insect Physiology 33:153–191

Juliani L（1962）O aprisionamento de rainhas virgens en colonias de Trigonini（Hymenoptera - Apoidea）. Boletim da Universidade do Paraná 20:1–11

Kajobe R, Roubik DW（2006）Honey-making bee colony abundance and predation by apes and humans in a Uganda forest reserve. Biotropica 38:210–218

Kennedy CM, Lonsdorf E, Neel MC, Williams NM, Ricketts TH, Winfree R, Bommarco R, Brittain C, Burley AL, Cariveau D（2013）A global quantitative synthesis of local and landscape effects on wild bee pollinators in agroecosystems. Ecology Letters 16（5）:584–599

Kerr WE（1948）Estudos sobre o gênero *Melipona*. Anais da Escola Superior de Agricultura Luiz de Queiroz 5:181–276

Kerr WE（1972）Effect of low temperature on male meiosis in *Melipona marginata*. Journal of Apicultural Research 11（2）:95–99

Kerr WE, Carvalho GA, Nascimento VA（1996）Abelha Uruçu: biologia, manejo e conservação, vol 2. Fundação Acangaú, Belo Horizonte, Minas Gerais

Kerr WE, Sakagami SF, Zucchi R, Portugal-Araújo Vd, Camargo Jd (1967) Observações sobre a arquitetura dos ninhos e comportamento de algumas espécies de abelhas sem ferrão das vizinhanças de Manaus, Amazonas (Hymenoptera, Apoidea). In: Atas do Simpósio sobre a biota Amazônica, 1967. Conselho Nacional de Pesquisas Rio de Janeiro, pp 255–309

Khuong A, Gautrais J, Perna A, Sbaï C, Combe M, Kuntz P, Jost C, Theraulaz G (2016) Stigmergic construction and topochemical information shape ant nest architecture. Proceedings of the National Academy of Sciences of the United States of America 113 (5):1303–1308

Kiatoko N, Langevelde FV, Raina SK (2018) Forest degradation influences nesting site selection of Afro-tropical stingless bee species in a tropical rain forest, Kenya. African Journal of Ecology 56:669–674

Klein AM, Steffan-Dewenter I, Buchori D, Tscharntke T (2002) Effects of land-use intensity in tropical agroforestry systems on coffee flower-visiting and trap-nesting bees and wasps. Conservation Biology 16 (4):1003–1014

Koedam D, Jungnickel H, Tentschert J, Jones G, Morgan E (2002) Production of wax by virgin queens of the stingless bee *Melipona bicolor* (Apidae, Meliponinae). Insectes Sociaux 49 (3):229–233

Kolmes S, Sommeijer M (1992) Ergonomics in stingless bees: changes in intranidal behavior after partial removal of storage pots and honey in *Melipona favosa* (Hym. Apidae, Meliponinae). Insectes Sociaux 39 (2):215–232

Lavinas FC, Macedo EHBC, Sá GBL, Amaral ACF, Silva JRA, Azevedo MMB, Vieira BA, Domingos TFS, Vermelho AB, Carneiro CS, Rodrigues IA (2019) Brazilian stingless bee propolis and geopropolis: promising sources of biologically active compounds. Revista Brasileira de Farmacognosia 29: 389–399

Leonhardt SD, Jung L-M, Schmitt T, Blüthgen N (2010) Terpenoids tame aggressors: role of chemicals in stingless bee communal nesting. Behavioral Ecology and Sociobiology 64:1415–1423

Lichtenberg EM, Mendenhall CD, Brosi B (2017) Foraging traits modulate stingless bee community disassembly under forest loss. Journal of Animal Ecology 86 (6):1404–1416

Lindauer M (1954) Temperaturregulierung und Wasserhaushalt im Bienenstaat. Zeitschrift für vergleichende Physiologie 36:391–432

Machado J (1971) Simbiose entre as abelhas sociais brasileiras (Meliponinae, Apidae) e uma espécie de bactéria. Ciencia e Cultura 23 (5):625–633

Macías-Macías JO, Quezada-Euán JJG, Contreras-Escareño F, Tapia-Gonzalez JM, Moo-Valle H, Ayala R (2011) Comparative temperature tolerance in stingless bee species from tropical highlands and lowlands of Mexico and implications for their conservation (Hymenoptera: Apidae: Meliponini). Apidologie 42 (6):679–689

Martins CF, Laurino MC, Koedam D, Fonseca VLI (2004) Tree species used for nidification by stingless bees in the Brazilian Caatinga (Seridó, PB; João Câmara, RN). Biota Neotropica 4:1–8

Massaro CF, Katouli M, Grkovic T, Vu H, Quinn RJ, Heard TA, Carvalho C, Manley-Harris M, Wallace HM, Brooks P (2014) Anti-staphylococcal activity of C-methyl flavanones from propolis of Australian stingless bees (*Tetragonula carbonaria*) and fruit resins of *Corymbia torelliana* (Myrtaceae). Fitoterapia 95:247–257

Massaro FC, Brooks PR, Wallace HM, Russell FD (2011) Cerumen of Australian stingless bees (*Tetragonula carbonaria*): gas chromatography-mass spectrometry fingerprints and potential anti-inflammatory properties. Naturwissenschaften 98:329–337

Melo GA (1996) Notes on the nesting biology of *Melipona capixaba* (Hymenoptera, Apidae). Journal of the Kansas Entomological Society 69:207–210

Menegatti C, Da Paixão Melo WG, Carrão DB, De Oliveira ARM, Do Nascimento FS, Lopes NP, Pupo MT (2018) *Paenibacillus polymyxa* Associated with the Stingless Bee *Melipona scutellaris* Produces Antimicrobial Compounds against Entomopathogens. J Chem Ecol 44:1158–1169

Menezes C, Colleto-Silva A, Gazeta G, Kerr W (2009a) Infestation by *Pyemotes tritici* (Acari, Pyemotidae) causes death of stingless bee colonies (Hymenoptera: Meliponina). Genetics and Molecular Research 8:630–634

Menezes C, Hrncir M, Kerr W (2009b) A mixed colony of *Scaptotrigona depilis* and *Nannotrigona testaceicornis* (Hymenoptera, Apidae, Meliponina). Genetics and Molecular Research 8 (2):507–514

Menezes C, Vollet-Neto A, Contrera FAFL, Venturieri GC, Imperatriz-Fonseca VL (2013) The role of useful microorganisms to stingless bees and stingless beekeeping.

In: Vit P, Pedro SRM, Roubik DW (eds) Pot-Honey. Springer, pp 153–171

Menezes C, Vollet-Neto A, Marsaioli AJ, Zampieri D, Fontoura IC, Luchessi AD, Imperatriz-Fonseca VL (2015) A Brazilian social bee must cultivate fungus to survive. Current Biology 25 (21):2851–2855

Messer AC (1985) Fresh dipterocarp resins gathered by megachilid bees inhibit growth of pollen-associated fungi. Biotropica 17:175–176

Michener CD (1946) Notes on the habits of some Panamanian stingless bees (Hymenoptera, Apidae). Journal of the New York Entomological Society 54:179–197

Michener CD (1961) Observations on the nests and behavior of *Trigona* in Australia and New Guinea (Hymenoptera, Apidae). American Museum Novitates 2026:1–46

Michener CD (1964) Evolution of the nests of bees. American Zoologist 4:227–239

Michener CD (1974) The Social Behavior of the Bees. Harvard University Press, Cambridge

Michener CD (2007) The bees of the world, 2nd edn. The Johns Hopkins University Press, Baltimore

Milborrow B, Kennedy J, Dollin A (1987) Composition of wax made by the Australian stingless bee *Trigona australis*. Australian Journal of Biological Sciences 40 (1):15–26

Moo-Valle H, Quezada-Euán JJG, Navarro J, Rodriguez-Carvajal LA (2000) Patterns of intranidal temperature fluctuation for *Melipona beecheii* colonies in natural nesting cavities. Journal of Apicultural Research 39 (1-2):3–7

Morais PB, Calaça PSST, Rosa CA (2013) Microorganisms associated with stingless bees. In: Vit P, Pedro SRM, Roubik DW (eds) Pot-Honey. Springer, pp 173–186

Moritz R, Crewe R (1988) Air ventilation in nests of two African stingless bees *Trigona denoiti* and *Trigona gribodoi*. Experientia 44 (11-12):1024–1027

Muli EM, J. M. Maingi, Macharia J (2008) Antimicrobial Properties of Propolis and Honey from the Kenyan Stingless bee, *Dactylurina schimidti*. Apiacta 49–61

Namu FN (2008) The possible role of stingless bees in the spread of Banana Xanthomonas Wilt in Uganda and the nesting biology of *Plebeina hildebrandti* and *Hypotrigona gribodoi* (Hymenoptera-Apidae-Meliponini). University of Bonn, PhD Thesis

Namu FN, Wittmann D (2016) An African stingless bee *Plebeina hildebrandti* Friese nest size and design (Apidae, Meliponini). African Journal of Ecology 55:111–114

Ndungu NN, Yusuf AA, Raina SK, Masiga DK, Pirk CWW, Nkoba K (2019) Nest Architecture as a Tool for Species Discrimination of *Hypotrigona* Species (Hymenoptera: Apidae: Meliponini). African Entomology 27:25–35

Neumann P, Elzen P (2004) The biology of the small hive beetle (*Aethina tumida*, Coleoptera: Nitidulidae): Gaps in our knowledge of an invasive species. Apidologie 35 (3):229–247

Njoya M (2009) Diversity of stingless bees in Bamenda Afromontane Forests-Cameroon: nest architecture, behaviour and labour calendar. PhD thesis: Wilhelms Universität Bonn-Institut für Nutzpflanzenwissenschaften und Ressourcenschutz Rheinische Friedrich (Deutschland)

Njoya MTM, Wittmann D, Ambebe TF (2016) Nest Architecture of *Dactylurina staudingeri* (Hymenoptera, Apidae, Meliponini) in Cameroon. International Journal of Research in Agricultural Sciences 3:335–338

Njoya MTM, Wittmann D, Azibo BR (2017) Subterranean Nest Architecture and Colony Characteristics of *Meliponula* (*Meliplebeia*) *becarii* (Hymenoptera, Apidae, Meliponini) in Cameroon. Journal of Chemical, Biological and Physical Sciences 7:220–233

Nogueira-Neto P (1948) Notas bionomicas sobre meliponineos. I. Sobre ventilação dos ninhos e as constucões com ela relationadas. Revista Brasileira de Biologia 8:465–488

Nogueira-Neto P (1954) Notas bionômicas sobre meliponíneos: III – Sobre a enxameagem. Arquivos do Museu Nacional 42:419–451

Nogueira-Neto P (1962) The scutellum nest structure of *Trigona* (*Trigona*) *spinipes* Fab. (Hymenoptera: Apidae). Journal of the New York Entomological Society 70:239–264

Nogueira-Neto P (1997) Vida e Criação de Abelhas Indígenas Sem Ferrão. Editora Nogueirapis, São Paulo

Oldroyd BP, Pratt SC (2015) Comb architecture of the eusocial bees arises from simple rules used during cell building. Advances in Insect Physiology 49:101–121

Oliveira ML, Morato EF, Garcia MV (1995) Diversity of species and density of stingless bee social nests (Hymenoptera, Apidae, Meliponinae) in terra firme forest in central Amazônia. Revista Brasileira de Zoologia 12 (1):13–24

Paludo CR, Menezes C, Silva-Junior EA, Vollet-Neto A, Andrade-Dominguez A, Pishchany G, Khadempour L, Nascimento FS, Currie CR, Kolter R (2018) Stingless bee larvae require fungal steroid to pupate. Scientific Reports 8:1122

Pedro SR, Cordeiro GD (2015) A new species of the stingless bee *Trichotrigona* (Hymenoptera: Apidae, Meliponini). Zootaxa 3956 (3):389–402

Perna A, Theraulaz G (2017) When social behaviour is moulded in clay: on growth and form of social insect nests. Journal of Experimental Biology 220:83–91

Pioker-Hara FC, Drummond MS, Kleinert AdMP (2014) The influence of the loss of Brazilian savanna vegetation on the occurrence of nests of stingless bees (Apidae: Meliponini). Sociobiology 61 (4):393–400

Portugal-Araujo V (1963) Subterranean nests of two African stingless bees (Hymenoptera: Apidae). Journal of the New York Entomological Society 71:130–141

Promnuan Y, Kudo T, Chantawannakul P (2009) *Actinomycetes* isolated from beehives in Thailand. World Journal of Microbiology and Biotechnology 25 (9):1685–1689

Ramírez VM, Calvillo LM, Kevan PG (2013) Effects of Human Disturbance and Habitat Fragmentation on Stingless Bees. In: Vit P, Pedro SRM, Roubik D (eds) Pot-Honey: A legacy of stingless bees. Springer, New York, NY, pp 269–282

Rasmussen C (2004) A stingless bee nesting with a paper wasp (Hymenoptera: Apidae, Vespidae). Journal of the Kansas Entomological Society 77 (4):593–601

Rasmussen C, Camargo JMF (2008) A molecular phylogeny and the evolution of nest architecture and behavior in *Trigona s.s.* (Hymenoptera: Apidae: Meliponini). Apidologie 39:102–118

Rasmussen C, Cameron S (2010) Global stingless bee phylogeny supports ancient divergence, vicariance, and long distance dispersal. Biological Journal of the Linnean Society 99:206–232

Ribeiro MF, Wenseleers T, Santos Filho PS, Alves DA (2006) Miniature queens in stingless bees: basic facts and evolutionary hypotheses. Apidologie 37 (2):191–206

Roldão-Sbordoni YS, Gomes G, Mateus S, Nascimento FS (2019) Scientific Note: Warming Nurses, a New Worker Role Recorded for the First Time in Stingless Bees. Journal of Economic Entomology 112:1485–1488

Rosa CA, Lachance M-A, Silva JO, Teixeira ACP, Marini MM, Antonini Y, Martins RP (2003) Yeast communities associated with stingless bees. FEMS Yeast Research 4

(3):271–275

Roubik DW (1981) A natural mixed colony of *Melipona* (Hymenoptera: Apidae). Journal of the Kansas Entomological Society 54:263–268

Roubik DW (1983) Nest and colony characteristics of stingless bees from Panama (Hymenoptera: Apidae). Journal of the Kansas Entomological Society 56:327–355

Roubik DW (1989) Ecology and Natural History of Tropical Bees. Cambridge University Press, New York

Roubik DW (2006) Stingless bee nesting biology. Apidologie 37:124–143

Roubik DW, Patiño JEM (2009) *Trigona corvina*: an ecological study based on unusual nest structure and pollen analysis. Psyche 2009:268756

Roubik DW, Peralta FJA (1983) Thermodynamics in nests of two *Melipona* species in Brasil. Acta Amazonica 13:453–466

Roubik DW, Wheeler QD (1982) Flightless beetles and stingless bees: phoresy of Scotocryptine beetles (Leiodidae) on their meliponine hosts (Apidae). Journal of the Kansas Entomological Society 55:125–135

Sakagami S, Inoue T, Yamane S, Salmah S (1983) Nest architecture and colony composition of the Sumatran stingless bee *Trigona* (*Tetragonula*) *laeviceps*. Kontyu 51 (1):100–111

Sakagami SF (1982) Stingless bees. In: Hermann HR (ed) Social Insects III. Academic Press, New York, pp 361–423

Sakagami SF, Camilo C, Zucchi R (1973) Oviposition behavior of a Brazilian stingless bee, *Plebeia* (*Friesella*) *schrottkyi*, with some remarks on the behavioral evolution in stingless bees. Journal of the Faculty of Science Hokkaido University Series VI, Zoology 19 (1):163–189

Sakagami SF, Inoue T, Yamane S, Salmah S (1989) Nests of the Myrmecophilous Stingless Bee, *Trigona moorei*: How do Bees Initiate Their Nest Within an Arboreal Ant Nest? Biotropica 21:265–274

Sakagami SF, Oniki Y (1963) Behavior studies of the stingless bees, with special reference to the oviposition process. I: *Melipona compressipes manaosensis* Schwarz. Journal of the Faculty of Science Hokkaido University Series VI, Zoology 15 (2):300–318

Sakagami SF, Zucchi R (1967) Behavior Studies of the Stingless Bees, with Special

Reference to the Oviposition Process: Ⅵ. *Trigona* (*Tetragona*) *clavipes*. (With 6 Text-figures and 2 Tables). Journal of the Faculty of Science Hokkaido University Series Ⅵ Zoology 16 (2): 292–313

Sakagami SF, Zucchi R (1974) Oviposition behavior of two dwarf stingless bees, *Hypotrigona* (*Leurotrigona*) *muelleri* and *H.* (*Trigonisca*) *duckei*, with notes on the temporal articulation of oviposition process in stingless bees Journal of the Faculty of Science Hokkaido University Series Ⅵ, Zoology 19 (2): 361–421

Sakagami SF, Zucchi R, Araujo VdP (1977) Oviposition behavior of an aberrant african stingless bee *Meliponula bocandei*, with notes on the mechanism and evolution of oviposition behavior in stingless bees. Journal of the Faculty of Science Hokkaido University Series Ⅵ, Zoology 20 (4): 647–690

Salt G (1929) A contribution to the ethology of the Meliponinae. Ecological Entomology 77 (2): 431–470

Samejima H, Marzuki M, Nagamitsu T, Nakasizuka T (2004) The effects of human disturbance on a stingless bee community in a tropical rainforest. Biological Conservation 120 (4): 577–587

Schwarz HF (1945) The wax of stingless bees (Meliponidae) and the uses to which it has been put. Journal of the New York Entomological Society 53: 137–144

Schwarz HF (1948) Stingless Bees (Meliponidae) of the Western Hemisphere. Bulletin of the American Museum of Natural History 90: 1–546

Segers FHID, Menezes C, Vollet-Neto A, Lambert D, Grüter C (2015) Soldier production in a stingless bee depends on rearing location and nurse behaviour. Behavioral Ecology and Sociobiology 69: 613–623

Segers FHID, Von Zuben LG, Grüter C (2016) Local differences in parasitism and competition shape defensive investment in a polymorphic eusocial bee. Ecology 97: 417–426

Shackleton K, Balfour NJ, Toufailia HA, Alves DA, Bento JM, Ratnieks FLW (2019) Unique nest entrance structure of *Partamona helleri* stingless bees leads to remarkable 'crash-landing' behaviour. Insect Soc 66: 471–477

Silva MD, Ramalho M (2014) Tree species used for nesting by stingless bees (Hymenoptera: Apidae: Meliponini) in the Atlantic Rain Forest (Brazil): Availability or Selectivity. Sociobiology 61 (4): 415–422

Simone-Finstrom M, Spivak M (2010) Propolis and bee health: the natural history and significance of resin use by honey bees. Apidologie 41 (3):295–311

Siqueira ENL, Bartelli BF, Nascimento ART, Nogueira-Ferreira FH (2012) Diversity and nesting substrates of stingless bees (Hymenoptera, Meliponina) in a forest remnant. Psyche: A Journal of Entomology 2012:370895

Slaa EJ (2006) Population dynamics of a stingless bee community in the seasonal dry lowlands of Costa Rica. Insectes Sociaux 53:70–79

Smith TJ, Mayfield MM (2018) The effect of habitat fragmentation on the bee visitor assemblages of three Australian tropical rainforest tree species. Ecol Evol 8:8204–8216

Starr C, Sakagami S (1987) An extraordinary concentration of stingless bee colonies in the Philippines, with notes on nest structure (Hymenoptera: Apidae: *Trigona* spp.). Insectes Sociaux 34 (2):96–107

Sung I-H, Yamane S, Hozumi S (2008) Thermal characteristics of nests of the Taiwanese stingless bee *Trigona ventralis hoozana* (Hymenoptera: Apidae). Zoological Studies 47 (4):417–428

Sung I-H, Yamane S, Lu S-S, Ho K-K (2011) Climatological influences on the flight activity of stingless bees (*Lepidotrigona hoozana*) and honeybees (*Apis cerana*) in Taiwan (Hymenoptera, Apidae). Sociobiology 58 (3):835–850

Tautz J, Maier S, Groh C, Rossler W, Brockmann A (2003) Behavioral performance in adult honey bees is influenced by the temperature experienced during their pupal development. Proceedings of the National Academy of Sciences of the United States of America 100 (12):7343–7347

Tornyie F, Kwapong PK (2015) Nesting ecology of stingless bees and potential threats to their survival within selected landscapes in the northern Volta region of Ghana. African Journal of Ecology 53 (4):398–405

Torres AR, Sandjo LP, Friedemann MT, Tomazzoli MM, Maraschin M, Mello CF, Santos ARS (2018) Chemical characterization, antioxidant and antimicrobial activity of propolis obtained from *Melipona quadrifasciata quadrifasciata* and *Tetragonisca angustula* stingless bees. Brazilian Journal of Medical and Biological Research 51:e7118

van Benthem FDJ, Imperatriz-Fonseca VL, Velthuis HHW (1995) Biology of the stingless bee *Plebeia remota* (Holmberg): observations and evolutionary implications. Insectes Sociaux 42:71–87

Velez-Ruiz RI, Gonzales VH, Engel MS (2013) Observations on the urban ecology of the Neotropical stingless bee *Tetragonisca angustula* (Hymenoptera: Apidae: Meliponini). Journal of Melittology 1:1–8

Velikova M, Bankova V, Marcucci MC, Tsvetkova I, Kujumgiev A (2000) Chemical composition and biological activity of propolis from Brazilian meliponinae. Zeitschrift für Naturforschung 55:785–789

Venturieri G (2009) The impact of forest exploitation on Amazonian stingless bees (Apidae, Meliponini). Genetics and Molecular Research 8 (2):684–689

Vijayakumar K, Jayaraj R (2013) Infestation of *Pyemotes* sp. (Acari, Pyemotidae) on *Tetragonula iridipennis* (Hymenoptera: Meliponinae) colonies. International Journal for Life Sciences and Educational Research 1:120–122

Villegas-Plazas M, Figueroa-Ramírez J, Portillo C, Monserrate P, Tibatá V, Sánchez OA, Junca H (2018) Yeast and Bacterial Composition in Pot-Pollen Recovered from Meliponini in Colombia: Prospects for a Promising Biological Resource. In: Vit P, Pedro SRM, Roubik DW (eds) Pot-Pollen in Stingless Bee Melittology. Springer International Publishing, Cham, pp 263–279

Vit P, Pedro SR, Roubik D (eds) (2013) Pot-honey: a legacy of stingless bees. Springer, New York

Vit P, Pedro SR, Roubik D (eds) (2018) Pot-Pollen in Stingless Bee Melittology. Springer, New York

Vollet-Neto A, Maia-Silva C, Menezes C, Imperatriz-Fonseca VL (2017) Newly emerged workers of the stingless bee *Scaptotrigona aff. depilis* prefer stored pollen to fresh pollen. Apidologie 48:204–210

Vollet-Neto A, Menezes C, Imperatriz-Fonseca VL (2015) Behavioural and developmental responses of a stingless bee (*Scaptotrigona depilis*) to nest overheating. Apidologie 46 (4):455–464

Vossler FG, Tellería MC, Cunningham M (2010) Floral resources foraged by *Geotrigona argentina* (Apidae, Meliponini) in the Argentine Dry Chaco forest. Grana 49:142–153

Wenseleers T, Ratnieks FL, de F Ribeiro M, de A Alves D, Imperatriz-Fonseca V-L (2005) Working-class royalty: bees beat the caste system. Biology letters 1 (2):125–128

Wille A (1966) Notes on two species of ground nesting stingless bees (*Trigona mirandula* and *T. buchwaldi*) from the pacific rain forest of Costa Rica. Revista de Biologia Tropical 14:251–277

Wille A (1976) Las abejas jicótes del género *Melipona* (Apidae: Meliponini) de Costa Rica. Revista de Biologia Tropical 24 (1):123–147

Wille A (1983) Biology of the stingless bees. Annual Review of Entomology 28:41–64

Wille A, Michener C (1973) The nest architecture of stingless bees with special reference to those of Costa Rica. Revista de Biologia Tropical 21:9–278

Wille A, Orozco E (1975) Observations on the founding of a new colony by *Trigona cupira* (Hymenoptera: Apidae) in Costa Rica. Revista de Biologia Tropical 22 (2):253–287

Winston ML (1987) The biology of the honey bee. Harvard University Press, Cambridge

Wittmann D (1989) Nest architecture, nest site preferences and distribution of *Plebeia wittmanni* (Moure & Camargo, 1989) in Rio Grande do Sul, Brazil (Apidae: Meliponinae). Studies on Neotropical Fauna and Environment 24 (1):17–23

Zucchi R, Sakagami S (1972) Capacidade termoreguladora em *Trigona spinipes* e em algumas outras espécies de abelhas sem ferrão (Hymenoptera: Apidae: Meliponinae). Paper presented at the Homenagem à WE Kerr, Rio Claro

4 无刺蜂分蜂及婚飞

要建立一个新的蜂群，无刺蜂蜂王需要工蜂的帮助。这些工蜂的首要任务是寻找和探查潜在的巢址，在选定的巢址建造蜂巢、收集食物和保卫新家。我们对这个过程的大部分细节仍然知之甚少，Paulo Nogueira-Neto（1954）最早进行了一些开创性工作，而后才有研究者开展无刺蜂分蜂和集体选择巢址的相关研究。因此，导致分蜂的内部和外部条件以及协调这一过程的信号在很大程度上仍然不清楚。然而现有的研究表明无刺蜂和蜜蜂采用了不同的方式来建立子蜂群，主要差异体现在3个方面：一是离巢蜂王的繁殖状况和年龄；二是蜂群建立的时间动态；三是巢址选择中所涉及的沟通过程。在蜂群建立的时间动态上，无刺蜂需要较长时间逐渐形成新蜂群基础（蜜蜂则多是分蜂群成群的单次迁徙），无刺蜂巢址选择中也缺乏明显的招募信号（蜜蜂则采用摇摆舞招募），这些特征使无刺蜂分蜂和巢址选择的研究具有挑战性。

4.1 无刺蜂分蜂

西方蜜蜂蜂群中的工蜂开始饲养替代蜂王后，母蜂王和数千名工蜂就会搬到新家（Winston，1987；Seeley，2010）。而对于无刺蜂来说，离开蜂巢的是未交配的（"处女"）蜂王，而母蜂王则留在母巢里，Hockings（1883）可能是第一个发现这一现象的人（Nogueira-Neto，1954）。造成这种差异的一个原因是无刺蜂王交配后腹部（和卵巢）的尺寸会增加，进而在很大程度上不能飞行（Hockings，1883），这种现象被称为胃囊症或膨腹（第1章；图4.1）。例如 *Scaptotrigona postica* 无刺蜂的蜂王的总体重在完全膨腹时会增加到出生体重的250%（Engels，1987）。膨腹蜂王飞行能力差的另一个原因是因为老蜂王的翅膀经常严重受损（Michener，1974），这可能是由于在幼虫食物供应和产卵过程（Provisioning and Oviposition Process，POP）中蜂王频繁地拍打翅膀造成的（第5章）。这也意味着，如果旧的蜂巢突然变得不适合居住，无刺蜂蜂群不能放弃它们的蜂巢迁移到一个新的地方，而这一弃旧换新的过程在蜜蜂中被称为飞逃。蜜蜂的蜂群飞逃行为是蜜蜂应对天敌攻击、资源稀缺、恶劣天气条件或火灾暴露等因素的一个重要策略（Winston，1987; Oldroyd and Wongsiri，2006）。然而无刺蜂也有例外，Inoue 等（1984a）观察到亚洲 Tetragonula laeviceps 无刺蜂中一个蜂群包括蜂王在内成功飞逃。还有研究者观察到无刺蜂工蜂会不带蜂王飞逃以应对刺激（Portugal-Araujo，1963; Ribeiro and Bego，1994）[1]。Ribeiro 和 Bego（1994）则观察到 Frieseomelitta silvestrii 无刺蜂飞逃的两个案例，在这两个案例中飞逃的工蜂加入了附近的同种蜂群。

图 4.1　不同新热带区无刺蜂的膨腹蜂王

注：A 为 Frieseomelitta flavicornis。B 为 F.varia。C 为 Plebeia minima。D 为 Tetragonisca angustula。E 为 Leurotrigona mualleri。F 为 Melipona flavolineata。G 为 Scaptotrigona depilis。H 为 Friesella schroukyi（摄影：Cristiano Menezes）。

---

[1] Schwarz（1948）提到 Goudot 的报告，他从哥伦比亚的土著人那里得知，有一个可能属于 Tetragona 属的无刺蜂有飞逃习性，土著人因而获得了这种无刺蜂的蜂巢。

另一种假说则是交配过的无刺蜂王因为其他原因才不离开蜂巢，这使得物理学界对无刺蜂有了更多的了解（Peters et al., 1999）。例如 Peters 等（1999）提出，蜜蜂蜂群中老蜂王在子蜂王出现之前离开有利，这能够避免了母女蜂王之间潜在的致命冲突。而与蜜蜂相比，无刺蜂雌性之间的亲缘关系更高（见下文），因此在无刺蜂中母蜂王可能更能容忍子蜂王的存在（然而需要注意的是无刺蜂对处女蜂王的容忍度因蜂种而异，见下文）。这反过来又为无刺蜂蜂群允许子王出房并离开蜂巢建立新蜂群的情况创造了条件。

4.1.1 分蜂阶段

分蜂主要发生在洞巢型无刺蜂蜂种中，因此下文也将集中在这类蜂种上。分蜂的过程大致可以分为 4 个相互重叠的阶段。

（1）侦察和准备阶段

侦察蜂通常花费几天的时间来视察新巢址及其周围环境（Nogueira-Neto, 1954; Inoue et al., 1984b; van Veen and Sommeijer, 2000 a）。在 *Partamona* 属无刺蜂蜂群（最初被称为 *Trigona cupira*，但可能是 *P. peckolti* 或 *P. bilineata*）分蜂的一个案例中，研究者观察到 100 多个侦察蜂 1 d 内不断进出一个潜在巢址，而后才开始物资的运输转移（Wille and Orozco, 1975）。这些侦察蜂很可能会基于诸如入口和洞穴大小或是否有天敌等信息来明确巢址是否合适。尽管侦察蜂对有些洞穴进行了最初的侦察活动，但蜂群最终并不一定搬入其中（Nogueira-Neto, 1954; Inoue et al., 1993），并且研究者观察到蜂群同时侦察了两个洞穴，而后放弃了其中一个（van Veen and Sommeijer, 2000 a）。这表明在分蜂之前，侦察蜂会评估潜在巢址的适宜性，在蜜蜂研究中也详细记录到了这种现象（Lindauer, 1955; Visscher, 2007; Seeley, 2010）。然而无刺蜂研究中尚未实验性地探索出工蜂是采用什么标准来确定一个巢址是否合适的。一旦无刺蜂工蜂认定某一个巢址更好，它们就会开始清理新家（Nogueira-Neto, 1954; van Veen and Sommeijer, 2000 a）。无刺蜂工蜂可能会在巢址附近留下化学标记（在 *Scaptotrigona postica* 和 *Trigona fulviventris* 蜂种中观察到这种情况），从而帮助同伴确定新巢址（Kerr et al., 1962; Hubbell and Johnson, 1977）。来自不同蜂群的侦察蜂偶尔会为了同一个巢穴打架（Hubbell and Johnson, 1977）。生活在白蚁或蚂蚁巢穴中的无刺蜂蜂种，如 *Scura* spp.，工蜂会通过在白蚁巢穴外建立一个蜂巢入口来开始分蜂过程，然后向宿主巢穴里挖掘进而创造一个洞穴（第 3 章）（Roubik, 1989; Sakagami et al., 1989）。

（2）筑巢材料和食物运输阶段

首先无刺蜂工蜂们会使用树脂、黏土或泥浆来填补洞内的裂缝。而后它们会建造一个入口管（Ihering，1903；Michener，1946；Nogueira-Neto，1954；Wille and Orozco，1975），这个入口管可能会作为同巢工蜂的视觉标志（Sakagami et al.，1989；Camargo and Pedro，2003）。守卫蜂接近蜂巢入口可以很快被看到（*Partamona bilineata*，Wille and Orozco，1975；*Tetragonisca angustula*，pers. observation）。然后，工蜂会制作第一个食物罐并在里面装满蜂蜜。这一过程需要源源不断的工蜂从母巢运送巢蜡和蜂蜜（Nogueira-Neto，1954）。Nogueira-Neto（1954）为了确认蜂蜜确实是从母巢带来的，他在母巢中放置了添加了人工色素的蜂蜜。很快他就在分蜂蜂巢里发现了带有颜色的蜂蜜。同样子分蜂蜂巢中的花粉也是工蜂采用花粉筐或者蜜囊以半液体的状态从母巢中携带过去的（Kerr，1951；Nogueira-Neto，1954；Wille and Orozco，1975）。蜜囊携带方式可能需要将花粉与蜂蜜或花蜜混合。这一阶段还包括一些其他活动，如建造内部入口管、巢质片和一个早期的包壳（第3章）。在建造包壳期间或之后，工蜂就会开始建造子脾巢室（Nogueira-Neto，1954）。

无刺蜂采用逐步建造的方式来建造分蜂蜂巢，这就需要工蜂往返于两个蜂巢之间，从而使母群和分蜂群保持着联系。这种社会联系的持续时间在不同的物种和蜂群中差异很大：*Frieseomelitta varia* 无刺蜂为2～3 d、*Tetragonula laeviceps* 无刺蜂为7～20 d、*lebeia mosquito* 和 *Tetragonisca angustula* 无刺蜂约为1个月、*Melipona orbignyi* 和 *Plebeia poecilochroa* 无刺蜂约为2个月，*Trigonisca* sp. 无刺蜂为3～4个月，*Partamona bilineata* 无刺蜂和另一种 *Tetragonisca angustula* 无刺蜂约为6个月（Nogueira-Neto，1954；Terada，1972；Wille and Orozco，1975；Sakagami，1982；Inoue et al.，1984 b；Drumond et al.，1995；van Veen and Sommeijer，2000 a）。而即使分蜂群已经开始产卵以后，工蜂仍然会经常继续从母蜂群中带回资源（Nogueira-Neto，1954）。这就导致分蜂巢中的生物量增加，而母巢的生物量减少（Darchen，1977）。然而，当分蜂群的采集蜂开始从附近的植物上收集花蜜、花粉和树脂时，分蜂群即使与母巢保持着联系也不完全依赖于母巢的资源（Wille and Orozco，1975；Inoue et al.，1984 b；van Veen and Sommeijer，2000 a）。例如，Wille 和 Orozco（1975）估算 *Partamona bilineata* 无刺蜂的分蜂蜂群在资源需求方面对母群的依赖度低于50%，*Tetragonula laeviceps* 无刺蜂则只从母巢运输相对少量的食物到分蜂群中

（Inoue et al.，1984 b）。目前还不清楚为什么这种联系最终会终止，但通过了解两个巢穴的工蜂的死亡，可能会导致两个巢穴的社会完全分离。

大量研究指出无刺蜂蜂群更喜欢住被其他蜂群使用过的洞穴，这种洞穴仍然有来自前居住者的建筑材料和巢结构（Nogueira-Neto，1954;Wille and Orozco，1975）。目前还不清楚蜂群喜欢这种洞穴是因为这种洞穴含有有价值的资源因此更加经济高效，还是因为洞本身质量更好因此被不同的蜂群先后选择。还有一种可能是这种洞穴中已经存在的筑巢材料发出吸引性的气味，从而使这种洞穴更容易被侦察蜂找到（Michener 1946;Nogueira-Neto，1954）。蜜蜂分蜂群也更喜欢搬入已经含有蜡的洞穴（Seeley，2010）。一个极端的例子是无刺蜂会侵占有蜂巢穴，这需要杀死这种巢穴中的原住蜂。研究者在澳洲的 *Tetragonula hockingsi* 无刺蜂蜂群就观察到这种侵占行为，它们侵占了 *T. carbonaria* 无刺蜂的蜂巢（Cunningham et al., 2014），也有研究观察到美洲 *Scaptotrigona postica* 蜂侵占了 *Frieseomelitta freiremaiai* 无刺蜂的蜂巢（Kerr et al., 1962），*Friesella schrottkyi* 蜂侵占了同类蜂的蜂巢（Nogueira-Neto，1954）（第 7 章）。侵占式分蜂在 *Tetragonisca angustula* 无刺蜂中似乎更常见，这种方式甚至可能是 *Lestrimelitta* 强盗蜂唯一的分蜂模式（Sakagami et al., 1993; Grüter et al., 2016）。

无刺蜂分蜂群从选择巢址到筑巢完成且新蜂王产卵之间的时间是长短不一的。*Tetragonisca angustula* 和 *Scaptotrigona postica* 无刺蜂从开始分蜂到产卵只需短短 2 周（Kerr，1951; Kerr et al., 1962; Roubik，1989）。巴西南部的 *Frieseomelitta varia* 无刺蜂完成一次分蜂则只需要 3 d（Terada，1972）。但是 Camargo（引用 Roubik，1989）提到 *Ptilotrigona* sp. 无刺蜂在硬沙基质建筑蜂巢花费了 3 年时间。

一个逐渐形成的蜂群基础可能会导致新生蜂群的生存机会更高，因为这样新蜂群可以不断获得母群的食物供应。事实上，Slaa（2006）发现，对于哥斯达黎加几种无刺蜂来说，新建立的蜂群和已建立的蜂群的生存机会没有差别。而亚洲 *Tetragonula minangkabau* 无刺蜂新建蜂群的存活机则较低（第 1 年存活率为 40%）（Inoue et al., 1993; 更多关于蜂群死亡率的信息见第 1 章），但是这个存活率也仍然远高于已报道的蜜蜂新建蜂群的存活率（Seeley，1978, 2017）。另外，无刺蜂群的繁殖频率低于蜜蜂蜂群，通常一年不到一次（第 1 章）（Roubik，1989; van Veen and Sommeijer, 2000 a）。例如，*Tetragonula Minangkabau* 无刺蜂蜂群每年以约 0.7 的频率繁殖（Inoue et al., 1993）。哥斯达黎加的 *Scaptotrigona pectoralis* 无刺蜂蜂群则每年只繁殖约 0.06 群（Slaa, 2006）。而新热带区的非洲蜜蜂通常每年分蜂几次，欧洲蜜蜂每年分蜂 1～3.6 次（Roubik, 1989; Winston, 1992）。这些估测值结合蜂群死亡率数据（第 1 章）表明，无刺蜂遵循高群体存活率和低群体繁殖率的生活史策略（Slaa, 2006）。

那么无刺蜂分蜂时有多少工蜂会跟随蜂王搬到新蜂巢呢？亚洲 *Tetragonula laeviceps* 无刺蜂蜂群中有 800～1 000 只工蜂（约占工蜂总数的 30%）在分蜂过程中会迁往分蜂巢（Inoue et al., 1984 b）。美国的 *Partamona bilineata* 蜂群工蜂迁移数也与之相似，为 700～800 只（Wille and Orozco, 1975）。而 *Tetragonisca angustula* 无刺蜂分蜂时迁往新巢的工蜂比例则较低，大约为工蜂总数的 10%（500～1 000 只）（van Veen and Sommeijer, 2000 a）。这些数据表明，无刺蜂分蜂时不到一半的工蜂会离开蜂巢去新巢蜂群[2]。非洲 *Hypotrigona* 无刺蜂工蜂颜色会随日龄变化，Darchen（1977）根据 *Hypotrigona* 无刺蜂工蜂年龄相关颜色变化（第 6 章）确定该种无刺蜂分蜂时新老工蜂的参与情况，起初主要是老工蜂参与准备新巢，但很快就有年轻的工蜂也参与进来。我们猜想有很大一部分年轻工蜂参与建巢，因为新蜂群需要相当长的时间才能繁殖自己的工蜂：从处女蜂王到达到第一批工蜂繁殖育成的时间通常会超过 50 d（工蜂发育时间见表 5.4），这一过程所耗时间甚至比很多蜂种中的工蜂的平均寿命还要长（表 1.1）。因此，分蜂群繁殖育成自己的第一代新工蜂之前，分蜂群与母巢保持联系并从中接收年轻工蜂对于分蜂蜂群避免工蜂短缺至关重要。此外，新巢中幼虫食物的准备和供应也需要年轻的工蜂（第 5 章，第 6 章）。

（3）蜂王到达阶段

在最初的准备工作开始后，工蜂们开始建造巢状建筑，一只处女蜂王就会在更多工蜂的陪同下到达新巢址（Wille and Orozco, 1975; Darchen, 1977; Engels and Imperatriz-Fonseca, 1990; Imperatriz-Fonseca and Zucchi, 1995）。一些研究者将此现象描述为蜂团迁移（Schwarz, 1948; Darchen, 1977; Inoue et al., 1984 b; Engels and Imperatriz-Fonseca, 1990）。例如，有研究者观察到 *Tetragonisca angustula* 无刺蜂的处女蜂王从母巢飞到分蜂巢时，有数百只工蜂紧随其后（几秒钟到几分钟后）（van Veen and Sommeijer, 2000 a）。还不清

---

[2] 西方蜜蜂分蜂时大约 2/3 的工蜂会离开蜂群，只留下 1/3 在原蜂巢（Seeley, 2010）。

楚蜂王是如何找到新巢穴的，但化学信号可能在这一过程中发挥了重要作用。然而，在其他蜂种分蜂期间并没有观察到这样的工蜂团迁移现象（Nogueira-Neto，1954）。*Paratrigona subnuda*、*Tetragonisca angustula* 和 *Scaptotrigona postica* 无刺蜂分蜂时则不止一只处女蜂王飞到新蜂巢（Kerr et al., 1962; Imperatriz-Fonseca，1977; Engels and Imperatriz-Fonseca，1990）。其中备用的处女蜂王要么被处决，要么被关在"监禁室"（第3章）（Kerr et al., 1962）。

（4）雄蜂到来阶段

大多数无刺蜂新建蜂巢后，雄性（或雄蜂）就会聚集在新建蜂巢外，或者当蜂巢有新蜂王后，雄性（或雄蜂）也会聚集在蜂巢外（Michener，1946; Nogueira-Neto，1954; Kerr et al., 1962; Engels and Engels，1984; Engels and Imperatriz-Fonseca，1990; Imperatriz-Fonseca and Zucchi，1995; Galindo López and Kraus，2009; Fierro et al., 2011）。雄蜂通常在分蜂开始后的几天内到达，抵达时间通常早于食物罐等巢结构建成之时（Michener，1946; Nogueira-Neto，1954; Kerr et al., 1962）。多种无刺蜂蜂巢入口处附近可以看到雄蜂在停留或飞翔（图4.2），这些雄蜂彼此之间没有太多互动（Roubik，1990; von Zuben，2017，但也有例外，见下文）。

图 4.2 *Tetragonisca angustula*（A）和 *Scaptotrigona depilis*（B）无刺蜂的雄蜂集群

注：照片是这两种蜂的几百只雄蜂正在新建蜂群附近的植物上休息的情况（摄影：C. Grüter）。

无刺蜂雄蜂通常"早出晚归"（Michener，1946; Engels and Engels，1984; Roubik，1990; Cortopassi-Laurino，2007），其活动在午后达到峰值（Kerr et al., 1962）。无刺蜂雄蜂可以在几天内快速聚集并且规模不断增大（Michener，1946; Kerr et al., 1962; Wille and Orozco，1975; Engels and Engels，1984; Roubik，1990）。*Partamona Bilineata* 无刺蜂分蜂时，侦察蜂刚探测

一个新巢穴后的 3 d 内就出现一个由约 800 只雄蜂组成的雄蜂集群，而后在接下来的 4 d 内这群雄蜂的数量才慢慢减少。无刺蜂雄蜂集群中雄蜂的数量从几十只到几千只不等［Kerr 等（1962）和 Fierro 等（2011）发现 *Tetragonisca angustula* 无刺蜂雄蜂集群中的雄蜂数量高达 3 000 只，Bänziger 和 Khamyotchai（2014）则发现 *Tetragonula laeviceps* 无刺蜂雄蜂集群中的雄蜂数量高达 7 000 只］。一只雄蜂会在一个雄蜂集群中逗留几天（Cortopassi-Laurino，2007; Koffler et al.，2016）。有研究发现雄蜂花费在雄蜂集群中的时间与其体型呈负相关，*Scaptotrigona depilis* 无刺蜂雄蜂在雄蜂集群中的时间还与雄蜂的精子特征有关（Koffler et al.，2016）。因此 Koffler 等（2016）推测雄蜂在雄蜂集群中的时长可能与雄蜂质量呈正相关。

  一个雄蜂集群中的雄蜂是来自许多不同的蜂群的，这些蜂群通常位于几百米以外（Kerr et al.，1962; Paxton，2000）。在一项研究中，巴西的 *Tetragonisca angustula* 无刺蜂雄蜂从它们的出生蜂巢飞到雄蜂集群地的距离平均约为 600 m，甚至有一只雄蜂是来自雄蜂集群地约 1 600 m 以外的一个蜂巢（Santos et al.，2016），这对于一只小小的无刺蜂来说是一个相当远的距离[3]。亚洲的 *Tetragonilla collina* 无刺蜂的一个大雄蜂集群是由来自大约 132 个蜂群的雄蜂组成的（Cameron et al.，2004）。*Scaptotrigona mexicana* 无刺蜂的雄蜂集群中则是来自 15～60 个蜂群生产的雄蜂（Kraus et al.，2008; Müller et al.，2012; Sánchez et al.，2018）。由于无刺蜂雄蜂集群中的雄蜂没有一只来自附近的蜂群，由此表明雄蜂会避免在自己的出生巢附近聚集成集群（Müller et al.，2012）。其他研究中也发现了类似的结果，即雄蜂集群中基本上没有处女蜂王的兄弟雄蜂（Cameron et al.，2004; Santos et al.，2016），而且雄蜂集群中的雄蜂之间通常没有血缘关系（Paxton，2000）。相对较短的分蜂距离会有蜂群近亲繁殖的风险，无刺蜂雄蜂这种避免在附近集群的策略则可以降低这种风险（见下文）。有时，雄蜂集群会在蜂王交配后迅速解散（Kerr et al.，1962），而在其他情况下，即使蜂王受精后雄蜂集群也会持续存在许多天（Galindo López and Kraus，2009; Fierro et al.，2011）。吸引无刺蜂雄蜂的信号具有高度的物种特异性，因为在无刺蜂雄蜂集群中只发现了非常小部分的异种雄蜂（*Scaptotrigona postica* 无刺蜂雄蜂集群中异种雄蜂约为 1%，*Tetragonula laeviceps* 无刺蜂雄蜂集群中异种雄蜂约为 1.4%）（Kerr et al.，1962; Bänziger

---

[3] 根据无刺蜂工蜂的体重，雄蜂约为 4mg（Grüter et al.，2012）。

and Khamyotchai，2014）。

　　雄蜂会偶尔暂时离开雄蜂集群去采花进食（例如，*Scaptotrigona postica*、*Melipona favosa* 和 *M. quadrifasciata* 无刺蜂雄蜂）（Kerr et al., 1962; Sakagami, 1982; Sommeijer et al., 2004; Boongird and Michener，2010）。*Nannotrigona* 和 *Scaptotrigona* 属无刺蜂雄蜂会被 *Mormolyca ringens* 兰花吸引，且这些雄蜂会试图与兰花的雌蜂拟态唇瓣交配，在这一过程中雄蜂会为兰花授粉（Singer et al., 2004）。在晚上，无刺蜂雄蜂会躲在植物里或花上，而不是回到它们的出生蜂群（Roubik，1990; Santos et al., 2014 b）。有人观察到无刺蜂雄蜂偶尔会在树枝上成群休息（"睡觉栖架"）（Santos et al., 2014 b）。

　　目前尚未最终确定吸引无刺蜂雄蜂到来的信号的本质和来源。一些人认为是处女蜂王产生了化学信号导致雄蜂集群（Engels and Engels，1984; Engels，1987; Fierro et al., 2011; Verdugo-Dardon et al., 2011），而另一些人则认为无刺蜂工蜂在进行采集等活动时会将一种化学信号带入环境中（Roubik，1990; von Zuben，2017）。研究者采用生物学研究时发现当把雄蜂和蜂王加入交配箱时，处女蜂王对雄蜂具有吸引力，这一现象支持了蜂王气味导致雄蜂群聚的观点（de Camargo，1972; Engels and Engels，1988; Fierro et al., 2011; Veiga et al., 2017）。Engels 和他的同事（Engels et al., 1990,1997）在 *Scaptotrigona postica* 无刺蜂蜂王的头部分泌物中鉴定出了对雄蜂有吸引力的挥发物。他们发现蜂王产生的不同化合物在不同的距离上对雄蜂具有吸引力：仲醇在几米的距离外都具有吸引力，而 2-酮类化合物则仅在几厘米的距离以内才具有吸引力并诱导雄蜂尝试交配。*S. mexicana* 无刺蜂蜂王对雄蜂吸引力似乎也是由蜂王头部产生的物质引起的（Verdugo-Dardon et al., 2011）。特别是仲醇 2-壬醇很可能是吸引雄蜂的关键物质。而 *Tetragonisca angustula* 无刺蜂中一种吸引雄蜂的化合物——己酸异丙酯，则主要来源于蜂王的腹部（Fierro et al., 2011）。

　　也有一些观察结果对于这些信号负责远距离吸引雄蜂这一观点提出了质疑。通常，测试蜂王的吸引力的生物实验是在相对较短的距离（从几厘米到几米）范围内，Fierro 等（2011）则发现蜂王对雄蜂的吸引力会随着距离的增加迅速下降，且不会超过 20 m。此外，还有一些报道提出在未交配的蜂王迁入新巢时，雄蜂就已经开始到达新巢附近集群了（Kerr et al., 1962; Roubik，1990; Imperatriz-Fonseca and Zucchi，1995; Velthuis et al., 2005）。例如，Kerr 等（1962）提到一群 *Scaptotrigona postica* 无刺蜂雄蜂在蜂群分蜂 3 d 后就到达这个新巢。处女蜂王则是之后两天才到达这里。因此，研究者认为无刺蜂工蜂会产生一种远距离吸引雄蜂的信号，一旦第一批雄蜂到达，这些雄蜂自己就会吸引更多的雄蜂（Roubik，1990; Imperatriz-Fonseca and Zucchi，1995）。研究者发现如果阻止 *Scaptotrigona depilis* 无刺蜂工蜂出外采集花蜜，就不会有雄蜂被吸引到这些独立蜂巢（von Zuben，2017），这表明了工蜂具有重要作用。在没有蜂王的蜂群中，采集蜂可能自己会产生一个信号，或者通过与一个能吸引雄蜂的处女蜂王接触来获得信号，然后把这种信号带到环境中去（Engels and Imperatriz-Fonseca，1990; von Zuben，2017）。在花丛中巡逻或觅食的雄蜂能够发现这种带有信号的工蜂，并跟着它们到达蜂巢（Roubik，1990; von Zuben，2017）。*Melipona favosa* 无刺蜂的雄蜂就有趋向工蜂的行为，这进一步表明工蜂信息素在雄蜂集群建立中起着重要作用（Sommeijer et al., 2004）。然而研究者尚未识别出这样的工蜂信号，*Scaptotrigona mexicana* 无刺蜂的工蜂甚至无法吸引雄蜂[4]，但它们会被其他雄蜂所吸引（Galindo López and Kraus，2009）。这些雄性信号似乎会留在雄蜂集群的表面，一旦雄蜂集群开始形成就会提供额外的正反馈效应（Galindo López and Kraus，2009）。

Melipona 属无刺蜂的一些蜂种则是例外，它们的雄蜂集群是根据蜂巢而定的（Michener，1946; van Veen et al., 1997; Sommeijer et al., 2004; Cortopassi-Laurino，2007）。但是 *M. favosa* 中雄蜂集群由几百只雄蜂组成，这又似乎并不与特定巢相关（Sommeijer and De Bruijn，1995; Sommeijer et al., 2004）。有些蜂王显然是被雄蜂（也可能是工蜂）产生的气味所吸引，它们会在特定的一天造访雄蜂集群（Sommeijer et al., 2004）。这种情况与蜜蜂的情况相似，雄蜂和蜂王在雄蜂集群区相遇，这一位置可能距离雄蜂和蜂王的本源蜂群几千米远（Winston，1987; Ruttner，1988）。在这些 *Melipona* 属无刺蜂的雄蜂集群中，雄蜂相互之间斗争激烈，但也经常会进行交哺活动。飞翔的雄蜂和蜂王间的吸引力以及蜂王着陆后的行为表明，雄蜂"休息处"上覆盖着有吸引力的信息素（Sommeijer and De Bruijn，1995）。Michener（1946）在巴拿马观察到 *Melipona phenax* 无刺蜂类似的雄蜂集群点（drone aggregation sites，DAS），但 DAS 在 *Melipona* 属无刺蜂的普遍程度仍然未知。例如 *Melipona*

---

[4] 应该注意的是这里提到的工蜂并不是专门指接受新蜂王的蜂群中的工蜂。那么工蜂可能只有当蜂群有新蜂王时才能产生或携带对雄蜂有吸引力的气味。

costaricensis 和 *M. flavolineata* 无刺蜂[5] 遵循更常见的依巢而定的雄蜂集群模式（van Veen et al., 1997; Veiga et al., 2018）。这种无刺蜂在非依巢而定的雄蜂集群中尚未观察到交配行为。

（5）婚飞

无刺蜂雄蜂集群中的雄蜂不会进入蜂群内部，而是在蜂群外面等待直到蜂王出巢婚飞（Roubik，1990）。Nogueira-Neto（1954）认为在无刺蜂中这是一种普遍存在的模式。这表明处女蜂王在最初的迁徙飞行中可能对雄蜂没有吸引力（Engels and Engels 1988; Veiga et al., 2017; van Veen and Sommeijer, 2000 b）。无刺蜂蜂王通常是在 1～14 日龄时婚飞（da Silva et al., 1972; Vollet-Neto et al., 2018）。

尽管人们很少观察到无刺蜂的婚飞行为，但研究者认为未交配的无刺蜂王只会进行一次婚飞，并其余生都是使用在这次飞行中获得的精子（第 4.6 节）（Engels and Imperatriz-Fonseca，1990; Imperatriz-Fonseca and Zucchi，1995; Vollet-Neto et al., 2018）。da Silva 等（1972）详细观察描述了 *Melipona quadrifasciata* 无刺蜂自然换王过程中（第 4.4 节）婚飞前的情况。他们观察到在 *Melipona quadrifasciata* 无刺蜂婚飞开始之前，蜂王的腹部开始变大，工蜂们则组成了一个蜂王护卫队（图 4.3）。蜂王婚飞前的时间段正好对应了卵黄原蛋白（卵黄原蛋白的前体）产量增加的时间段[6]（Engels and Engels，1977; Engels，1987）。*M. quadrifasciata* 无刺蜂婚飞前的一两天，可以在蜂巢入口或入口管附近看到蜂王，或者蜂王会在巢内从一个巢脾或食物罐短距离飞行到另一个巢脾或食物罐（da Silva et al., 1972）。然后，蜂王会离开蜂巢，并在离巢约半米远的地方转身面向蜂巢入口。接着蜂王会绕着越来越大的圆圈飞行，这可能是为了了解蜂巢入口和周围区域的视觉特征。研究者观察到 *Tetragonisca angustula* 无刺蜂有类似的处女蜂王定向飞行，但在 *Melipona beecheii* 无刺蜂中则没有观察到这种行为（van Veen and Sommeijer，2000 b）。*M. quadrifasciata* 无刺蜂婚飞行为会持续 30～100 min（da Silva et al., 1972），*M. quinquefasciata* 无刺蜂则约为 50 min（de Camargo，1972），*M. beecheii* 无刺蜂为

---

[5] 这个蜂种的独特之处在于它有两个不同颜色的雄性变种，浅色（约 80% 的新出房雄蜂）和深色（约 20% 雄蜂）（Veiga et al., 2018）。

[6] 关于卵黄原蛋白（vg）和无刺蜂繁殖之间的联系我们知之甚少。卵黄原蛋白似乎并不直接作用于无刺蜂繁殖，因为在无生育能力的工蜂中 vg 也有表达（Dallacqua et al., 2007），但在经常产卵的 *Melipona quadrifasciata* 无刺蜂工蜂中却没有发现 vg（Paes de Oliveira et al., 2012）。

5～40 min（van Veen and Sommeijer，2000 b），*Paratrigona subnuda* 无刺蜂约为 30 min（Imperatriz-Fonseca，1977），*Tetragonisca angustula* 无刺蜂为 2～20 min （van Veen and Sommeijer，2000 b），*Plebeia droryana* 无刺蜂约为 15 min（da Silva, 1972），*Melipona favosa* 无刺蜂则仅为 2～8 min（Sommeijer et al., 2003 b） [关于交配飞行时间的更多详细内容参见 Vollet-Neto 等（2018）]。

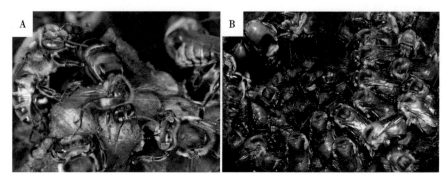

图 4.3 A. *Melipona beecheii* 无刺蜂工蜂正在攻击处女蜂王（Jarau et al., 2009）；
B. *Melipona flavolineata* 无刺蜂工蜂在其接受的蜂王周围形成蜂王护卫队
（摄影：Cristiano Menezes）

Tetragonisca angustula 无刺蜂婚飞行为通常发生在处女蜂王到达新巢的当天或第二天（van Veen and Sommeijer，2000）。研究者观察到当该蜂种的处女蜂王离开蜂群交配时，雄蜂扑向蜂王后，蜂王会降落到地面上，开始交配（Imperatriz-Fonseca et al., 1998）。Boongird 和 Michener（2010）研究了亚洲 *Tetragonula* 和 *Heterotrigona* 属无刺蜂的婚飞，他们发现这两种无刺蜂交配通常发生在处女蜂王降落到蜂巢入口附近之后。在他们的研究中，在黎明之后会发生交配行为，蜂王在交配期间会经常进入蜂巢中，有时一些其他雄蜂会跟随其后。这些观察结果似乎与新热带区无刺蜂种中得到的观察结果不同，但尚不清楚这些观察结果是否表明了不同支系中存在着不同的交配模式。

目前我们尚不清楚在无刺蜂的交配中配偶选择是否有作用。无刺蜂蜂王在交配过程中会试图用力甩脱雄蜂，从而避免选择到条件较差的雄蜂（Smith，2020），而雄蜂也会更强烈地被质量较高的蜂王吸引，例如雄蜂会通过化学信号来衡量蜂王质量。例如，*Scaptotrigona depili* 无刺蜂雄蜂会在具有大量生物量蜂群的独立蜂巢前聚集更多（von Zuben，2017），这意味着群体和（或）蜂王的整体质量都会影响对雄蜂的吸引力。因为无刺蜂雄蜂生殖器在交配过程中会被撕裂，因此雄蜂仅能交配一次（Engels and Imperatriz-Fonseca，1990; Veiga et al., 2018; Smith, 2020）。

Melipona quadrifasciata 无刺蜂蜂王婚飞归来时，雄蜂生殖器官仍附着在蜂王上（Kerr et al., 1962），而蜂王自己通常能够在 1 h 内（Vollet-Neto et al., 2018）用后肢移除雄蜂的交配器官（"交配标志"）（da Silva et al., 1972）。而其他无刺蜂蜂种的蜂王则需要工蜂的帮助来移除雄蜂的交配标志（Colonello and Hartfelder, 2005）。交配标志是由雄蜂破裂的生殖器囊构成的（Colonello and Hartfelder, 2005）。*Tetragonisca angustula* 无刺蜂蜂王交配后需要 2～7 d 才会产卵，*Melipona quadrifasciata* 无刺蜂需要 6 d，*Melipona rufiventris* 无刺蜂需要 16 d，*Melipona bicolor schenckis* 无刺蜂需要 30 d，*Plebeia droryana* 无刺蜂需要 40 d（Kerr, 1948, 1951; Kerr et al., 1962; Terada et al., 1975; da Silva, 1972; van Veen and Sommeijer, 2000 a; Vollet-Neto et al., 2018）。然而，应该注意的是，即使是同一蜂种的蜂王从交配到产卵所需要的时间也有相当大的差异。以 *M. quadrifasciata* 无刺蜂为例，蜂王从交配到产卵开始的时间可以是 3～20 d 不等（Kerr, 1951; Silva et al., 1972）。

Lestrimelitta ehrhardti 和 *Schwarziana* sp. 无刺蜂蜂王即使交配后仍然能够吸引雄蜂，当蜂巢被打开时雄蜂会试图与膨腹蜂王交配（Sakagami and Laroca, 1963; Sakagami, 1982）。同样，研究者生物实验发现 *Tetragonisca angustula* 无刺蜂的膨腹蜂王与处女蜂王对雄蜂的吸引力相同，且这两种类型的蜂王具有相似的化学特征（Fierro et al., 2011）。然而对于 *Scaptotrigona mexicana* 无刺蜂而言，处女蜂王对雄蜂的吸引力是强于膨腹蜂王的（Verdugo-Dardon et al., 2011）。在其他社会性昆虫中，通常雌性蜂王在交配后就不能再吸引雄性（Oppelt and Heinze, 2009）。这表明一些无刺蜂的蜂王会持续产生性信息素，这可能是因为性信息素在蜂群组织中有其他功能。

4.2 分蜂距离

无刺蜂的分蜂距离对于觅食竞争、种群遗传和无刺蜂对栖息地变化的反应能力都有重要意义（第 2 章）。无刺蜂的分蜂距离往往比蜜蜂短得多。蜜蜂的分蜂蜂群经常会迁移到 1 000 m 开外的新巢，有时甚至会迁移几千米以外（Lindauer, 1955; Schmidt, 1995; Seeley and Buhrman, 1999）。而 *Tetragonisca angustula* 无刺蜂的分蜂距离仅为 2～280 m（Kerr, 1951; van Veen and Sommeijer, 2000 a）。*Tetragonula laeviceps* 无刺蜂蜂群甚至只迁徙约 20 m

就到新巢址（Inoue et al., 1984），而研究者观察到的 *Partamona bilineata*（Wille and Orozco，1975）、*Melipona orbignyi*（Nogueira-Neto，1954）和 *Plebeia poecilochroa*（Drumond et al., 1995）无刺蜂的分蜂距离分别为 100 m、75 m 和 40 m。前一章中提到在同一株树或同一白蚁穴中有时会发现多个同种蜂的蜂巢（第 3 章）的现象，那么这就证实分蜂距离低于几米对无刺蜂来说也并不罕见。研究者发现 *Plebeia* sp. 无刺蜂的一个蜂群在母巢的 365 m 外建立新巢穴，这是无刺蜂研究中发现的最长分蜂距离（Nogueira-Neto，1954）。对于无刺蜂来说，分蜂距离再远一点可能是不现实的，这是因为能量消耗的原因（因为工蜂会在新巢和旧巢之间来回飞行，所以能量消耗会随着时间的推移不断积累），或者由于附近有大量的可用巢穴，分蜂距离再远一点也没有必要。当然，也有可能是因为研究者难以观察无刺蜂向更遥远的洞穴分蜂，因此文献中对远距离分蜂的案例报道不足（van Veen and Sommeijer，2000 a）。

4.3 分蜂时期

对于无刺蜂来说，热带和亚热带环境在气候和觅食条件上具有季节性，因此人们猜测无刺蜂的分蜂时期也遵循着一定程度的季节性。例如，von Ihering（1903）在巴西圣保罗州观察到无刺蜂分蜂群在夏秋季节更多。这两个季节的雨水也较多。在巴拿马和哥斯达黎加这两个雨季和旱季比较明显的国家，无刺蜂分蜂行为在旱季比在雨季更频繁（Michener，1946; Slaa，2006）。Michener（1946）认为发生这种现象是因为蜂群更倾向于在阳光明媚的时候分蜂，而 Slaa（2006）则认为是因为旱季有更多的食物来源才导致这种现象。有趣的是，Peckolt [1894，引自 Schwarz（1948）] 在巴西南部观察到了 *Trigona spinipes* 无刺蜂[7]在夏末的雷暴期间分蜂的案例。印度南部的 *Tetragonula iridipennis* 无刺蜂蜂群通常在雨季结束时（8—9 月）分蜂（Schwarz，1948）。在赤道非洲加蓬的 *Hypotrigona* 属无刺蜂蜂群中也发现了明显的季节性分蜂，蜂群通常在 11 月至翌年 5 月开始分蜂，在 12 月和翌年 1 月达到高峰，这一时间段刚好对应于雨季结束和旱季开始（Darchen，1977）。这段时间有良好的觅食条件，蜂群利用这段时间进行大量的食物储备。这个区域的意大利蜜蜂（*Apis mellifera*）也在这个时间段分蜂，表明这两个类群需

---

[7] 该蜂种等同于 *Trigona ruficrus* 无刺蜂，但是根据 Camargo 和 Pedro 研究（2013, Moure's Bee Catalogue）该蜂种尚未被鉴别。

要相似的生态条件来支持分蜂（Darchen，1977）。

相反，Nogueira-Neto（1954）在圣保罗州观察到不同的无刺蜂蜂种在不同的月份分蜂，并没有特定的分蜂季节。例如，*Nannotrigona testaceicornis* 和 *Plebeia mosquito* 无刺蜂在四季都会分蜂。只有在冬天最冷的月份（6月）才没观察到这些无刺蜂。苏门答腊岛的 *Tetragonula minangkabau* 无刺蜂蜂群分蜂似乎也没有表现出很强的季节性，但在非常干燥或非常多雨的天气里较少观察到这种无刺蜂分蜂（Inoue et al.，1993）。

目前研究者尚未深入研究出促使无刺蜂分蜂的因素，但是 van Veen 和 Sommeijer（2000 a）观察到分蜂的 *Tetragonisca angustula* 蜂群比未分蜂的蜂群群势更强。食物储存量可能是触发无刺蜂蜂群分蜂的因素之一（Roubik，1982），但需要更多的研究来证实这个科学猜想。蜜蜂的繁殖性分蜂通常发生强群增长和相应的食物储备积累之后（Seeley and Visscher，1985; Winston，1992）。工蜂可以通过工蜂密度增加、蜂王信号的减少或直接检查大食物罐来感知蜂群强度（Seeley，1979）。

4.4 换王

当年老的蜂王将死之时会产生二倍体雄蜂（见下文），待精子耗尽或生产力不高时，老蜂王就会被一个新蜂王取代。没有生产力的蜂王通常会被工蜂杀死，工蜂杀王有时是在处女蜂王的帮助下完成的（Imperatriz-Fonseca，1978; Engels and Imperatriz-Fonseca，1990; Imperatriz-Fonseca and Zucchi，1995）。例如，研究者观察到 *Paratrigona subnuda* 无刺蜂的处女蜂王会通过在膨腹蜂王上释放成分未知的下颚腺分泌物来启动换王（Imperatriz-Fonseca，1977, 1978）。这会引发工蜂去舔触蜂王并随后清除处女蜂王或膨腹蜂王。研究者还观察到 *Plebeia droryana*、*P. remota* 和 *Frieseomelitta languida* 无刺蜂的处女蜂王会通过攻击膨腹蜂王来促进换王（da Silva，1972; Imperatriz-Fonseca and Zucchi，1995; van Benthem et al.，1995）。新蜂王偶尔会在老蜂王去世之前就被工蜂所接纳（da Silva et al.，1972; Imperatriz-Fonseca，1978），这种情况导致无刺蜂蜂群中可以短时间内有一个以上的产卵蜂王（图4.4）（Engels and Imperatriz-Fonseca，1990; Alves et al.，2011）。Imperatriz-Fonseca（1978）观察到无刺蜂换王现象在分蜂过程中更为常见。这可能是由于在分蜂过程中膨腹蜂王和处女蜂王之间的竞争加剧，或者是由于工蜂在分蜂过程中对处女蜂王的容忍度更高。在自然条件下换王的发生频率尚未被量化[8]。

图 4.4　有两个膨腹蜂王的 *Melipona flavolineata* 无刺蜂蜂群
（摄影：Cristiano Menezes）

注：照片上也展示了空的打开的巢室和一个装有育虫食物的打开的巢室。

4.5　蜂王的产生和选择

无刺蜂和蜜蜂的一个主要区别是无刺蜂会常年繁育蜂王，而西方蜜蜂只在需要时才繁育蜂王。特别是 *Melipona* 属无刺蜂就以繁育大量蜂王而广为人知，这些繁育出的蜂王的大多数在它们出房后不久就会被工蜂杀死（Kerr，1946；Kerr et al., 1962；Imperatriz-Fonseca and Zucchi, 1995；Koedam et al., 1995；Wenseleers et al., 2004；Wenseleers and Ratnieks, 2004；Ratnieks et al., 2006）。其他无刺蜂蜂种也会在一年中大部分时间里都在巢中保留备用的处女蜂王（图 3.14；第 3.2.4 节）（Kerr et al., 1962；Inoue et al., 1984 b；Engels and Imperatriz-Fonseca, 1990；Imperatriz-Fonseca and Zucchi, 1995）。无刺蜂蜂群中繁育的蜂王数量取决于蜂种、蜂群大小、觅食条件或膨腹蜂王的繁殖状况（第 5.4.3 节）（Engels and Imperatriz-Fonseca, 1990；Imperatriz-Fonseca and Zucchi, 1995；van Veen et al., 2004）。研究者在一个强大 *Scaptotrigona postica* 无刺蜂蜂群中观察到了多达 13 只处女蜂王（Engels and Imperatriz-Fonseca, 1990），而 von Ihering（1903）报道在一个未知蜂种的无刺蜂蜂群中出现了 24 只处女蜂王。*Trigona crassipes* 无刺蜂的一个蜂群中甚至在巢室中繁育了 132

---

[8] 雄蜂群出现在蜂群前的频率可以作为估测换王频率的一种间接方法。

只蜂王（Camargo and Roubik，1991），但尚不清楚这些蜂王中有多少能存活下来。

无刺蜂蜂群中持续存在备用蜂王的一个原因可能是无刺蜂繁育新蜂王所需时间比蜜蜂更长（第5章；表5.4）。因此，拥有备用蜂王可能是一种防止失王的保险策略（Engels and Imperatriz-Fonseca，1990）。然而有报道称一些无刺蜂蜂种能够在相对较短的时间内繁育出应急蜂王（Faustino et al., 2002; Nunes et al., 2015; Luz et al., 2017）。*Frieseomelitta varia* 无刺蜂蜂种是一种建簇状巢的蜂种（第3章），这种无刺蜂独立蜂群的工蜂会在装有晚期幼虫的巢室旁边构建含有食物的辅助巢室，然后它们会在辅助巢室和含有幼虫的巢室之间建立连接，使得幼虫能够享用额外份额的食物（第5章）（Faustino et al., 2002; Luz et al., 2017）。关于无刺蜂多个蜂王同群存在还有另一个假设，即多王存在可以使得无刺蜂工蜂能够不断比较不同的蜂王，来保证一个高质量蜂王的存在（Koedam et al., 1995）。

新蜂王出房对工蜂来说是一件潜在的重大事件。新出房蜂王能够成为生殖蜂王，从而有助于工蜂的进化适应性（第4.5节）。相反，在蜂群不需要蜂王时也会有新蜂王出生。那么这时候新出房蜂王是如何被工蜂对待的呢？在一些蜂种（一些 *Melipona* 无刺蜂）中工蜂最初对新出房蜂王没有什么兴趣，但在这些雌蜂出房几天后工蜂就开始有反应（Imperatriz-Fonseca and Zucchi，1995; Souza et al., 2017）。而在其他蜂种中，如 *Meliponula bocandei*、*Tetragonula laeviceps* 和 *Plebeia remota* 无刺蜂的处女蜂王一出房就会立即引起工蜂的反应（友好或攻击）（Sakagami et al., 1977; Inoue et al., 1984b; van Benthem et al., 1995）。*Melipona beecheii* 无刺蜂是一种繁育许多蜂王的无刺蜂蜂种，这种无刺蜂的处女蜂王还在尝试出巢室时就开始受到攻击（van Veen et al., 1999），而且据估测新出房的 *M. beecheii* 蜂王平均存活时间为 1 d（Wenseleers et al., 2004）。*Plebeia remota* 无刺蜂中多余的蜂王通常在出房后几小时内就被工蜂杀死（van Benthem et al., 1995）。工蜂可能是通过蜂表皮上特定的碳氢化合物混合物来识别蜂王的（Abdalla et al., 2003; Kerr et al., 2004; Nunes et al., 2009b; Ferreira-Caliman et al., 2013; Souza et al., 2017）。蜂王和工蜂的表皮碳氢谱（CHC）差异会随出房时间的延长而增加，例如 *Melipona bicolor* 和 *M. scutellaris* 无刺蜂就是如此（Abdalla et al., 2003; Souza et al., 2017），这种现象可以用来解释工蜂对蜂王行为的变化。

当无刺蜂蜂群需要一个新王时，工蜂对新出房蜂王的容忍度就会上升，此时新出房蜂王的生存机会就会增加（Kärcher et al., 2013），尽管如此，独立蜂群仍然可能会在一段时间内继续杀死处女蜂王（da Silva et al., 1972; van Veen et al., 1999; Kärcher et al., 2013）。鉴于无刺蜂蜂群中通常有好几个处女蜂王可供选择，那么工蜂是如何选择出它们的下一任蜂王呢？Imperatriz-Fonseca（1977）提到在某些蜂种中新出房蜂王会散发出一种强烈的气味，甚至连人类观察者都能闻到，若通过处女蜂王周围工蜂的数量来衡量处女蜂王的吸引力，则发现这些新出房的处女蜂王对工蜂的吸引力差异很大。这些蜂王的气味可能传达了关于处女蜂王品质的信息，但研究者尚未鉴别出这其中所涉及的信息素（Jarau et al., 2009）。Cruz-Landim 等（1980）发现蜂王对工蜂的吸引力与其背腺的分泌活动相关，特别是蜂王背部两侧的腺体。这表明这些腺体产生的化学物质可能具有信息素功能（Cruz-Landim et al., 2006）。有报道称年轻的蜂王在走动的时候会膨胀腹部并暴露背囊，这与上述这一假说相一致，这可能是为了释放一种信息素（Engels and Imperatriz-Fonseca，1990; Imperatriz-Fonseca and Zucchi, 1995; van Veen et al., 1999; Jarau et al., 2009; Kärcher et al., 2013）。*Melipona beecheii* 无刺蜂中最终被工蜂接纳的蜂王通常腹部膨胀的时间更长（van Veen et al., 1999）。有研究观察到工蜂经常攻击未被选中的蜂王的腹部（图 4.3），这进一步证明腹腺产生的化学物质在工蜂选王过程中发挥了作用（Engels and Imperatriz-Fonseca，1990; van Veen et al., 1999; Jarau et al., 2009）。

也有研究者猜测无刺蜂蜂王的表皮碳氢化合物会影响工蜂对蜂王的行为（Abdalla et al., 2003; Souza et al., 2018）。例如 *Friesella schrottki* 无刺蜂的蜂王体表的碳氢化合物具有信息素功能，能抑制工蜂卵巢，从而调节该物种的生殖劳动分工（Nunes et al., 2014）。Nunes 等（2014）接着提供了进一步的证据，即腹腺（可能是杜福尔斯腺）释放了蜂王信息素。蜂王的腹部运动似乎是与工蜂的仪式化互动中的重要部分。然而，他们的数据也表明来自不同腺体和身体部位的复杂的混合化学成分影响了工蜂的行为和生理。综上所述，这些研究表明腹腺产生的信息素在蜂王选择过程中发挥了重要作用，但这些研究也暴露出我们对介导蜂王选择的信号的理解尚不充分（关于其他社会性昆虫群体中蜂王信号的信息参见 van Oystaeyen et al., 2014; Oi et al., 2015; Oliveira et al., 2015）。

Schwarziana quadripunctata 无刺蜂是一种会同时繁育大体型蜂王和小体型蜂王的蜂种（第 5 章），工蜂似乎更喜欢大蜂王，而小蜂王则占所有新出

房蜂王的 80% 以上，新出房的小蜂王大部分会被工蜂杀死，而大蜂王则被关在"监禁室"里（Engels and Imperatriz-Fonseca，1990; Wenseleers et al., 2005; Ribeiro et al., 2006）。因此，该蜂种中只有大约 20% 的蜂群是以小蜂王为首的（第 5 章）（Wenseleers et al., 2005）。由于小蜂王的繁殖力较低，因此工蜂可能会对蜂王的化学、行为或形态上的生育指标作出应对（Camargo，1974; Wenseleers et al., 2005）。

处女蜂王的行为也会影响蜂群对蜂王的选择。*Melipona beecheii* 无刺蜂工蜂不攻击刚被杀的蜂王或假死的蜂王（即六足紧贴身体躺着不动），这表明蜂王过激行为可能会引发工蜂的攻击行为（van Veen et al., 1999; Jarau et al., 2009）。处女蜂王经常试图通过奔跑和在受到攻击时快速转身来避免被工蜂杀死。最重要的是它们会藏在蜂巢的外围或空的储物罐中（Engels and Imperatriz-Fonseca，1990; van Veen et al., 1999; Jarau et al., 2009）。Kärcher 等（2013）发现 *Melipona quadrifasciata* 无刺蜂的蜂王生存和工蜂接受度相关性最相关的行为是处女蜂王的躲藏时间和避免战斗的能力。在失王后几小时到几天里是工蜂接受新蜂王的关键时期。上述这些因素增加了处女蜂王在这个时期里的存活机会（da Silva et al., 1972; Imperatriz-Fonseca，1977; van Veen et al., 1999; Sommeijer et al., 2003b; Kärcher et al., 2013）。研究者还观察到处女蜂王也会形成小群体并建立统治等级（Engels and Imperatriz-Fonseca，1990）。

近期研究发现，*Melipona* 属无刺蜂新出房蜂王会通过离巢来躲避工蜂的攻击。离开蜂群的 *M. favosa* 无刺蜂新出房处女蜂王有 57% 得以存活（Sommeijer et al., 2003 b）。这些处女蜂王还会试图潜入同种蜂群去取代产卵蜂王（Sommeijer et al., 2003 b; Sommeijer et al., 2003 a; Wenseleers et al., 2011; van Oystaeyen et al., 2013）。如果在母巢中，一个处女蜂王取代产卵蜂王的机会是非常小的，那么从蜂群和蜂王的角度来看这种策略是有益的。有证据表明，这些离开蜂群飞散的处女蜂王会优先入侵无蜂王的蜂群（据估测，在一个区域内就有 8.6% 的 *M. scutellaris* 无刺蜂蜂群为无蜂王群，Wenseleers et al., 2011），且它们会在晚上蜂巢入口守卫不那么森严时尝试进入其他巢穴（van Oystaeyen et al., 2013）。这看起来是一个相对成功的策略，因为在 Wenseleers 等（2011）和 van Oystaeyen 等（2013）的研究观察中发现，25%～37.5% 的 *M. scutellaris* 无刺蜂蜂王在换王过程中被外来蜂王所取代。Sommeijer 等（2003 a，2003 b）认为这种社会寄生的现象可以部分解释为什么 *Melipona* 属无刺蜂蜂群会产生如此多的蜂王（补充解释见第 5 章）。处女蜂王在寻找无蜂王的蜂巢入侵的同时，能够通过采花进食（在 *M. favosa* 无刺蜂中观察到该现象）而存活很长一段时间（Sommeijer et al., 2003 a）。

一旦无刺蜂蜂王被蜂群接纳，该蜂王自己就可能加入攻击其他处女蜂王的行列（da Silva et al., 1972）。被选中的蜂王会频繁从工蜂那里获得食物，而在几个无刺蜂蜂种中也观察到蜂王向工蜂提供未知的黄色液体，这反过来增加了工蜂对新选蜂王的饲喂（da Silva et al., 1972; Engels and Imperatriz-Fonseca, 1990）。当 *Melipona quadrifasciata* 无刺蜂蜂群明显选定了蜂王后，工蜂会聚集在被选中的蜂王周围形成蜂王护卫队（Kärcher et al., 2013）。然后蜂王会表现出一种类似舞蹈的行为（类似于蜜蜂的"圆舞"，见 von Frisch, 1967）。在舞蹈中，蜂王会摇动膨胀的腹部并用其触碰工蜂，这可能是在向工蜂发出蜂王存在的信号。这场"登基舞会"后不久，蜂王和工蜂们都会冷静下来。*Scaptotrigona postica* 无刺蜂选定的蜂王则是在子巢区顶部表演这种"登基舞"（Engels and imperiz-Fonseca, 1990）。*S. postica* 无刺蜂蜂王的"登基舞"会表演 1～2d，并引起工蜂的兴奋。有人推测蜂王会通过信息素来塑造其在蜂群中的存在感，这些信息素存在于蜂王的粪便里并被工蜂吃掉（Sakagami et al., 1977; Engels and Imperatriz-Fonseca, 1990; Kleinert, 2005）。

4.6　无刺蜂中的单雄交配

无刺蜂蜂王通常只与一只雄蜂交配，或者只使用一只雄蜂的精子（Kerr et al., 1962; da Silva et al., 1972; Peters et al., 1999; Strassmann, 2001; Green and Oldroyd, 2002; Palmer et al., 2002; Tóth et al., 2002, 2004; Jaffé et al., 2014; Vollet-Neto et al., 2018, 2019）。因此无刺蜂中工蜂间的亲缘关系估算值通常与膜翅目昆虫预测的姐妹亲缘关系值（$r = 0.75$）相差无几（图 4.5）（Ratnieks et al., 2006）。

单雄交配最早是根据精子数量推测出来的：Kerr 等（1962）发现 *Melipona quadrifasciata* 无刺蜂的雄蜂平均有 1 156 850 个精子细胞，而刚交配的蜂王输卵管和受精囊中约有 100 万个精子细胞[9]。然而精子计数本身并不准确，已有人应用分子工具证明 20 多种无刺蜂的有效亲子关系值都接近于 1（Peters et al., 1999; Strassmann, 2001; Palmer et al., 2002; Tóth et al., 2002,

---

[9] 这显著低于 *Apis mellifera* 蜜蜂（其受精囊中约有 500 万个精子，Baer, 2005）。

2004; Jaffé et al., 2014）。因此，在无刺蜂群中雌蜂之间的亲缘关系较蜜蜂更为密切，蜜蜂蜂王则会与许多雄蜂交配且蜜蜂工蜂之间的亲缘关系约为 0.3（Strassmann，2001; Ratnieks et al., 2006）。

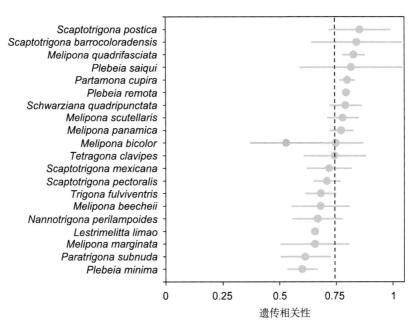

图 4.5　20 种无刺蜂的同巢蜂间的平均亲缘关系和 95% 置信区间

注：由于 *Lestrimelitta limao* 无刺蜂的样本量较小，因而该蜂种没有标准差。灰色的点代表了 *Melipona bicolor* 蜂群中的多蜂王同巢蜂群（数据源自 Paxton et al., 1999; Peters et al., 1999; Palmer et al., 2002; Tóth et al., 2002; Reis et al., 2011）。

当然也偶有报道发现无刺蜂工蜂间亲缘关系值较低的现象（图 4.5）（Peters et al., 1999; Paxton et al., 1999; Paxton，2000），而且 Imperatriz–Fonseca 等（1998）已经观察到一只 *Tetragonisca angustula* 蜂王在与两只雄蜂交配。也有研究观察到同蜂种的蜂王进行了两次婚飞（van Veen and Sommeijer，2000b）。这表明在无刺蜂中多雄交配也会偶尔发生（Engels and Engels，1988; Paxton et al., 1999; Paxton，2000; Francini et al., 2012; Viana et al., 2015）。然而也有人对于无刺蜂中发现的较低亲缘关系值做出了其他解释，如工蜂换群、蜂群换王、无效等位基因或被分析的样本中存在雄蜂污染（Paxton，2000; Palmer et al., 2002; Vollet-Neto et al., 2018）。那么多雄交配是否在特定的无刺蜂种中更常发生呢？这一问题还需要进一步的研究。

无刺蜂较之蜜蜂的高度近亲缘关系可能会对无刺蜂蜂群的社会组织和整

体表现产生重要影响（Peters et al., 1999; Ratnieks et al., 2006）。例如，由于无刺蜂的亲缘关系背景，工蜂生产雄蜂是有益的，因为工蜂与其姐妹的儿子的亲缘关系（$r = 0.375$）比与其母亲蜂王的儿子的亲缘关系（$r = 0.25$）更密切（更多细节见第5章）[Ratnieks等（2006）对此进行了综述]。这可能有助于解释为什么在多种无刺蜂蜂种中很大一部分雄蜂是由工蜂繁育的（Tóth et al., 2004; Ratnieks et al., 2006; Grüter, 2018）。更高亲缘关系也可以增加工蜂对处女蜂王的容忍度，从而解释为什么一些无刺蜂蜂王被关在"监禁室"内（第3章）（Peters et al., 1999）。然而，事实上多雄交配可以增加工蜂之间的遗传多样性，进而改善蜂群健康（Baer and Schmid-Hempel，1999）、采集行为（Mattila and Seeley，2007）和环境变化后的蜂群恢复力（Jones et al., 2004; Oldroyd and Fewell，2007）。那么为什么无刺蜂蜂王不像蜜蜂蜂王那样与多个雄蜂交配呢？

这可能是由于多雄交配也可能会付出较高的代价，因为蜂王暴露于天敌的风险会增加，且受到攻击性雄蜂伤害和被传染疾病的风险也会增加（Strassmann，2001）。但是蜜蜂的蜂王也会面临相同的代价，蜜蜂蜂王却仍然会与多个雄蜂交配。二倍体雄蜂假说认为无刺蜂的单雄交配是得益于自然选择的结果，因为无刺蜂会在蛰饲且密封的巢室中繁育幼虫，这一特殊的生物学特征与膜翅目蜂的性别决定系统相结合，促成了自然选择，形成了无刺蜂单雄交配（第5章）（Ratnieks，1990）。膜翅目蜂的性别是由互补性别决定机制（Complementary Sex Determination，CSD）决定的（Ratnieks，1990; Cook and Crozier, 1995; Heimpel and de Boer, 2008; Zayed and Packer, 2005; Hasselmann et al., 2008）。在互补性别决定位点上杂合的是雌性，而只有一个拷贝（半合子/单倍体）或有相同拷贝（纯合子）是雄性。在互补性别决定位点上有相同拷贝的情况下就会产生二倍体雄蜂，这种二倍体雄蜂通常是不育的或不能存活的（Cook and Crozier，1995; Zayed and Packer，2005; Heimpel and de Boer，2008）。但是，至少在诸如 *Melipona quadrifasciata*、*Tetragonula carbonaria* 和 *Tetragonisca angustula* 等无刺蜂蜂种中二倍体雄蜂是可存活的，并且在雄蜂集群中发现了二倍体雄蜂的存在（Santos et al., 2013; Vollet-Neto et al., 2015），但无刺蜂的二倍体雄蜂似乎寿命更短，而且也可能无法生育（de Camargo，1982; Green and Oldroyd，2002; Tavares et al., 2003）。因此，对于一个蜂群来说二倍体雄蜂是一项失败的投资。

如果无刺蜂能够像采用渐进式喂养幼虫的蜜蜂那样，工蜂在二倍体雄蜂发育早期就能识别并将其杀死，那么蜂群损失就会很低（Ratnieks，1990）。然而，在无刺蜂中巢室在蜂王产卵后会被大宗供应食物然后立即封盖（第5章）（Michener，1974;Sakagami，1982;Wille，1983），这意味着二倍体雄蜂会被养到其出巢室为止（Carvalho，2001；Francini et al., 2012）。这种情况则会导致巨大的成本消耗，因为单雄交配的蜂王如果繁育了二倍体雄性，就意味着有50%的潜在工蜂幼虫会损失掉[10]（de Camargo，1979，1982；Ratnieks，1990；Green and Oldroyd，2002；Cameron et al., 2004；Vollet-Neto et al., 2018, 2019），进而导致蜂群增长减缓和蜂群大小降低，从而可能会对蜂群健康产生极大的影响（Ratnieks，1990）。即使雄蜂在幼虫中所占比例较小，二倍体雄蜂也可能会对蜂群增长产生负面影响，因为会造成工蜂的缺失和繁育工蜂空间的减少（Ratnieks，1990；Vollet-Neto et al., 2017）。因此，正在繁育二倍体雄蜂的蜂王的适应度较低，并有人猜测这种蜂王会被工蜂杀死（de Camargo，1979；Ratnieks，1990；Carvalho，2001；Alves et al., 2011；Francini et al., 2012；Vollet-Neto et al., 2017, 2019）。

据估计，不同的无刺蜂蜂王与具有相同性别等位基因的雄性交配，产生二倍体雄蜂（称为配对交配）的概率为5%～20%不等（Vollet-Neto et al., 2017）。根据二倍体雄性假说，在多雄交配的情况下配对交配的概率会增加，蜂王被工蜂处死的风险也会增加。近期研究者发现 *Scaptotrigona depilis* 无刺蜂的交配频率、二倍体雄蜂产量和蜂王被处死情况之间确实存在上述的关系。*Scaptotrigona depilis* 无刺蜂中繁育大量二倍体雄蜂（25%及以上）的蜂王，或者在二倍体雄蜂出房时引入的蜂王都会被工蜂杀掉（图4.6）（Vollet-Neto et al., 2017, 2019）。无刺蜂工蜂杀死蜂王的时间也与二倍体雄蜂分化出特有的表皮碳氢谱的时间相吻合，这表明工蜂可以通过化学方法识别二倍体雄蜂（Vollet-Neto et al., 2017）。有趣的是，如果二倍体雄性只占到出房蜂的12.5%及以下，就很少出现杀死蜂王的现象（Vollet-Neto et al., 2019）。这表明，如果蜂王与4只或4只以上的雄性交配，如果其中只有一只是配对交配，那么蜂王就不会被处死。然而，双雄交配的蜂王被处死的风险较高，这似乎阻止了无刺蜂进行多雄交配的进化（Vollet-Neto et al., 2019）。

---

[10] 如果蜂王与具有匹配性等位基因的雄蜂交配，意味着50%的受精卵（通常会发育成工蜂）会发育成二倍体雄蜂。

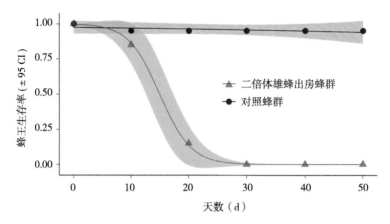

图 4.6　*Scaptotrigona depilis* 无刺蜂蜂王的存活率对比

注：图中对比的是被放入过量产生二倍体雄蜂的蜂群的蜂王和放入无二倍体雄蜂产生蜂群的蜂王（Vollet-Neto et al., 2017）。

4.7　无刺蜂的一夫一妻制

无刺蜂主要是一妻制的，这意味着蜂群中只有一个产卵蜂王（Sakagami, 1982; Nogueira-Neto, 1997）。如前所述，无刺蜂蜂群也会出现短暂的一夫多妻现象，且至少在 6 种无刺蜂中观察到了这种情况（Vollet-Neto et al., 2018）。甚至在一个 *Melipona quadrifasciata* 蜂群中出现过 8 个同胞膨腹蜂王共存的现象（Alves et al., 2011）。然而，只有一种已知的无刺蜂是稳定的一夫多妻制，这就是 *Melipona bicolor* 无刺蜂。在 *M. bicolor* 蜂群中会有多达 5 只蜂王在蜂群中繁殖，因此工蜂亲缘关系低于其他单蜂王蜂群（图 4.5）（Michener, 1974; Bego, 1983; Velthuis et al., 2001, 2006; Reis et al., 2011）。这些蜂王之间是有血缘关系的，它们要么是母女要么是姐妹（Velthuis et al., 2006），但外来蜂王的入侵似乎也偶有发生（Reis et al., 2011）。这些蜂王之间并不表现出攻击性（Engels and Imperatriz-Fonseca, 1990），但这些蜂王的繁殖量表现出相当大的个体倾斜，可能是因为对工蜂越有吸引力的蜂王会获得越多的食物，例如营养卵（Velthuis et al., 2001, 2006）。实验室蜂群中人工喂养似乎有利于一夫多妻制的形成（Imperatriz-Fonseca and Zucchi, 1995），但是一夫多妻制的蜂群似乎并不比一夫一妻的蜂群更多产。因此，一夫多妻制对 *M. bicolor* 无刺蜂的好处尚不清楚（Velthuis et al., 2006）。

参考文献

Abdalla FC, Jones GR, Morgan ED, da Cruz-Landim C (2003) Comparative study of the cuticular hydrocarbon composition of *Melipona bicolor* Lepeletier, 1836 (Hymenoptera, Meliponini) workers and queens. Genetics and Molecular Research 2:191–199

Alves DA, Menezes C, Imperatriz-Fonseca VL, Wenseleers T (2011) First discovery of a rare polygyne colony in the stingless bee *Melipona quadrifasciata* (Apidae, Meliponini). Apidologie 42:211–213

Araújo ED, Costa M, Chaud-Netto J, Fowler HG (2004) Body size and flight distance in stingless bees (Hymenoptera: Meliponini): Inference of flight range and possible ecological implications. Brazilian Journal of Biology 64:563–568

Baer B, Schmid-Hempel P (1999) Experimental variation in polyandry affects parasite loads and fitness in a bumble-bee. Nature 397 (6715):151–154

Bänziger H, Khamyotchai K (2014) An unusually large and persistent male swarm of the stingless bee *Tetragonula laeviceps* in Thailand (Hymenoptera: Apidae: Meliponini). Journal of Melittology 32:1–5

Bego L (1983) On some aspects of bionomics in *Melipona bicolor bicolor* Lepeletier (Hymenoptera, Apidae, Meliponinae). Revista Brasileira de Entomologia 27:211–224

Boongird S, Michener CD (2010) Pollen and propolis collecting by male stingless bees (Hymenoptera: Apidae). Journal of the Kansas Entomological Society 83(1):47–50

Camargo JMD (1974) Notas sobre a morfologia e biologia de *Plebeia* (*Schwarziana*) *quadripunctata quadripunctata* (Hymenoptera, Apidae, Meliponinae). Studia Entomologica 17:433–470

Camargo JMD, Pedro SR (2003) Neotropical Meliponini: the genus *Partamona* Schwarz, 1939 (Hymenoptera, Apidae, Apinae)-bionomy and biogeography. Revista Brasileira de Entomologia 47 (3):311–372

Camargo JMD, Roubik DW (1991) Systematics and bionomics of the apoid obligate necrophages: the *Trigona hypogea* group (Hymenoptera: Apidae; Meliponinae). Biological Journal of the Linnean Society 44 (1):13–39

Camargo JMF, Pedro SRM (2013) Meliponini Lepeletir, 1836. In: Moure JS, Urban

D, Melo GAR (eds) Catalogue of Bees (Hymenoptera, Apoidea) in the Neotropical Region-online version. Available at http: //www.moure. cria. org. br/catalogue

Cameron EC, Franck P, Oldroyd BP (2004) Genetic structure of nest aggregations and drone congregations of the southeast Asian stingless bee *Trigona collina*. Molecular Ecology 13:2357–2364

Carvalho G (2001) The number of sex alleles (CSD) in a bee population and its practical importance (Hymenoptera: Apidae). Journal of Hymenoptera Research 10 (1):10–15

Colonello NA, Hartfelder K (2005) She's my girl-male accessory gland products and their function in the reproductive biology of social bees. Apidologie 36 (2):231–244

Cook JM, Crozier RH (1995) Sex determination and population biology in the Hymenoptera. Trends in Ecology & Evolution 10 (7):281–286

Cortopassi-Laurino M (2007) Drone congregations in Meliponini: what do they tell us? Bioscience Journal 23:153–160

Cruz-Landim C, Abdalla FC, Gracioli-Vitti LF (2006) Class III glands in the abdomen of Meliponini. Apidologie 37 (2):164–174

Cruz-Landim Cd, Hofling M, Imperatriz-Fonseca VL (1980) Tergal and mandibular glands in queens of *Paratrigona subnuda* (Moure) (Hymenoptera: Apidae). Morphology and associated behaviour. Naturalia 58 (5):121–133

Cunningham JP, Hereward JP, Heard TA, De Barro PJ, West SA (2014) Bees at War: Interspecific Battles and Nest Usurpation in Stingless Bees. American Naturalist 184 (6):777

da Silva DLN, Zucchi R, Kerr WE (1972) Biological and behavioural aspects of the reproduction in some species of *Melipona* (Hymenoptera, Apidae, Meliponinae). Animal Behaviour 20 (1):123–132

da Silva LN (1972) Considerações em Torno de um Caso de Substituição de Rainha em *Plebeia* (*Plebeia*) *droryana* (Friese, 1900). In: Homenagem ao 50º aniversário de WE Kerr. Rio Claro, pp 267–273

Dallacqua R, Simões Z, Bitondi M (2007) Vitellogenin gene expression in stingless bee workers differing in egg-laying behavior. Insectes Sociaux 54 (1):70–76

Darchen R (1977) L'essaimage chez les *Hypotrigones* au Gabon: dynamique de quelques populations. Apidologie 8 (1):33–59 de Camargo CA (1972) Mating of the social bee *Melipona quadrifasciata* under controlled conditions (Hymenoptera, Apidae). Journal of the Kansas Entomological Society 45:520–523

de Camargo CA (1979) Sex determination in bees. XI Production of diploid males and sex determination in *Melipona quadrifasciata*. Journal of Apicultural Research 18 (2):77–84

de Camargo CA (1982) Longevity of diploid males, haploid males, and workers of the social bee *Melipona quadrifasciata* Lep. (Hymenoptera: Apidae). Journal of the Kansas Entomological Society 55:8–12

Drumond P, Bego L, Melo GAR (1995) Nest architecture of the stingless bee *Plebeia poecilochroa* Moure and Camargo, 1993 and related considerations (Hymenoptera, Apidae, Meliponinae). Iheringia, Ser Zool, Porto Alegre 79:39–49

Engels E, Engels W (1984) Drohnen-Ansammlungen bei Nestern der stachellosen Biene *Scaptotrigona postica*. Apidologie 15 (3):315–328

Engels E, Engels W (1988) Age-dependent queen attractiveness for drones and mating in the stingless bee, *Scaptotrigona postica*. Journal of Apicultural Research 27 (1):3–8

Engels W (1987) Pheromones and reproduction in Brazilian stingless bees. Memórias do Instituto Oswaldo Cruz 82:35–45

Engels W, Engels E, Francke W (1997) Ontogeny of cephalic volatile patterns in queens and mating biology of the neotropical stingless bee, *Scaptotrigona postica*. Invertebrate Reproduction & Development 31 (1-3):251–256

Engels W, Engels E, Lübke G, Schröder W, Francke W (1990) Volatile cephalic secretions of drones, queens and workers in relation to reproduction in the stingless bee, *Scaptotrigona postica* (Hymenoptera: Apidae: Trigonini). Entomologia Generalis 15 (2):91–101

Engels W, Imperatriz-Fonseca VL (1990) Caste development, reproductive strategies, and control of fertility in honey bees and stingless bees. In: Engels W (ed) Social Insects. Springer, Heidelberg, pp 167–230

Engles W, Engels E (1977) Vitellogenin und Fertilität bei stachellosen Bienen. Insectes Sociaux 24 (1):71–94

Faustino C, Silva-Matos E, Mateus S, Zucchi R (2002) First record of emergency queen rearing in stingless bees (Hymenoptera, Apinae, Meliponini). Insectes

Sociaux 49（2）:111–113

Ferreira-Caliman MJ, Falcón T, Mateus S, Zucchi R, Nascimento FS（2013）Chemical identity of recently emerged workers, males, and queens in the stingless bee *Melipona marginata*. Apidologie 44（6）:657–665

Fierro MM, Cruz-López L, D. S, Villanueva-Gutiérrez R, Vandame R（2011）Queen volatiles as a modulator of *Tetragonisca angustula* drone behavior. Journal of Chemical Ecology 37:1255–1262

Francini IB, Nunes-Silva CG, Carvalho-Zilse GA（2012）Diploid male production of two Amazonian *Melipona* bees（Hymenoptera: Apidae）. Psyche: A Journal of Entomology 2012:484618

Galindo López JC, Kraus FB（2009）Cherchez la femme? Site choice of drone congregations in the stingless bee *Scaptotrigona mexicana*. Animal Behaviour 77:1247–1252

Green C, Oldroyd BP（2002）Queen mating frequency and maternity of males in the stingless bee *Trigona carbonaria* Smith. Insectes Sociaux 49（3）:196–202

Grüter C, Menezes C, Imperatriz-Fonseca VL, Ratnieks FLW（2012）A morphologically specialized soldier caste improves colony defence in a Neotropical eusocial bee. Proceedings of the National Academy of Sciences of the United States of America 109:1182–1186

Grüter C, von Zuben LG, Segers FHID, Cunningham J（2016）Warfare in stingless bees. Insectes Sociaux 63:223–236

Grüter C（2018）Repeated switches from cooperative to selfish worker oviposition during stingless bee evolution. Journal of Evolutionary Biology 31:1843–1851

Hasselmann M, Gempe T, Schiøtt M, Nunes-Silva CG, Otte M, Beye M（2008）Evidence for the evolutionary nascence of a novel sex determination pathway in honeybees. Nature 454（7203）:519–522

Heimpel GE, de Boer JG（2008）Sex determination in the Hymenoptera. Annual Review of Entomology 53:209–230

Hockings HJ（1883）Notes on two Australian species of *Trigona*. Transactions of the Entomological Society of London 2:149–157

Hubbell SP, Johnson LK（1977）Competition and nest spacing in a tropical stingless bee community. Ecology 58:949–963

Imperatriz-Fonseca VL (1978) Studies on *Paratrigona subnuda* (Moure) (Hymenoptera, Apidae, Meliponinae) III. Queen supersedure. Boletim de Zoologia, Universidade de São Paulo 3:153–162

Imperatriz-Fonseca VL, Matos E, Ferreira F, Velthuis HHW (1998) A case of multiple mating in stingless bees (Meliponinae). Insectes Sociaux 45 (2):231–233

Imperatriz-Fonseca VL, Zucchi R (1995) Virgin queens in stingless bee (Apidae, Meliponinae) colonies: a review. Apidologie 26 (3):231–244

Imperatriz-Fonseca VL (1977) Studies on *Paratrigona subnuda* (Moure) Hymenoptera, Apidae, Meliponinae-II. Behaviour of the virgin queen. Boletim de Zoologia, Universidade de São Paulo 2:169–182

Inoue T, Nakamura K, Salmah S, Abbas I (1993) Population dynamics of animals in unpredictably-changing tropical environments. Journal of Biosciences 18 (4):425–455

Inoue T, Sakagami SF, Salmah S, Nukmal N (1984 a) Discovery of successful absconding in the stingless bee *Trigona* (*Tetragonula*) *laeviceps*. Journal of Apicultural Research 23 (3):136–142

Inoue T, Sakagami SF, Salmah S, Yamane S (1984 b) The process of colony multiplication in the Sumatran stingless bee *Trigona* (*Tetragonula*) *laeviceps*. Biotropica 16:100–111

Jaffé R, Pioker-Hara FC, Santos CF, Santiago LR, Alves DA, Kleinert AMP, Francoy TM, Arias MC, Imperatriz-Fonseca VL (2014) Monogamy in large bee societies: a stingless paradox. Naturwissenschaften 101:261–264

Jarau S, van Veen JW, Aguilar I, Ayasse M (2009) Virgin queen execution in the stingless bee *Melipona beecheii*: The sign stimulus for worker attacks. Apidologie 40 (4):496–507

Jones JC, Myerscough MR, Graham S, Oldroyd BP (2004) Honey bee nest thermoregulation: diversity promotes stability. Science 305:402–404

Kärcher M, Menezes C, Alves DA, Beveridge OS, Imperatriz-Fonseca VL, Ratnieks FLW (2013) Factors influencing survival duration and choice of virgin queens in the stingless bee *Melipona quadrifasciata*. Naturwissenschaften 100:571–580

Kerr WE, Jungnickel H, Morgan ED (2004) Workers of the stingless bee *Melipona scutellaris* are more similar to males than to queens in their cuticular compounds.

Apidologie 35（6）:611–618

Kerr WE（1948）Estudos sobre o gênero *Melipona*. Anais da Escola Superior de Agricultura Luiz de Queiroz 5:181–276

Kerr WE（1951）Bases para o estudo da genética de populações dos Hymenoptera em geral e dos Apinae sociais em particular. Anais da Escola Superior de Agricultura Luiz de Queiroz 8:219–354

Kerr WE, Zucchi R, Nakadaira JT, Butolo JE（1962）Reproduction in the social bees （Hymenoptera: Apidae）. Journal of the New York Entomological Society 70:265–276

Kleinert A（2005）Colony strength and queen replacement in *Melipona marginata* （Apidae: Meliponini）. Brazilian Journal of Biology 65（3）:469–476

Koedam D, Monge IA, Sommeijer M（1995）Social interactions of gynes and their longevity in queenright colonies of *Melipona favosa*（Apidae: Meliponinae）. Netherlands Journal of Zoology 45（3）:480–494

Koffler S, Meneses HM, Kleinert AMP, Jaffé R（2016）Competitive males have higher quality sperm in a monogamous social bee. BMC Evolutionary Biology 16（1）:195

Kraus F, Weinhold S, Moritz RF（2008）Genetic structure of drone congregations of the stingless bee *Scaptotrigona mexicana*. Insectes Sociaux 55（1）:22–27

Lindauer M（1955）Schwarmbienen auf Wohnungssuche. Zeitschrift für vergleichende Physiologie 37:263–324

Luz GF, Campos LA de O, Zanuncio JC, Serrão JE（2017）Auxiliary brood cell construction in nests of the stingless bee *Plebeia lucii*（Apidae: Meliponini）. Apidologie 48:681–691

Mattila HR, Seeley TD（2007）Genetic diversity in honey bee colonies enhances productivity and fitness. Science 317:362–364

Michener CD（1946）Notes on the habits of some Panamanian stingless bees （Hymenoptera, Apidae）. Journal of the New York Entomological Society 54:179–197

Michener CD（1974）The Social Behavior of the Bees. Harvard University Press, Cambridge, Massachusetts

Müller MY, Moritz RF, Kraus FB（2012）Outbreeding and lack of temporal genetic structure in a drone congregation of the neotropical stingless bee *Scaptotrigona mexicana*. Ecology and Evolution 2（6）:1304–1311

Nogueira-Neto P（1954）Notas bionômicas sobre meliponíneos: Ⅲ – Sobre a enxameagem. Arq Mus Nac 42:419–451

Nogueira-Neto P（1997）Vida e Criação de Abelhas Indígenas Sem Ferrão. Editora Nogueirapis, São Paulo

Nunes TM, Mateus S, Favaris AP, Amaral MF, von Zuben LG, Clososki GC, Bento JM, Oldroyd BP, Silva R, Zucchi R（2014）Queen signals in a stingless bee: suppression of worker ovary activation and spatial distribution of active compounds. Scientific Reports 4:7449

Oi CA, van Zweden JS, Oliveira RC, Van Oystaeyen A, Nascimento FS, Wenseleers T（2015）The origin and evolution of social insect queen pheromones: novel hypotheses and outstanding problems. BioEssays 37:808–821

Oldroyd BP, Wongsiri S（2006）Asian Honey Bees: Biology, Conservation, and Human Interactions. Harvard University Press, Cambridge

Oldroyd BP, Fewell J（2007）Genetic diversity promotes homeostasis in insect colonies. Trends in Ecology & Evolution 22:408–413

Oliveira RC, Oi CA, do Nascimento MMC, Vollet-Neto A, Alves DA, Campos MC, Nascimento FS, Wenseleers T（2015）The origin and evolution of queen and fertility signals in Corbiculate bees. BMC Evolutionary Biology 15（1）:254

Oppelt A, Heinze J（2009）Mating is associated with immediate changes of the hydrocarbon profile of *Leptothorax gredleri* ant queens. Journal of Insect Physiology 55（7）:624–628

Paes de Oliveira VT, Berger B, Cruz–Landim C, Simões ZLP（2012）Vitellogenin content in fat body and ovary homogenates of workers and queens of *Melipona quadrifasciata* anthidioides during vitellogenesis. Insect Science 19（2）:213–219

Palmer KA, Oldroyd BP, Quezada-Euán JJ, Paxton RJ, May-Itza WJ（2002）Paternity frequency and maternity of males in some stingless bee species. Molecular Ecology 11:2107–2113

Paxton R（2000）Genetic structure of colonies and a male aggregation in the stingless bee *Scaptotrigona postica*, as revealed by microsatellite analysis. Insectes Sociaux 47（1）:63–69

Paxton RJ, Weißschuh N, Engels W, Hartfelder K, Quezada-Euán JJG（1999）Not only single mating in stingless bees. Naturwissenschaften 86（3）:143–146

Peters JM, Queller DC, Imperatriz-Fonseca VL, Roubik DW, Strassmann JE（1999）Mate number, kin selection and social conflicts in stingless bees and honeybees. Proceedings of the Royal Society of London Series B: Biological Sciences 266（1417）:379–384

Portugal-Araujo V（1963）Subterranean nests of two African stingless bees（Hymenoptera: Apidae）. Journal of the New York Entomological Society 71:130–141

Ratnieks FLW（1990）The evolution of polyandry by queens in social Hymenoptera: the significance of the timing of removal of diploid males. Behavioral Ecology and Sociobiology 26:343–348

Ratnieks FLW, Foster KR, Wenseleers T（2006）Conflict resolution in insect societies. Annual Review of Entomology 51:581–608

Reis EP, Campos LAO, Tavares MG（2011）Prediction of social structure and genetic relatedness in colonies of the facultative polygynous stingless bee *Melipona bicolor*（Hymenoptera, Apidae）. Genetics and Molecular Biology 34（2）:338–344

Ribeiro MF, Bego L（1994）Absconding in the Brazilian stingless bee *Frieseomelitta silvestrii languida* Moure（Hymenoptera: Apidae: Meliponinae）. Anais Sociedade Entomologica do Brasil 23（2）:355–358

Ribeiro MF, Wenseleers T, Santos-Filho PS, Alves DA（2006）Miniature queens in stingless bees: basic facts and evolutionary hypotheses. Apidologie 37（2）:191–206

Roubik DW（1982）Seasonality in colony food storage, brood production and adult survivorship: studies of *Melipona* in tropical forest（Hymenoptera: Apidae）. Journal of the Kansas Entomological Society 55:789–800

Roubik DW（1989）Ecology and Natural History of Tropical Bees. Cambridge University Press, New York

Roubik DW（1990）Mate location and mate competition in males of stingless bees（Hymenoptera: Apidae: Meliponinae）. Entomologia Generalis 15（2）:115–120

Ruttner F（1988）Biogeography and taxonomy of honeybees Springer, Berlin, Germany

Sakagami SF, Roubik DW, Zucchi R（1993）Ethology of the robber stingless bee, *Lestrimelitta limao*（Hymenoptera: Apidae）. Sociobiology 21:237–277

Sakagami SF（1982）Stingless bees. In: Hermann HR（ed）Social Insects Ⅲ. Academic

Press, New York, pp 361–423

Sakagami SF, Inoue T, Yamane S, Salmah S (1989) Nests of the Myrmecophilous Stingless Bee, *Trigona moorei*: How do Bees Initiate Their Nest Within an Arboreal Ant Nest? Biotropica 21:265–274

Sakagami SF, Laroca S (1963) Additional observations on the habits of the cleptobiotic stingless bees, the genus *Lestrimelitta* Friese (Hymenoptera, Apoidea). Journal of the Faculty of Science Hokkaido University 15:319–339

Sakagami SF, Zucchi R, Araujo VP (1977) Oviposition behavior of an aberrant african stingless bee *Meliponula bocandei,* with notes on the mechanism and evolution of oviposition behavior in stingless bees. Journal of the Faculty of Science Hokkaido University Series VI, Zoology 20 (4):647–690

Sánchez D, Vandame R, Kraus FB (2018) Genetic analysis of wild drone congregations of the stingless bee *Scaptotrigona mexicana* (Hymenoptera: Apidae) reveals a high number of colonies in a natural protected area in Southern Mexico. Revista Mexicana de Biodiversidad 89:226–231

Santos CF, Imperatriz-Fonseca VL, Arias MC (2016) Relatedness and dispersal distance of eusocial bee males on mating swarms. Entomological Science 19 (3):245–254

Santos CF, Menezes C, Imperatriz-Fonseca VL, Arias MC (2013) A scientific note on diploid males in a reproductive event of a eusocial bee. Apidologie 44:519–521

Schmidt JO (1995) Dispersal distance and direction of reproductive European honey bee swarms (Hymenoptera: Apidae). Journal of the Kansas Entomological Society 68:320–325

Schwarz HF (1948) Stingless Bees (Meliponidae) of the Western Hemisphere. Bulletin of the American Museum of Natural History 90:1–546

Seeley TD (1978) Life history strategy of the honey bee, *Apis mellifera*. Oecologia 32 (1):109–118

Seeley TD (1979) Queen substance dispersal by messenger workers in honeybee colonies. Behavioral Ecology and Sociobiology 5 (4):391–415

Seeley TD (2010) Honeybee Democracy. Princeton University Press, Princeton

Seeley TD (2017) Life-history traits of wild honey bee colonies living in forests around Ithaca, NY, USA. Apidologie 48:743–754.

Seeley TD, Buhrman SC (1999) Group decision making in swarms of honey bees. Behavioral Ecology and Sociobiology 45:19–31

Seeley TD, Visscher PK (1985) Survival of honeybees in cold climates: the critical timing of colony growth and reproduction. Ecological Entomology 10:81–88

Singer RB, Flach A, Koehler S, Marsaioli AJ, Amaral MCE (2004) Sexual mimicry in *Mormolyca ringens* (Lindl.) Schltr. (Orchidaceae: Maxillariinae). Annals of Botany 93 (6):755–762

Slaa EJ (2006) Population dynamics of a stingless bee community in the seasonal dry lowlands of Costa Rica. Insectes Sociaux 53:70–79

Smith TJ (2020) Evidence for male genitalia detachment and female mate choice in the Australian stingless bee *Tetragonula carbonaria*. Insectes Sociaux 67:189–193

Sommeijer MJ, De Bruijn L (1995) Drone congregations apart from the nest in *Melipona favosa*. Insectes Sociaux 42 (2):123–127

Sommeijer MJ, de Bruijn LL, Meeuwsen F (2003 a) Reproductive behaviour of stingless bees: solitary gynes of *Melipona favosa* (Hymenoptera: Apidae, Meliponini) can penetrate existing nests. Entomologische Berichten-Nederlandsche Entomologische Vereenigung 63 (2):31–35

Sommeijer MJ, de Bruijn LL, Meeuwsen F, Slaa EJ (2003 b) Reproductive behaviour of stingless bees: nest departures of non-accepted gynes and nuptial flights in *Melipona favosa* (Hymenoptera: Apidae, Meliponini). Entomologische Berichten-Nederlandsche Entomologische Vereenigung 63:7–13

Sommeijer MJ, de Bruijn LLM, Meeuwsen FJAJ (2004) Behaviour of males, gynes and workers at drone congregation sites of the stingless bee *Melipona favosa*(Apidae: Meliponini). Entomologische Berichten 64:10–15

Souza EA, Trigo JR, Santos DE, Vieira CU, Serrão JE (2017) The relationship between queen execution and cuticular hydrocarbons in stingless bee *Melipona scutellaris* (Hymenoptera: Meliponini). Chemoecology 27:25–32

Strassmann J (2001) The rarity of multiple mating by females in the social Hymenoptera. Insectes Sociaux 48:1–13

Tavares M, Irsigler AST, de Oliveira Campos LA (2003) Testis length distinguishes haploid from diploid drones in *Melipona quadrifasciata* (Hymenoptera: Meliponinae). Apidologie 34 (5):449–455

Terada Y (1972) Enxameagem em *Frieseomelitta varia* Lep. (Hymenoptera, Apidae). In: Homenageam a W.E. Kerr, por ocasião de seu 50° aniversário. Rio Claro

Terada Y, Garofalo CA, Sakagami SF (1975) Age-survival curves for workers of two eusocial bees (*Apis mellifera* and *Plebeia droryana*) in a subtropical climate, with notes on worker polyethism in *P. droryana*. Journal of Apicultural Research 14 (3-4):161–170

Tóth E, Queller DC, Dollin A, Strassmann JE (2004) Conflict over male parentage in stingless bees. Insectes Sociaux 51:1–11

Tóth E, Strassmann JE, Nogueira-Neto P, Imperatriz-Fonseca VL, Queller DC (2002) Male production in stingless bees: variable outcomes of queen–worker conflict. Molecular Ecology 11:2661–2667

van Benthem FDJ, Imperatriz-Fonseca VL, Velthuis HHW (1995) Biology of the stingless bee *Plebeia remota* (Holmberg): observations and evolutionary implications. Insectes Sociaux 42:71–87

van Oystaeyen A, Alves DA, Oliveira RC, Nascimento DL, Nascimento FS, Billen J, Wenseleers T (2013) Sneaky queens in *Melipona* bees selectively detect and infiltrate queenless colonies. Animal Behaviour 86:603–609

van Oystaeyen A, Oliveira RC, Holman L, van Zweden JS, Romero C, Oi CA, D'Ettorre P, Khalesi M, Billen J, Wäckers F, Millar JG, Wenseleers T (2014) Conserved class of queen pheromones stops social insect workers from reproducing. Science 343:287–290

van Veen J, Arce Arce H, Sommeijer M (2004) Production of queens and drones in *Melipona beecheii* (Meliponini) in relation to colony development and resource availability. Proceedings of the Netherlands Entomological Society 15:35–39

van Veen J, Sommeijer M, Aguilar Monge I (1999) Behavioural development and abdomen inflation of gynes and newly mated queens of *Melipona beecheii* (Apidae, Meliponinae). Insectes Sociaux 46 (4):361–365

van Veen J, Sommeijer MJ, Meeuwsen F (1997) Behaviour of drones in *Melipona* (Apidae, Meliponinae). Insectes Sociaux 44 (4):435–447

van Veen JW, Sommeijer MJ (2000 a) Colony reproduction in *Tetragonisca angustula* (Apidae, Meliponini). Insectes Sociaux 47:70–75

van Veen JW, Sommeijer MJ (2000 b) Observations on gynes and drones around nuptial flights in the stingless bees *Tetragonisca angustula* and *Melipona beechii* (Hymenoptera, Apidae, Meliponinae). Apidologie 31:47–54

Veiga JC, Menezes C, Contrera FAL (2017) Insights into the role of age and social interactions on the sexual attractiveness of queens in an eusocial bee, *Melipona flavolineata* (Apidae, Meliponini). The Science of Nature 104 (3–4):31

Veiga JC, Leão KL, Coelho BW, Queiroz ACM de, Menezes C, Contrera FAL (2018) The Life Histories of the "Uruçu Amarela" Males (*Melipona flavolineata*, Apidae, Meliponini). Sociobiology 65:780–783

Velthuis HHW, De Vries H, Imperatriz-Fonseca VL (2006) The polygyny of *Melipona bicolor*: scramble competition among queens. Apidologie 37 (2):222

Velthuis HHW, Koedam D, Imperatriz-Fonseca VL (2005) The males of *Melipona* and other stingless bees, and their mothers. Apidologie 36 (2):169–185

Velthuis HHW, Roeling A, Imperatriz-Fonseca VL (2001) Repartition of reproduction among queens in the polygynous stingless bee *Melipona bicolor*. Proceedings of the Section Experimental and Applied Entomology - Netherlands Entomological Society 12:45–50

Verdugo-Dardon M, Cruz-Lopez L, Malo E, Rojas J, Guzman-Diaz M (2011) Olfactory attraction of *Scaptotrigona mexicana* drones to their virgin queen volatiles. Apidologie 42:543–550

Viana M, de Carvalho C, Sousa H, Francisco A, Waldschmidt A (2015) Mating frequency and maternity of males in *Melipona mondury* (Hymenoptera: Apidae). Insectes sociaux 62 (4):491–495

Visscher PK (2007) Group decision making in nest-site selection among social insects. Annual Review of Entomology 52:255–275

Vollet-Neto A, dos Santos CF, Santiago LR, de Araujo Alves D, de Figueiredo JP, Nanzer M, Arias MC, Imperatriz-Fonseca VL (2015) Diploid males of *Scaptotrigona depilis* are able to join reproductive aggregations (Apidae, Meliponini). Journal of Hymenoptera Research 45:125–130

Vollet-Neto A, Koffler S, Santos CF, Menezes C, Nunes FMF, Hartfelder K, Imperatriz-Fonseca VL, Alves DA (2018) Recent advances in reproductive biology of stingless bees. Insectes Sociaux 65:201–212

Vollet-Neto A, Oliveira RC, Schillewaert S, Alves DA, Wenseleers T, Nascimento FS,

Imperatriz-Fonseca VL, Ratnieks FLW (2017) Diploid male production results in queen death in the stingless bee *Scaptotrigona depilis*. Journal of Chemical Ecology 43 (4):403–410

Vollet-Neto A, Imperatriz-Fonseca VL, Ratnieks FLW (2019) Queen execution, diploid males, and selection for and against polyandry in the Brazilian stingless bee *Scaptotrigona depilis*. The American Naturalist 194:725–735

von Frisch K (1967) The dance language and ovientation of bees. Harverd University Press, Cambridge

von Ihering H (1903) Biologie der stachellosen Honigbienen Brasiliens. Zoologische Jahrbücher Abteilung für Systematik Ökologie und Geographie der Tiere 19:179–287

von Zuben LG (2017) The mating communication of stingless bees (Hymenoptera: Apidae, Meliponini). PhD Thesis, University of São Paulo, Ribeirão Preto, Brazil

Wenseleers T, Alves DA, Francoy TM, Billen J, Imperatriz-Fonseca VL (2011) Intraspecific queen parasitism in a highly eusocial bee. Biology Letters 7:173–176

Wenseleers T, Hart AG, Ratnieks FLW, Quezada-Euán JJ (2004) Queen execution and caste conflict in the stingless bee *Melipona beecheii*. Ethology 110 (9):725–736

Wenseleers T, Ratnieks FLW (2004) Tragedy of the commons in *Melipona* bees. Proceedings of the Royal Society of London Series B: Biological Sciences 271(Suppl 5):S310–S312

Wenseleers T, Ratnieks FLW, de F Ribeiro M, Alves DA, Imperatriz-Fonseca VL (2005) Working-class royalty: bees beat the caste system. Biology letters 1 (2):125–128

Wille A (1983) Biology of the stingless bees. Annual Review of Entomology 28:41–64

Wille A, Orozco E (1975) Observations on the founding of a new colony by *Trigona cupira* (Hymenoptera: Apidae) in Costa Rica. Revista de Biologia Tropical 22 (2):253–287

Winston ML (1987) The biology of the honey bee. Harvard University Press, Cambridge

Winston ML (1992) The biology and management of Africanized honey bees. Annual Review of Entomology 37 (1):173–193

Zayed A, Packer L (2005) Complementary sex determination substantially increases extinction proneness of haplodiploid populations. Proceedings of the National Academy of Sciences of the United States of America 102 (30):10742–10746

5 无刺蜂幼蜂繁育

无刺蜂养育幼蜂是一个复杂的过程，需要工蜂与蜂王协调合作完成许多不同的工作。一个新的幼虫巢室建造完成后，几只工蜂会将幼虫食物反刍到这个巢房中，这一过程被称为"瓮饲"（mass provisioning）。不久之后蜂王会将卵产在幼虫食物上，随后该巢室会被立即封盖（Sakagami, 1982; Engels and Imperatriz-Fonseca, 1990; Zucchi et al., 1999）。由于大多数独居型的胡蜂和蜂类中都发现存在瓮饲、产卵和封盖的生活方式，因此研究者认为这种生活方式是一种蜂类的原始特性（Sakagami and Zucchi, 1974; Sommeijer, 1985; Roubik, 1989）。然而无刺蜂的巢室建造、供给、产卵和封盖是一个高度整合的社会过程，蜂王扮演着领导者的角色。这一过程与蜜蜂哺育幼虫的过程有很大的不同。蜜蜂的蜂王将卵产在可重复使用的空巢室中，然后工蜂会每天哺育幼虫，最后才将巢室封盖。无刺蜂这种独特的育虫方式会影响蜂群健康、繁殖冲突（第 5.6 节）和交配系统（第 4 章）。

5.1 食物供给和产卵过程

当无刺蜂蜂王和工蜂选择一个空巢室后[1]，工蜂就会将幼虫食物填满这个巢室，然后蜂王在里面产卵。产卵后年轻工蜂就会立即将巢室封盖。这种看似简单的食物供给和产卵过程（Provisioning and Oviposition Process，POP）却引起了研究人员相当大的兴趣，一部分原因是这一过程中工蜂和蜂王有显著的行为表现，且蜂种间存在显著的行为差异（Sakagami and Oniki,

---

[1] 在 POP 的早期定义中这种选择巢室的行为也常被称为"固巢"。

1963; Sakagami et al., 1973, 1977; Sakagami and Zucchi，1974; Sakagami，1982; Bego，1990; Bego et al., 1999; Zucchi et al., 1999; Pereira et al., 2009）。研究者将这一过程中无刺蜂工蜂和蜂王的行为描述为激动或兴奋，并且认为蜂王和工蜂的互动是习惯性行为。这些描述有主观性的成分，但研究人员一致认为在许多无刺蜂蜂种中蜂王和工蜂之间的互动是习惯性的（例如 *Melipona compressipes*、*Plebeia droryana*、*Scaptotrigona postica*、*Trigonisca duckei*，而 *Tetragona clavipes* 和 *Friesella schrottkyi* 无刺蜂中则较少发现），并且参与这一过程的无刺蜂会表现出焦躁不安甚至是明显的攻击性（Sakagami and Oniki, 1963; Sakagami and Zucchi，1967, 1974; Sakagami et al., 1977; Sakagami，1982; Bego，1990; Drumond et al., 1999; Drumond et al., 1996; Bego et al., 1999; Chinh et al., 2003）。

5.1.1 幼虫食物供给前阶段

无刺蜂蜂王一到子脾上就会引发工蜂兴奋，它们快速在其周围移动和频繁地检查打开的空巢室。蜂王在这个过程中也表现得很兴奋，它会频繁地振动翅膀（Sakagami and Oniki，1963; Sakagami and Zucchi，1967; Bego，1990; Drumond et al., 1996; Bego et al., 1999），这可能有助于传播蜂王信息素（Sommeijer，1985）。*Friesella schrottkyi* 和 *Mourella caerulea* 无刺蜂的蜂王甚至会冲向正在检查空巢室的工蜂（Sakagami et al., 1973; Wittmann et al., 1991），而 *Melipona quinquefasciata* 和 *Trigonisca duckei* 无刺蜂的蜂王则会用触角和前腿攻击在巢室附近等待着的工蜂（Sakagami and Zucchi，1974; Sakagami，1982）。一些无刺蜂蜂种的蜂王还会反刍出一种不明来源和成分的淡黄色液体（第 4.5 节）（Sakagami，1982）。Sakagami（1982）认为这种液体可以刺激工蜂将准备好的幼虫食物反刍到空巢室中去。除了非洲的 *Meliponula bocandei* 和亚洲的 *Lepidotrigona hoozana* 无刺蜂，大多数无刺蜂都是只有当蜂王等在一个巢室旁时，工蜂才会开始给这个巢室供给幼虫食物（Sakagami et al., 1977; Sakagami and Yamane，1987）。*Mourella caerulea* 无刺蜂则可能还需要蜂王攻击，巢室附近的工蜂才会启动幼虫食物供给过程（Wittmann et al., 1991）。在蜂王到子脾区后，第一个巢室（蜂王产第一"批"卵用的第一个巢室，见下文）的幼虫食物的供给前阶段时间通常较长一些（Sakagami and Zucchi，1967; Sakagami et al., 1973）。

5.1.2 幼虫食物供给和产卵阶段

目前研究过的一些无刺蜂蜂种工蜂供给幼虫食物的行为模式是相似的：

一只工蜂头朝前进入巢室，然后进行一次快速的腹部收缩（Sakagami and Oniki，1963；Sakagami et al.，1973）。因此，每只工蜂负责1次食物供给。每个巢室由2～19只不等的工蜂供应幼虫食物，负责单个巢室供给食物的工蜂只数随蜂种不同有所差异（表5.1），工蜂们会不断去巢室填充直到食物填满巢室的2/3为止（图5.1 A）。*Leurotrigona muelleri* 无刺蜂的每个巢室则基本只需两只工蜂供应，这是研究者记录到的无刺蜂单巢室食物供给工蜂的最低值（Sakagami and Zucchi，1974）。工蜂完成巢室内的食物供给后就会迅速离开巢室（"放完就撤"）（Sakagami and Zucchi，1974；Sakagami et al.，1977；Sommeijer et al.，1984；Yamane et al.，1995；Drumond et al.，1999）。而无刺蜂中，至少某些无刺蜂种中，工蜂供应的幼虫食物是通过交哺行为从其他工蜂那里获得的，而不是自己制作的（Sommeijer et al.，1984）。

表5.1 无刺蜂不同蜂种蜂巢内每个巢室的填装次数、蜂王日产卵速率（与巢室建造速率一致）以及蜂群规模

无刺蜂蜂种	幼虫食物填装次数	蜂王产卵速率	蜂群规模	参考文献
Austroplebeia australis		40～50	1 000～5 000	Drumond et al.（1999）
Austroplebeia cassiae		40～50	1 000～5 000	Drumond et al.（1999）
Friesella schrottkyi	3～5	40～65	300～2 500	Sakagami et al.（1973）
Frieseomelitta nigra	2～5			Sommeijer et al.（1984）
Hypotrigona gribodoi	4～6	20～80	100～750	Bassindal（1955）；Darchen（1972）
Lestrimelitta limao		600～740	2 000～7 000	Wittmann et al.（1991）
Leurotrigona muelleri	2	高达100	500～1 000	Sakagami and Zucchi（1974）
Melipona		高达50	150～3 000	Sakagami and Oniki（1963）；Chinh et al.（2003）
Melipona beecheii	6～19			van Veen（2000）
Melipona compressipes	6～15			Sakagami and Oniki（1963）
Meliponula bocandei	4～9			Sakagami et al.（1977）
Mourella caerulea	5～7			Wittmann et al.（1991）

续表

无刺蜂蜂种	幼虫食物填装次数	蜂王产卵速率	蜂群规模	参考文献
Plebeia droryana	4～7	～80	2 000～3 000	Drumond et al.（1996）
Plebeia juliani	4～12	13～30	300	Drumond et al.（1998）
Plebeia remota	4～12	高达 220	2 000～5 000	van Benthem et al.（1995）
Tetragona clavipes		150～500	7 000～29 000	Sakagami and Zucchi（1967）
Tetragonisca angustula	4～14			Segers et al.（2015）
Tetragonula carbonaria		～250	10000	Yamane et al.（1995）
Trigona recursa		～720		Yamane et al.（1995）
Trigonisca duckei	4～6			Sakagami and Zucchi（1974）

图 5.1 无刺蜂子脾及不同日龄的幼虫

注：A 为 *Tetragonisca angustula* 无刺蜂的子脾，如图所示，每个巢室内装入了约 2/3 体积的幼虫食物（白色箭头所示）。其中一个巢室是打开的，有典型的蜡质边沿，即巢柱（摄影：C. Grüter）。B 为 *Melipona fasciculata* 无刺蜂蜂巢中幼虫食物上的营养卵（摄影：Cristiano Menezes）。C 为 *Scaptotrigona depilis* 巢室中有真菌生长，这些真菌会被幼虫食用，图中幼虫的日龄和体型从左到右不断递增。日龄最大的幼虫已经吃掉了所有的真菌了（摄影：Cristiano Menezes）。

Tetragonisca angustula 无刺蜂蜂巢子脾靠中心的巢室中供给的幼虫食物量多于外围巢室，很可能是因为中心巢室体积稍大一些（Segers et al., 2015）。从这些巢室中育出的工蜂是体型较大的兵蜂（第 6 章）。群势较弱的

Geotrigona mumbuca 无刺蜂尽管巢室体积更小且从这些巢室中出房的工蜂也更小，但弱群中工蜂向每个巢室内供应幼虫食物的次数却多于强群（Lacerda et al., 1991）。因此很可能是弱群中单个工蜂向巢室中供应的食物量更少[2]。

无刺蜂蜂王在产卵之前会通过插入触角来短暂地检查巢室（Sakagami and Oniki, 1963; Sakagami et al., 1973; Sakagami and Zucchi, 1974; Drumond et al., 1996）[3]。多种无刺蜂蜂王在产卵前会吃掉一些幼虫食物（Sakagami and Oniki, 1963; Sakagami et al., 1977; Sommeijer, 1985; Wittmann et al., 1991; Drumond et al., 1996）。这些幼虫食物和营养卵（见下文）可能是蜂王的主要能量来源，在大多数无刺蜂中（除了 *Lepidotrigona hoozana* 无刺蜂，工蜂经常喂养蜂王）很少看见工蜂直接通过交哺方式喂养蜂王（Sakagami et al., 1977; Sakagami and Yamane, 1987; Wittmann et al., 1991; Yamane et al., 1995; Drumond et al., 1998）。蜂王产卵后该巢室就会立即被封盖（Sakagami et al., 1973）。

无刺蜂的幼虫食物供给和产卵有的是成批进行，有的则是以更连续的方式进行，具有明显的区分（Sakagami and Zucchi, 1974）。成批进行是指一批巢室被装好了幼虫食物，然后蜂王在里面产卵后就会离开子区。例如 *Scaptotrigona postica* 无刺蜂就可在短时间内装好 30 个以上巢室的幼虫食物并且完成蜂王产卵过程（Bego, 1990）。其他无刺蜂蜂种成批产卵时的批量大小参见 Sakagami 和 Zucchi（1967, 1973, 1974）、Sakagami 等（1973）、Sakagami 和 Yamane（1987）、Bego 等（1999）、Segers 等（2015）。一些无刺蜂蜂种的巢室是多个同时被装好幼虫食物（例如，*Frieseomelitta nigra*、*Plebeia droryana*、*P. remota*、*Scaptotrigona postica*、*Tetragonisca angustula* 和 *Tetragonula carbonaria* 无刺蜂），而另一些无刺蜂蜂种的产卵批次里的巢室是一个接一个地被装好幼虫食物的（例如，*Tetragona clavipes* 和 *Cephalotrigona*）（Sakagami and Zucchi, 1967; Sommeijer et al., 1984; Sommeijer, 1985; Yamane et al., 1995; Drumond et al., 1996; Zucchi et al., 1999; Segers et al., 2015）。有证据表明，无刺蜂成批产卵的批次大小和批次之间的时间间隔与蜂群大小和可利用食物率有关（Sakagami and Zucchi, 1967）。例如，van Benthem 等（1995）

---

[2] 无刺蜂的蜂群规模和工蜂体型大小之间呈正相关（Segers et al., 2016），这表明工蜂群体是决定巢室体积大小、幼虫食物供给量和工蜂体型大小的重要因素。

[3] *Plebeia remota* 和 *P. juliani* 无刺蜂蜂王则例外，它们在产卵前通常不会检查巢室（van Benthem et al., 1995; Drumond et al., 1998）。

发现强群产卵的批量规模更大。*Melipona* 属无刺蜂是一个以小规模蜂群为主的属（表1.3）（Michener，1974; Roubik，1983），该属和 *Meliponula bocandei* 无刺蜂蜂种的POP则不是成批的，而是依次连续的供应每个巢室的幼虫食物和产卵的（Sakagami and Oniki，1963; Sakagami and Zucchi，1967; Sakagami et al.，1977; Sommeijer，1985）。当然，无刺蜂的成批POP行为和连续POP行为也不能完全区分开来（例如，*Melipona quadrifasciata*）（Sakagami and Zucchi，1974）。

总的来说，小规模蜂群无刺蜂蜂种的蜂王基本每天只产几个卵，而大规模蜂群的无刺蜂蜂种的蜂王每天可产上百个卵（表5.1）。与产卵模式相似的是，无刺蜂的巢室建设也分持续建设（一个接一个地，例如 *Hypotrigona gribodoi* 和 *Schwarziana quadripunctata* 无刺蜂）和同时按批次建设（例如 *Lepidotrigona ventralis* 和 *Plebeia remota* 无刺蜂）（Sakagami，1982）。当生存条件不利时，例如在食物短缺期间，无刺蜂工蜂甚至可能打开幼虫巢室去食用幼虫食物（Sakagami and Zucchi，1974; Roubik，1982）。

蜂王—工蜂的互动

一些无刺蜂蜂种（例如 *Austroplebeia* spp.，*Melipona* spp.，*Plebeia droryana*，*Scaptotrigona postica* 或 *Tetragonisca angustula*）即将或已经被供应了幼虫食物的巢室附近的工蜂会对蜂王表现出"猛冲"或"摇摆"行为，这可能导致蜂王更加躁动，甚至是逃跑或攻击工蜂（Sakagami and Oniki，1963; Bego，1990; Bego et al.，1999; Drumond et al.，1996，1999，2000）。一些研究者观察到一些诸如 *Meliponula bocandei* 无刺蜂蜂种的工蜂即使是在POP阶段以外也会偶尔向蜂王"猛冲"（Sakagami et al.，1977）。*Trigonisca duckei* 无刺蜂的工蜂弯曲身体向蜂王展示自己的胸部，研究者将这种行为解释为一种工蜂用来阻止蜂王靠近巢室的惯性"屏障"（Sakagami and Zucchi，1974）。*Plebeia remota* 无刺蜂的工蜂似乎也会在蜂王产卵前表现出"保护"巢室的行为来避免巢室被蜂王破坏（van Benthem et al.，1995）。*Plebeia droryana* 无刺蜂工蜂还会展现出一种被研究者称为"催眠转弯"的行为（Sakagami，1982; Drumond et al.，1996）。*Lepidotrigona hoozana* 无刺蜂中则是蜂王会在POP期间会表现出独特的转弯行为（Sakagami and Yamane，1987）。

一些研究者提出，无刺蜂蜂王—工蜂在POP期间的惯性且看似对抗的互动，是工蜂阻止蜂王在空巢室中产卵的原始本能和衍生的拒王行为之间的一种矛盾的行为表达。研究者也经常在工蜂遇到漫游蜂王时或在供给完幼虫食物后观察到工蜂的这种拒王行为（Sakagami et al., 1973, 1977; Sakagami and Zucchi, 1974; Sakagami, 1982; Sommeijer, 1985; Drumond et al., 1996, 2000）[4]。换句话说，这种互动可能是无刺蜂蜂群中的一种原始的（且仍持续存在）的关于谁生产雄蜂的冲突的外在表现（第5.8节）（Hamilton, 1972; Crespi, 1992; Drumond et al., 1999; Grüter, 2018）。Crespi（1992）以及Oldroyd和Pratt（2015）也认为无刺蜂工蜂和蜂王之间的惯性的对抗互动是蜂群中工蜂能够繁殖蜂种的原始特性。但是值得注意的是，一些无刺蜂蜂种尽管蜂群中有大量工蜂也能生育后代，例如 *Melipona favosa*（见下文），这些蜂种的蜂王—工蜂互动中却并未表现出丝毫的对抗攻击性（Chinh et al., 2003; Velthuis et al., 2005）。相反，*Frieseomelitta paupera* 无刺蜂蜂群中的工蜂是不能生育后代的，该蜂种的工蜂却也有习惯性的摇摆行为和猛冲向蜂王的行为（Sommeijer et al., 1984）。因此，蜂群中工蜂能否生育和蜂王—工蜂互动之间似乎没有直接的联系，但需要进一步地比较分析来探索生殖矛盾（第5.8节）和惯性互动之间的联系。

无蜂王的无刺蜂蜂群中也会有POP过程，但是与有蜂王存在的蜂群相比更加无规律且具有不协调性，这凸显出在POP阶段中无刺蜂蜂王作为领导者起到了重要作用（Sakagami, 1982; Engels and Imperatriz-Fonseca, 1990）。

幼虫食物组成

无刺蜂的幼虫食物呈半流质，主要由蜂蜜、花粉、工蜂咽下腺和下颌腺分泌物组成（Quezada-Euán et al., 2011）。一些无刺蜂蜂种的幼虫食物中腺体分泌物的添加量似乎并不高（Quezada-Euán et al., 2011），这和蜜蜂有很大的不同，蜜蜂的幼虫食物中大部分源自工蜂的腺体分泌物（Winston, 1987）。无刺蜂和蜜蜂的幼虫食物的另一个区别是无刺蜂幼虫食物含水量（40%～60%）低于蜜蜂幼虫食物（60%～70%），然而两种幼虫食物中的糖含量（5%～20%）和游离氨基酸量（0.2%～1.3%）则是相似的（Hartfelder and Engels, 1989; Velthuis et al., 2003; Wang et al., 2016）。花粉粒是无刺蜂主要的蛋白质来源（Velthuis et al., 2003; Menezes et al., 2007）。无刺蜂幼虫食物之所以含水量低可能是因为无刺蜂幼虫食物需要形成一种高度黏稠的液体来让卵和幼虫漂浮在幼虫食物上面并防止其下沉（图5.1 B）（Hartfelder and

---

[4] 当蜂王停留一定时间或当其到达一个巢室去产卵前，就会形成蜂王的工蜂护卫圈（Sakagami, 1982; Yamane et al., 1995）。

Engels, 1989; Velthuis and Velthuis, 1998）[5]。无刺蜂幼虫食物的物质组成会随着季节改变，特别是蛋白质含量（Quezada-Euán et al., 2011），而幼虫食物的蛋白质含量则影响着新出房工蜂的体型大小（Quezada-Euán et al., 2011）。

Scaptotrigona depilis 无刺蜂的幼虫还会食用生长在巢室内壁和幼虫食物表面上的真菌（*Zygosaccharomyces* 属）（图 5.1 C）（Menezes et al., 2015; Paludo et al., 2018）。没有食用真菌的幼虫则会死亡。这些真菌可能为无刺蜂幼虫提供了关键化合物——类固醇前体，这是幼虫变态发育所必需的，而这些蜂种的无刺蜂则无法从头合成类固醇前体（Paludo et al., 2018）。但是幼虫并不依赖真菌提供的营养物质。无刺蜂—真菌的关系是如何进化的尚不清楚，但是最初建立关联可能是由于真菌具有营养或激素补充的功能。即使 *Tetragona clavipes* 和 *Melipona flavolineata* 无刺蜂在系统发育关系上相距较远，但这两种蜂的巢室内也都观察到了真菌的生长（Menezes et al., 2013）。这些观察结果表明无刺蜂中的蜂—真菌互利共生的关系很可能比目前所知的更为普遍存在。

5.2 工蜂产卵

许多无刺蜂和其他社会性昆虫的一个显著区别就是无刺蜂工蜂的卵巢处于激活状态且会频繁产卵，而其他社会性蜂种中蜂王的存在会导致工蜂不孕（Hoover et al., 2003; Matsuura et al., 2010; van Oystaeyen et al., 2014; Oi et al., 2015; Grüter and Keller, 2016; Grüter, 2018）[6]。当巢室被无刺蜂工蜂供给完幼虫食物后，偶尔可以看到工蜂将卵产在供给好的巢室的侧壁上，或者直接产在幼虫食物上（Bassindale, 1955; Sakagami and Oniki, 1963; Beig, 1972; Sakagami and Yamane, 1987; Wittmann et al., 1991; Bego et al., 1999; Koedam et al., 1996; Segers et al., 2015）。一些 *Plebeia* 属无刺蜂蜂种的工蜂甚至会在巢脾里随地产卵，而且通常是在给巢室供给幼虫食物之前，但这些工蜂产卵是随着靠近蜂王而被激发的（Drumond et al., 1996, 1998, 2000）。工蜂产卵的时间和地点的不同就引出这样一个问题：这些工蜂产的卵是用来喂养

---

[5] 此外，无刺蜂卵表面的理化性质也有助于其漂浮在幼虫食物上（Velthuis and Velthuis, 1998）。

[6] Silva-Matos 等（2006）发现在一个 *Trigona cilipes* 无刺蜂蜂群中工蜂体内分化出了卵巢并且有模仿产卵的行为，但是这些工蜂并没有产下卵。这是不是这种无刺蜂中的普遍情况？这种行为是否发挥了一些交流的作用或者是蜂王和工蜂间冲突的一种信号？这都不得而知。

蜂王的小能量包？还是工蜂自己想要生育雄蜂[7]。

Beig（1972）进行了一项首创性研究，在该研究中他观察到当 *Scaptotrigona postica* 无刺蜂蜂群中某一时期雄蜂数量较多时，许多巢室内会含有一个以上的卵。通常青年蜂就已经有发育的卵巢，并且可以看到它们产两种不同类型的卵，这两种卵都比雌蜂卵大：一种是细长的雄蜂卵，这种卵通常是在蜂王产卵后不久和巢室封盖之前被产在幼虫食物上，另一种是形状相对更圆的球形卵，这种卵则被产在巢室内侧上部边缘，会被作为蜂王的食物，即营养卵（图 5.2）。繁殖卵主要是由青年工蜂产下的，而营养卵则是中年工蜂产的（Bego et al., 1983; Bego, 1990）。工蜂的解剖结果显示从第 10 天到第 25 天之间它们的卵巢就已经发育完全。在工蜂的生命后期，即在外出采集阶段，工蜂的卵巢会退化（Bego et al., 1983）。*Melipona subnitida* 无刺蜂工蜂则是在相同日龄（约 15 d）产营养卵和繁殖卵，但是它们开始产繁殖卵的日龄就更小（Koedam et al., 1999）[8]。

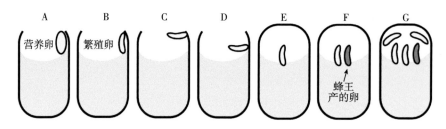

图 5.2 *Scaptotrigona postica* 无刺蜂工蜂产的不同类型的卵

注：A 为正常位置的营养卵。B 为了饲喂蜂王而产的繁殖卵的位置。C、D 为工蜂产下的卵所在的其他位置。E 为蜂王产卵前就封盖的蜂子巢室。F 为蜂王产卵后封盖的蜂子巢室。G 为蜂王产卵后几种工蜂又在该巢室内产卵（修改自 Bego,1990）。

一些其他无刺蜂蜂种的工蜂会既产繁殖卵又产营养卵（表 5.2），而蜂王（偶尔也有工蜂）则会食用这两种卵（Beig, 1972; Sommeijer, 1985; Bego, 1990; van Benthem et al., 1995; Koedam et al., 1999, 2001; Velthuis et al., 2002; Tanaka et al., 2009）。在像美洲的 *Geotrigona mombuca*、*Tetragonisca angustula*

---

[7] 由于膜翅目昆虫的单倍二倍体性别决定系统，无刺蜂工蜂能产单倍体雄蜂卵，但是不能生育二倍体雌蜂，因为它们无法与雄蜂交配（第 1 章）。

[8] *Melipona bicolor* 无刺蜂的工蜂中 30% ~ 40% 的工蜂在产卵，且产繁殖卵的工蜂与不产卵或产营养卵的工蜂相比会更多地参与到 POP 中（Cepeda, 2006; Koedam and Imperatriz-Fonseca, 2012）。*Tetragonisca angustula* 无刺蜂蜂种的工蜂在有王群中只产营养卵，这种无刺蜂中参与到 POP 中的工蜂中约 35% 都具有活跃的卵巢，该比例与 *Melipona bicolor* 无刺蜂类似（Koedam et al., 1997）。

和澳洲的 *Austroplebeia cassiae* 、*A. australis* 无刺蜂等其他无刺蜂蜂种中工蜂只在蜂王存在时才会产营养卵（表 5.2）（Drumond et al., 1999; Koedam et al., 1996; Bego et al., 1999; Lacerda and Zucchi，1999; Grosso et al., 2000）。在这些案例中蜂王信号很可能会导致工蜂不孕（Nunes et al., 2014）。然而在蜂王不在时，工蜂就会开始产卵（Sakagami et al., 1973; Koedam et al., 1996; Drumond et al., 1999; Nunes et al., 2014）。*Geotrigona mombuca* 无刺蜂则是个例外，这种无刺蜂即使失去蜂王，蜂群中的工蜂也只会继续产营养卵（Lacerda and Zucchi，1999）。

表 5.2　蜂群在有无蜂王情况下不同无刺蜂蜂种的工蜂产卵情况

工蜂产卵情况	参考文献
工蜂不产卵	
Duckeola ghiliani	Sakagami（1982）
Frieseomelitta languida	Engels and Imperatriz-Fonseca（1990）
Frieseomelitta paupera	Sommeijer et al.（1984）
Frieseomelitta silvestrii	Cruz-Landim（2000）
Frieseomelitta varia	Boleli et al.（1999）
Tetragonula carbonaria	Gloag et al.（2007）
Tetragonula minangkabau	Suka and Inoue（1993）
蜂群失王时工蜂主要产繁殖卵	
Leurotrigona muelleri	Sakagami and Zucchi（1974）
Friesella schrottkyi*	Nunes et al.（2014）
Plebeia lucii	Nunes et al.（2017）
蜂群有蜂王时工蜂也产卵的情况	
工蜂既产营养卵也产繁殖卵	
Melipona asilvai	Pereira et al.（2006）
Melipona bicolor	Koedam et al.（2001）
Melipona fasciculata	Pereira et al.（2006）
Melipona quadrifasciata	Cruz-Landim（2000）
Melipona subnitida	Hartfelder et al.（2006）

续表

工蜂产卵情况	参考文献
Melipona favosa	Sommeijer et al.（1999）
Melipona scutellaris	Tóth et al.（2002a）; Pereira et al.（2006）
Paratrigona subnuda	Tóth et al.（2002a）
Plebeia droryana	Drumond et al.（1996）
Plebeia remota	van Benthem et al.（1995）
Scaptotrigona postica	Beig（1972）
Scaura sp.	Sakagami（1982）
工蜂主要产繁殖卵	
Friesella schrottkyi*	Imperatriz-Fonseca and Kleinert（1998）
工蜂主要产营养卵	
Austroplebeia australis	Drumond et al.（1999）
Austroplebeia cassiae	Drumond et al.（1999）
Geotrigona mombuca**	Lacerda and Zucchi（1999）
Hypotrigona gribodoi	Sakagami（1982）
Lestrimelitta sp.	Sakagami（1982）
Plebeia minima	Drumond et al.（2000）
Plebeia saiqui	Tóth et al.（2004）; Drumond et al.（2000）
Tetragonisca angustula	Grosso et al.（2000）
Schwarziana quadripunctata	Tóth et al.（2003）

注：*当蜂群有蜂王时工蜂产繁殖卵速率较低。
**该蜂种即使失王也只产营养卵。

最后，有一些无刺蜂蜂种，如美洲的 *Duckeola ghilianii*、*Frieseomelitta* spp.、*Leurotrigona muelleri*、*Trigonisca duckei* 和亚洲的 *Tetragonula minangkabau* 无刺蜂，当蜂群中存在产卵蜂王时工蜂不会产任何类型的卵（表5.2）（Sakagami et al., 1973; Sakagami and Zucchi, 1974; Sommeijer et al., 1984; Suka and Inoue, 1993），但有些工蜂可能会在蜂群失去蜂王时卵巢激活（例如

Friesella schrottkyi、*Leurotrigona muelleri* 无刺蜂）（Tóth et al., 2004; Nunes et al., 2014）。其他无刺蜂蜂种的工蜂则不能产卵（表 5.2）（Sakagami et al., 1964; Sakagami et al., 1973; Sakagami and Zucchi, 1974; Suka and Inoue, 1993; Cruz-Landim, 2000）。例如，*Frieseomelitta varia* 和 *F. silvestrii* 无刺蜂工蜂的卵巢在其幼虫期时就被重吸收了（Boleli et al., 1999; Cruz-Landim, 2000; Luna-Lucena et al., 2018）。

在无刺蜂蜂王产卵前后，通常是由负责封盖工蜂封盖同时产下繁殖卵（Beig, 1972; Koedam, 1999; Koedam et al., 2001, 2005; Velthuis et al., 2005; Koedam and Imperatriz-Fonseca, 2012）。*Melipona bicolor* 无刺蜂的繁殖工蜂会相互激烈打斗来争夺一个空巢室（Velthuis et al., 2002）。研究者观察一些无刺蜂蜂种的多个工蜂会在同一个巢室中各产一个卵（*Hypotrigona gribodoi* 和 *Melipona compressipes* 无刺蜂）（Bassindale, 1955; Sakagami and Oniki, 1963），并且 Nogueira-Neto（1963）观察到 *Melipona quadrifasciata* 无刺蜂的一个巢室中有不止一只幼虫，这种情况导致了幼虫之间的竞争。Beig（1972）发现在有雄性和雌性幼虫共存的巢室中，雄性幼虫通常更大更灵活，并且最终会杀死它的同巢室幼虫。

无刺蜂工蜂有时会在 POP 后再次打开巢室，并吃掉蜂王产下的卵，然后把自己的卵产在幼虫食物上（Imperatriz-Fonseca and Kleinert, 1998; Tóth et al., 2002 a; Velthuis et al., 2002, 2005; Koedam et al., 2007; Koedam and Imperatriz-Fonseca, 2012）。蜂王遇到这样的工蜂时，可能会试图把工蜂从这个巢室推开，但这似乎并不能阻止工蜂稍后返回这个巢室产卵（Tóth et al., 2002 a）。而研究者观察到 *Melipona trinitatis* 无刺蜂的一只蜂王会用它的上颚打开了一个巢室并且吃掉工蜂产的卵（Sommeijer et al., 1984），但这种 POP 后的蜂王监管行为似乎并不常见。事实上，无刺蜂巢室产卵后就被直接封盖了，这可能会使这种被产入了繁殖卵的巢室难以被蜂王或其他工蜂识别，那么其中的繁殖卵也就不会被吃掉（即蜂王/工蜂的监察行为，这种行为在许多其他社会性昆虫中很常见；第 5.6 节）。尽管如此，这些行为表明蜂王和工蜂之间以及工蜂之间存在产卵冲突（第 5.6 节）。

无刺蜂的营养卵和繁殖卵在大小、形状和表面形态上都有所不同（营养卵缺乏网状绒毛膜）（图 5.3）（Sakagami, 1982; Sommeijer, 1985; Bego, 1990; Wittmann et al., 1991; Koedam et al., 1996, 1997, 2001; Pereira et al., 2006;

Tanaka et al., 2009）[9]。此外，营养卵还缺少细胞核（Koedam et al., 1996），而且要么没有要么只有一个高度退化的卵膜孔，卵膜孔是穿过绒毛膜的小开口通道，精子就是通过卵膜孔进入卵子的（图 5.3 E，图 5.3 F）（Koedam et al., 1996）.

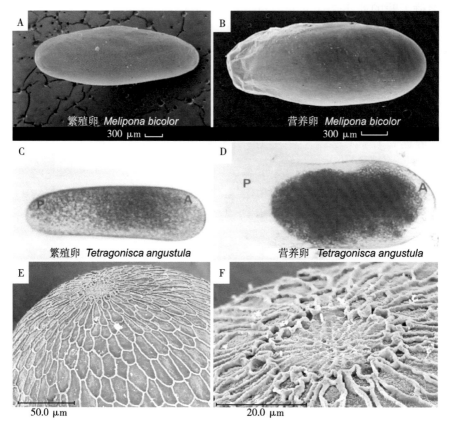

图 5.3 *Melipona bicolor* 和 *Tetragonisca angustula* 无刺蜂工蜂产的繁殖卵（A、C）和营养卵（B、D）的扫描电镜图（A、B）与光学显微镜图（C、D）

注：P = 后部，A = 前部。图 A 显示的网状绒毛膜最终过渡到一个光滑的基底顶点（Koedam et al.,1996, 2001）。E 为繁殖卵的网状绒毛膜（最外层膜）的扫描电镜图，此处展示的为 *Scaptotrigona pectoralis* 无刺蜂的繁殖卵（Rozen et al. ,2019）。F 为卵膜孔的扫描电镜图特写（Rozen et al. ,2019）。

营养卵富含脂质和蛋白质，这使它们成了蜂王的一种营养食物来源（Hartfelder et al., 2006; Tanaka et al., 2009; Luna-Lucena et al., 2018）。营养卵

---

[9] 根据无刺蜂蜂种不同，工蜂产的营养卵会大于（*Scaptotrigona* 无刺蜂）或小于（一些 *Melipona* 无刺蜂）繁殖卵（Beig 1972; Koedam et al., 2001; Pereira et al., 2006）。*Melipona* 属无刺蜂工蜂产下的繁殖卵比蜂王产下的繁殖卵还要小（Pereira et al., 2006）。

的另一种或者说是附加功能是它保持了工蜂卵巢的功能，这可能有利于工蜂在蜂王死亡后能够产卵繁殖（West-Eberhard，1981）。营养卵通常在POP期间产在开口的巢室中，这也同样值得注意。营养卵仅仅是蜂王的食物还是为了维持工蜂的卵巢功能？我们可以预测即使不在POP阶段或者是在其他地方，一旦工蜂遇到蜂王就会产营养卵。有人认为营养卵是关于远古蜂群关于谁产卵的冲突的残存性状，在某些蜂种的祖先中胜出的是蜂王（Peters et al.，1999）。然而研究者最近对无刺蜂一项祖先状态测定中发现，无刺蜂工蜂产营养卵是祖先状态的可能性更大，而蜂王统治下工蜂产繁殖卵的情况的出现时间则相对较晚（图5.4）（Grüter，2018）。营养卵和繁殖卵之间似乎有很强的关联性，因为在蜂王存在的情况下会产繁殖卵的蜂种大多也是会产营养卵的蜂种（表5.2）。因此，无刺蜂祖先的工蜂产卵可能是一种利他功能，但这种情况可能随后促进了某些蜂种工蜂的利己性繁殖进化（Grüter，2018）。当然也并不能排除另一种可能性，即繁殖卵是从无刺蜂先祖蜂种的营养卵中衍生出来的，而无刺蜂先祖蜂种有可能就是已灭绝的真社会性的 *Meliponini* 和 *Melikertini* 属蜂，它们是无刺蜂近亲（第2章）。

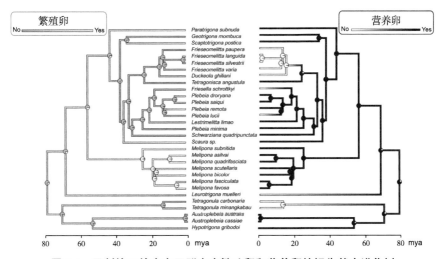

图5.4　无刺蜂工蜂在有王群中产繁殖卵和营养卵的祖先状态进化树

注：该进化树中的颜色是采用MCMC方法计算的祖先状态情况。饼图是基于连续时间马尔可夫链模型计算而来（更多细节参见Grüter,2018）。

5.3 幼蜂发育阶段

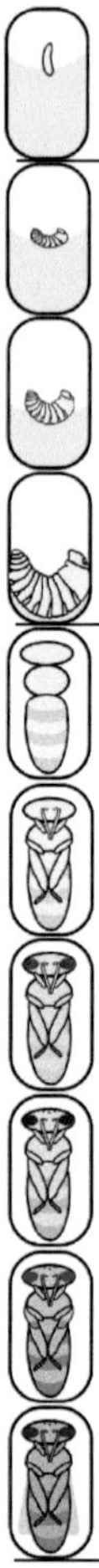

无刺蜂的幼蜂的发育通常分为10个阶段：卵期、3个幼虫期、1个蛹前期和5个蛹期（图5.5）（Nates-Parra et al., 1989; Salmah et al., 1996; Moo-Valle et al., 2004; Amaral et al., 2010; Cardoso-Júnior et al., 2017）。无刺蜂在3个幼虫期期间的大小和颜色都有不同，对应着无刺蜂幼虫的3个不同龄期（Nates-Parra et al., 1989; Salmah et al., 1987; Salmah et al., 1996; Amaral et al., 2010）。无刺蜂幼虫末期的特征是在胸部节段和一些腹部节段上存在一对背外侧圆锥状结节（Rozen et al., 2019）。Amaral等（2010）根据幼虫食物的浓度和是否存在粪便等特征进一步将3龄幼虫期划分成5个阶段。而到了蛹前期，无刺蜂幼蜂的头、胸和腹部都变得可以辨别，但看不到腿（Salmah et al., 1987, 1996）。紧随其后便是5个蛹期，分别是带有肉眼可见的腿和触角的白眼蛹（蛹1期）、粉色眼蛹（蛹2期）、带有色素斑的棕色眼蛹（蛹3期）、有折叠翅膀的褐色眼蛹（蛹4期）以及身体颜色更深且翅膀变大的棕色眼蛹（蛹5期）。研究者目前只测定了少数无刺蜂蜂种的不同幼蜂发育阶段的时长（图5.5）（Nates-Parra et al., 1989; Salmah et al., 1987; Salmah et al., 1996; Moo-Valle et al., 2004）。

发育阶段	雄蜂	蜂王	工蜂
卵期	8	8	8 (5.5)
幼虫期1期	5	5	5 (1)
幼虫期2期	4	4	4 (2)
幼虫期3期	10	8	9 (7)
蛹前期	2	2	2 (1)
蛹1期	3	3	3 (5.3)
蛹2期	8	6	8 (15.7)
蛹3期	6	4	6 (4)
蛹4期	3	4	3 (3)
蛹5期	6	7	5 (2)
总天数：	53.5	50.8	52.7(46.5)

图5.5 无刺蜂幼蜂的发育阶段

注：图中展示的是美洲 *Melipona beecheii* 无刺蜂的雄蜂（M）、蜂王（Q）、工蜂（W）幼蜂在各个发育阶段的天数（Moo-Valle et al., 2004）。括号中的值则是亚洲 *Sundatrigona moorei* 无刺蜂工蜂幼蜂在各个发育阶段的天数（Salmah et al., 1987）。其他蜂种的相关数据参见 Salmah 等（1996）。

5.4 幼蜂的繁育

无刺蜂繁殖幼蜂通常有一定程度的季节性，这与可利用的食物源情况和食物储存量有关（Terada et al., 1975; Roubik，1982; Bego，1990; van Benthem et al., 1995; Ribeiro et al., 2003a; van Veen et al., 2004; Chinh and Sommeijer，2005; Alves et al., 2009 a; Halcroft et al., 2013; Prato and Soares，2013; Maia-Silva et al., 2016）。经历了寒冷冬季或极端季节性降雨的无刺蜂蜂种可能在最冷或最潮湿/最干燥的时期完全停止育虫（Santos et al., 2014, 2015）。这种繁殖滞育情况主要出现在新热带区的几个 *Plebeia* 和 *Melipona* 属无刺蜂蜂种以及中国台湾的 *Lepidotrigona hoozan* 无刺蜂中（Terada et al., 1975; van Benthem et al., 1995; Ribeiro et al., 2003 a; Borges and Blochtein，2006; Sung et al., 2008; Alves et al., 2009 a; Nascimento and Nascimento，2012; Santos et al., 2014; Ferreira-Caliman et al., 2017）。这种繁殖滞育通常持续1～4个月，但也会持续长达5～6个月（如 *Plebeia saiqui* 和 *Melipona marginata* 无刺蜂就有这种情况）（Pick and Blochtein, 2002; Borges and Blochtein, 2006）。在滞育开始之前，无刺蜂工蜂通常会减缓巢室建造速度，并通过增加额外的包壳层来增加包壳的强度（第3章）（Borges and Blochtein，2006; Santos et al., 2014）。工蜂也会减少对花粉的采集（Pick and Blochtein, 2002），且工蜂的寿命延长了（van Benthem et al., 1995），并且 *P. remota* 无刺蜂的蜂王在繁殖滞育期间会出现体重下降并且其膨腹表现出一定程度减小的情况（Ribeiro et al., 2003 a）。

5.4.1 工蜂幼蜂的繁育

无刺蜂蜂巢中繁育的幼蜂65%～99%都是工蜂幼蜂（表5.3）。工蜂幼蜂的繁育量与食物储备特别是花粉储备呈正相关（Roubik, 1982; Chinh and Sommeijer, 2005; Maia-Silva et al., 2016），也与蜂群规模和觅食条件呈正相关（Santos-Filho et al., 2006; Prato and Soares, 2013; Maia-Silva et al., 2015）[10]。无刺蜂工蜂幼蜂从蜂王产卵到幼蜂出房成为成年蜂需要35～55 d（图5.5，表5.4）。这种时间长短似乎与无刺蜂的体型大小无关（图5.6 A）。因此，无刺蜂的工蜂幼蜂生长发育所需的平均时间大约是温带蜜蜂（约21 d）

---

[10] 当食物储存量匮乏时，工蜂偶尔会动用巢内已经存在的幼虫及其幼虫食物来作为食物（Roubik，1982）。

的 2 倍（Winston，1987）。其中一个原因可能是蜜蜂相比于一些无刺蜂蜂种更善于在育虫区域维持较高且恒定的温度（第 3 章）。这种保持蜂巢温度恒定的能力在无刺蜂中相对退化，这也可以用来解释为什么无刺蜂的发育时间较之蜜蜂差异更大。例如，van Veen（2000）观察到在哥斯达黎加的 *Melipona beecheii* 无刺蜂工蜂幼蜂需 43 d 出房，而在墨西哥的同种无刺蜂工蜂幼蜂则需要 50 d 以上的发育时间（Moo-Valle et al., 2004）。无刺蜂幼蜂发育时间较长的另一个原因可能是由于无刺蜂幼虫食物中含有较高含量的花粉和较低比例的腺体分泌物，因而更难被幼虫消化（Moo-Valle et al., 2004）。然而，这不能完全解释为什么无刺蜂的蛹期要长于蜜蜂。

表 5.3 不同无刺蜂蜂种蜂巢内繁育的幼蜂中蜂王、工蜂和雄蜂的百分比

无刺蜂蜂种	蜂王	工蜂	雄蜂	是否有蜂王巢室	参考文献
Apotrigona nebulata	0%	89.5%	10.5%		Darchen（1969）
Lepidotrigona ventralis	0.08%	96.8%	3.1%	是	Chinh and Sommeijer（2005）
Lisotrigona carpenteri	7.90%			是	Chinh et al.（2005）
Melipona asilvai	6.2%	76.2%	17.6%	否	Santos-Filho et al.（2006）
Melipona beecheii	5.3%～11.3%	65.8%～89%	5.7%～22.9%	否	Moo-Valle et al.（2001）; van Veen et al.（2004）
Melipona bicolor	6.2%	87.6%	6.2%	否	Santos-Filho et al.（2006）
Melipona fasciata**	10.30%	79.30%	10.40%	否	Kerr（1948）
Melipona favosa	4.2%～7.7%	71%～78.5%	17%	否	Koedam（1999）; Sommeijer et al.（2003）
Melipona flavipennis***	5.10%	75.50%	19.40%	否	Kerr（1951）
Melipona interrupta***	13.70%	86.30%	0%	否	Kerr（1951）
Melipona marginata	19.1%	68.20%	12.70%	否	Kerr（1948
Melipona melanoventer***	6.30%	62.10%	31.60%	否	Kerr（1951）
Melipona quadrifasciata	10.30%	78.2%	11.50%	否	Kerr（1951）
Melipona subnitida	7.8%	85.1%	7.1%	否	Santos-Filho et al.（2006）
Melipona trinitatis	5.20%	87.10%	7.70%	否	Sommeijer et al.（2003）
Nannotrigona testaceicornis	0.01%	98.3%	1.7%	是*	Eterovic et al.（2009）

续表

无刺蜂蜂种	蜂王	工蜂	雄蜂	是否有蜂王巢室	参考文献
Plebeia remota	0.1%	73.4%	26.5%	是 *	Alves et al.（2009a）
Schwarziana quadripunctata	0.60%	81.20%	18.2%	是 *	Wenseleers et al.（2005） Santos-Filho et al.（2006）
Tetragonisca angustula	0.16%	86.7%	13.1%	是	Prato and Soares（2013）
Trichotrigona extranea			0%～18%		Camargo and Pedro（2007）

注：* 繁育小型蜂王的蜂种。
** 该蜂种未确定，根据 Moure's 的蜂分类可能是 *M. rufiventris*（Camargo and Pedro，2013）。
*** 样本量较小。

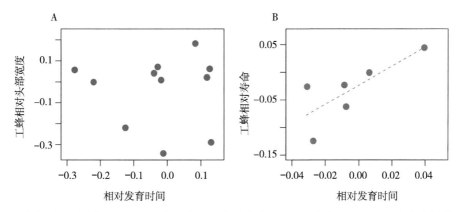

图 5.6 A 为 12 个无刺蜂蜂种的工蜂发育时间（从产卵到出房）和平均工蜂大小［工蜂体型大小数据主要来自 Grüter et al.（2017）］的相关性。B 为 7 个无刺蜂蜂种的工蜂发育时间与工蜂平均寿命（数据来自表 5.4）的相关性

注：图中使用了系统发育独立对比和物种起源线性回归（Felsenstein,1985; Garland et al.,1992）。系统发育关系和分支长度是基于一个已发表的系统发育树（Rasmussen and Cameron,2010）。发育时间和工蜂大小之间不存在显著相关性（t 值 =0.02；p=0.98），但发育时间与工蜂寿命间存在显著的正相关关系（t 值 =2.83，p=0.048）（数据和进化树参见：www.socialinsect-research.com/book.php）。发育时间与工蜂寿命的正相关可能是由各种生态因素驱动的，如气候。

无刺蜂工蜂的寿命往往比蜜蜂工蜂长（表 1.1，表 5.4），这可能是因为无刺蜂工蜂幼蜂需要的发育时间更长。有证据表明，7 个无刺蜂蜂种的数据显示工蜂幼蜂发育时间与其寿命之间存在正相关（图 5.6 B）。*Tetragonisca angustula* 无刺蜂幼蜂的发育时间还取决于幼蜂巢室靠近巢中心的距离（Nates-

Parra et al., 1989），这可能是因为随着巢室与子脾中心的距离增加，巢室中的食物量会随之减少，其内幼蜂的体型大小也随之减小（Segers et al., 2015）。

5.4.2 雄蜂幼蜂的繁育

无刺蜂蜂巢中繁育幼蜂的 1% ～ 30% 是雄蜂幼蜂（表 5.3）。雄蜂幼蜂的发育时长与工蜂幼蜂的发育时长相似（表 5.4），这也是合情合理的，因为雄蜂和工蜂是在相同的巢室中繁育出来的，而且这些巢室中的食物量也相差无几[11]。

表 5.4　不同无刺蜂蜂种的工蜂、雄蜂和蜂王幼蜂的发育时间（即从产卵到出房的时长）

发育时间（d）	工蜂	雄蜂	蜂王	工蜂平均寿命	参考文献
Austroplebeia australis	～55				Halcroft et al.（2013）
Friesella schrottkyi	35～36		39～48	30.1	Sakagami（1982）; Giannini（1997）
Frieseomelitta varia			25		Faustino et al.（2002）
Geotrigona mombuca	52±1.5				Lacerda et al.（1991）
Heterotrigona itama	46.5				Salmah et al.（1996）
Hypotrigona gribodoi	～35	～32			Bassindale（1955）
Melipona beecheii	52±1.3	53.4±1.12	50.8±1.5	51	Biesmeijer and Toth（1998）; Moo-Valle et al.（2004）
Melipona quadrifasciata	34～37		27～31		Salmah et al.（1987）; da Silva et al.（1972）
Melipona scutellaris	49			43.8	Moo-Valle et al.（2004）; Santos（2013）
Nannotrigona perilampoides	36～38				Quezada-Euan et al.（2011）

---

[11] *Scaptotrigona postica* 无刺蜂中装工蜂产的雄蜂卵的巢室中幼虫食物量（37.3 μL）低于装蜂王产的雄蜂卵的巢室中幼虫食物量（38.9 μL）（Beig et al., 1982），两者之间尽管差异较小但是存在显著差异。

续表

发育时间（d）	工蜂	雄蜂	蜂王	工蜂平均寿命	参考文献
Plebeia remota	40～45	40～45	48～54	67.7	van Benthem et al.（1995）; Grosso and Bego（2002）
Scaptotrigona depilis	～35				Vollet-Neto et al.（2017）
Sundatrigona moorei	46.5				Salmah et al.（1987）
Tetragonisca angustula	33.5～40.3			27	Nates-Parra et al. 1989; Hammel et al.（2016）
Tetragonula laeviceps	～50			45	Inoue et al.（1984）
Tetragonula minangkabau	～42			37	Salmah et al.（1987）; Inoue et al.（1996）

注：以天（d）为单位展示了工蜂的平均寿命，也见表1.1。

蜂群繁育的雄蜂幼虫量具有季节性，这种季节性则与外部因素（如降水、温度和食物供应）和内部因素（如蜂群规模和食物储存量）有关（第4章）（Kerr，1948; Bego，1990; van Veen et al., 2004; Chinh and Sommeijer，2005; Koedam et al., 2005; Velthuis et al., 2005; Prato and Soares，2013; Becker et al., 2018）。巴西南部的 *Tetragonisca angustula* 无刺蜂蜂群繁育雄蜂就是季节性的，该蜂种中大多数雄蜂幼蜂是在夏末出房，这个时间与该地区最潮湿炎热的时期结束的时间相对应（图5.7）（Prato and Soares，2013）。*Lepidotrigona ventralis* 无刺蜂则更多是在雨季繁育雄蜂，蜂群在这个时期也有更多的食物储备（Chinh and Sommeijer，2005）。*Scaptotrigona postica* 无刺蜂繁育雄蜂的时期则是与蜂群群体生产力最强的时期相吻合（Bego，1990）。研究者发现，如果实验性地调整 *Melipona beecheii* 无刺蜂蜂群中食物储备量，就会影响蜂群中繁育的雄蜂幼虫比例，这一发现进一步证实了蜂群的食物储备会影响蜂群中雄蜂繁育量的观点（Moo-Valle et al., 2001）。往 *M. subnitida* 无刺蜂蜂群中添加花粉还会刺激工蜂和蜂王产卵（Koedam et al., 2005）。由于蜂群中的食物储备量和蜂群的规模还可能促进分蜂（第4章）（Darchen，1977），因此我们可以预测繁育雄蜂和蜂群分蜂之间具有一定程度的时间轴相关性（Moo-Valle et al., 2001）。

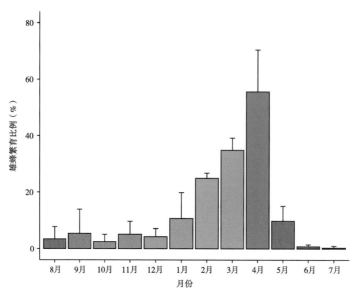

图 5.7　2008 年 8 月至 2009 年 7 月,圣保罗州 *Tetragonisca angustula* 无刺蜂 3 个有王群的雄蜂繁育情况(平均值 ± 标准差,Prato and Soares,2013)

一些无刺蜂蜂种的蜂群还会扎堆繁育雄蜂幼蜂(雄蜂繁育期;male producing periods,MPP 期),在这一时期相同蜂种和相同地区的蜂群繁育雄蜂的时间是不同步的。存在这些 MPP 期的一种解释是无刺蜂工蜂会定期繁育雄蜂(Chinh et al., 2003; Koedam et al., 2005)。一些无刺蜂蜂种(*Melipona beecheii*、*Scaptotrigona postica* 和 *Tetragonisca angustula* 无刺蜂)繁育的雄蜂幼虫分散分布于整个巢脾的巢室中(图 5.8)(Beig, 1972; Moo-Valle et al., 2004; Segers et al., 2015),但还有些无刺蜂蜂种(*M. quadrifasciata*、*M. favosa* 和 *Nannotrigona testaceicornis*)的雄蜂幼虫巢室则分布得比较集中且通常是在子脾的中心位置(Sakagami et al., 1965; Koedam, 1999; Eterovic et al., 2009)。雄蜂幼虫巢室是否扎堆集中分布可能取决于这些雄蜂主要是由蜂王生育的还是由工蜂生育的。如果主要是工蜂生育的雄蜂幼虫,则雄蜂幼虫分布比较集中,这是因为诸如 *Melipona favosa* 等无刺蜂蜂种的工蜂会在某些特定时期产卵,也可能是因为诸如 *M. subnitida* 等无刺蜂蜂种的工蜂喜欢在新建的巢脾巢室中产卵(因此这些卵都产在靠近巢脾中心的巢室中)(Koedam et al., 1999; Koedam et al., 2005; Chinh et al., 2003; Moo-Valle et al., 2004)。

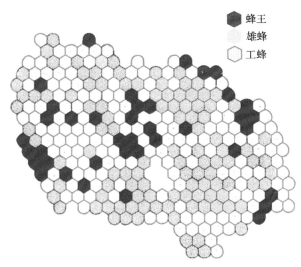

图 5.8 *Melipona fulva* 无刺蜂的一个子脾中不同级型和性别的幼蜂的巢室分布情况（Kerr et al.,1967）

蜂群如果繁育的雄蜂幼蜂越多，那么蜂群繁育工蜂资源和空间就更少，这就导致蜂群需要进行能量分配的权衡（Velthuis et al., 2005）。这种情况也导致蜂种繁育的雄蜂和工蜂量之间经常呈负相关（Moo-Valle et al., 2004; Santos-Filho et al., 2006; Prato and Soares, 2013）。当食物充足时（见上文），蜂群可能会将繁育的重心更多地转向繁育雄蜂，这可能是因为在这种情况下蜂群能够耐受工蜂数量较少的情况。此外，当觅食条件有利时，工蜂利己性的产卵繁殖对于蜂群来说可能更经济节约（第5.6节）。

巨型雄蜂

无刺蜂蜂群偶尔会在蜂王巢室中繁育雄蜂[12]。研究者已经在 *Friesella schrottkyi*（Camillo, 1971）、*Geotrigona mombuca*、*Nannotrigona testaceicornis*、*Paratrigona subnuda*（Engels and Imperatriz-Fonseca, 1990）、*S. postica*（Bego and de Camargo, 1984）、*Schwarziana quadripunctata*（Santos-Filho et al., 2006）和几个 *Plebeia* 属（Engels and Imperatriz-Fonseca, 1990; Alves et al., 2009 a）无刺蜂蜂种中发现了这样的巨型雄蜂。巨型雄蜂的发育时间更长，而且会产生更多的精子细胞（Camillo, 1971），但是这种巨型雄蜂是否是

---

[12] 在一些极为稀少的情况下，无刺蜂蜂群还会在蜂王巢室中繁育巨型工蜂，而这种情况是令人费解的，因为当雌性幼虫被给予大量蜂王食物时会发育成为蜂王（Imperatriz-Fonseca and Zucchi, 1995）。这种巨型工蜂可能是因为发育错误，但是它们仍然会执行普通工蜂的任务（Engels and Imperatriz-Fonseca, 1990）。

功能性的且能否使蜂王受精则不得而知。

5.4.3 蜂王幼蜂的繁育

尽管无刺蜂全年都会繁育蜂王幼蜂（第4章），但是一些无刺蜂蜂种繁育蜂王还是表现出一定的季节性。例如哥斯达黎加的 *Melipona beecheii* 无刺蜂6月蜂群中的蜂王幼虫占到蜂群中所有幼虫的15%，而到了8月蜂王幼虫则只占3%（van Veen et al., 2004）。蜂王繁育量与雄蜂繁育量呈正相关，与蜂巢中的花粉储藏量也呈正相关（van Veen et al., 2004; Moo-Valle et al., 2001）。Kerr（1948）在巴西观察到 *Melipona* 属不同蜂种中蜂王的繁育也呈现了类似的季节性。在Kerr（1948）的研究中，蜂王繁育量在9月至翌年4月最高，这与研究区域内出现有利于无刺蜂觅食的条件的时间段相吻合。由于有利觅食条件似乎既有利于蜂群繁育雄蜂，也有利于繁育蜂王，因此不难推断蜂群中雄蜂的繁育量和蜂王的繁育量是呈正相关的（Kerr, 1948; Moo-Valle et al., 2001; van Veen et al., 2004）。

大多数无刺蜂蜂种的蜂王幼蜂只占所有蜂子的一小部分（表5.3）。因此，具有生育后代能力的蜂的性别比例严重偏向雄性，这在通过分蜂来实现再生产的蜂种中是可想而知的，因为每个分蜂群（只包含一个或几个蜂王）都代表着母蜂群的巨大投资（Cronin et al., 2013）。当然也有例外，*Melipona* 属无刺蜂就是最典型的，该属无刺蜂蜂群中出房幼虫的 10%±4.5%（平均 ±SD, N = 12）都是蜂王（表5.3）。这意味着 *Melipona* 属蜂群中繁育的蜂王与雄蜂幼蜂的比例接近1（约13%为雄蜂）。然而在非 *melipona* 属无刺蜂蜂种中新出房蜂王幼蜂的占比则仅为约0.2%[13]。*Melipona* 属无刺蜂工蜂会杀死大部分新出房蜂王幼蜂（第4章），这意味着蜂群损失了大约10%（最高约25%）的繁育投入。

Melipona 属无刺蜂如此过量繁育蜂王似乎是非常浪费的，对此现象已经有了几种解释。第一种解释是 *Melipona* 属无刺蜂未交配的蜂王有时会试图进入其他蜂群并接管该蜂群的繁殖工作，这可能会增加那些产生大量处女蜂王的母蜂群的繁殖成功率（第4章）。然而，由于大多数新出房蜂王都会被工蜂杀死，蜂群仍然面临着相当大的投资损失（Wenseleers et al., 2004 b; Jarau et al., 2009; Kärcher et al., 2013）。第二种解释是 *Melipona* 属无刺蜂产生过多的蜂王更有可能是出于其他原因，且一些 *Melipona* 蜂王通过进化出的寄生繁

---

[13] *Lisotrigona carpenteri* 无刺蜂则不是该百分比，这很有趣（Chinh et al., 2005），但是这里并未将这种蜂计算在内，因为其工蜂幼蜂的比例尚不清楚。

殖作为一种替代策略而获得直接的适应性。*Melipona* 属无刺蜂蜂群繁育很多蜂王的原因可能是因为这样工蜂就可以不断地比较处女蜂王和膨腹蜂王，以确保高质量的蜂王统领蜂群（第 4 章）（Roubik，1989; Imperatriz-Fonseca and Zucchi，1990）。这对于 *Melipona* 属无刺蜂可能特别重要，因为该属蜂群只会繁育"微型"蜂王（见下文），这种蜂王的繁殖力较低。因此我们推测，无刺蜂蜂群繁育蜂王幼蜂的量与蜂巢中现有蜂王的繁殖力之间存在负相关关系，因为如果当前蜂王的繁殖力很强，工蜂就会减少在养育候补蜂王方面的投入。然而，蜂群群势强弱和繁育蜂王幼蜂量之间则不是负相关的（见下文），而且持续对比蜂王的好处也不能弥补持续淘汰蜂王的投入。

第三种解释是这样过量繁育蜂王还可能是保障母蜂王突然死亡的一种应急策略（Michener，1974）。然而其他无刺蜂蜂种也会一直在蜂群中保持处女蜂王的存在（例如，在"监禁室"养育着处女蜂王，见第 4 章），但是这些蜂种繁育的蜂王幼蜂量却不会过量。第四种解释则是基于蜂群的社会等级矛盾：许多 *Melipona* 幼虫会因为其个体利益和该属蜂种特有的蜂王决定系统而成为蜂王，这意味着工蜂是不能阻止这一进程的（这个假设详细讨论见第 5.6 节）。

5.5 无刺蜂的蜂王级型确立机制

社会性昆虫的蜂王级型确立机制通常分为两大类，环境决定型和遗传决定型蜂王级型。一些无刺蜂蜂种的蜂王级型确立则是两类因素的共同作用（Schwander et al., 2010）。影响蜂王级型确立的主要环境因素是食物量和/或质量，但诸如温度等其他因素对于蜂王发育也很重要。营养和遗传因素也是无刺蜂蜂王发育的关键（Velthuis，1976; Hartfelder et al., 2006）。在大多数无刺蜂蜂种中，一些幼虫获得了更多的食物从而触发了内分泌系统的响应，进而引发了蜂王级型发育（Hartfelder et al., 2006）。而 *Melipona* 属无刺蜂蜂王级型确立很可能是取决于遗传和营养因素的共同作用。

5.5.1 营养确立蜂王级型

无刺蜂中营养型蜂王级型确立存在两种不同的模式。第一种是在大多数无刺蜂属中都有发现的情况，即蜂王是从体积更大的蜂王巢室中出房的，这种巢室比工蜂和雄蜂巢室含有更多的幼虫食物（图 5.9 A，图 5.9 B）。尽管研究者没有进行过准确测定，但他们发现 *Scaptotrigona depilis* 和 *S. postica* 无刺蜂的蜂王巢室中的幼虫食物大约是其工蜂巢室的 4 倍，而养育一只 *Tetragonisca angustula* 无刺蜂蜂王所用的幼虫食物是一只工蜂的 6～7 倍

（Engels and Imperatriz-Fonseca, 1990; Menezes, 2010; Cham et al., 2019）。因此，蜂王巢室中的幼虫可以吃得更久、长得更大（图 1.6）。采用这种蜂王级型确立模式的蜂王的蜕变时间滞后于工蜂，而蜂王幼蜂却拥有了更长的发育时间（表 5.3 中的 *Friesella schrottkyi* 和 *Plebeia remota* 无刺蜂）[14]。无刺蜂的蜂王巢室通常位于子脾的外围（图 5.9 A，图 5.9 B），但偶尔也会位于子脾较中心的位置（Sakagami，1982）。

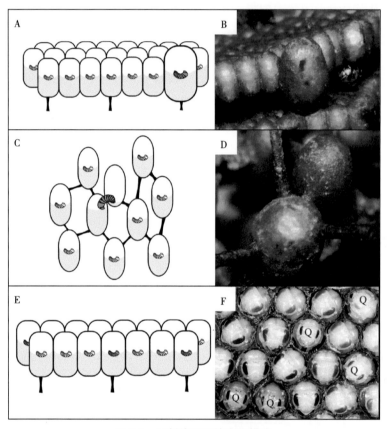

图 5.9　无刺蜂不同的育王模式

注：大多数无刺蜂蜂种在较大的蜂王巢室中养育蜂王，这种巢室中含有更多的幼虫食物（A、B）。蜂王巢室通常位于蜂巢的边缘。构建簇状巢的无刺蜂蜂种的一些幼虫在吃完自己的食物后会去吃相邻的辅助巢室中的食物（C、D）。*Melipona* 无刺蜂则是在相同大小的巢室中养育蜂王、工蜂和雄蜂（E、F）。在同时繁殖大型蜂和微小型蜂王的无刺蜂蜂种中也发现这种现象（B 显示的是 *Tetragonisca angustula*，C 显示的是 *Frieseomelitta longipes*，照片由 Cristiano Menezes 拍摄；F 显示的是 *Melipona subnitida*，源自 Wenseleers and Ratnieks, 2004）。

---

[14] 研究者观察到 *Trigona hypogea* 无刺蜂蜂王幼虫在其巢室中保持一种休眠状态长达 11 个月（Imperatriz-Fonseca and Zucchi，1995）。

无刺蜂幼虫使用的食物量似乎是其社会级型确立的主要因素（Velthuis，1976; Hartfelder and Engels，1989）。例如，de Camargo（1972 b）在实验室用不同的食物量喂养 *Scaptotrigona postica* 幼虫后发现，当给幼虫饲喂了较多的食物时就会养育成蜂王，而只有当幼虫获得了与工蜂巢室中平均食物量相当的食物时幼虫才会长成工蜂（Velthuis，1976）。Darchen 和 Delage-Darchen（1971）甚至可以用异种幼虫养育出蜂王。这与蜜蜂的情况不同，蜜蜂的 3 日龄幼虫从食用蜂王浆转变为食用工蜂浆后就会发育成工蜂（Haydak，1970; Mao et al., 2015）。然而导致无刺蜂工蜂建造新蜂王巢室的激素因素仍有待研究。

在一些（可能是大多数）建造簇状蜂巢的无刺蜂蜂种中发现了营养确立蜂王级型的第二种模式（例如 *Austroplebeia cassia*、*Celetrigona longicornis*、*Frieseomelitta* spp.、*Leurotrigona muelleri*、*Plebeia minima* 和 *P. lucii* 无刺蜂），这些蜂种的所有幼虫都是在常规巢室中繁育的，但有些幼虫可以通向相邻巢室从而获得了额外份额的幼虫食物进而发育成蜂王（图 5.9 C，图 5.9 D）（Sakagami，1982; Engels and Imperatriz-Fonseca，1990; Faustino et al., 2002; Ribeiro et al., 2006b; Luz et al., 2017）。例如，*Leurotrigona muelleri* 无刺蜂一些末日龄期的幼虫在吃完它们巢室内的所有食物后，会咬穿巢室壁到后期建造的相邻巢室中去吃其中的幼虫食物（Engels and Imperatriz-Fonseca，1990）。工蜂经常会将两个巢室连在一起为正在发育的蜂王建造成一个大巢室。这种蜂王幼蜂养育模式有助于育成"紧急蜂王"：如果突然需要新蜂王，工蜂就在含有晚期雌幼虫的巢室旁紧挨着建造一个"辅助"巢室并装入幼虫食物（Faustino et al., 2002; Nunes et al., 2015; Luz et al., 2017）[15]。这些辅助巢室中不含钾却装满了幼虫食物，通常建造在含有幼虫的巢室旁边（Luz et al., 2017），因此，蜂王消耗的幼虫食物大约是工蜂的两倍（*Frieseomelitta varia* 无刺蜂工蜂消耗的幼虫食物为约 27 μL，蜂王则是 50～65 μL，Baptistella et al., 2012）。

目前还不清楚为什么有些幼虫会咬穿它们的巢室壁并吃掉相邻巢室中的食物，而有些幼虫则不会，很可能是工蜂通过在两个巢室之间建立连接才触发了这个过程（Faustino et al., 2002）。这就提出了一个问题，即工蜂是如何选择巢室来养育蜂王的。

---

[15] *Tetragonula carbonaria* 无刺蜂在饲养应急蜂王时结合了蜂王饲养的两种模式；蜂王通常饲养在较大的巢室中，但是一旦出现蜂王损失，急需一个新蜂王时，两个工蜂巢室会互相融合成一个较大的巢室（Nunes et al., 2015）。

5.5.2　遗传确立蜂王级型

Melipona 是一种新热带属无刺蜂，其中包含了 70 多个蜂种，*Melipona* 属无刺蜂的蜂王和工蜂是在大小和幼虫食物量都几乎没有差异的巢室中养育而成的（图 5.9 E，图 5.9 F）。这使得食物量不太可能是决定这种蜂级型分化命运的关键因素（Kerr，1946，1948; Sakagami，1982）。而 Kerr（1946, 1948, 1950 a, 1950 b）则提出了一种决定了级型分化的遗传基础。他认为无刺蜂存在两个遵循孟德尔遗传定律的双等位基因，这两个基因决定了雌蜂的级型。他提出只有在这两个基因位点都杂合（AaBb）的雌蜂才能发育成蜂王，这导致在单雄交配的蜂群中繁育出的雌蜂分化成工蜂与蜂王的比例为 3∶1[16]。研究者观察发现在自然条件下多达 25% 的雌性幼虫会分化成蜂王（Kerr，1948，1951; de Camargo et al.，1976; Velthuis，1976），上述假设就是基于这些观察发现得来的。尽管人们已经广泛认可了这一假设，但这种假设仍然缺乏确凿的证据，如尚未鉴别出级型分化确立的基因位点[17]。事实上，大多数无刺蜂蜂种的雌性后代中只有不到 25% 能分化成蜂王（平均 10%，表 5.3）（Kerr，1948; Koedam，1999; Moo-Valle et al.，2001; Sommeijer et al.，2003; van Veen et al.，2004; Wenseleers and Ratnieks，2004; Santos-Filho et al.，2006），这表明其他因素也起着重要作用。目前科学家推测，在食物质量和/或数量不佳的情况下一些具有蜂王基因型的幼虫会分化成工蜂（Kerr，1948; Kerr and Nielsen，1966; Darchen and Delage-Darchen，1975; Koedam，1999; van Veen，2000; Hartfelder et al.，2006）。有迹象表明，具有蜂王基因型而表型为工蜂的无刺蜂在幼虫蜕变期会有融合的蜂王样腹神经节［Kerr and Nielsen，1966; Darchen 和 Delage-Darchen（1975）则并未在 *Melipona beecheii* 无刺蜂中确切观察到这种情况］，基于这种特征以及蜂王基因型工蜂的背部腺体数量（Cruz-Landim et al.，1980）仍然可以从工蜂基因型的工蜂中识别出蜂王基因型工蜂，但这需要进一步研究。因此如果生存条件不利，例如食物储备不足时蜂群中育成的蜂王就会更少。然而关于食物储备是否真的起到了重要作用的相关证据则是混杂

---

[16] 早期 Kerr（1946, 1948, 1950 a）也针对一些 *Melipona* 属无刺蜂提出过三基因位点模型来解释为什么这些蜂种中蜂王的产生比例低于预估的 25%。三基因位点模式预测出的蜂王级型的比率为 12.5%，但是后来 Kerr 自己否定了这种解释并认为营养因素更大程度的作用于蜂王级型分化。

[17] 育种试验也提供了一些证据。例如采用保幼激素处理的大多数幼虫会发育成蜂王。这意味着具有工蜂基因型的蜂王是可以被育成的。这种蜂王的人工交配理论上可以为我们测试由"工蜂"蜂王领导的蜂群繁育出的蜂王／工蜂幼虫比例是否有所不同。然而这个试验至今尚未成功，因为这种方式育成的蜂王死亡率很高（Engels and Imperatriz-Fonseca，1990）。

的。Moo-Valle 等（2001）通过在蜂群中添加或移除食物发现，即使是高食物供应的蜂群中育成的蜂王数量也只是轻微地增多（13.5% 相对于 10.5%），而在低食物供应的蜂群雄蜂幼蜂的繁育则几乎完全停滞。Maciel-Silva 和 Kerr（1991）发现对于 *Melipona fasciculata* 无刺蜂而言，如果巢室中食物不足则其中的幼虫不太可能分化成蜂王，而当巢室中供应了超过平均量的食物时分化成蜂王的雌蜂则接近工蜂数量的 1/3，这证明了单个巢室的食物量也起到了一定作用。而巢室内食物量高于这个阈值时则不再能进一步影响分化成蜂王的雌蜂比例。他们的研究数据也提到注定分化成蜂王的幼虫要重于注定分化成工蜂的幼虫，这表明蜂王级型的幼虫吃食物的速度更快。这也就解释了为什么直至至今的相关研究中都发现 *Melipona* species 无刺蜂蜂王发育时间比工蜂发育时间要短 2～10 d（表 5.3）。研究者通过在单个巢室中补充幼虫食物发现，*Melipona* 属无刺蜂的 4 个其他蜂种产生了类似的效果（de Camargo et al., 1976）。有趣的是 Darchen 和 Delage-Darchen（1975）发现往 *Melipona beecheii* 无刺蜂巢室中添加食物，其内雌性幼虫发育成蜂王概率会增加到 40%～60%（甚至高达约 70%）[18]，这个比率值则引发了对 Kerr 提出的两基因/双等位基因决定级型分化机制的质疑，因为该机制预测蜂王的比例上限是 25%[19]。应该注意的是，该属无刺蜂蜂王和工蜂通常是用等量的食物养育而成的且出房时体重相同（Wenseleers et al., 2004 b）。Kerr 和 Nielsen（1966）以及 van Veen（2000）都报道了一个有趣的观察结果，即工蜂向养育蜂王的巢室内排入食物的次数较少，这表明个别工蜂会向养育蜂王的巢室中一次排入较大量的幼虫食物。研究者推测这些大于平均排入量的幼虫食物在物质组成上与普通幼虫食物有所不同，例如里面含有腺体分泌物的花粉与蜂蜜的比例可能有所不同。

也有研究者猜测幼虫食物中的腺体分泌物决定了无刺蜂雌蜂的级型分化（Darchen and Delage-Darchen, 1975; Jarau et al., 2010）。Jarau 等（2010）提出无刺蜂哺育蜂唇腺中产生的香叶醇是诱发蜂王级型分化的一个关键因素。香叶醇可能参与了保幼激素（juvenile hormone，JH）生物合成的激活通路，而

---

[18] 研究者发现向巢室中添加或移除食物会影响出房蜂的大小，且研究者已经发现了小型蜂王、巨型蜂王和中型蜂王的存在（Darchen and Delage-Darchen, 1975; Sakagami, 1982; Engel and Imperatriz-Fonseca, 1990）。

[19] 这项研究（Darchen and Delage-Darchen, 1975）因没有提供实验的实际样本量而受到质疑（Velthuis, 1976），这使得我们很难明确该数据的统计意义。

保幼激素则会触发无刺蜂雌蜂分化成蜂王：JH 处理雌蜂幼虫会增加其分化发育成蜂王的概率，有时甚至接近 100%（Velthuis，1976; Campos and Coelho，1993; Bonetti et al., 1995; Pinto et al., 2002; Hartfelder et al., 2006; Cardoso-Júnior et al., 2017）。此外，*Melipona quadrifasciata*（Kerr et al., 1975）无刺蜂的咽下腺的大小（JH 的合成部位）能预测该蜂种蜂王的级型分化。能够决定无刺蜂雌蜂级型分化的 JH 敏感期正好是末龄幼虫期的"翻身"阶段（Pinto et al., 2002; Hartfelder et al., 2006; Luna-Lucena et al., 2018）。

Jarau 等（2010）猜测，哺育蜂会通过只向部分巢室内加入足量的香叶醇从而实现对雌性蜂幼虫级型分化命运的控制。然而这个假说又引发了新的问题：为什么有些无刺蜂工蜂要养育这么多的蜂王幼蜂，然后在这些蜂王幼蜂出房后又立即杀死其中的大多数蜂王幼蜂（第 4 章；第 5.6 节）。Maciel-Silva 和 Kerr（1991）研究认为 *Melipona fasciculata* 无刺蜂的幼虫食物组成和幼蜂的级型分化方向之间并没有直接的关系。他们的研究中均质化了所有的幼虫食物以确保所有的幼虫都得到相同的食物，然后发现雌性幼虫中约 18% 分化发育成了蜂王。因此，这就可以排除 *Melipona fasciculata* 无刺蜂的幼虫食物组成因素参与其蜂王级型分化确立，那么我们又回到这个问题：到底是什么因素决定了 *Melipona fasciculata* 无刺蜂的蜂王级型分化确立。尽管目前存在这些开放性的问题，但是目前最合理解释仍然是遗传和营养因素的组合作用影响了 *Melipona* 属无刺蜂的蜂王级型分化确立，这两种因素共同作用影响了工蜂咽下腺中保幼激素的产生。然后保幼激素会诱发蜂王幼虫的蜕皮激素水平的变化，还会引起驱动级型分化的相关基因表达的变化（Judice et al., 2004, 2006; Hartfelder et al., 2006）。最近 Brito 等（2015）就发现了一个这样的受保幼激素影响的基因。他们发现这个雌性化基因在 *Melipona interrupta* 无刺蜂幼蜂的胚胎期高度表达，该基因在蜂王中的表达量高于在工蜂中的表达量，因此这个基因可能在雌蜂分化发育成蜂王过程中起到了重要作用。

5.5.3 非 *Melipona* 属无刺蜂的微小型蜂王

一些无刺蜂蜂种不仅会在蜂王巢室中繁育大型蜂王，还会在工蜂巢室中繁育微小型（"侏儒"）蜂王。目前研究者已经发现在 *Cephalotrigona capitata*、*C. femorata*、*Nannotrigona testaceicornis*、*Plebeia droryana*、*P. emerina*、*P. julianii*、*P. mosquito*、*P. pugnax*、*P. remota* 和 *Schwarziana quadripunctata* 无刺蜂蜂群中存在微小型蜂王（Camargo，1974; Engels and Imperatriz-Fonseca，1990; Imperatriz-Fonseca and Zucchi，1995; Imperatriz-Fonseca et al., 1997;

Bourke and Ratnieks, 1999; Wenseleers et al., 2005; Ribeiro et al., 2006 b; Alves et al., 2009 a)。而且，*Plebeia mosquito*、*Schwarziana quadripunctata* 和 *Nannotrigona testaceicornis* 无刺蜂的新出房蜂王中大多数都是微小型蜂王，而 *Plebeia remota* 无刺蜂的大多数新出房蜂王则基本都是大型蜂王（Imperatriz-Fonseca et al., 1997; Ribeiro and Alves, 2001; Wenseleers et al., 2005; Ribeiro et al., 2006 b; Santos-Filho et al., 2006）。然而，无刺蜂不同蜂种繁育微小型蜂王的情况各不相同，而且尚不清楚雌蜂幼蜂是如何启动蜂王分化发育的。例如，我们并不知道香叶醇是否与这些蜂种微小型蜂王的繁育有关，就像在 *Melipona* 属无刺蜂中发现的那样（Jarau et al., 2010）。

研究者发现约 20% 的 *S. quadripunctata* 无刺蜂蜂群和 14% 的 *P. remota* 无刺蜂蜂群由微小型蜂王统领的（Ribeiro and Alves, 2001; Wenseleers et al., 2005; Ribeiro et al., 2006 b），这表明微小型蜂王是具有生物学功能的。然而有证据表明微小型蜂王的卵巢重量较低，卵巢管数量较少，产卵率也较低，且它们产下的卵体积更小（Imperatriz-Fonseca et al., 1997; Cruz-Landim, 2000; Ribeiro and Alves, 2001; Wenseleers et al., 2005; Ribeiro et al., 2006 b; Ribeiro et al., 2006 a）。Ribeiro 等（2003 b）猜测，稍微增加一点 *Plebeia remota* 无刺蜂幼虫的食物，就足以增加该幼虫长成蜂王的可能性。之前的研究也发现一些 *Plebeia remota* 无刺蜂蜂群中的微小型蜂王的体重介于工蜂和正常蜂王之间（Ribeiro et al., 2006 b）。因此，微小型蜂王可能是幼虫在巢室内被意外过度喂养的结果。然而，*Nannotrigona testaceicornis* 和 *S. quadripunctata* 无刺蜂（还包括盛产微小型蜂王的 *P. remota* 无刺蜂）的微小型蜂王的体重则与工蜂体重没有显著差异（Imperatriz-Fonseca et al., 1997; Wenseleers et al., 2005; Ribeiro et al., 2006 b），这表明在很多情况下，微小型蜂王并不是过度喂食幼蜂造成的结果（Bourke and Ratnieks, 1999; Wenseleers et al., 2005; Ribeiro et al., 2006 b）。研究者还提出另一种解释，他们认为微小型蜂王是一种发育紊乱的蜂王（Camargo, 1974; Ribeiro et al., 2006 b）。然而，微小型蜂王能够统领蜂群，这一事实表明微小型蜂王是无刺蜂的一种替代性的繁殖策略。无刺蜂蜂群中的一些幼虫似乎能够逃避工蜂的社会控制（工蜂通过控制幼虫食物实现这种控制），并通过成为蜂群中有繁殖能力的蜂王而不是不能生育的工蜂，来提高它们的适应性（Wenseleers et al., 2005; Ribeiro et al., 2006 b）。而无刺蜂工蜂则更喜欢体型较大的蜂王是因为这种蜂王的繁殖力更高，且工蜂更倾向于杀死体型较小的蜂王（第 4 章）（Engels and Imperatriz-Fonseca, 1990; Ribeiro et al., 2006 b）。这可能是为什么无刺蜂中能产卵的微小型蜂王却不是主流蜂王类型的原因，即使蜂群繁育微小型蜂王速度更快（Ribeiro et al., 2006 b）。例如，*Schwarziana quadripunctat* 无刺蜂的新出房蜂王中 80%～90% 都是微小型蜂王，但是该蜂种仅有大约 20% 的蜂群是由微小型蜂王统领的（Wenseleers et al., 2005; Ribeiro et al., 2006 b）。微小型蜂王没能成为主流蜂王类型的原因还有另一种解释：微小型蜂王不太擅长飞行，这可能会降低它们在分蜂或婚飞中的生存概率，也可能是微小型蜂王对雄蜂的吸引力较小（Ribeiro et al., 2006 b）。

无刺蜂这些微小型蜂王与 *Melipona* 属无刺蜂繁育的蜂王有一些相似之处。这两种蜂王都是在工蜂巢房中育成的，而且通常繁育量都很大。此外，这两种蜂王中"过量"的蜂王通常都会被工蜂杀死（Ribeiro et al., 2006 b）。在大巢室中育成的蜂王很可能代表了祖先状态，但 *Melipona* 属无刺蜂丢失了这种祖先状态。有研究者提出 *Melipona* 属无刺蜂的祖先可能会同时繁育正常体型蜂王和微小型蜂王，但是大型蜂王表型在进化过程中丢失了，因为微小型蜂王数量较多，这意味着拥有大型表型的蜂王统领蜂群的概率较低（Ribeiro et al., 2006 b）。因此，*Melipona* 属无刺蜂工蜂就完全停止了建造大的蜂王巢室。这种情况就需要 *Melipona* 祖先蜂中微小型蜂王相比大型蜂王繁殖力更强。Ribeiro 等（2006 b）也讨论提出了一种可能性，即对于 *Melipona* 属无刺蜂和繁育微小型蜂王的蜂种而言，微小型蜂王级型确立机制是类似的。具体来说，他们认为遗传因素决定了无刺蜂雌蜂分化发育成一只微小型蜂王。

5.6　无刺蜂的繁殖矛盾

人们称赞社会性昆虫是因为它们会相互合作且极具利他主义精神（Wilson, 1971; Hölldobler and Wilson, 2009; Bourke, 2011），但是其实社会性昆虫的群体内部是存在矛盾的，并且这种矛盾在无刺蜂蜂群中尤为明显。特别是在繁殖这一问题上，一个无刺蜂蜂群内的成员并不总是齐心协力的（Hamilton,1972; Ratnieks,1988; Ratnieks and Reeve,1992; Queller and Strassmann,1998; Bourke and Franks,1995; Bourke,2011; Ratnieks et al. ,2006; Ratnieks and Helanterä, 2009）。

亲缘选择理论则可以用来解释无刺蜂家族群体的利他行为和矛盾冲突。亲缘选择是指一个生物体会通过帮助近亲来提高双方的适应性，即使这样会降低帮助者的生存率或繁殖成功率。这是因为帮助者和受助者有一些相同的基因（Hamilton,1972; West-Eberhard,1975;Dawkins,1976; Bourke and Franks,1995; Queller and Strassmann,1998; Ratnieks et al. ,2006; Bourke,2011）。

随着二者间亲缘度（r）的增加，帮助者的潜在生物适应性也会有所增加。然而，亲缘选择也会导致矛盾冲突。因为单雄交配的无刺蜂蜂王统领了无刺蜂蜂群（第4章），而由于单倍二倍性理论（雌性来自受精的二倍体卵，雄性来自未受精的单倍体卵）（Wilson，1971），无刺蜂姊妹工蜂间的亲缘度（$r=0.75$，图4.5）是工蜂与其兄弟雄蜂（即蜂王的儿子）亲缘度（$r=0.25$）的3倍。无刺蜂蜂王与子女间的亲缘度（$r=0.5$）则是相等的。因此，工蜂与群内其他工蜂繁殖的雄蜂即与其子侄间的亲缘度（$r=0.375$）比与其兄弟雄蜂间的亲缘度（$r=0.25$）更大一些。

这些亲缘关系的不对称性导致了无刺蜂蜂群内出现了一些潜在的矛盾冲突（Ratnieks et al.，2006；Bourke，2011）。例如，工蜂养育自己繁殖的雄蜂（$r=0.5$）和姊妹工蜂繁殖的雄蜂（$r=0.375$）时比养育蜂王繁殖的雄蜂（$r=0.25$）获益更多。然而，蜂王从繁殖雄蜂中得到的好处则要多于养育工蜂繁殖的雄蜂（也就是她的"孙子"）。因此，蜂王倾向于阻止工蜂繁殖雄蜂。另一个潜在矛盾冲突是级型分化：尽管不育的工蜂能够通过帮助蜂王繁殖来获得间接的生物适应性，但如果它们自己本身就是具有繁殖力的蜂王，它们的生物适应性会更高，与自己后代间的亲缘度也更高（Bourke and Ratnieks，1999；Ratnieks，2001）。无刺蜂工蜂不能成为有繁殖力的蜂王是因为它们不能交配（Luna-Lucena et al.，2018），但是雌性幼虫可以尝试分化发育成蜂王而不是工蜂。当然，这些亲缘关系的不对称性还导致了一些其他方面潜在矛盾冲突（Ratnieks et al.，2006），但笔者将把重点放在繁育雄蜂的矛盾和雌蜂级型分化的矛盾上，因为它们是无刺蜂蜂群中最直接的矛盾冲突，而且二者间很可能是相互关联的（Ratnieks，2001；Wenseleers and Ratnieks，2004；Tóth et al.，2004；Wenseleers et al.，2013）。

5.6.1 繁育雄蜂的矛盾冲突

由于无刺蜂工蜂与其自己生殖的雄蜂（$r=0.5$）及其姊妹工蜂的儿子（$r=0.375$）间的亲缘度要比工蜂与蜂王生殖的雄蜂间的亲缘度（$r=0.25$）更高，因此无刺蜂蜂群中工蜂生殖应该会是广泛存在的现象。事实上，对于一些无刺蜂种而言，蜂群中有很大比例的雄蜂都是工蜂生殖的。例如 *Melipona favosa*、*M. mondury*、*M. quadrifasciata*、*Paratrigona subnuda* 和 *Tetragona clavipes* 无刺蜂蜂群中超过50%的雄蜂通常是由工蜂生殖的（Sommeijer et al.，1999；Tóth et al.，2002 a，2002 b；Viana et al.，2015）。而且，近一半的无刺蜂蜂种的蜂群中工蜂负责生殖了蜂群中至少10%的雄蜂（Tóth et al.，2004）。而在蜜蜂和其他社会性昆虫中，工蜂生殖则是非常罕见，它们产的卵通常会被其他工蜂（或蜂王）吃掉，这与无刺蜂的情况存在很大差异（Ratnieks and Wenseleers,2005; Ratnieks et al. ,2006; Wenseleers and Ratnieks,2006; Zanette et al. ,2012）。这种所谓的工蜂监管的行为因亲缘度而在蜜蜂蜂群中得以执行，因为对于蜜蜂而言，蜂王是多雄交配的，这样就降低了蜜蜂工蜂之间的亲缘度，也就是说工蜂与蜂王的儿子间的亲缘度高于工蜂与姊妹工蜂（蜜蜂姊妹可能是同母异父的）儿子间的亲缘度。因此，在由多雄交配的蜂王统领的蜂群中，工蜂会更倾向于养育它们的兄弟而不是它们的子侄（Ratnieks,1988; Ratnieks and Visscher,1989）。

无刺蜂工蜂偶尔也会吃工蜂产下的卵（包括营养卵和繁殖卵，见上文），但这是否会显著降低蜂群中由工蜂生殖的雄蜂在所有雄蜂中的占比？这个问题目前尚不清楚。此外，还没有研究者比较过无刺蜂工蜂或蜂王吃掉工蜂产的和蜂王产的雄性卵的比例（Ratnieks and Visscher,1989 年在蜜蜂上开展过这种研究）。但是已有证据表明，无刺蜂工蜂能够识别工蜂产的卵。Koedam 等（2007）发现，*Melipona bicolor* 无刺蜂工蜂常常会将已经含有另一只工蜂卵的巢室内的卵吃掉并且将自己的卵产入其中，却不会去试图取代蜂王产的卵，而蜂王产的卵则很可能是雌性卵。研究者发现其他一些社会性蜂的工蜂可能是基于化学线索来识别卵是蜂王产的还是工蜂产的（Ratnieks and Visscher,1989; Endler et al. ,2004）。

研究者还观察到无刺蜂蜂群中存在"蜂王监管"的现象（见上文），但他们认为"蜂王监管"比"工蜂监管"效率要低一些（Tóth et al. ,2004; Ratnieks et al.,2006），因为在大蜂群中，单只蜂王很难全面监控工蜂的产卵行为。即使是一些蜂群规模较小的蜂种中，蜂王监管行为也并不常见。例如，*Melipona* 属无刺蜂的蜂群规模就普遍相对较小（表 1.3），但该属无刺蜂的工蜂生殖后代的现象却是很常见的。*Melipona favosa* 无刺蜂蜂王也不会对正在繁殖的工蜂表现出任何攻击性（Chinh et al.,2003）。综上所述，研究者认为无刺蜂蜂王监管不是解释工蜂生殖力的重要因素（Tóth et al.,2004）。

如果无刺蜂蜂群中已有的蜂王被新蜂王取代（例如由于换王，第 4 章），那么蜂群成员之间的亲缘度就会发生变化。新蜂王繁殖的工蜂与旧工蜂间的亲缘度就会降低。这种情况就会增加老工蜂繁殖后代的动机，因为后续成本会更多地由与他们亲缘度较低的工蜂（侄女）承担（Hamilton,1964; Ratnieks et al.,2006）。Alves 等（2009 b）研究发现，*Melipona scutellaris* 无刺蜂换王后，工蜂产的雄性卵中有 80% 是由前蜂王的女儿产下的，这也与上述猜想一致。

他们的数据发现这些老工蜂只占工蜂总量的一小部分，但它们的寿命却非常长（高达 110 d），它们会把精力都用于繁殖，而不是从事通常由老工蜂承担的且风险更大的任务，如采集工作和守卫工作（第 6 章）。

研究者目前研究过的无刺蜂蜂种中，有一半蜂种的工蜂只要蜂群中存在产卵蜂王便不会产卵繁殖。Duckeola、Frieseomelitta、Geotrigona 和 Tetragonula（表 5.2）属的无刺蜂蜂种的工蜂就从来都不产卵繁殖。这些蜂种的工蜂要么卵巢没有繁殖功能，要么是即便失王也只会产营养卵。很难从亲缘关系的角度来解释这种现象，那么蜂群中为什么这些工蜂不能生育而让蜂王产所有的雄性卵呢？一种解释是蜂王能够以某种方式阻止工蜂繁殖。无刺蜂蜂王的体型比工蜂更大，蜂王可以利用体型优势控制参与 POP 的工蜂。然而无刺蜂蜂王嘴部则相对脆弱（Schwarz,1948），这降低了蜂王咬击工蜂的可能性（Tóth et al.,2003），而且当前无刺蜂蜂王通常不使用明显的统治行为来垄断繁殖（Sakagami et al.,1977）[20]。无刺蜂蜂王还可能是用化学方法抑制了工蜂繁殖，但这种情况好像在进化过程中不太稳定，目前也没有经验证据证实这种假设（Keller and Nonacs,1993; Oi et al.,2015; Grüter and Keller,2016; Van Oystaeyen et al.,2014）。与之相对的是，Nunes 等（2017）提供证据表明，无刺蜂工蜂不育不是蜂王化学性抑制的结果，而更像是进化出来的特征，因为这种特征对无刺蜂工蜂也有一定的好处（Grüter,2018）。换句话说，只要有一个可生育的蜂王存在，蜂群的工蜂们生育受限是有好处的。

无刺蜂工蜂这种生殖自限（以及工蜂监管行为）的另一种解释是对蜂群来说，工蜂繁殖需要付出高昂的代价（Ratnieks and Reeve,1992; Tóth et al.,2003; Ratnieks et al., 2006）。首先，工蜂产卵繁殖则其用于其他任务的时间就会更少，或者就会使用更多的蜂群资源（Tóth et al., 2003; Ratnieks et al., 2006; Alves et al.,2009 b）。其次，工蜂产卵繁殖雄蜂就减少了蜂群中繁育的工蜂幼蜂量，这可能会降低整个蜂群的生产力（Koedam et al.,1999; Tóth et al.,2002 a; Wenseleers et al.,2013）。最后，蜂王与工蜂对抗来竞争繁殖雄蜂的机会可能会伤害到蜂王，这对整个蜂群来说都是一种危险（Ratnieks and Reeve,1992）。因此这就提示，无刺蜂蜂群中工蜂是否生殖后代的决策应该是工蜂自身繁殖的个体利益和工蜂繁殖产生的集体成本之间相互平衡得到的

---

[20] 然而某些无刺蜂蜂种的蜂王索求食物的表现十分明显，出现蜂王压住工蜂进行"勒索"的现象（Sakagami et al., 1977; Sakagami and Yamane, 1987）。

结果。如果工蜂生殖雄蜂的群体成本是巨大的，那么蜂群就会进化出工蜂监管或者工蜂的生殖自限（即工蜂不育）（Ratnieks et al. ,2006）。当蜂群中的工蜂监管十分有效地抑制了工蜂的生殖欲望，工蜂也能间接地进化出生殖自限（Wenseleers et al.,2004 a; Wenseleers and Ratnieks,2006）。无刺蜂蜂群中工蜂产卵繁殖的程度在不同蜂种中差异极大，即使这些蜂种有相似的亲缘关系结构，这一事实表明不同蜂种工蜂产卵生殖的群体成本投入也有所不同（Wenseleers et al. ,2013）。例如 Melipona 属无刺蜂的工蜂产卵的成本投入就比较低，因为工蜂产的卵取代了很多蜂王产的卵，而这些蜂王产的雌性卵发育成的蜂王对于蜂群来说是多余的（Tóth et al. ,2004）。这就可以解释在这个属的无刺蜂蜂群中工蜂繁殖十分普遍的原因。

5.6.2 雌蜂级型分化的矛盾冲突

对于无刺蜂雌性幼虫个体来说，分化发育成蜂王的好处远大于分化发育成工蜂，但是蜂群水平则会产生一些成本投入（Bourke and Ratnieks,1999; Ratnieks,2001; Wenseleers et al. ,2003）。蜂群中许多或大多数的雌性幼虫分化发育成蜂王就会影响整个蜂群的继代繁衍，也就是所谓的"公共悲剧"（Hardin,1968; Ratnieks et al. ,2006;Ratnieks et al. ,2006）。这种情况不符合蜂群中成年工蜂和蜂王的利益。那么结果就是幼雌蜂和成年雌蜂之间会产生矛盾冲突。然而，实际上无刺蜂蜂群中很少发生这种矛盾冲突，因为营养决定蜂王级型分化的策略阻止了大多数雌性幼虫分化发育成蜂王（第 5.7 节）（Bourke and Ratnieks,1999; Ratnieks et al. ,2006）。而 Melipona 属无刺蜂蜂王级型分化则不依赖于食物量，这就为幼雌蜂分化发育成蜂王提供了机会。由此不难推测 Melipona 属无刺蜂较之其他无刺蜂蜂种可能会有更多的雌性幼虫分化发育成蜂王，而实际观察到现象也正是如此（表 5.3，表 5.5）。Melipona 属无刺蜂的雌性幼虫中蜂王占比（约 10%）略低于 Ratnieks（2001）和 Wenseleers 等（2003）的预测值（他们的模型预测为 14%~20%）。这表明蜂群过量繁育蜂王对群体生产力的影响要大于模型中预测值。

Melipona 属无刺蜂蜂种中几乎所有新出房的蜂王都被工蜂杀死，这一观察结果进一步证实了工蜂和幼虫在级型上存在矛盾冲突的假设。幼虫成为蜂王的比例也可能取决于工蜂是否能产卵繁殖。如果蜂群中普遍存在工蜂产卵繁殖的情况，那么幼虫级型分化成蜂王的就更少，因为工蜂产卵繁殖会增加雌性幼虫分化发育成工蜂的个体利益（Ratnieks,2001）。因此我们推测 Melipona 属无刺蜂的工蜂产卵繁殖和雌性幼虫级型分化成蜂王之间存在负相关关系。但这并不容易测定，因为即使一个蜂群繁育出的蜂王幼蜂和雄蜂幼蜂的量都是变化不定的，但 Wenseleers 和 Ratnieks（2004）对此提供了一些初步试验证据，例如他们发现 4 种 *Melipona* 属无刺蜂蜂群中存在这种负相关性。

值得注意的是，这样用生物适应性来解释 *Melipona* 无刺蜂过量繁育蜂王幼蜂现象的假说不能完全替代基因决定蜂王级型分化的假说，也不能完全替代幼虫食物中的化合物触发蜂王级型发育的假说（第 5.5 节）。相反，可以用雌蜂的级型分化矛盾和蜂个体的自然选择来解释为什么 *Melipona* 无刺蜂属的蜂王确立机制会进化成允许大量幼虫分化发育成蜂王的情况（Bourke and Ratnieks,1999）。

构建簇状蜂巢的无刺蜂蜂种的蜂群会利用融合巢室来培育蜂王，幼虫可以通过吃掉邻近巢室中的幼虫食物而分化发育成蜂王。然而，实际上这样的幼虫几乎无法控制自己的分化级型，因为它们需要另一个紧挨着自己巢室的巢室，且这个相邻巢室需要在关键的时间窗口期内含有足够的幼虫食物。只有这样，末龄幼虫才能获得另一个巢室中全部的幼虫食物。这意味着在这些蜂种中可能是建造巢室的工蜂控制着蜂王的产生。

上述例子表明，无刺蜂蜂群的成员之间会经历激烈的繁殖矛盾冲突。在某些情况下矛盾冲突被抑制了是因为一方能够迫使另一方按其利益行事，而在某些情况下蜂群内成员间的冲突及其代价是无法避免的。这表明无刺蜂个体的利他主义一部分是个体自愿的，一部分则是被强迫的（Ratnieks et al., 2006; Ratnieks and Helanterä,2009）。

参考文献

Alves DA, Imperatriz-Fonseca V, Santos-Filho P（2009 a）Production of workers, queens and males in *Plebeia remota* colonies（Hymenoptera, Apidae, Meliponini）, a stingless bee with reproductive diapause. Genetics and Molecular Research 8（2）:672–683

Alves DA, Imperatriz-Fonseca VL, Francoy TM, Santos Filho PS, Nogueira-Neto P, Billen J, Wenseleers T（2009 b）The queen is dead—long live the workers: intraspecific parasitism by workers in the stingless bee *Melipona* scutellaris. Molecular Ecology 18（19）:4102–4111

Amaral IMR, Neto JFM, Pereira GB, Franco MB, Beletti ME, Kerr WE, Bonetti AM,

Ueira-Vieira C (2010) Circulating hemocytes from larvae of *Melipona scutellaris* (Hymenoptera, Apidae, Meliponini): Cell types and their role in phagocytosis. Micron 41 (2):123–129

Baptistella AR, Souza CCM, Santana WC, Soares AEE (2012) Techniques for the In Vitro Production of Queens in Stingless Bees (Apidae, Meliponini). Sociobiology 59:297–310

Bassindale R (1955) The biology of the Stingless Bee *Trigona* (*Hypotrigona*) *gribodoi* Magretti (Meliponidae). Proceedings of the Zoological Society of London 125 (1):49–62

Becker T, Pequeno PACL, Carvalho-Zilse GA (2018) Impact of environmental temperatures on mortality, sex and caste ratios in *Melipona interrupta* Latreille (Hymenoptera, Apidae). Science of Nature 105:55

Bego LR (1990) On social regulation in *Nannotrigona* (*Scaptotrigona*) *postica* Latreille, with special reference to productivity of colonies (Hymenoptera, Apidae, Meliponinae). Revista Brasileira de Entomologia 34 (4):721–738

Bego LR, de Camargo CA (1984) The occurrence of giant males in Nannotrigona (Scaptotrigona) postica Latreille (Hymenoptera, Apidae, Meliponinae). Boletim de Zoologia, Universidade de São Paulo 8:11–16

Bego LR, Grosso AF, Zucchi R, Sakagami SF (1999) Oviposition behavior of the stingless bees XXIV. Ethological relationships of *Tetragonisca angustula angustula* to other Meliponinae taxa (Apidae: Meliponinae). Entomological Science 2 (4):473–482

Bego LR, Simões D, Zucchi R (1983) On some structures in *Nannotrigona* (*Scaptotrigona*) *postica* Latreille (Hymenoptera, Meliponinae). Boletim de Zoologia, Universidade de São Paulo 6 (6):79–88

Beig D (1972) The production of males in queenright colonies of *Trigona* (*Scaptotrigona*) *postica*. Journal of Apicultural Research 11 (1):33–39

Beig D, Bueno OC, da Cunha RA, de Moraes HJ (1982) Differences in quantity of food in worker and male brood cells of *Scaptotrigona postica* (Latr. 1807) (Hymenoptera, Apidae). Insectes Sociaux 29 (2):189–194

Biesmeijer J, Tóth E (1998) Individual foraging, activity level and longevity in the stingless bee *Melipona beecheii* in Costa Rica (Hymenoptera, Apidae,

Meliponinae). Insectes Sociaux 45(4):427–443

Boleli IC, Paulino-Simões ZL, Gentile Bitondi MM (1999) Cell death in ovarioles causes permanent sterility in *Frieseomelitta varia* worker bees. Journal of Morphology 242(3):271–282

Bonetti AM, Kerr WE, Matusita SH (1995) Effects of juvenile hormones Ⅰ, Ⅱ and Ⅲ, in single and fractionated dosage in *Melipona* bees. Revista Brasileira de Biologia 55:113–120

Borges FVB, Blochtein B (2006) Seasonal variations in the internal conditions of colonies in *Melipona marginata obscurior* Moure, at Rio Grande do Sul, Brazil. Revista Brasileira de Zoologia 23(3):711–715

Bourke AF (2011) Principles of social evolution. Oxford University Press, Oxford

Bourke AF, Franks NR (1995) Social evolution in ants. Princeton University Press, Princeton

Bourke AF, Ratnieks FLW (1999) Kin conflict over caste determination in social Hymenoptera. Behavioral Ecology and Sociobiology 46(5):287–297

Brito DV, Silva CGN, Hasselmann M, Viana LS, Astolfi-Filho S, Carvalho-Zilse GA (2015) Molecular characterization of the gene feminizer in the stingless bee *Melipona interrupta* (Hymenoptera: Apidae) reveals association to sex and caste development. Insect Biochemistry and Molecular Biology 66:24–30

Camargo JMF (1974) Notas sobre a morfologia e biologia de *Plebeia* (*Schwarziana*) *quadripunctata quadripunctata* (Hymenoptera, Apidae, Meliponinae). Studia Entomologica 17:433–470

Camargo JMF, Pedro SR (2007) Notes on the bionomy of *Trichotrigona extranea* Camargo & Moure (Hymenoptera, Apidae, Meliponini). Revista Brasileira de Entomologia 51(1):72–81

Camillo C (1971) Estudos adicionais sobre zangoes de *Trigona* (*Friesella*) *schrottkyi* Friese (Hym., Apidae). Ciência e Cultura 23:273

Campos LAO, Coelho C (1993) Determinação de sexo em abelhas. XXX. Influência da quantidade de alimento e do hormônio juvenil na determinação das castas em *Partamona cupira helleri* (Hymenoptera, Apidae, Meliponinae). Revista Brasileira de Zoologia 10(3):449–452

Cardoso-Júnior CAM, Silva RP, Borges NA, de Carvalho WJ, Walter SL, Simões ZLP,

Bitondi MMG, Vieira CU, Bonetti AM, Hartfelder K（2017）Methyl farnesoate epoxidase（*mfe*）gene expression and juvenile hormone titers in the life cycle of a highly eusocial stingless bee, *Melipona scutellaris*. Journal of Insect Physiology 101:185–194

Cepeda OI（2006）Division of labor during brood production in stingless bees with special reference to individual participation. Apidologie 37（2）:175–190

Cham KO, Nocelli RCF, Borges LO, Viana-Silva FEC, Tonelli CAM, Malaspina O, Menezes C, Rosa-Fontana AS, Blochtein B, Freitas BM, Pires CSS, Oliveira FF, Contrera FAL, Torezani KRS, Ribeiro M de F, Siqueira MAL, Rocha MCLSA(2019) Pesticide Exposure Assessment Paradigm for Stingless Bees. Environal Entomology 48:36–48

Chinh TX, Sommeijer MJ, Boot W, Michener C（2005）Nest and colony characteristics of three stingless bee species in Vietnam with the first description of the nest of *Lisotrigona carpenteri*（Hymenoptera: Apidae: Meliponini）. Journal of the Kansas Entomological Society 78（4）:363–372

Chinh TX, Grob GB, Meeuwsen FJ, Sommeijer MJ（2003）Patterns of male production in the stingless bee *Melipona favosa*（Apidae, Meliponini）. Apidologie 34（2）:161–170

Chinh TX, Sommeijer MJ（2005）Production of sexuals in the stingless bee *Trigona*（*Lepidotrigona*）*ventralis flavibasis* Cockerell（Apidae, Meliponini）in northern Vietnam. Apidologie 36（4）:493–503

Crespi BJ（1992）Cannibalism and trophic egg in subsocial and eusocial insects. In: Elgar MA, Crespi BJ（eds）Cannibalism: ecology and evolution among diverse taxa. Oxford University Press, Oxford, pp 176–213

Cronin AL, Molet M, Doums C, Monnin T, Peeters C（2013）Recurrent evolution of dependent colony foundation across eusocial insects. Annual Review of Entomology 58:37–55

Cruz-Landim C（2000）Ovarian development in Meliponine bees（Hymenoptera: Apidae）: the effect of queen presence and food on worker ovary development and egg production. Genetics and Molecular Biology 23（1）:83–88

Cruz-Landim C, Höfling M, Imperatriz-Fonseca VL（1980）Tergal and mandibular glands in queens of *Paratrigona subnuda*（Moure）（Hymenoptera: Apidae）.

Morphology and associated behaviour. Naturalia 58（5）:121–133

da Silva DLN, Zucchi R, Kerr WE（1972）Biological and behavioural aspects of the reproduction in some species of *Melipona*（Hymenoptera, Apidae, Meliponinae）. Animal Behaviour 20（1）:123–132

Darchen R（1969）Sur la biologie de *Trigona*（*Apotrigona*）*nebulata* komiensis Cock. I. Biologia Gabonica 5:151–183

Darchen R（1972）Ecologie de quelques trigones（*Trigona* sp.）de la savane de Lamto（Cote D'Ivoire）. Apidologie 3（4）:341–367

Darchen R（1977）L'essaimage chez les *Hypotrigones* au Gabon: dynamique de quelques populations. Apidologie 8（1）:33–59

Darchen R, Delage-Darchen B（1971）Le déterminisme des castes chez les Trigones（Hyménoptères Apidés）. Insectes Sociaux 18（2）:121–134

Darchen R, Delage-Darchen B（1975）Contribution a l'étude d'une abeille du mexique *Melipona beecheii* b.（Hymenoptère : Apide）. Le déterminisme des castes chez les mélipones. Apidologie 6（4）:295–339

Dawkins R（1976）The selfish gene. Oxford University Press, Oxford

de Camargo CA（1972 a）Mating of the social bee *Melipona quadrifasciata* under controlled conditions（Hymenoptera, Apidae）. Journal of the Kansas Entomological Society 45:520–523

de Camargo CA（1972 b）Produção "in vitro" de intercastes em *Scaptotrigona postica* Latreille. Paper presented at the Homenagem à Warwick E.. Kerr, Rio Claro

de Camargo CA, de Almeida MG, Parra MGN, Kerr WE（1976）Genetics of sex determination in bees. IX. Frequencies of queens and workers from larvae under controlled conditions（Hymenoptera: Apoidea）. Journal of the Kansas Entomological Society 49:120–125

Drumond PM, Zucchi R, Oldroyd BP（2000）Description of the cell provisioning and oviposition process of seven species of *Plebeia* Schwarz（Apidae, Meliponini）, with notes on their phylogeny and taxonomy. Insectes Sociaux 47（2）:99–112

Drumond PM, Oldroyd BP, Dollin AE, Dollin LJ（1999）Oviposition behaviour of two Australian stingless bees, *Austroplebeia symei* Rayment and *Austroplebeia australis* Friese（Hymenoptera: apidae: Meliponini）. Austral Entomology 38（3）:234–241

Drumond PM, Zucchi R, Mateus S, Bego LR, Yamane S, Sakagami SF（1996）

Oviposition behavior of the stingless bees, Ⅷ.: *Plebeia* (*Plebeia*) *droryana* and an ethological comparison with other Meliponine taxa (Hymenoptera, Apidae). Japanese Journal of Entomology 64 (2):385–400

Drumond PM, Zucchi R, Yamane S, Sakagami SF (1998) Oviposition behavior of the stingless bees: XX. *Plebeia* (*Plebeia*) *julianii* which forms very small brood batches (Hymenoptera: Apidae, Meliponinae). Entomological Science 1 (2):195–205

Endler A, Liebig J, Schmitt T, Parker JE, Jones GR, Schreier P, Hölldobler B (2004) Surface hydrocarbons of queen eggs regulate worker reproduction in a social insect. Proceedings of the National Academy of Sciences of the United States of America 101 (9):2945–2950

Engels W, Imperatriz-Fonseca VL (1990) Caste development, reproductive strategies, and control of fertility in honey bees and stingless bees. In: Engels W (ed) Social Insects. Springer, Heidelberg, pp 167–230

Eterovic A, Cabral G, Oliveira A, Imperatriz-Fonseca V, Santos-Filho PS (2009) Spatial patterns in the brood combs of *Nannotrigona testaceicornis* (Hymenoptera: Meliponinae): male clusters. Genetics and Molecular Research 8 (2):577–588

Faustino C, Silva-Matos E, Mateus S, Zucchi R (2002) First record of emergency queen rearing in stingless bees (Hymenoptera, Apinae, Meliponini). Insectes Sociaux 49 (2):111–113

Felsenstein J (1985) Phylogenetics and the comparative method. American Naturalist 125:1–15

Ferreira-Caliman M, Galaschi-Teixeira J, do Nascimento F (2017) A scientific note on reproductive diapause in *Melipona marginata*. Insectes Sociaux 64 (2):297–301

Garland T, Harvey PH, Ives AR (1992) Procedures for the analysis of comparative data using phylogenetically independent contrasts. Systematic Biology 41:18–32

Giannini KM (1997) Labor division in *Melipona compressipes fasciculata* Smith (Hymenoptera: Apidae: Meliponinae). Anais da Sociedade Entomológica do Brasil 26 (1):153–162

Gloag RS, Beekman M, Heard TA, Oldroyd BP (2007) No worker reproduction in the Australian stingless bee *Trigona carbonaria* Smith (Hymenoptera, Apidae). Insectes Sociaux 54:412–417

Grosso AF, Bego LR (2002) Labor division, average life span, survival curve, and nest architecture of *Tetragonisca angustula angustula*（Hymenoptera, Apinae, Meliponini）. Sociobiology 40:615–637

Grosso AF, Bego LR, Martinez AS（2000）The production of males in queenright colonies of *Tetragonisca angustula angustula*（Hymenoptera, Meliponinae）. Sociobiology 35（3）:475–485

Grüter C, Keller L（2016）Inter-caste communication in social insects. Current Opinion in Neurobiology 38:6–11

Grüter C, Segers FHID, Menezes C, Vollet-Neto A, Falcon T, von Zuben L, Bitondi MMG, Nascimento FS, Almeida EAB（2017）Repeated evolution of soldier sub-castes suggests parasitism drives social complexity in stingless bees. Nature Communications 8:4

Halcroft M, Haigh AM, Spooner-Hart R（2013）Ontogenic time and worker longevity in the Australian stingless bee, *Austroplebeia australis*. Insectes Sociaux 60:259–264

Hamilton WD（1964）The genetical evolution of social behaviour. I. Journal of Theoretical Biology 7:1–16

Hamilton WD（1972）Altruism and related phenomena, mainly in social insects. Annual Review of Ecology and Systematics 3（1）:193–232

Hammel B, Vollet-Neto A, Menezes C, Nascimento FS, Engels W, Grüter C（2016）Soldiers in a stingless bee: work rate and task repertoire suggest guards are an elite force. The American Naturalist 187:120–129

Hardin G（1968）The tragedy of the commons. Science 162:1243–1248

Hartfelder K, Engels W（1989）The composition of larval food in stingless bees: evaluating nutritional balance by chemosystematic methods. Insectes Sociaux 36（1）:1–14

Hartfelder K, Makert GR, Judice CC, Pereira GA, Santana WC, Dallacqua R, Bitondi MM（2006）Physiological and genetic mechanisms underlying caste development, reproduction and division of labor in stingless bees. Apidologie 37（2）:144–163

Haydak MH（1970）Honey bee nutrition. Annual Review of Entomology 15（1）:143–156

Hölldobler B, Wilson EO（2009）The Superorganism: The Beauty, Elegance, and Strangeness of Insect Societies. W. W. Norton, New York

Hoover SE, Keeling CI, Winston ML, Slessor KN（2003）The effect of queen pheromones on worker honey bee ovary development. Naturwissenschaften 90 (10):477–480

Imperatriz-Fonseca VL, Cruz-Landim C, de Moraes RS (1997) Dwarf gynes in *Nannotrigona testaceicornis* (Apidae, Meliponinae, Trigonini). Behaviour, exocrine gland morphology and reproductive status. Apidologie 28 (3-4):113–122

Imperatriz-Fonseca V, Zucchi R (1995) Virgin queens in stingless bee (Apidae, Meliponinae) colonies: a review. Apidologie 26 (3):231–244

Imperatriz-Fonseca VL, Kleinert A (1998) Worker reproduction in the stingless bee species *Friesella schrottkyi* (Hymenoptera: Apidae: Meliponinae). Entomologia Generalis 23:169–175

Inoue T, Sakagami SF, Salmah S, Yamane S (1984) The process of colony multiplication in the Sumatran stingless bee *Trigona* (*Tetragonula*) *laeviceps*. Biotropica 16:100–111

Inoue T, Salmah S, Sakagami SF (1996) Individual variations in worker polyethism of the Sumatran stingless bee, *Trigona* (*Tetragonula*) *minangkabau* (Apidae, Meliponinae). Japanese Journal of Entomology 64:641–668

Jarau S, van Veen JW, Aguilar I, Ayasse M (2009) Virgin queen execution in the stingless bee *Melipona beecheii*: The sign stimulus for worker attacks. Apidologie 40:496–507

Jarau S, van Veen JW, Twele R, Reichle C, Gonzales EH, Aguilar I, Francke W, Ayasse M (2010) Workers make the queens in *Melipona* bees: identification of geraniol as a caste determining compound from labial glands of nurse bees. Journal of Chemical Ecology 36 (6):565–569

Judice C, Carazzole M, Festa F, Sogayar M, Hartfelder K, Pereira G (2006) Gene expression profiles underlying alternative caste phenotypes in a highly eusocial bee, *Melipona quadrifasciata*. Insect Molecular Biology 15 (1):33–44

Judice C, Hartfelder K, Pereira GAG (2004) Caste-specific gene expression in the stingless bee *Melipona quadrifasciata* - Are there common patterns in highly eusocial bees? Insectes Sociaux 51:352–358

Kärcher M, Menezes C, Alves DA, Beveridge OS, Imperatriz-Fonseca VL, Ratnieks FLW (2013) Factors influencing survival duration and choice of virgin queens in the stingless bee *Melipona quadrifasciata*. Naturwissenschaften 100:571–580

Keller L, Nonacs P (1993) The role of queen pheromones in social insects: queen control or queen signal? Animal Behaviour 45 (4):787–794

Kerr WE (1946) Formação das castas no gênero *Melipona* (Illiger, 1806). Anais da Escola Superior de Agricultura Luiz de Queiroz 3:288–312

Kerr WE (1948) Estudos sobre o gênero *Melipona*. Anais da Escola Superior de Agricultura Luiz de Queiroz 5:181–276

Kerr WE (1950 a) Evolution of the mechanism of caste determination in the genus *Melipona*. Evolution 4 (1):7–13

Kerr WE (1950 b) Genetic determination of castes in the genus *Melipona*. Genetics 35 (2):143–152

Kerr WE (1951) Bases para o estudo da genética de populações dos Hymenoptera em geral e dos Apinae sociais em particular. Anais da Escola Superior de Agricultura Luiz de Queiroz 8:219–354

Kerr WE, Akahira Y, Camargo CA (1975) Sex determination in bees. IV. Genetic control of juvenile hormone production in *Melipona quadrifasciata* (Apidae). Genetics 81 (4):749–756

Kerr WE, Nielsen RA (1966) Evidences that genetically determined *Melipona* queens can become workers. Genetics 54 (3):859–866

Kerr WE, Sakagami SF, Zucchi R, Portugal-Araújo Vd, Camargo JMF (1967) Observações sobre a arquitetura dos ninhos e comportamento de algumas espécies de abelhas sem ferrão das vizinhanças de Manaus, Amazonas (Hymenoptera, Apoidea). In: Atas do Simpósio sobre a biota Amazônica. Conselho Nacional de Pesquisas Rio de Janeiro, pp 255–309

Koedam D (1999) Production of queens, workers and males in the stingless bee *Melipona favosa* (Apidae: Meliponinae): patterns in time and space. Netherlands Journal of Zoology 49 (4):289–302

Koedam D, Cepeda-Aponte OI, Imperatriz-Fonseca VL (2007) Egg laying and oophagy by reproductive workers in the polygynous stingless bee *Melipona bicolor* (Hymenoptera, Meliponini). Apidologie 38 (1):55–66

Koedam D, Broné M, van Tienen P (1997) The regulation of worker-oviposition in the stingless bee *Trigona* (*Tetragonisca*) *angustula* Illiger (Apidae, Meliponinae). Insectes Sociaux 44 (3):229–244

Koedam D, Contrera FAL, Fidalgo AdO, Imperatriz-Fonseca VL (2005) How queen and workers share in male production in the stingless bee *Melipona subnitida* Ducke (Apidae, Meliponini). Insectes Sociaux 52 (2):114–121

Koedam D, Contrera FAL, Imperatriz-Fonseca V (1999) Clustered male production by workers in the stingless bee *Melipona subnitida* Ducke (Apidae, Meliponinae). Insectes Sociaux 46 (4):387–391

Koedam D, Imperatriz-Fonseca VL (2012) Cell-sealing efficiency and reproductive workers in the species *Melipona bicolor* (Hymenoptera, Meliponini): double standard and possible rogue conduct. Apidologie 43 (4):371–383

Koedam D, Velthausz P, Dohmen M, Sommeijer M (1996) Morphology of reproductive and trophic eggs and their controlled release by workers in *Trigona* (*Tetragonisca*) *angustula* Illiger(Apidae, Meliponinae). Physiological Entomology 21(4):289–296

Koedam D, Velthuis HHW, Dohmen M, Imperatriz-Fonseca VL (2001) The behaviour of laying workers and the morphology and viability of their eggs in *Melipona bicolor bicolor*. Physiological Entomology 26 (3):254–259

Lacerda LM, Simões ZP, Velthuis HHW (2010) The sharing of male production among workers and queens in *Scaptotrigona depilis* (Moure, 1942) (Apidae, Meliponini). Insectes Sociaux 57 (2):185–192

Lacerda LM, Zucchi R (1999) Behavioral alterations and related aspects in queenless colonies of *Geotrigona mombuca* (Hymenoptera, Apidae, Meliponinae). Sociobiology 33:277–288

Lacerda LdM, Zucchi R, Zucoloto F (1991) Colony condition and bionomic alterations in *Geotrigona inusitata*(Apidae, Meliponinae). Acta Biológica Paranaense 20(1):2

Luna-Lucena D, Rabico F, Simões ZL (2018) Reproductive capacity and castes in eusocial stingless bees (Hymenoptera: Apidae). Current Opinion in Insect Science 31:20–28

Luz GF, Campos LA de O, Zanúncio JC, Serrão JE (2017) Auxiliary brood cell construction in nests of the stingless bee *Plebeia lucii* (Apidae: Meliponini). Apidologie 48:681–691

Maciel-Silva V, Kerr WE (1991) Sex determination in bees. XXVII. Castes obtained from larvae fed homogenized food in *Melipona compressipes* (Hymenoptera, Apidae). Apidologie 22 (1):15–19

Maia-Silva C, Hrncir M, da Silva CI, Imperatriz-Fonseca VL (2015) Survival strategies of stingless bees (*Melipona subnitida*) in an unpredictable environment, the Brazilian tropical dry forest. Apidologie 46:631–643

Maia-Silva C, Hrncir M, Imperatriz-Fonseca VL, Schorkopf DLP (2016) Stingless bees (*Melipona subnitida*) adjust brood production rather than foraging activity in response to changes in pollen stores. Journal of Comparative Physiology A 202:723–732

Mao W, Schuler MA, Berenbaum MR (2015) A dietary phytochemical alters caste-associated gene expression in honey bees. Science Advances 1 (7):e1500795

Matsuura K, Himuro C, Yokoi T, Yamamoto Y, Vargo EL, Keller L (2010) Identification of a pheromone regulating caste differentiation in termites. Proceedings of the National Academy of Sciences of the United States of America 107 (29):12963–12968

Menezes C (2010) A produção de rainhas ea multiplicação de colônias em *Scaptotrigona aff. depilis* (Hymenoptera, Apidae, Meliponini). PhD Thesis, University of São Paulo, Ribeirão Preto

Menezes C, Bonetti AM, Amaral IMR, Kerr WE (2007) Alimentação larval de *Melipona* (Hymenoptera, Apidae): estudo individual das células de cria. Bioscience Journal 23:70–75

Menezes C, Vollet-Neto A, Contrera FAFL, Venturieri GC, Imperatriz-Fonseca VL (2013) The role of useful microorganisms to stingless bees and stingless beekeeping. In: Vit P, Pedro SRM, Roubik DW (eds) Pot-Honey. Springer, pp 153–171

Menezes C, Vollet-Neto A, Marsaioli AJ, Zampieri D, Fontoura IC, Luchessi AD, Imperatriz-Fonseca VL (2015) A Brazilian social bee must cultivate fungus to survive. Current Biology 25 (21):2851–2855

Michener CD (1974) The Social Behavior of the Bees. Harvard University Press, Cambridge

Moo-Valle H, Quezada-Euán JJG, Wenseleers T (2001) The effect of food reserves on the production of sexual offspring in the stingless bee *Melipona beecheii* (Apidae, Meliponini). Insectes Sociaux 48 (4):398–403

Moo-Valle H, Quezada-Euán JJG, Canto-Martín J, Gonzalez-Acereto JA (2004) Caste ontogeny and the distribution of reproductive cells on the combs of *Melipona beecheii*（Apidae: Meliponini）. Apidologie 35（6）:587–594

Nascimento DL, Nascimento FS（2012）Extreme effects of season on the foraging activities and colony productivity of a stingless bee（*Melipona asilvai* Moure, 1971）in Northeast Brazil. Psyche: A Journal of Entomology 2012:267361

Nates-Parra G, Lopera AV, Briceño CV（1989）Ciclo de desarrollo de *Trigona* (*Tetragonisca*)*angustula*, latreille 1811（hymenoptera, trigonini）. Acta Biológica Colombiana 1（5）:91–98

Nogueira-Neto P（1963）A arquitetura das celulas de cria dos meliponideos（Apoidea, Hym.）. Universidade de São Paulo

Nunes TM, Heard TA, Venturieri GC, Oldroyd BP（2015）Emergency queens in *Tetragonula carbonaria*（Smith, 1854）（Hymenoptera: Apidae: Meliponini）. Austral Entomology 54（2）:154–158

Nunes TM, Mateus S, Favaris AP, Amaral MF, von Zuben LG, Clososki GC, Bento JMS, Oldroyd BP, Silva R, Zucchi R（2014）Queen signals in a stingless bee: suppression of worker ovary activation and spatial distribution of active compounds. Scientific Reports 4:7449

Nunes TM, Oldroyd BP, Elias LG, Mateus S, Turatti IC, Lopes NP（2017）Evolution of queen cuticular hydrocarbons and worker reproduction in stingless bees. Nature Ecology & Evolution 1（7）:0185

Oi CA, Van Zweden JS, Oliveira RC, Van Oystaeyen A, Nascimento FS, Wenseleers T（2015）The origin and evolution of social insect queen pheromones: novel hypotheses and outstanding problems. BioEssays 37:808–821

Oldroyd BP, Pratt SC（2015）Comb architecture of the eusocial bees arises from simple rules used during cell building. Advances in Insect Physiology 49:101–121

Paludo CR, Menezes C, Silva-Junior EA, Vollet-Neto A, Andrade-Dominguez A, Pishchany G, Khadempour L, Nascimento FS, Currie CR, Kolter R（2018）Stingless bee larvae require fungal steroid to pupate. Scientific Reports 8（1）:1122

Pereira R, Morais M, Bego LR, Nascimento FS（2009）Intrinsic colony conditions affect the provisioning and oviposition process in the stingless bee *Melipona scutellaris*. Genetics and Molecular Research 8（2）:725–729

Pereira RA, Morais MM, Gioli LD, Nascimento FS, Rossi MA, Bego LR（2006）Comparative morphology of reproductive and trophic eggs in *Melipona* bees（Apidae,

Meliponini). Brazilian Journal of Morphological Sciences 23 (3-4):349–354

Peters JM, Queller DC, Imperatriz-Fonseca VL, Roubik DW, Strassmann JE (1999) Mate number, kin selection and social conflicts in stingless bees and honeybees. Proceedings of the Royal Society of London Series B 266 (1417):379–384

Pick RA, Blochtein B (2002) Atividades de coleta e origem floral do pólen armazenado em colônias de *Plebeia saiqui* (Holmberg) (Hymenoptera, Apidae, Meliponinae) no sul do Brasil. Revista Brasileira de Zoologia 19:289–300

Pinto LZ, Hartfelder K, Bitondi MMG, Simões ZLP (2002) Ecdysteroid titers in pupae of highly social bees relate to distinct modes of caste development. Journal of Insect Physiology 48:783–790

Prato M, Soares AEE (2013) Production of sexuals and mating frequency in the stingless bee *Tetragonisca angustula* (Latreille) (Hymenoptera, Apidae). Neotropical Enomology 42:474–482

Queller DC, Strassmann JE (1998) Kin selection and social insects. Bioscience 48 (3):165–175

Quezada-Euán JJG, López-Velasco A, Pérez-Balam J, Moo-Valle H, Velazquez-Madrazo A, Paxton RJ (2011) Body size differs in workers produced across time and is associated with variation in the quantity and composition of larval food in *Nannotrigona perilampoides* (Hymenoptera, Meliponini). Insectes Sociaux 58:31–38

Rasmussen C, Cameron S (2010) Global stingless bee phylogeny supports ancient divergence, vicariance, and long distance dispersal. Biological Journal of the Linnean Society 99:206–232

Ratnieks FLW (1988) Reproductive harmony via mutual policing by workers in eusocial Hymenoptera. American Naturalist 132:217–236

Ratnieks FLW, Helanterä H (2009) The evolution of extreme altruism and inequality in insect societies. Philosophical Transactions of the Royal Society B: Biological Sciences 364 (1533):3169–3179

Ratnieks FLW, Reeve HK (1992) Conflict in single-queen hymenopteran societies: the structure of conflict and processes that reduce conflict in advanced eusocial species. Journal of Theoretical Biology 158 (1):33–65

Ratnieks FLW (2001) Heirs and spares: caste conflict and excess queen production in

Melipona bees. Behavioral Ecology and Sociobiology 50 (5):467–473

Ratnieks FLW, Foster KR, Wenseleers T (2006) Conflict resolution in insect societies. Annual Review of Entomology 51:581–608

Ratnieks FLW, Visscher PK (1989) Worker policing in the honeybee. Nature 342 (6251):796–797

Ratnieks FLW, Wenseleers T (2005) Policing insect societies. Science 307:54–56

Ribeiro MF, Imperatriz-Fonseca VL, Santos-Filho PS, Melo GAR, Alves-dos-Santos I (2003 a) A interrupção da construção de células de cria e postura em *Plebeia remota* (Holmberg) (Hymenoptera, Apidae, Meliponini). Apoidea Neotropica: homenagem aos 90 Anos de Jesus Santiago Moure 90:177–188

Ribeiro MF, Alves DA (2001) Size variation in *Schwarziana quadripunctata* queens (Hymenoptera, Apidae, Meliponini). Revista de Etologia 3 (1):59–65

Ribeiro MF, Santos-Filho P, Imperatriz-Fonseca VL (2006 a) Size variation and egg laying performance in *Plebeia remota* queens (Hymenoptera, Apidae, Meliponini). Apidologie 37 (6):653–664

Ribeiro MF, Imperatriz-Fonseca VL, Santos-Filho PS (2003 b) Exceptional high queen production in the Brazilian stingless bee *Plebeia remota*. Studies on Neotropical Fauna and Environment 38 (2):111–114

Ribeiro MF, Wenseleers T, Santos Filho PS, Alves DA (2006 b) Miniature queens in stingless bees: basic facts and evolutionary hypotheses. Apidologie 37 (2):191–206

Roubik DW (1982) Seasonality in colony food storage, brood production and adult survivorship: studies of *Melipona* in tropical forest (Hymenoptera: Apidae). Journal of the Kansas Entomological Society 55:789–800

Roubik DW (1983) Nest and colony characteristics of stingless bees from Panama (Hymenoptera: Apidae). Journal of the Kansas Entomological Society 56:327–355

Roubik DW (1989) Ecology and Natural History of Tropical Bees. Cambridge University Press, New York

Rozen JG, Quezada-Euán JJG, Roubik DW, Smith CS (2019) Immature Stages of Selected Meliponine Bees (Apoidea: Apidae). American Museum Novitates 2019:1–28

Sakagami SF (1982) Stingless bees. In: Hermann HR (ed) Social Insects Ⅲ. Academic Press, New York, pp 361–423

Sakagami SF, Beig D, Akahira Y (1964) Behavior studies of the stingless bees, with special reference to the oviposition process Ⅲ .: appearance of laying workers in an orphan colony of *Partamona* (*Partamona*) *testacea testacea* (Klug). Japanese Journal of Ecology 14:50–57

Sakagami SF, Camilo C, Zucchi R (1973) Oviposition behavior of a Brazilian stingless bee, *Plebeia* (*Friesella*) *schrottkyi*, with some remarks on the behavioral evolution in stingless bees. Journal of the Faculty of Science Hokkaido University Series Ⅵ, Zoology 19 (1):163–189

Sakagami SF, Montenegro MJ, Kerr WE (1965) Behavior studies of the stingless bees, with special reference to the oviposition process: V. *Melipona quadrifasciata anthidioides* Lepeletier. Journal of the Faculty of Science Hokkaido University Series Ⅵ, Zoology 15 (4):578–607

Sakagami SF, Oniki Y (1963) Behavior studies of the stingless bees, with special reference to the oviposition process. I: *Melipona compressipes manaosensis* Schwarz. Journal of the Faculty of Science Hokkaido University Series Ⅵ, Zoology 15 (2):300–318

Sakagami SF, Yamane S (1987) Oviposition behavior and related notes of the Taiwanese stingless bee *Trigona* (*Lepidotrigona*) *ventralis hoozana*. Journal of Ethology 5 (1):17–27

Sakagami SF, Zucchi R (1967) Behavior Studies of the Stingless Bees, with Special Reference to the Oviposition Process: Ⅵ. *Trigona* (*Tetragona*) *clavipes* (With 6 Text-figures and 2 Tables). Journal of the Faculty of Science Hokkaido University Series Ⅵ Zoology 16 (2):292–313

Sakagami SF, Zucchi R (1974) Oviposition behavior of two dwarf stingless bees, *Hypotrigona* (*Leurotrigona*) *muelleri* and *H.* (*Trigonisca*) *duckei*, with notes on the temporal articulation of oviposition process in stingless bees Journal of the Faculty of Science Hokkaido University Series Ⅵ, Zoology 19 (2):361–421

Sakagami SF, Zucchi R, Araujo VP (1977) Oviposition behavior of an aberrant african stingless bee *Meliponula bocandei,* with notes on the mechanism and evolution of oviposition behavior in stingless bees. Journal of the Faculty of Science Hokkaido University Series Ⅵ, Zoology 20 (4):647–690

Salmah S, Inoue T, Mardius P, Sakagami SF (1987) Incubation period and post-emergence pigmentation in the Sumatran stingless bee, *Trigona* (*Trigonella*) *moorei*. Kontyu 55 (3):383–390

Salmah S, Inoue T, Sakagami SF (1996) Incubation period and post-emergence pigmentation in the Sumatran stingless bee *Trigona* (*Heterotrigona*) *itama* (Apidae, Meliponinae). Japanese Journal of Entomology 64 (2):401–411

Santos-Filho PS, Alves DA, Eterovic A, Imperatriz-Fonseca VL, Kleinert AMP (2006) Numerical investment in sex and caste by stingless bees (Apidae: Meliponini): a comparative analysis. Apidologie 37 (2):207–221

Santos CF, Nunes-Silva P, Blochtein B (2015) Temperature rise and its influence on the cessation of diapause in *Plebeia droryana*, a eusocial bee (Hymenoptera: Apidae). Annals of the Entomological Society of America 109 (1):29–34

Santos CF, Nunes-Silva P, Halinski R, Blochtein B (2014) Diapause in Stingless Bees (Hymenoptera: Apidae). Sociobiology 61 (4):369–377

Santos DCdJ (2013) Divisão de trabalho e sua relação com a dinâmica dos hidrocarbonetos cuticulares em *Melipona scutellaris* (Hymenoptera, Apidae, Meliponini). MSc Thesis, University of São Paulo, Ribeirão Preto

Schwander T, Lo N, Beekman M, Oldroyd BP, Keller L (2010) Nature versus nurture in social insect caste differentiation. Trends in Ecology & Evolution 25:275–282

Schwarz HF (1948) Stingless Bees (Meliponidae) of the Western Hemisphere. Bulletin of the American Museum of Natural History 90:1–546

Segers FHID, Menezes C, Vollet-Neto A, Lambert D, Grüter C (2015) Soldier production in a stingless bee depends on rearing location and nurse behaviour. Behavioral Ecology and Sociobiology 69:613–623

Segers FHID, Von Zuben LG, Grüter C (2016) Local differences in parasitism and competition shape defensive investment in a polymorphic eusocial bee. Ecology 97:417–426

Silva-Matos EV, Noll FB, Mateus S, Zucchi R (2006) Non-ovipositing nurse workers with developed ovaries in *Trigona cilipes cilipes* (Hymenoptera, Meliponini). Brazilian Journal of Morphological Sciences 23:343–347

Sommeijer M, Chinh TX, Meeuwsen F (1999) Behavioural data on the production of males by workers in the stingless bee *Melipona favosa* (Apidae, Meliponinae). Insectes Sociaux 46 (1):92–93

Sommeijer M, De Bruijn L, Meeuwsen F, Martens E (2003) Natural patterns of caste and sex allocation in the stingless bees *Melipona favosa* and *M. trinitatis* related to worker behaviour. Insectes Sociaux 50 (1):38–44

Sommeijer M, Houtekamer J, Bos W (1984) Cell construction and egg-laying in *Trigona nigra paupera* with a note on the adaptive significance of oviposition behaviour of stingless bees. Insectes Sociaux 31 (2):199–217

Sommeijer MJ (1985) The social behavior of *Melipona favosa*: some aspects of the activity of the queen in the nest. Journal of the Kansas Entomological Society 58:386–396

Suka T, Inoue T (1993) Nestmate recognition of the stingless bee *Trigona* (*Tetragonula*) *minangkabau* (Apidae: Meliponinae). Journal of Ethology 11 (2):141–147

Sung I-H, Yamane S, Hozumi S (2008) Thermal characteristics of nests of the Taiwanese stingless bee *Trigona ventralis hoozana* (Hymenoptera: Apidae). Zoological Studies 47 (4):417–428

Tanaka ÉD, Santana WC, Hartfelder K (2009) Ovariole structure and oogenesis in queens and workers of the stingless bee *Melipona quadrifasciata* (Hymenoptera: Apidae, Meliponini) kept under different social conditions. Apidologie 40 (2):163–177

Terada Y, Garofalo CA, Sakagami SF (1975) Age-survival curves for workers of two eusocial bees (*Apis mellifera* and *Plebeia droryana*) in a subtropical climate, with notes on worker polyethism in *P. droryana*. Journal of Apicultural Research 14 (3-4):161–170

Tóth E, Queller DC, Dollin A, Strassmann JE (2004) Conflict over male parentage in stingless bees. Insectes Sociaux 51:1–11

Tóth E, Queller DC, Imperatriz-Fonseca VL, Strassmann JE (2002 a) Genetic and behavioral conflict over male production between workers and queens in the stingless bee *Paratrigona subnuda*. Behavioral Ecology and Sociobiology 53 (1):1–8

Tóth E, Strassmann JE, Imperatriz-Fonseca VL, Queller DC (2003) Queens, not workers, produce the males in the stingless bee *Schwarziana quadripunctata quadripunctata*. Animal Behaviour 66 (2):359–368

Tóth E, Strassmann JE, Nogueira-Neto P, Imperatriz-Fonseca VL, Queller DC (2002 b) Male production in stingless bees: variable outcomes of queen–worker conflict.

Molecular Ecology 11:2661–2667

van Benthem FDJ, Imperatriz-Fonseca VL, Velthuis HHW (1995) Biology of the stingless bee *Plebeia remota* (Holmberg): observations and evolutionary implications. Insectes Sociaux 42:71–87

van Oystaeyen A, Oliveira RC, Holman L, van Zweden JS, Romero C, Oi CA, D'Ettorre P, Khalesi M, Billen J, Wäckers F, Millar JG, Wenseleers T (2014) Conserved class of queen pheromones stops social insect workers from reproducing. Science 343:287–290

van Veen J, Arce Arce H, Sommeijer MJ (2004) Production of queens and drones in *Melipona beecheii* (Meliponini) in relation to colony development and resource availability. Proceedings of the Netherlands Entomological Society 15:35–39

van Veen JW (2000) Cell provisioning and oviposition in *Melipona beecheii* (Apidae, Meliponinae), with a note on caste determination. Apidologie 31 (3):411–419

Velthuis B-J, Velthuis HHW (1998) Columbus Surpassed: Biophysical Aspects of How Stingless Bees Place an Egg Upright on Their Liquid Food. Naturwissenschaften 85:330–333

Velthuis HHW, Koedam D, Imperatriz-Fonseca VL (2005) The males of *Melipona* and other stingless bees, and their mothers. Apidologie 36 (2):169–185

Velthuis HHW, Alves DA, Imperatriz-Fonseca VL, Duchateau MJ (2002) Worker bees and the fate of their eggs. Proceedings of the Section Experimental and Applied Entomology of the Netherlands Entomological Society 13:97–102

Velthuis HHW, Cortopassi-Laurino M, Pereboom Z, Imperatriz-Fonseca VL (2003) Speciation, development, and the conservative egg of the stingless bee genus *Melipona*. Proceedings of the Section Experimental and Applied Entomology - Netherlands Entomological Society 14:53–58

Velthuis HHW (1976) Environmental, genetic and endocrine influences in stingless bee caste determination. Paper presented at the XV International Congress of Entomology, Washington DC

Viana M, de Carvalho C, Sousa H, Francisco A, Waldschmidt AM (2015) Mating frequency and maternity of males in *Melipona mondury* (Hymenoptera: Apidae). Insectes Sociaux 62 (4):491–495

Vollet-Neto A, Oliveira RC, Schillewaert S, Alves DA, Wenseleers T, Nascimento FS,

Imperatriz-Fonseca VL, Ratnieks FLW（2017）Diploid male production results in queen death in the stingless bee *Scaptotrigona depilis*. Journal of Chemical Ecology 43（4）:403–410

Wang Y, Ma L, Zhang W, Cui X, Wang H, Xu B（2016）Comparison of the nutrient composition of royal jelly and worker jelly of honey bees（*Apis mellifera*）. Apidologie 47（1）:48–56

Wenseleers T, Hart AG, Ratnieks FLW（2004 a）When resistance is useless: policing and the evolution of reproductive acquiescence in insect societies. The American Naturalist 164（6）:E154–E167

Wenseleers T, Hart AG, Ratnieks FLW, Quezada-Euán JJG（2004 b）Queen execution and caste conflict in the stingless bee *Melipona beecheii*. Ethology 110（9）:725–736

Wenseleers T, Helanterä H, Alves DA, Dueñez-Guzmán E, Pamilo P（2013）Towards greater realism in inclusive fitness models: the case of worker reproduction in insect societies. Biology Letters 9（6）:20130334

Wenseleers T, Ratnieks FLW（2004）Tragedy of the commons in *Melipona* bees. Proceedings of the Royal Society of London Series B: Biological Sciences 271(Suppl 5）:S310–S312

Wenseleers T, Ratnieks FLW, Billen J（2003）Caste fate conflict in swarm-founding social Hymenoptera: an inclusive fitness analysis. Journal of Evolutionary Biology 16（4）:647–658

Wenseleers T, Ratnieks FLW, Ribeiro MF, Alves DA, Imperatriz-Fonseca VL（2005）Working-class royalty: bees beat the caste system. Biology letters 1（2）:125–128

Wenseleers T, Ratnieks FLW（2006）Enforced altruism in insect societies. Nature 444:50

West-Eberhard M（1981）Intragroup selection and the evolution of insect societies. In: Alexander RD, Tinkle DW（eds）Natural selection and social behavior: recent research and new theory. Chiron Press, New York and Concord, pp 3–17

West-Eberhard MJ（1975）The evolution of social behavior by kin selection. Quarterly Review of Biology 50:1–33

Wheeler WM（1911）The ant-colony as an organism. Journal of Morphology 22（2）:307–325

Wilson EO（1971）The insect societies. Harvard University Press, Cambridge

Winston ML (1987) The biology of the honey bee. Harvard University Press, Cambridge

Wittmann D, Bego LR, Zucchi R, Sakagami SF (1991) Oviposition behavior and related aspects of the stingless bees: XIV. *Plebeia* (*Mourella*) *caerulea*, with comparative notes on the evolution of the oviposition patterns (Apidae, Meliponinae). Japanese Journal of Entomology 59 (4):793–809

Yamane S, Heard TA, Sakagami SF (1995) Oviposition behavior of the stingless bees (Apidae, Meliponinae) XVI. *Trigona* (*Tetragonula*) *carbonaria* endemic to Australia, with a highly integrated oviposition process. Japanese Journal of Entomology 63 (2):275–296

Zanette LR, Miller SD, Faria C, Almond EJ, Huggins TJ, Jordan WC, Bourke AF (2012) Reproductive conflict in bumblebees and the evolution of worker policing. Evolution 66 (12):3765–3777

Zucchi R, Silva-Matos EV da, Nogueira Ferreira FH, Garcia Azevedo G (1999) On the cell provisioning and oviposition process (POP) of the stingless bees - nomenclature reappraisal and evolutionary considerations (Hymenoptera, Apidae, Meliponinae). Sociobiology 34:65–86

6 无刺蜂蜂群的劳动分工

  Adam Smith（1776）在他著名的"大头针工厂"案例中计算发现，一群分工明确的工人每天生产的大头针数量是没有分工的工人的240倍以上。他将生产力的显著提升归因于几个因素。首先，劳动分工使工人因重复任务而在特定任务上变得更加熟练；其次，工人可以避免频繁切换任务造成的时间损耗；最后，工人可以专注于他们更擅长的任务。这3个因素可能也同样适用于社会性昆虫群体中的劳动者，并且研究者提供了一些实践经验和理论依据来证实这些因素确实对劳动分工产生的效果有利（Julian and Cahan,1999; Duarte et al.,2012; Goldsby et al., 2012; Leighton et al., 2017; Dornhaus,2008）[1]。因此截至目前，研究者顺理成章地发现他们研究过所有的社会性昆虫物种几乎都表现出了某种程度的劳动分工情况（Hölldobler and Wilson,1990, 2009）。不同社会性昆虫物种、群体内工种和工作任务间，任务行为专业化程度则各不相同（Wilson,1984; Robinson and Page,1995; Sempo and Detrain,2004; Mertl and Traniello,2009）。大家普遍认为分工是社会性昆虫生态成功的重要原因（Wilson,1971; Oster and Wilson,1978; Hölldobler and Wilson,1990）。虽然试验性地量化"劳动分工的益处"具有一定的挑战性（Ulrich et al.,2018），但是毫无疑问的是劳动分工确实能显著提升昆虫群体的表现。

---

[1] 笔者这里使用了"分工"的广义定义，即工人执行任务上的非随机性差异（Jeanne,2016）。长期个体专业化（Gordon,2016）则不完全属于本文所使用的劳动分工定义，它代表的是劳动分工的一个特殊和极端情况。

6.1 无刺蜂的时间阶段性职级分工

最先研究无刺蜂劳动分工的研究者是Bassindale（1955）以及Kerr和Santos Neto（1956）。这两项研究都发现，随着无刺蜂工蜂年龄的增长，它们会执行不同的任务，后来研究者也发现其他十几种无刺蜂存在这种模式（表6.1）。例如，Bassingdale（1955）就发现在非洲的 *Hypotrigona gribodoi* 无刺蜂蜂群中年轻工蜂负责蜂巢建造工作和为幼虫巢室供应食物，而年长的工蜂则负责守卫蜂巢入口或采集花粉、花蜜、树脂和其他筑巢材料（第3章，第8章）。然而这不代表一个蜂群内的所有工蜂都是在同样的年龄担任同样的任务（Inoue et al.,1996; Hammel et al.,2016），但是工蜂们确实会在特定的年龄表现出从事特定工作任务的倾向。这种阶段性分工（或者年龄分工）模式在蚂蚁、熊蜂、蜜蜂和黄蜂中也同样常见（Rösch,1925; Lindauer,1952; Wilson,1971; Seeley,1982; Cameron,1989; Hölldobler and Wilson,1990; Johnson,2005; Yerushalmi et al.,2006; Shorter and Tibbetts,2009）。其一般规律是年轻工蜂更倾向于照顾幼虫，而年长工蜂则更倾向于保卫蜂群或者寻找食物。从群体水平来说，这个任务顺序是十分合理的，因为这种模式中生命接近尾声的工蜂才去执行最危险的任务，而剩余寿命更长的工蜂则会去执行更加安全的任务，这样最大程度地保证工蜂为蜂群工作更久的时间（Tofts and Franks, 1992; Moroń et al., 2012）[2]。研究者对于工蜂这样的阶段性分工行为提出了一个非工蜂自适应性的解释，即工蜂由于在幼虫巢室区出房，因此工蜂更可能首先受到哺育工作的相关刺激信号。换句话说，它们一出生就处于第一项工作任务环境中（Tofts and Franks,1992），因此它们的工作更有可能从哺育幼虫这项任务开始（Tofts and Franks,1992; Beshers and Fewell,2001）。随着新工蜂不断出房，幼虫区的年长工蜂发现自己难以找到工作后，就离开幼虫区去寻找新的任务（表6.1）。研究者提出的这种"为了工作而去采集觅食"模式的灵感是来自对蚂蚁和蜜蜂的观察，但这种模式似乎与无刺蜂的劳动分工不太一致，因为无刺蜂工蜂最先从事的工作任务是在幼虫区之外进行蜂巢建造活动，而后才会参与巢室建造、食物供给和产卵过程。

---

[2] Ravaiano等（2018）还发现在 *Melipona quadrifasciata* 无刺蜂中觅食蜂的血细胞数量低于哺育蜂。悬浮在血淋巴中的血细胞在免疫防御中发挥着重要作用。因此，觅食蜂也可能比年轻的哺育蜂在细胞免疫反应方面投入更少。

表 6.1 研究者对蜜蜂随年龄变化行为随之改变的特性在社会学、内分泌、神经化学和分子学基础上进行了广泛的研究（Winston,1987; Robinson, 1992, 2002; Ben-Shahar et al.,2002; Schulz et al.,2002; Whitfield et al., 2003; Leoncini et al., 2004; Herb et al.,2012）。但是关于无刺蜂的劳动分工和行为成长过程的最直接的原因，我们仍然知之甚少。例如，保幼激素（JH）在蜜蜂的行为成长过程中起着重要作用（Fahrbach,1997; Sullivan et al.,2000; Schulz et al.,2002）。但是我们却不明确 JH 在无刺蜂劳动分工中的作用。近期的一项研究发现了一个有趣的现象，*Melipona scutellaris* 无刺蜂的采集蜂的 JH 滴度比哺育蜂低，这与在蜜蜂中发现的情况相反（Cardoso-Júnior et al.,2017）。无刺蜂劳动分工另一个悬而未决的问题是，如果情况突然改变，蜂群如何灵活地将工蜂重新分配到不同的岗位中［也就是说，当需要更多地采集蜂时幼蜂是否可以成为采集蜂，或者当照料幼蜂的工蜂短缺时采集蜂是否可以恢复哺育幼虫的能力；见 Robinson（1992）关于蜜蜂的文章］。有一些证据证明 *Melipona quadrifasciata* 无刺蜂的行为是灵活可变的（Waldschmidt et al., 1998），因为即使蜂群中只有年轻工蜂，该蜂群仍然表现出正常的采集活动。而如果从 *Tetragonisca angustula* 无刺蜂蜂群中移走守卫蜂，几天后该蜂群守卫蜂的数量就会达到移走守卫蜂之前的水平（Baudier et al.,2019）。此外，社会因素是否以及如何加速或减缓无刺蜂的行为成长过程也仍有待研究。

6.2 无刺蜂主要工作任务的顺序和工作内容

多项研究发现，无刺蜂存在相对连续的基于时间的行为多态性。笔者在下文中将这些不同的行为归纳到 4 个阶段（表 6.1，表 6.2）。许多无刺蜂蜂种随着承担这些阶段性任务会出现色素沉积增加（图 6.1）（Bassindale,1955; Salmah et al.,1987; Salmah et al.,1996）、蕈形体和触角叶体积增加，而蕈形体和触角叶这两种脑组织对感官感知和认知是很重要的（*Melipona quadrifasciata*, Tomé et al., 2014）。而与蜜蜂相比，无刺蜂工蜂需要更长的时间才能色素沉着完全。

表 6.1 17 种无刺蜂工蜂执行任务的顺序

无刺蜂蜂种	第二阶段				第三阶段				第四阶段			参考文献
	用巢质和巢蜡进行蜂巢建设工作	建造幼虫巢室	工蜂产卵	养育蜂王过程	处理蜂巢垃圾	花蜜脱水	给采集蜂卸下负载物	出巢扔蜂巢垃圾	守卫蜂巢	采集工作	
Apotrigona nebulata	1	2		2				3	3	3	Darchen (1969)
Frieseomelitta varia	1	2	2	2	3				4	4	Cardoso (2010)
Hypotrigona gribodoi	1	1	1		2	2		3	4	4	Bassindale (1955)
Meliplebeia beccarii	1	2			3				4	4	Mogho and Wittmann (2013)
Melipona eburnea	1	1	1	1	2	2			3	4	Bustamante (2006)
Melipona fasciculata	1	2		3						5	Giannini (1997)
Melipona favosa	1	1	2		3	4	4		4	4	Sommeijer (1984)
Melipona lateralis	1	1	1	1	2	3			3	4	Bustamante (2006)
Melipona quadrifasciata	1	2	3	4	4	5	5	6	6	7	Kerr and Santos-Neto (1956)
Melipona scutellaris	2	1		1	3	3	3			4	Santos (2013)
Melipona seminigra	1	2	2	1	3	3	3		3	4	Bustamante (2006)
Plebeia droryana	2	1		2		3				3	Terada et al. (1975)
Plebeia remota	1	2	2	2	4	3	4			5	van Benthem et al. (1995)

续表

无刺蜂蜂种	第一阶段 用巢质和巢蜡进行蜂巢建设工作	第二阶段 建造幼虫巢室	第二阶段 工蜂产卵	第二阶段 养育蜂王	第二阶段 过程	第三阶段 处理蜂巢垃圾	第三阶段 花蜜脱水	第三阶段 给采集蜂卸下负载物	第三阶段 出巢扔巢垃圾	第四阶段 守卫蜂巢	第四阶段 采集工作	参考文献
Tetragonisca angustula	1	1	1		1	2			3	4	4	Grosso and Bego (2002); Hammel et al. (2016)
Trigona pallens	1	2			2	3					4	Silva (2008)
Scaptotrigona postica	1	1			1	2	3			4	5	Simões and Bego (1979)
Scaptotrigona xanthotricha	1	1	1		1	2	2		2	3	4	Hebling et al. (1964)

注：每个阶段的相关内容请参阅文本。表中数字是相关研究中观察的蜂种的工蜂执行任务的时间顺序。并非所有的研究都研究过工蜂每个阶段的行为。第一阶段未显示，因为该阶段工蜂的特征是自我梳理和静止状态。

表 6.2　无刺蜂的劳动分工研究中经常观察到的任务内容和行为清单

工蜂工作阶段	行为内容	具体行为清单
第一阶段		自我梳理
		接收食物
第二阶段	蜂巢建造	建造蜂巢包壳
		产蜡
		建造储物罐
		建造幼虫巢室
		建造蜂巢入口
	POP 和哺育工作	进入食物罐吃花粉和蜂蜜
		检查空的幼虫巢室
		向幼虫巢室反刍幼虫食物
		饲喂蜂王
		产营养卵
		为幼虫巢室封盖
		打开巢室帮助工蜂出房
	调控蜂巢气候	为蜂巢通风
第三阶段	处理食物	为采蜜蜂卸货
		催熟蜂蜜
	处理非食物材料	将蜂蜡放在蜡堆上
		处理树脂和泥浆
		用树脂或泥浆封住蜂巢缝隙
		将蜂巢废弃物运往蜂巢垃圾堆
		制作垃圾球并将其扔出蜂巢
第四阶段	守卫蜂巢	巡逻侦察入口管
		在蜂巢入口处守卫
		审查入巢蜂
		打击入侵者
		开闭蜂巢入口

续表

工蜂工作阶段	行为内容	具体行为清单
第四阶段	采集活动	采集花蜜
		采集花粉
		采集树脂或泥浆
		采水
		将花粉储存进花粉罐
		将泥浆或树脂成堆储存
		卸载花蜜至花蜜加工
		招募同巢蜂去食物源
各个阶段	多项内容	交哺作用
		游走状态
		静默状态
		梳理同巢蜂
		触角感触

注：相关行为源自 Bassindale（1955）、Inoue 等（1996）、Cardoso（2010）和 Hammel 等（2016）的相关研究。

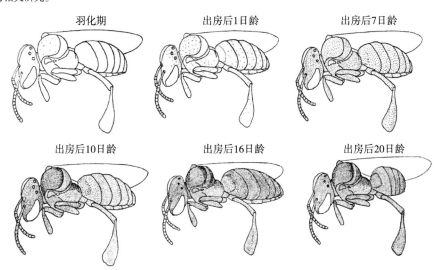

图 6.1　马来西亚 *Sundatrigona moorei* 无刺蜂的工蜂的色素沉积情况

注：该蜂种在出房后 30 d 左右开始采集工作（Salmah et al., 1987）。

第一阶段

无刺蜂工蜂出房后最初几天都在梳理自己和接收食物。工蜂活跃度低且颜色较浅（Hebling et al., 1964; Bassindale, 1955; Kerr and Santos Neto, 1956; Bustamante, 2006; Hammel et al., 2016）。

第二阶段

无刺蜂工蜂这一阶段的主要任务是建造巢室，其次是与幼虫食物供应和产卵过程（POP）有关的任务（第5章；图6.2）。这一阶段，无刺蜂工蜂主要在包壳内或在包壳上工作（子区周围的巢质层，第3章）。这一阶段的无刺蜂工蜂会不断用上颚打磨它们的筑巢材料（Bassindale, 1955）。这一阶段中一个幼虫巢室的食物供应需要由几只无刺蜂工蜂在产卵前将幼虫食物吐入其内（第5章）（Bassindale, 1955; van Benthem et al., 1995; Segers et al., 2015）。产卵后，第二阶段的工蜂会立即对该巢室进行封盖。工蜂偶尔也会给蜂王喂食，有的是通过交哺给蜂王供应液体食物，有的则是通过产营养卵的方式（第5章）（Sommeijer, 1984），当蜂王不移动时，一些工蜂还会在蜂王周围形成蜂王护卫队。一些无刺蜂蜂种的工蜂还会在这一阶段相应的时间段内（1~3周）产繁殖卵，繁殖卵可繁育出雄蜂幼蜂（第5章）（Bego,1990; Koedam et al.,1999）。而这一时期也正好是工蜂卵黄蛋白的前体——卵黄原蛋白产生的高峰期（Engles and Engels,1977; Engels,1987; Hartfelder et al., 2006）。巢室封盖几天后（当巢室中的幼虫构建由丝制成的茧时），第二阶段的工蜂才会去这些巢室上移除一些巢质，并将其重新用于建筑其他蜂巢区域（Michener,1974）。也有研究观察到，无刺蜂工蜂最先是在幼虫区包壳外工作，例如建造储物罐（Bassindale,1955; Hammel et al., 2016）。而老年工蜂（第三阶段）也参与了建造储物罐的工作（Bassindale,1955; Mateus et al., 2019）。第二阶段的工蜂位于第4至第7节（Justino et al., 2018）或第3至第6节或第3至第5节（Cruz–Landim,1963）的背甲的蜡腺会开始分泌蜡质。不同无刺蜂蜂种的蜡腺的确切位置也不同，且不同无刺蜂蜂种是否在连续3~4个背甲中含有蜡腺方面存在差异。虽然在这个阶段里无刺蜂工蜂的蜡产量最高，但是工蜂其实直至出巢外勤时仍可以持续产蜡（Kerr and Santos Neto, 1956; Terada et al., 1975; Sakagami, 1982; Bego et al., 1983; Ferreira–Caliman et al., 2010; Justino et al., 2018）。例如，*Melipona marginata* 无刺蜂工蜂从第5天开始直至第24天会分泌蜡质，而第24天也是最早一批工蜂开始从事采集工作的时间（Ferreira–Caliman et al., 2010）。*Friesella schrottkyi* 无刺蜂工蜂的蜡产量是在第13天左右达到顶峰，然后迅速下降（Justino et al., 2018）。不是所有的无刺蜂工蜂都能产蜡：研究者检测了 *F. schrottkyi* 无刺蜂工蜂发现其中大约 30% 的工蜂似乎根本不分泌蜡质，这表明可能存在生理上和行为上不同的职级分工，但这需要进一步研究（Justino，2018）。无刺蜂工蜂们是从其基腹节的末端将蜡鳞从腹部拖拽出来的（图 3.7）（Cruz-Landim,1963）。工蜂偶尔会将这些蜡质堆积在蜂巢内活跃的建筑区附近（Bassindale,1955; Hebling et al., 1964）。

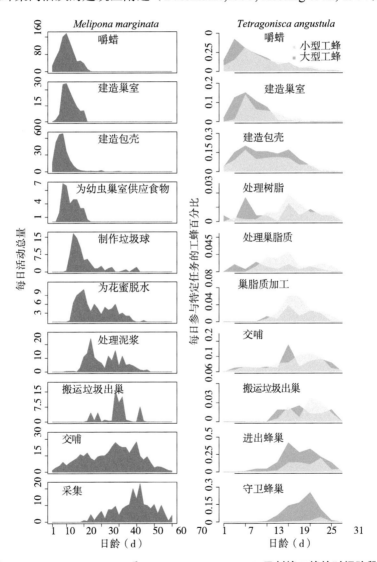

图 6.2　*Melipona marginata* 和 *Tetragonisca angustula* 无刺蜂工蜂的时间阶段性劳动分工情况（Hammel et al.,2016; Mateus et al.,2019）

注：图中展示的 *T. angustula* 无刺蜂的数据是来自小型工蜂和大型工蜂（多数）的所有数据。需要注意的是图中涉及的两项研究中每天的活动以不同的方式测量的（y 轴）。

第三阶段

这一阶段里无刺蜂工蜂会改变工作位置并开始在包壳之外进行许多活动。它们会卸下返巢采集蜂带回的液体食物（例如花蜜、蜜露或果汁），并为之脱水后将其加工成蜂蜜（第1章）。正在从事脱水工作的工蜂会站着不动，并且有节奏地移动它们的口器，来扩大和缩小口器上的小液滴的大小（这就是所谓"舌击打"）（Sakagami and Oniki,1963）。该阶段无刺蜂工蜂的另一个常见活动就是在蜂群中界定的垃圾堆中处理废弃物。这种废物主要包括了粪便、死亡工蜂的尸体和旧的幼虫巢室（Bassindale,1955; Medina et al.,2014）。无刺蜂与蜜蜂不同，无刺蜂工蜂是在巢内排便的（蜂王除外，蜂王的粪便会被工蜂吃掉）（da Silva et al.,1972; Sakagami et al.,1977）。然后无刺蜂工蜂会将垃圾堆中的物质做成垃圾球并随后将其运送到蜂巢外，在蜂巢附近扔下（图6.3）。一些 *Melipona* 属无刺蜂甚至会飞行近50 m 去扔垃圾球，而 *Tetragonisca angustula* 无刺蜂工蜂则只飞行很短的距离就会将垃圾球扔下[3]（Darchen,1969; Kerr and Kerr,1999; Medina et al.,2014）。一些无刺蜂蜂种，例如美洲的 *Leurotrigona muelleri* 和 *Lestrimelitta limao* 无刺蜂，搬运垃圾的工蜂会走到蜂巢入口的边缘将垃圾球直接丢到地上，导致其蜂巢入口正下方会出现小型垃圾堆（Sakagami and Zucchi,1974）。在无刺蜂蜂群中，在巢内将废物堆积到垃圾堆上的工蜂的日龄比把废物搬出蜂巢的工蜂日龄要小几天（图6.2）（Bassindale,1955; Hammel et al.,2016）。例如，*T. angustula* 无刺蜂在巢内垃圾堆工作的工蜂平均日龄为19 d，它们比将垃圾搬运到蜂巢外的工蜂小6 d。而在巢内垃圾堆处工作的工蜂后期去巢外扔垃圾的可能性是从未在垃圾堆工作过工蜂的4倍[4]，这表明一些无刺蜂工蜂对处理垃圾有趋向性。第三阶段的工蜂也会大量参与处理其他材料，如树脂或泥浆（图6.2）。

---

[3] 据 Roubik 和 Patiño（2009）报道，*Cephalotrigona* 属无刺蜂是唯一已知的将废弃物堆积在洞穴底部而不是移出巢外的无刺蜂属。笔者猜测这种习性最终将因耗尽空间而导致蜂群的死亡。

[4] 研究发现，前期在垃圾堆工作的无刺蜂工蜂的44.4%或54只中的24只会在后期去巢外扔垃圾；从未出现在垃圾堆处的工蜂的12.3%或138只中的17只会在后期去巢外扔垃圾。卡方检验 $X^2=23.8$, $df=1$, $p<0.001$。数据源自 Hammel 等（2016）的补充材料信息。

图6.3 一只 *Scaptotrigona bipunctata* 无刺蜂工蜂正带着垃圾球出巢

注：照片在巴西圣保罗拍摄（摄影：C. Grüter）。

研究者还观察到第三阶段的无刺蜂工蜂会打开巢室，帮助它们的姐妹工蜂出房（Bassindale,1955; Hammel et al. ,2016）。该阶段的工蜂还会负责在入口管内扇风以产生气流（第3章）（Bassindale,1955）。工蜂在此阶段为了准备采集工作也会开始进行短暂的定向飞行（5～10 min）来了解蜂巢的周围环境（Biesmeijer and Tóth,1998）。

第四阶段

第四阶段里的老无刺蜂工蜂就开始在蜂巢外执行任务，主要是守卫任务和采集食物。花粉采集蜂会将它们负载的花粉直接放入花粉罐中，在此之前它们偶尔会在蜂巢入口附近逗留一段时间以求得一些食物（Sommeijer et al., 1985）。当采集蜂的花粉筐装满花粉时，其他工蜂则会用它们的触角触碰载满花粉的花粉筐，或轻轻啃食上面的花粉（Bassindale，1955; Sommeijer et al., 1983）。花蜜采集蜂则是将带回的食物传递给花蜜接收工蜂（第三阶段），这种行为被称为交哺。有时，它们会同时将食物传递给多个花蜜接收工蜂（关于交哺在采集食物和招募过程中的作用的进一步讨论见第8章和第10章）（Sommeijer et al.,1983; Hart and Ratnieks,2002）。因此，许多无刺蜂蜂群中花蜜采集是一项分割的任务，而花粉的收集和储存是由同一只采集蜂完成的，就像蜜蜂一样（von Frisch,1967; Ratnieks and Anderson,1999）。然而这似乎不是一个普遍规律，Sommeijer等（1983）发现 *Melipona favosa* 无刺蜂归巢的采集蜂在向其他工蜂提供花蜜后，通常还会在蜜囊中保留大量的花蜜，然后这些采集蜂进行一次腹部收缩将剩余的花蜜反哺到一个打开的储蜜罐中。非洲 *Plebeina armata* 无刺蜂采集蜂也偶尔会直接将液体食物放入储藏罐中，而不是将其传递给另一只工蜂（Krausa et al.,2017 b, 文中写的是 *P. hildebrandti* 无刺蜂）。Darchen（1969）还报道了另一种非洲 *Apotrigona nebulata* 无刺蜂的采集蜂会自己将食物放入储藏罐中。因此，无刺蜂是否将花蜜采集任务进行分割的程度似乎因蜂种而异。从更普遍性的问题来看，我们对无刺蜂的交哺（及其所含物质，见 LeBoeuf et al., 2016）在蜂群分工组织中的重要性和作用仍知之甚少。

采集蜂将树脂和其他黏性植物材料装在花粉筐上搬运，通常（但不总是）它们会在没有其他工蜂帮助的情况下自己卸下这些物质，一些无刺蜂蜂种将树脂和其他黏性植物材料储存在特定的树脂堆中（图 3.7）（dos Santos et al., 2010）。采集蜂卸下树脂只需要几秒钟，因为它们会直接将带回的树脂按压在巢内树脂堆或巢壁上。树脂有许多不同的功能（第 3 章，第 7 章），例如密封巢腔中的裂缝和开口，第四阶段的工蜂也负责完成密封巢腔中的裂缝和开口的任务（Bassindale,1955）。然而，尚不清楚处理树脂的任务是否主要由采集树脂的工蜂完成。一些无刺蜂蜂种会收集泥土作为筑巢材料，这些泥土被采集蜂堆在巢内的泥土储存区（Sidnei Mateus）。

无刺蜂采集蜂通常会专性收集某一种类型的资源（第 8 章）：*Melipona beecheii*、*M. marginata*、*M. favosa* 和 *Plebeia tobagoensis* 无刺蜂的采集蜂大部分表现出采集花蜜、花粉、水、树脂或者泥土的倾向（Sommeijer et al., 1983; Biesmeijer and Tóth,1998; Hofstede and Sommeijer,2006; Mateus et al., 2019）。Hofstede 和 Sommeijer（2006）还观察到，专性采集程度较高的采集蜂比可变性采集的采集蜂采集效率更高，这进一步证实了行为专业化与效率提高相关的观点（见上文）。相反，Biesmeijer 和 Tóth（1998）则发现 *Melipona beecheii* 无刺蜂蜂群中专性采集蜂工作效率和不断切换采集资源的采集蜂相比并没有差异。蜜蜂采集蜂也表现出偏好性采集花粉、花蜜、水或树脂，这些偏好可以通过蔗糖敏感性来测定（Pankiw and Page,2000; Scheiner et al., 2004; Simone-Finstrom et al., 2010）。研究者尚未在无刺蜂上开展蔗糖敏感性和采集偏好之间关系的深入研究，但 Balbuena 和 Farina（2020）最新研究发现，*Tetragonisca angustula* 无刺蜂的花粉采集蜂比非花粉采集蜂（可能是花蜜采集蜂）具有更高的蔗糖敏感性，结果与研究者在蜜蜂中发现的情况类似（Pankiw and Page,2000; Scheiner et al., 2004）。我们也不知道有多少工蜂在某个时刻专门从事采集工作，但是 Wille（1966）发现新热带区 *Tetragonisca buchwald* 无刺蜂蜂群中 3 d 内有大约 10% 的工蜂积极从事采集工作。

在大多数无刺蜂蜂种中，保卫蜂巢是一项过渡性的任务（表 6.1），这项任务由相对较少数量的工蜂来完成（Couvillon et al., 2008; Grüter et al., 2012; Segers et al., 2015），之后它们就会成为采集蜂。我们通常可以在无刺蜂蜂巢入口处的前面、上面或里面观察到守卫蜂（图 3.8）。一个值得注意的例外情况是 *Tetragonisca angustula* 无刺蜂，*Tetragonisca angustula* 无刺蜂的守卫工作是由体型较大的兵蜂承担的，这些兵蜂会执行守卫蜂巢的任务直至死亡（见下文）（Hammel et al., 2016）。一些无刺蜂品种的守卫蜂的守卫时间可长达 3 周，而无刺蜂守卫蜂的平均守卫时间约为 5 d（Grüter et al., 2011; Hammel et al., 2016）。无刺蜂工蜂守卫蜂巢通常是先飞悬在入口管附近同时面向气道，在几天后切换到站立在入口管之上或附近（飞悬守卫蜂）（图 6.4，见下文）（Baudier et al., 2019）。

图 6.4 *Tetragonisca angustula* 无刺蜂的两种守卫蜂

注：飞悬守卫蜂在入口管附近飞行，面对通向入口孔的飞行走廊。
站岗守卫蜂则站在入口管上或入口管里面（摄影：C. Grüter）。

无刺蜂工蜂的守卫和采集工作开始之后，工蜂个体的死亡风险也随之增高（图 6.5；第 1.2 节）（Giannini, 1997; Grosso and Bego, 2002; Njoya and Wittmann, 2013; Lopes et al., 2020），这很可能是因为这些任务与巢内工作相比在代谢上要求更高并且更危险（Wolf and Schmid-Hempel, 1989; Visscher and Dukas, 1997; Page and Peng, 2001）。有研究者观察发现 *Melipona fasciculata* 无刺蜂的采集蜂如果不每天去采集的话，采集蜂的寿命会更长（Gomes et al., 2015），而 *M.fulva* 和 *M.favosa* 无刺蜂采集活动较少的时期，采集蜂的寿命会延长两倍之多（Roubik, 1982），这都进一步证实了采集活动对工蜂寿命有显著的负面影响。*M. beecheii* 无刺蜂的花粉采集蜂每天的行程比花蜜采集蜂要短，其寿命也相应的比花蜜采集蜂要长得多（Biesmeijer and Toth, 1998）。

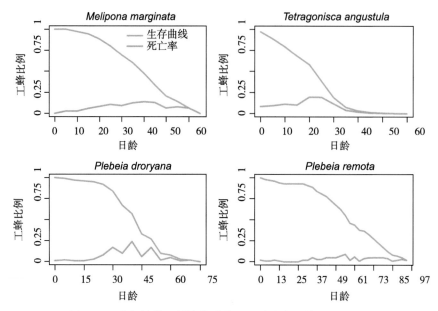

图 6.5　4 种新热带无刺蜂蜂种基于日龄的存活情况和死亡率

注：数据来自 Terada 等（1975）、van Benthem 等（1995）、Grosso 和 Bego（2002）、Mateus 等（2019）。x 轴为工蜂出房后的天数，即成年蜂日龄。

无刺蜂和蜜蜂在随日龄出现阶段性分工方面表现出许多相似之处（Kerr and Santos Neto,1956），但也有些差异值得注意（Sakagami,1982）：一是无刺蜂蜂群中非常年幼的工蜂承担了许多蜂巢建造任务，并且这些任务是先于哺育幼虫的任务的（表 6.1），而在蜜蜂中，年幼工蜂承担了哺育幼虫的任务，而蜂巢建造则是由中年蜜蜂承担的（Rösch,1927; Seeley,1982; Johnson and Frost,2012）。二是无刺蜂通常会比蜜蜂收集更多的树脂和其他筑巢材料。三是无刺蜂工蜂离巢时间晚于蜜蜂。一个原因可能是无刺蜂工蜂是在巢内排泄的，因此它们不需要离开蜂巢。例如 *Melipona Quadrifasciata* 无刺蜂工蜂在 27～30 日龄才第一次出远门去扔蜂巢垃圾，而后不久就开始守护蜂巢和进行采集（Kerr and Santos Neto,1956）。研究者在其他 *Melipona* 属无刺蜂中也发现了相似的规律（Bustamante,2006）。四是无刺蜂的幼虫食物供应和产卵过程与蜜蜂的幼虫食物供应和哺育幼虫的过程有很大不同（第 5 章）。五是无刺蜂工蜂在幼龄时就开始分泌蜂蜡并持续到采集年龄（例如，*Melipona seminigra* 无刺蜂的蜡分泌阶段为第 5～36 日龄）（Bustamante，2006；Sakagami，1982），而蜜蜂的蜂蜡主要是中年蜜蜂分泌的（Rösch，1925；Lindauer，1952）。

当然，并不是所有的无刺蜂都是基于时间段分工来组织劳动的。Inoue 等

（1996）对亚洲 *Tetragonula minangkabau* 无刺蜂进行个体标记并研究了其一生工作概况，他们发现不同工蜂个体的任务流程存在显著个体差异。通过聚类分析他们确定了 4 个行为上不同的工蜂群体，即核心哺育蜂、辅助哺育蜂、采集蜂和寿命短且长时间不活跃的工蜂。例如，哺育蜂很少从事户外工作，但比采集蜂存活时间长得多。*Melipona marginata* 无刺蜂不同日龄的工蜂会执行不同的任务（图 6.2），但 Mateus 等（2019）发现这些工蜂个体之间始终存在差异，这种差异与日龄无关。他们研究还发现大约 60% 的工蜂从未参与过采集工作，而其他部分工蜂则未参与 POP 工作（Mateus et al.,2019）。这表明不同的无刺蜂蜂种有一系列的劳动分工模式，只有单独标记工蜂个体并跟踪记录其一生劳动轨迹，才能探明其劳动分工模式。

6.3 无刺蜂的生理性次级分工

多种蚂蚁和白蚁群体是基于不同工种工蚁间的形态差异来区分劳动分工的（Wilson,1953; Oster and Wilson,1978; Hölldobler and Wilson,1990, 2009; Korb and Hartfelder,2008）。具有生理性次级分工的物种通常具有以下特征：一是不同身体部位尺寸间存在异速生长关系，即体型大小不同的工蚁在外形上也不同；二是多模态的工蚁体型大小—频率分布（Wilson,1953; Oster and Wilson,1978; Wheeler,1991）。研究者认为生理性次级分工可以进一步提高劳动分工的效率，因为工蜂可以在形态上适应特定的任务，例如守卫蜂利用其强壮的上颚来防御敌人。

虽然研究者发现蚂蚁上存在生理性次级分工的极端例子（例如 *Atta*、*Pheidole* 或 *Eciton*），但尚未在无刺蜂上发现这样的极端性生理性次级分工现象，只在几个无刺蜂蜂种中发现存在中等分化的生理性次级分工情况（Grüter et al., 2012, 2017 a; Wittwer and Elgar,2018）。目前生理性次级分工研究得最好的是小型 *Tetragonisca angustula* 无刺蜂（图 6.6 A）（Grüter et al., 2012; Segers et al., 2015; Segers et al., 2016; Hammel et al., 2016; Baudier et al., 2019），这也是新热带区最常见的无刺蜂蜂种之一（Freitas,2001; Slaa,2006; Velez-Ruiz et al., 2013）。*T. angustula* 无刺蜂蜂群中有两种不同的守卫蜂驻扎在蜂巢的入口管处：10～30 个站岗守卫蜂停留在蜂蜡入口管的里面、上面或附近，而 5～15 个飞悬守卫蜂则监视着通往入口管的飞行走廊（图 6.4，图 6.6 B）（Wittmann,1985; Wittmann et al., 1990; Bowden et al., 1994; Grüter et al., 2011; Kärcher and Ratnieks,2009; van Zweden et al., 2011; Baudier et al., 2019）。下午时

段，该蜂种蜂巢守卫蜂的数量达到顶峰（Grüter et al.,2011）。这些守卫蜂中的飞悬的守卫蜂[5]主要基于视觉信号攻击异种入侵者，而站岗守卫蜂则识别看起来相似但闻起来不同的同种不同巢蜂（Wittman,1985; Wittmann et al.,1990; Bowden et al.,1994; Kärcher and Ratnieks,2009; Jones et al.,2012; Couvillon et al., 2013）。该蜂种这两种类型的守卫蜂比普通采集蜂的体型大约30%（图6.7 A）。由于这两项任务中每个任务组内工蜂数量变化很小，因此蜂群中守卫蜂和采集蜂之间角色重叠的工蜂数量几乎为0。该蜂种蜂群中守卫蜂（或士兵蜂）占总劳动力的1%～6%（Grüter et al.,2012; Segers et al.,2015）。无刺蜂蜂群中负责垃圾清运的工蜂属于中等规模群体（Grüter et al.,2012），而 Hammel 等（2016）观察到一些负责垃圾清运的工蜂也会执行了一些守卫任务[6]。无刺蜂蜂群中体型较大的垃圾清运工蜂有可能成为守卫蜂，而体型较小的垃圾清运工蜂则会成为采集蜂。另外，由于清运蜂巢垃圾可能会使工蜂暴露于病原体，因此垃圾清运工蜂可能会成为另一类专业分工的工蜂，它们很可能会坚持这项任务直至死亡。而研究者对无刺蜂的垃圾清运活动仍然缺少深入研究。

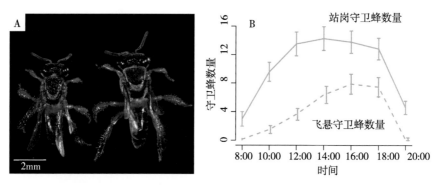

图6.6 *Tetragonisca angustula* 无刺蜂的采集蜂（A, 左）和守卫蜂（A, 右）

注：采集蜂胸部覆盖着一层薄薄的树脂（图A所示），其功能尚不清楚。守卫蜂腿上有树脂，但胸部没有。守卫蜂比采集蜂重30%（Grüter et al.,2012）。图B显示蜂巢入口处不同时间点飞悬守卫蜂数量和站岗守卫蜂的数量（Grüter et al.,2011）。

---

[5] 该蜂种因蜂巢入口附近有一群飞悬守卫蜂这种异常特征而被称为"angelitas"（西班牙语，小天使）和"us-kaab"（危地马拉玛雅语中的"飞行蜂"，Źrałka et al., 2018）。

[6] Goulson 等（2005）发现 *Tetragonisca angustula*（和 *Scaptotrigona mexicana*）无刺蜂工蜂体型大小在劳动分工中起到了一定作用，但是近期的多项研究结果与之不同，这些研究提出采集蜂比巢内工蜂体型更大。然而这种明显的矛盾结果最可能是因为这些研究中抓获工蜂方法不同而导致的。他们采用笼子抓捕巢外工蜂并认为这些工蜂就是采集蜂。然而考虑 *T. angustula* 无刺蜂蜂巢入口有大体型的飞悬守卫蜂守护在巢外，则 Goulson 等（2005）抓捕的巢外蜂很有可能是守卫蜂而不是采集蜂。

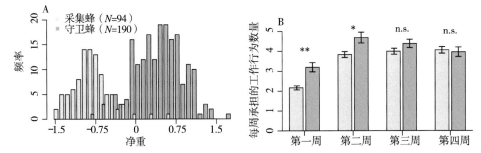

图 6.7　图 A 所示为来自巴西圣西芒 Aretuzina 农场的 *Tetragonisca angustula* 无刺蜂 20 个不同蜂群的采集蜂和守卫蜂的体重分布（每个蜂群取 5 个采集蜂，5 个飞悬守卫蜂和 5 个站岗守卫蜂）。该图中将飞悬守卫蜂和站岗守卫蜂的数据汇总在一起，然而飞悬守卫蜂的体重略高于站岗守卫蜂（+5%）（线性混合效应模型：飞悬守卫 vs 站岗守卫，z-value = 3.9, $p < 0.001$）（Segers et al., 2016）。图 B 显示 *T. angustula* 无刺蜂蜂群中小型工蜂和兵蜂体型工蜂在 4 周龄内担任过的不同工作行为的数量。**$p < 0.01$, *$p < 0.05$, *n.s.*= 不显著（Hammel et al., 2016）

Tetragonisca angustula 无刺蜂的守卫蜂和采集蜂在形态上也有所不同。总的来说，*Tetragonisca angustula* 无刺蜂的守卫蜂的头部相对较小（但从绝对尺寸来说头部较大），后肢则相对较大。由此可以推测这种形态上的差异可能会影响工蜂完成各自任务的能力，但这需要进一步的研究。无刺蜂这种身体大小和头部宽度之间的负异速生长是较为常见的现象。Grüter 等（2017 a）研究了 28 个新热带无刺蜂蜂种，发现其中 9 个蜂种的头部相对尺寸随着身体尺寸的增加而减小（负异速生长似乎与体型较大的守卫蜂尺寸或兵蜂的存在无关）。

Tetragonisca fiebrigi[7] 和 *Frieseomelitta longipes*[8] 无刺蜂亲缘关系较近，而研究者也发现了这两种无刺蜂的守卫蜂和采集蜂的大小呈双峰分布（Grüter et al., 2017 a），这表明这些蜂种为保卫蜂巢分化出了守卫蜂级型。*Frieseomelitta* 属无刺蜂蜂种的守卫蜂和采集蜂之间则表现出另一种有趣的差异。该属无刺蜂的守卫蜂往往颜色更深（图 6.8），尤其是面部的颜色差异特别明显，而守卫蜂在许多其他身体部位上的颜色也相对更暗一点（图 6.8 C）。我们对这种守卫蜂着色更深的原因仍然是未知的，但有研究者认为工蜂在守卫蜂巢入口

---

[7] 一些作者认为 *T. angustula* 和 *T. fiebrigi* 是一个物种形成过程中的两个亚种（Francisco et al., 2014）。这两种无刺蜂的明显差异主要包括独特的颜色（例如 *T. angustula* 无刺蜂有颜色更深的中轴线）、雄蜂的生殖器形态以及子脾的建筑结构（图 3.9 A，*T. fiebrigi* 的子脾更大）。

[8] 测定 *F. longipes* 无刺蜂工蜂体型分布的样本量相对较低，需要检测更大的样本量来确定或否定该蜂种工蜂体型呈双峰分布这一观点。

时颜色更深的守卫蜂可能会伪装的更好（图 6.8）（Grüter et al., 2017 a），这样就能使依赖于视觉线索的（例如蜘蛛等捕食者）天敌更难以找到蜂巢入口。守卫蜂上增加的黑化作用也可以增加其对病原体和物理损伤的抵抗力（Dubovskiy et al., 2013），这对于蜂巢入口的守卫蜂来说是十分有利的。

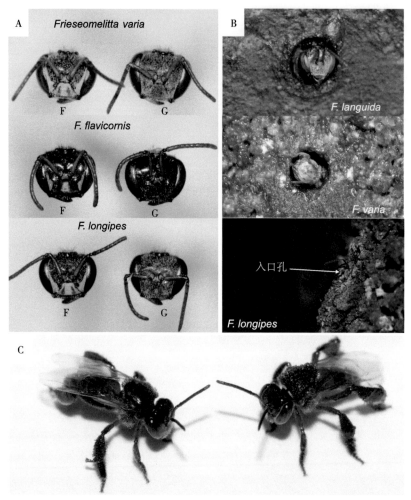

图 6.8　A 为 3 种不同 *Frieseomelitta* 属无刺蜂蜂种的守卫蜂和采集蜂的头部（摄影：C.Crüter）。B 为 *Frieseomelitta languida*，*F. varia* 和 *F. longipes* 无刺蜂蜂巢的入口孔。*Frieseomelitta* 属无刺蜂的入口孔是无刺蜂中最小的（Couvillon et al.,2008）。颜色较深的 *F. languida* 和 *F. silvestrii* 无刺蜂蜂巢的入口孔颜色比颜色更浅的无刺蜂蜂种（*F. varia* 或 *F. longipes*）蜂巢入口孔颜色更深。无刺蜂工蜂会用黏性物质覆盖住蜂巢入口周围的表面（摄影：前两张为 C.Crüter，后一张为 Cristiano Menezes）。C 为显示的是 *F. flavicornis* 无刺蜂的守卫蜂（左）和采集蜂（右）（摄影：Cristiano Menezes）

一些其他无刺蜂蜂种的守卫蜂虽然体型明显大于采集蜂，但工蜂体型大小频率分布是呈单峰的（Grüter et al., 2017 a），与在一些熊蜂中的发现类似（Goulson et al., 2002）。目前有一些原因可以解释为什么守卫蜂体型更大是有益的。例如，*Tetragonisca angustula* 无刺蜂在与 *Lestrimelitta limao* 强盗蜂战斗时，体型越大的守卫蜂能够坚持的时间更长（Grüter et al., 2012）。这样就可以给蜂群争取更多的时间来招募更多的防御力量。此外，研究者发现守卫蜂体型大小与其识别非同巢工蜂的能力呈正相关（Grüter et al., 2017 b）。这项研究中比较了来自 10 个蜂群的守卫蜂的识别准确度。守卫蜂体型最小的蜂群接受了大约 20% 的非同巢工蜂，也就是说，这些非同巢工蜂没有受到攻击，而守卫蜂体型最大的蜂群只接受了大约 2% 的非同巢工蜂。体型越大的守卫蜂的触角上有更多的感觉感受器，这可能与其触角知觉的增加有关（Spaethe et al., 2007; Gill et al., 2013）。工蜂触角上感受器的密度不依赖于触角的大小，但是体型越大的守卫蜂触角表面积更大，导致守卫蜂的体型大小和感受器数量之间呈正相关（Grüter et al., 2017 b）。*Tetragonula carbonaria* 无刺蜂蜂种的守卫蜂体型也较大，该蜂种守卫蜂感觉感受器的密度本身就高于采集蜂（图 6.9）（Wittwer and Elgar, 2018）。Month-Juris 等（2020）分析测定了两个 *Tetragonisca fiebrigi* 无刺蜂守卫蜂的触角，发现有迹象表明在该蜂种的守卫蜂和其他工蜂之间某些类型的感受器长度和数量也有所不同（Balbuena and Farina，2020）。

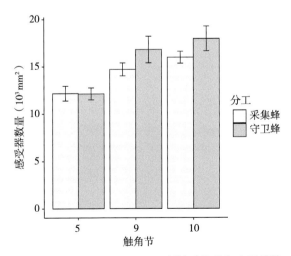

图 6.9 *Tetragonula carbonaria* 无刺蜂采集蜂和守卫蜂第 5、第 9 和第 10 节触角的感受器密度（平均 ± 标准误）

注：感受器密度向触角顶端方向不断增加，守卫蜂的感受器密度总体上高于采集蜂（$p = 0.036\ 9$, $N = 10$）（Wittwer and Elgar, 2018）。

拥有体型更大的守卫蜂来保卫蜂群确实具有很多潜在好处，这就引出了一个问题：为什么不是所有的无刺蜂蜂种都有守卫蜂？拥有专性工蜂次级级型的一个缺点就是蜂群应对突然的环境变化或内部需求变化的反应能力会变弱（Oster and Wilson，1978；Wheeler，1991）。专业化的工蜂在行为上可能不太灵活可变（Oster and Wilson,1978; Mertl and Traniello,2009）。Baudier 等（2019）通过抓捕 *Tetragonisca angustula* 无刺蜂的所有守卫蜂模拟了一场灾难性的天敌捕食事件。虽然蜂群能够通过重新分配体型稍小的工蜂去执行守卫任务来弥补部分损失，但蜂群需要几天时间才能完全取代失去的守卫蜂。因此蜂群面对失去一小群重要的工蜂（*T. angustula* 蜂群中仅有大约 1% 的工蜂是守卫蜂）采取的缓慢的集体应对可能会使蜂群付出高昂的代价。

Grüter 等（2017 a）猜测无刺蜂蜂群中存在体型较大的守卫蜂与蜂群受强盗蜂袭击的风险有关。*Lestrimelitta* 强盗蜂是 2 500 万—2 000 万年前分化出来的，这与无刺蜂蜂群中出现形态不同的守卫蜂的估算时间相对应（Grüter et al.,2017 a）。研究者认为 *Lestrimelitta* 强盗蜂的掠夺行为是新热带无刺蜂守卫蜂体型大小增加的重要驱动力，这一观点得到了以下观察的证实：受 *Lestrimelitta* 侵害的无刺蜂蜂种拥有体型较大的守卫蜂的可能性是非受害蜂种的 4 倍；约 70% 的受害蜂种都拥有体型较大的守卫蜂，而只有约 17% 的非受害蜂种拥有体型较大的守卫蜂。那么增加专业守卫蜂的益处是否超过了劳动力灵活度低造成的潜在损失就需要取决于蜂群面对的捕食者和寄生物威胁的情况。

生理性的劳动分工和时间阶段性的劳动分工之间并不是互斥的，二者经常是结合在一起的（蚂蚁：Wilson,1980; Seid and Traniello,2006; Camargo et al., 2007）。*Tetragonisca angustula* 无刺蜂蜂群中小型和大型工蜂年轻时都执行相似的任务，但大型工蜂（守卫蜂）从一个任务到另一个任务的过渡更快（图 6.2）（Hammel et al., 2016）。随着两种工蜂日龄的增长，两种工蜂掌握的行为技能似乎都在增加（蚂蚁：Seid and Traniello, 2006）。但是总的来说，守卫蜂大小的工蜂的工作任务更大，并且研究者观察到这些工蜂在出房后的头两周内执行了更多的工作任务（图 6.7 B）（Hammel et al., 2016）。这与在蚂蚁中观察到的情况形成鲜明对比，在蚂蚁中兵蚁掌握的行为技能比幼年蚁少，也就是说它们更专业，并且往往工作任务更少（Wilson,1984; Beshers and Fewell, 2001; Sempo and Detrain, 2004）。

6.4 工蜂工作活动的分布情况

无刺蜂蜂群的表现和适应力不仅取决于工蜂执行的任务类型和执行任务的工蜂数量，还取决于每只工蜂执行不同任务的强度。研究者对蚂蚁、蜜蜂和胡蜂开展的一些研究发现，工蜂的工作强度分布呈高度倾斜且具有非随机性，每项工作都是由极少部分的工蜂完成了大部分内容，而大部分其他工蜂则几乎没有参与这项工作（Robson and Traniello,1999; Hurd et al., 2003; Pinter-Wollman et al., 2012; Tenczar et al.,2014）。研究者发现 *Melipona marginata* 和 *Tetragonisca angustula* 无刺蜂也存在同样的情况（图 6.10）（Hammel et al., 2016; Mateus et al., 2019）。然而，如果我们考虑工蜂整个生命周期的工作表现，而不是专注于单项工作时，工蜂工作活动就不是那么不平衡了（图 6.10），剩下的不平衡大部分是由于早期死亡引起的，而不是由工蜂持续不工作引起的（Hammel et al., 2016）。3 个新热带无刺蜂蜂种的工蜂整个生命周期里表现的工作总量则基本是遵循高斯分布的（图 6.11）。

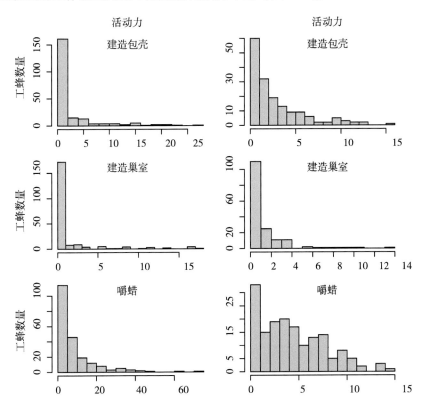

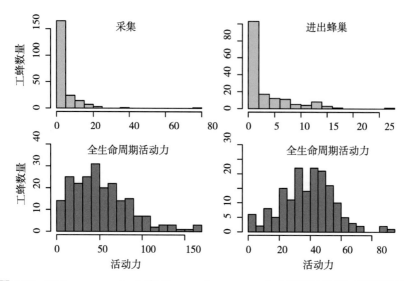

图 6.10 *Melipona marginata* 和 *Tetragonisca angustula* 无刺蜂工蜂在 4 项不同工作任务上的活动力（以观察到的工蜂进行这项工作活动行为的次数来衡量）分布情况以及其终身工作表现情况

注：终身工作表现是所有研究行为中的活动总数（Mateus et al., 2019；Hammel et al., 2016）。

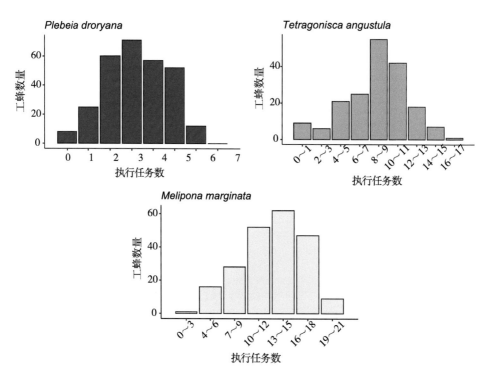

图 6.11 （观察到的）3 种新热带无刺蜂蜂种的工蜂在其成年生活中完成的任务总数（van Benthem et al.,1995; Hammel et al.,2016; Mateus et al.,2019）

6.5 工作的无刺蜂雄蜂和蜂王

群居膜翅目昆虫群体中的雄性（或雄蜂）通常不会在群体中进行利他性工作，如果研究者观察到在群体中工作的雄性通常也并不能明确这种行为是否真的是利他性的，还是说这种行为对雄性自身有直接好处（Lucas and Field,2011）。一些因素可以解释为什么无刺蜂雄蜂不工作：首先，单个无刺蜂雌蜂祖先为哺育幼虫而产生的行为上和形态学上的适应性可能会使雌性倾向于照顾后代（West–Eberhard,1975），而无刺蜂雄蜂没有螫刺可能会阻止雄蜂为蜂群防御做出贡献（Starr,1985）。其次，由于单倍二倍性，膜翅目昆虫的同胞姐妹（$r = 0.75$）比同胞兄弟（$r = 0.5$）亲缘度更高，从而导致最初就偏向于雌性利他主义（Reeve,1993）。最后，交配的可能性会激励雄蜂将大部分资源用于交配获得直接的适应性，而不是通过帮助他人获得间接的适应性。尽管如此，在一些胡蜂和熊蜂蜂群中一些雄蜂也会进行潜在利他工作，且可能会为其力所能及的工作做出贡献。

无刺蜂雄蜂需要 1～2 周的时间达到性成熟，之后它们就会离开蜂巢（*Tetragonisca angustula* 无刺蜂雄蜂 7～12 日龄出房，Santos, 2018; *Melipona* 无刺蜂雄蜂 10～20 日龄出房；van Veen et al.,1997; Velthuis et al., 2005; Bustamante,2006; Veiga et al.,2018）。一旦无刺蜂雄蜂离开蜂巢就基本就不会再回来了（第 4 章）（van Veen et al.,1997; Veiga et al., 2018），然而在雄蜂离开家园之前，雄蜂可能会在不牺牲交配机会的情况下为母蜂群做出贡献。研究者就曾在新热带区的 *Plebeia remota*（van Benthem et al.,1995）、*P. droryana*（Cortopassi–Laurino,1979）、*Schwarziana quadripunctata*（Imperatriz-Fonseca,1973）、*Melipona beecheii*、*M. favosa*、*M. flavolineata* 和 *M. marginata*（Roubik,1989; van Veen et al.,1997; Veiga et al.,2018）无刺蜂蜂群中发现雄蜂参与了蜂群中的蜂蜜催熟工作。甚至有研究者观察到一些无刺蜂雄蜂是归巢采集蜂的花蜜接收员（Kerr,1997），然而尚无研究观察到雄蜂将成熟的蜂蜜放入蜜罐中（Kerr,1997），因此目前还不清楚雄蜂参与催熟花蜜是利他行为还是利己行为。为蜂蜜脱水可能是无刺蜂雄蜂的一种减轻体重和增加其体能能量含量的策略，这种策略可能会增加试图与蜂王交配的雄蜂的灵活性（Cortopassi–Laurino,2007）。然而，已经有研究观察到雄蜂也会以其他看似利他的方式为蜂群做出贡献。例如有报道提到无刺蜂雄蜂会泌蜡（*Melipona compressipes, M. quadrifasciata, M. marginata, M. scutellaris, M. rufiventris, Nannotrigona testaceicornis*）和帮助蜂巢建筑工作（Sakagami,1982; Kerr,1997）。Kerr（1951）观察到 *M. marginata* 和 *Plebeia droryana* 无刺蜂蜂群中有 100 多只雄蜂能产生大量的蜡。Kerr（1997）还提到 *M. compressipes* 无刺蜂雄蜂的行为也会从它们的背板上取下鳞片状蜡质，并将这些蜡质放入蜂巢里的蜡堆中，该行为与工蜂的行为类似。他还观察到了 *M. rufiventris* 和 *M. compressipes* 无刺蜂的雄蜂用蜡来为蜂巢建造小柱巢、储物罐和包壳层，但不包括子室。目前还不清楚无刺蜂雄蜂产蜡现象是否普遍（蜜蜂雄蜂是否有这种情况也是未知的），而研究者对几只无刺蜂蜂种雄蜂的组织学研究中发现，仅美洲 *Melipona marginata* 和非洲 *Meliponula bocandei* 无刺蜂雄蜂存在蜡腺，这为无刺蜂雄蜂产蜡提供了证据（Cruz-Landim，1963, 1967）[9]。

无刺蜂上颚腺产生的警报信息素在蜂群防御中起着重要作用（第 7 章）（Cruz-López et al.,2007; Schorkopf et al.,2009; Nunes et al.,2014）。Schorkopf（2016）最近研究了 *Scaptotrigona depilis* 无刺蜂雄蜂的上颚腺，发现工蜂上颚腺中所含化合物也存在于雄蜂上颚腺中，但是雄蜂下上颚腺化合物的量较少一些。这些化合物中就有警报信息素 2- 正庚醇。在蜂巢入口处模拟对 *S. depilis* 和 *S. bipunctata* 无刺蜂雄蜂进行攻击会引起雄蜂张开上颚，守卫工蜂则会表现出防御行为，这表明受到攻击的雄蜂可能能够诱发守卫蜂的防御反应（Schorkopf,2016）。而雄蜂接触雄蜂上颚腺化合物后的反应则是逃跑。

无刺蜂雄蜂的其他潜在利他行为还包括养育蜂王（例如，*Plebeia droryana* 无刺蜂雄蜂）（Cortopasi-Laurino,1979; Kerr,1997）或者养育其他雄蜂（*P. droryana*、*Melipona compressipes*、*M. quadrifasciata*、*M. beecheii* 和 *Scaptotrigona postica* 无刺蜂雄蜂）（Cortopassi-Laurino,1979; Kerr,1997; van Veen et al.,1997），还可能参与幼虫的表面孵化（Kerr，1997）。这些观察大多是传闻，尚无关于无刺蜂雄蜂对蜂群总工作量的贡献的系统研究，但这些报告表明无刺蜂雄蜂比蜜蜂雄蜂执行了更多的工作（van Veen et al.,1997）。Kerr 等（1997，2004）指出由于无刺蜂的性别决定和哺育幼虫系统，无刺蜂雄蜂和雌性工蜂在行为上和形态上相对相似（第 1 章，第 5 章）。因此，无刺蜂的雄蜂可能比蜜蜂雄蜂更适合工作。而无刺蜂雄蜂的工作贡献也可以用来解释为什么它们似乎比蜜蜂雄蜂更少受到蜂群内工蜂的攻击（Sakagami,1982），即使在一些无刺蜂

---

[9] 研究者注意到 *Melipona* 属无刺蜂雄蜂与蜂王—工蜂中间级型外观上是相似的。因此，需要更仔细地观察来确定这些在工作的"雄蜂"是不是中间级型的雌蜂。

种中也观察到了工蜂驱逐甚至杀死雄蜂的现象（Kerr,1951）。研究者还观察到 *Melipona beecheii* 和 *M. rufiventris* 无刺蜂雄蜂会在特定时节被拖着翅膀丢到蜂巢入口处（von Ihering, 1903; Schwarz, 1948）。而 *Tetragonisca angustula* 无刺蜂工蜂对雄蜂的攻击则是随着雄蜂停留在蜂巢内时间的延长而不断增加的（Santos,2018）。

Koedam 等（2002）还研究了 *Melipona bicolor* 无刺蜂处女蜂王产蜡的情况。无刺蜂蜂王产的蜡与工蜂产的蜡成分相似，但蜂王产蜡的时间阶段短于工蜂。其他研究还观察到 *Scaptotrigona bipunctata* 无刺蜂的处女蜂王也会分泌蜡（Kerr and de Lello,1962），*Schwarziana quadripunctata* 无刺蜂处女蜂王甚至可以用蜡进行建筑工作（Imperatriz-Fonseca,1973）。

参考文献

Balbuena MS, Farina WM（2020）Chemosensory reception in the stingless bee *Tetragonisca angustula*. Journal of Insect Physiology 125:104076

Bassindale R（1955）The biology of the Stingless Bee *Trigona*（*Hypotrigona*）*gribodoi* Magretti（Meliponidae）. Proceedings of the Zoological Society of London 125（1）:49–62

Baudier KM, Ostwald MM, Grüter C, Segers FHID, Roubik DW, Pavlic TP, Pratt SC, Fewell JH（2019）Changing of the guard: mixed specialization and flexibility in nest defense（*Tetragonisca angustula*）. Behavioral Ecology 30:1041–1049

Bego LR（1990）On social regulation in *Nannotrigona*（*Scaptotrigona*）*postica* Latreille, with special reference to productivity of colonies（Hymenoptera, Apidae, Meliponinae）. Revista Brasileira de Entomologia 34（4）:721–738

Bego LR, Simões D, Zucchi R（1983）On some structures in *Nannotrigona*（*Scaptotrigona*）*postica* Latreille（Hymenoptera, Meliponinae）. Boletim de Zoologia, Universidade de São Paulo 6（6）:79–88

Ben-Shahar Y, Robichon A, Sokolowski MB, Robinson GE（2002）Influence of gene action across different time scales on behavior. Science 296:741–744

Beshers SN, Fewell JH（2001）Models of division of labor in social insects. Annual Review of Entomology 46:13–40

Biesmeijer JC, Tóth E（1998）Individual foraging, activity level and longevity in the stingless bee *Melipona beecheii* in Costa Rica（Hymenoptera, Apidae,

Meliponinae). Insectes Sociaux 45:427–443

Bowden RM, Garry MF, Breed MD (1994) Discrimination of con- and heterospecific bees by *Trigona* (*Tetragonisca*) *angustula* guards. Journal of the Kansas Entomological Society 67:137–139

Bustamante NCR (2006) Divisão de Trabalho em Três Espécies de Abelhas do Gênero *Melipona* (Hymenoptera, Apidae) na Amazônia, Brasileira. PhD Thesis, Universidade Federal do Amazonas, Manaus

Camargo RS, Forti LC, Lopes JFS, Andrade APP, Ottati ALT (2007) Age polyethism in the leaf-cutting ant *Acromyrmex subterraneus brunneus* Forel, 1911 (Hym., Formicidae). Journal of Applied Entomology 131:139–145

Cameron S (1989) Temporal patterns of division of labor among workers in the primitively eusocial bumble bee, *Bombus griseocollis* (Hymenoptera: Apidae). Ethology 80:137–151

Cardoso-Júnior CAM, Silva RP, Borges NA, de Carvalho WJ, Walter SL, Simões ZLP, Bitondi MMG, Vieira CU, Bonetti AM, Hartfelder K (2017) Methyl farnesoate epoxidase (*mfe*) gene expression and juvenile hormone titers in the life cycle of a highly eusocial stingless bee, *Melipona scutellaris*. Journal of Insect Physiology 101:185–194

Cardoso R (2010) Divisao etária de trabalho em operárias de *Frieseomelitta varia* (Hymenoptera, Apidae, Apini). BSc Thesis, University of São Paulo, Ribeirão Preto

Cortopassi-Laurino M (1979) Observações sobre atividades de machos de *Plebeia droryana* Friese (Hymenoptera, Apidae, Meliponinae). Revista Brasileira de Entomologia 23:177–191

Cortopassi-Laurino M (2007) Drone congregations in Meliponini: what do they tell us? Bioscience Journal 23:153–160

Couvillon MJ, Segers FHID, Cooper-Bowman R, Truslove G, Nascimento DL, Nascimento FS, Ratnieks FLW (2013) Context affects nestmate recognition errors in honey bees and stingless bees. Journal of Experimental Biology 216:3055–3061

Couvillon MJ, Wenseleers T, Imperatriz-Fonseca VL, Nogueira-Neto P, Ratnieks FLW (2008) Comparative study in stingless bees (Meliponini) demonstrates that nest entrance size predicts traffic and defensivity. Journal of Evolutionary Biology 21:194–201

Cruz-Landim C (1963) Evaluation of the wax and scent glands in the Apinae (Hymenoptera: Apidae). Journal of the New York Entomological Society 71:2–13

Cruz-Landim C (1967) Estudo comparativo de algumas glândulas das abelhas (Hymenoptera, Apoidea) e respectivas implicações evolutivas. Arquivos de Zoologia 15 (3):177–290

Cruz-López L, Aguilar S, Malo E, Rincón M, Guzman M, Rojas J (2007) Electroantennogram and behavioral responses of workers of the stingless bee *Oxytrigona mediorufa* to mandibular gland volatiles. Entomologia Experimentalis et Applicata 123 (1):43–47

da Silva DLN, Zucchi R, Kerr WE (1972) Biological and behavioural aspects of the reproduction in some species of *Melipona* (Hymenoptera, Apidae, Meliponinae). Animal Behaviour 20 (1):123–132

Darchen R (1969) Sur la biologie de *Trigona* (*Apotrigona*) *nebulata* komiensis Cock. I. Biologia Gabonica 5:151–183

Dornhaus A (2008) Specialization does not predict individual efficiency in an ant. PLoS Biology 6:e285

dos Santos CG, Blochtein B, Megiolaro FL, Imperatriz-Fonseca VL (2010) Age polyethism in *Plebeia emerina* (Friese) (Hymenoptera: Apidae) colonies related to propolis handling. Neotropical Entomology 39 (5):691–696

Duarte A, Pen I, Keller L, Weissing FJ (2012) Evolution of self-organized division of labor in a response threshold model. Behavioral Ecology and Sociobiology 66:947–957

Dubovskiy IM, Whitten MMA, Kryukov VY, Yaroslavtseva ON, Grizanova EV, Greig C, Mukherjee K, Vilcinskas A, Mitkovets PV, Glupov VV, Butt TM (2013) More than a colour change: insect melanism, disease resistance and fecundity. Proceedings of the Royal Society of London Series B-Biological Sciences 280:20130584

Engels W (1987) Pheromones and reproduction in Brazilian stingless bees. Memórias do Instituto Oswaldo Cruz 82:35–45

Engles W, Engels E (1977) Vitellogenin und Fertilität bei stachellosen Bienen. Insectes Sociaux 24 (1):71–94

Fahrbach SE (1997) Regulation of Age Polyethism in Bees and Wasps by Juvenile Hormone. In: Slater PJB, Rosenblatt JS, Snowdon CT, Milinski M (eds) Advances in the Study of Behavior. Academic Press, pp 285–316

Ferreira-Caliman M, Nascimento FS, Turatti I, Mateus S, Lopes N, Zucchi R (2010) The cuticular hydrocarbons profiles in the stingless bee *Melipona marginata* reflect task-related differences. Journal of Insect Physiology 56 (7):800–804

Francisco FO, Santiago LR, Brito RM, Oldroyd BP, Arias MC (2014) Hybridization and asymmetric introgression between *Tetragonisca angustula* and *Tetragonisca fiebrigi*. Apidologie 45:1–9

Freitas GS (2001) Levantamento de ninhos de meliponíneos (Hymenoptera, Apidae) em área urbana: Campus da USP, Ribeirao Preto-SP. Master Thesis, University of Sao Paulo, Brazil

Giannini KM (1997) Labor division in *Melipona compressipes fasciculata* Smith (Hymenoptera: Apidae: Meliponinae). Anais da Sociedade Entomológica do Brasil 26 (1):153–162

Gill KP, Van Wilgenburg E, Macmillan DL, Elgar MA (2013) Density of antennal sensilla efficacy of communication in a social insect. Am Nat 182:834–840

Goldsby HJ, Dornhaus A, Kerr B, Ofria C (2012) Task-switching costs promote the evolution of division of labor and shifts in individuality. Proceedings of the National Academy of Sciences of the United States of America 109:13686–13691

Gomes RLC, Menezes C, Contrera FAL (2015) Worker longevity in an Amazonian *Melipona* (Apidae, Meliponini) species: effects of season and age at foraging onset. Apidologie 46 (2):133–143

Gordon DM (2016) From division of labor to the collective behavior of social insects. Behavioral Ecology and Sociobiology 70:1101–1108

Goulson D, Derwent LC, Peat J (2005) Evidence for alloethism in stingless bees (Meliponinae). Apidologie 36:411–412

Goulson D, Peat J, Stout J, Tucker J, Darvill B, Derwent LC, Hughes WOH (2002) Can alloethism in workers of the bumblebee, *Bombus terrestris*, be explained in terms of foraging efficiency? Animal Behaviour 64:123–130

Grosso AF, Bego LR (2002) Labor division, average life span, survival curve, and nest architecture of *Tetragonisca angustula angustula* (Hymenoptera, Apinae, Meliponini). Sociobiology 40:615–637

Grüter C, Kärcher M, Ratnieks FLW (2011) The natural history of nest defence in a stingless bee, *Tetragonisca angustula* (Latreille) (Hymenoptera: Apidae), with two distinct types of entrance guards. Neotropical Entomology 40:55–61

Grüter C, Menezes C, Imperatriz-Fonseca VL, Ratnieks FLW (2012) A morphologically specialized soldier caste improves colony defence in a Neotropical eusocial bee. Proceedings of the National Academy of Sciences of the United States of America 109:1182–1186

Grüter C, Segers FHID, Menezes C, Vollet-Neto A, Falcon T, von Zuben LG, Bitondi MMG, Nascimento FS, Almeida EAB (2017 a) Repeated evolution of soldier sub-castes suggests parasitism drives social complexity in stingless bees. Nature Communication 8:4

Grüter C, Segers FHID, Santos LLG, Hammel B, Zimmermann U, Nascimento FS (2017 b) Enemy recognition is linked to soldier size in a polymorphic stingless bee. Biology Letters 13:20170511

Hammel B, Vollet-Neto A, Menezes C, Nascimento FS, Engels W, Grüter C (2016) Soldiers in a stingless bee: work rate and task repertoire suggest guards are an elite force. The American Naturalist 187:120–129

Hart AG, Ratnieks FLW (2002) Task-partitioned nectar transfer in stingless bees: work organisation in a phylogenetic context. Ecological Entomology 27:163–168

Hartfelder K, Makert GR, Judice CC, Pereira GA, Santana WC, Dallacqua R, Bitondi MMG (2006) Physiological and genetic mechanisms underlying caste development, reproduction and division of labor in stingless bees. Apidologie 37 (2):144–163

Hebling NJ, Kerr WE, Kerr F (1964) Divisão de trabalho entre operárias de *Trigona* (*Scaptotrigona*) *xanthotricha* Moure. Papéis Avulsos de Zoologia, São Paulo 16:115–127

Herb BR, Wolschin F, Hansen KD, Aryee MJ, Langmead B, Irizarry R, Amdam GV, Feinberg AP (2012) Reversible switching between epigenetic states in honeybee behavioral subcastes. Nature Neuroscience 15:1371–1373

Hofstede FE, Sommeijer MJ (2006) Effect of food availability on individual foraging specialisation in the stingless bee *Plebeia tobagoensis* (Hymenoptera, Meliponini). Apidologie 37 (3):387

Hölldobler B, Wilson EO (1990) The Ants. The Belknap Press of Harward University, Cambridge

Hölldobler B, Wilson EO (2009) The Superorganism: The Beauty, Elegance, and Strangeness of Insect Societies. W. W. Norton, New York

Hurd CR, Nordheim EV, Jeanne RL (2003) Elite workers and the colony-level pattern of labor division in the yellow jacket wasp, *Vespula germanica*. Behaviour 140:827–845

Imperatriz-Fonseca VL (1973) Miscellaneous observations on the behaviour of *Schwarziana quadripunctata* (Hym. Apidae, Meliponinae). Boletim de Zoologia e Biologia Marinha 30:633–640

Imperatriz-Fonseca VL (1977) Studies on *Paratrigona subnuda* (Moure) Hymenoptera, Apidae, Meliponinae-II. Behaviour of the virgin queen. Boletim de Zoologia 2(2):169–182

Inoue T, Salmah S, Sakagami SF (1996) Individual variations in worker polyethism of the Sumatran stingless bee, *Trigona* (*Tetragonula*) *minangkabau* (Apidae, Meliponinae). Japanese Journal of Entomology 64:641–668

Jeanne RL (2016) Division of labor is not a process or a misleading concept. Behavioral Ecology and Sociobiology 70:1109–1112

Johnson BR (2005) Limited flexibility in the temporal caste system of the honey bee. Behavioral Ecology and Sociobiology 58:219–226

Johnson BR, Frost E (2012) Individual-level patterns of division of labor in honeybees highlight flexibility in colony-level development mechanisms. Behavioral Ecology and Sociobiology 66:923–930

Jones SM, van Zweden JS, Grüter C, Menezes C, Alves D, Nunes-Silva P, Czaczkes TJ, Imperatriz-Fonseca VL, Ratnieks FLW (2012) The role of wax and resin in the nestmate recognition system of a stingless bee, *Tetragonisca angustula*. Behavioral Ecology and Sociobiology 66:1–12

Julian GE, Cahan S (1999) Undertaking specialization in the desert leaf-cutter ant *Acromyrmex versicolor*. Animal Behaviour 58(2):437–442

Justino CEL, Noll FB, Mateus S, Billen J (2018) Wax gland size according to worker age in *Friesella schrottkyi*. Apidologie 49:356–366

Kärcher M, Ratnieks FLW (2009) Standing and hovering guards of the stingless bee *Tetragonisca angustula* complement each other in entrance guarding and intruder recognition. Journal of Apicultural Research 48:209–214

Kerr A, Kerr W (1999) *Melipona* garbage bees release their cargo according to a Gaussian distribution. Revista Brasileira de Biologia 59 (1):119–123

Kerr WE, Jungnickel H, Morgan ED (2004) Workers of the stingless bee *Melipona scutellaris* are more similar to males than to queens in their cuticular compounds. Apidologie 35 (6):611–618

Kerr WE (1951) Bases para o estudo da genética de populações dos Hymenoptera em geral e dos Apinae sociais em particular. Anais da Escola Superior de Agricultura Luiz de Queiroz 8:219–354

Kerr WE (1997) Sex determination in honey bees (Apinae and Meliponinae) and its consequences. Brazilian Journal of Genetics 20 (4):601

Kerr WE, de Lello E (1962) Sting glands in stingless bees: a vestigial character (Hymenoptera: Apidae). Journal of the New York Entomological Society 70:190–214

Kerr WE, Santos Neto GR (1956) Contribuição para o conhecimento da bionomia dos Meliponini. 5. Divisão de trabalho entre as operarias de *Melipona quadrifasciata quadrifasciata*. Insectes Sociaux 3:423–430

Koedam D, Contrera FAL, Imperatriz-Fonseca VL (1999) Clustered male production by workers in the stingless bee *Melipona subnitida* Ducke (Apidae, Meliponinae). Insectes Sociaux 46 (4):387–391

Koedam D, Jungnickel H, Tentschert J, Jones G, Morgan E (2002) Production of wax by virgin queens of the stingless bee *Melipona bicolor* (Apidae, Meliponinae). Insectes Sociaux 49 (3):229–233

Korb J, Hartfelder K (2008) Life history and development-a framework for understanding developmental plasticity in lower termites. Biological Reviews 83 (3):295–313

LeBoeuf AC, Waridel P, Brent CS, Gonçalves AN, Menin L, Ortiz D, Riba-Grognuz O, Koto A, Soares ZG, Privman E, Miska EA, Benton R, Keller L (2016) Oral transfer of chemical cues, growth proteins and hormones in social insects. eLife 8:e51082

Leighton GM, Charbonneau D, Dornhaus A (2017) Task switching is associated with temporal delays in *Temnothorax rugatulus* ants. Behavioral Ecology 28:319–327

Leoncini I, Le Conte Y, Costagliola G, Plettner E, Toth AL, Wang M, Huang Z, Bécard JM, Crauser D, Slessor KN, Robinson GE (2004) Regulation of behavioral maturation by a primer pheromone produced by adult worker honey bees. Proceedings of the National Academy of Sciences of the United States of America 101:17559–17564

Lindauer M (1952) Ein Beitrag zur Frage der Arbeitsteilung im Bienenstaat. Zeitschrift für vergleichende Physiologie 34:299–345

Lopes BSC, Campbell AJ, Contrera FAL (2020) Queen loss changes behavior and increases longevity in a stingless bee. Behavioral Ecology and Sociobiology 74:35

Lucas ER, Field J (2011) Active and effective nest defence by males in a social apoid wasp. Behavioral Ecology and Sociobiology 65 (8):1499–1504

Mateus S, Ferreira-Caliman MJ, Menezes C, Grüter C (2019) Beyond temporal-polyethism: division of labor in the eusocial bee *Melipona marginata*. Insectes Sociaux 66:317–328

Medina LAM, Hart AG, Ratnieks FLW (2014) Waste management in the stingless bee *Melipona beecheii* Bennett (Hymenoptera: Apidae). Sociobiology 61:435–440

Mertl AL, Traniello JFA (2009) Behavioral evolution in the major worker subcaste of twig-nesting *Pheidole* (Hymenoptera: Formicidae): does morphological specialization influence task plasticity? Behavioral Ecology and Sociobiology 63:1411–1426

Michener CD (1974) The Social Behavior of the Bees. Harvard University Press, Cambridge

Month-Juris E, Ravaiano SV, Lopes DM, Salomão TMF, Martins GF (2020) Morphological assessment of the sensilla of the antennal flagellum in different castes of the stingless bee *Tetragonisca fiebrigi*. Journal of Zoology 310:110–125

Moroń D, Lenda M, Skórka P, Woyciechowski M (2012) Short-lived ants take greater risks during food collection. The American Naturalist 180 (6):744–750

Njoya MTM, Wittmann D (2013) Tasks partitioning among workers of *Meliponula* (*Meliplebeia*) *becarrii* (Meliponini) in Cameroon. International Journal of Environmental Sciences 3 (5):1796–1805

Nunes TM, von Zuben LG, Costa L, Venturieri GC (2014) Defensive repertoire of the stingless bee *Melipona flavolineata* Friese (Hymenoptera: Apidae). Sociobiology 61 (4):541–546

Oster GF, Wilson EO (1978) Caste and Ecology in the Social Insects. Princeton

University Press, Princeton

Page RE, Peng CY-S (2001) Aging and development in social insects with emphasis on the honey bee, *Apis mellifera* L. Experimental Gerontology 36 (4):695–711

Pankiw T, Page RE (2000) Response thresholds to sucrose predict foraging division of labor in honeybees. Behavioral Ecology and Sociobiology 47 (4):265–267

Pinter-Wollman N, Hubler J, Holley J-A, Franks NR, Dornhaus A (2012) How is activity distributed among and within tasks in *Temnothorax* ants? Behavioral Ecology and Sociobiology 66:1407–1420

Ratnieks FLW, Anderson C (1999) Task partintioning in insect societies. Insectes Sociaux 46:95–108

Ravaiano SV, Barbosa WF, Campos LA, Martins GF (2018) Variations in circulating hemocytes are affected by age and caste in the stingless bee *Melipona quadrifasciata*. The Science of Nature 105: 48

Reeve HK (1993) Haplodiploidy, eusociality and absence of male parental and alloparental care in Hymenoptera: a unifying genetic hypothesis distinct from kin selection theory. Philosophical Transactions of the Royal Society of London Series B: Biological Sciences 342 (1302):335–352

Robinson GE (1992) Regulation of division of labor in insect societies. Annual Review of Entomology 37:637–665

Robinson GE (2002) Genomics and integrative analyses of division of labor in the honeybee colonies. American Naturalist 160:S160–S172

Robinson GE, Page RE (1995) Genotypic constraints on plasticity for corpse removal in honey bee colonies. Animal Behaviour 49:867–876

Robson SK, Traniello JFA (1999) Key individuals and the organisation of labor in ants. In: Detrain C, Deneubourg JL, Pasteels JM (eds) Information Processing in Social Insects. Springer, Basel, pp 239–260

Rösch G (1927) Ueber die Bautätigkeit im Bienenvolk und das Alter der Baubienen. Zeitschrift für vergleichende Physiologie 6 (2):264–298

Rösch GA (1925) Untersuchungen über die Arbeitsteilung im Bienenstaat. Zeitschrift für vergleichende Physiologie 2:571–631

Roubik DW (1982) Seasonality in colony food storage, brood production and adult survivorship: studies of *Melipona* in tropical forest (Hymenoptera: Apidae). Journal of the Kansas Entomological Society 55:789–800

Roubik DW (1989) Ecology and Natural History of Tropical Bees. Cambridge University Press, New York

Roubik DW, Patiño JEM (2009) *Trigona corvina*: an ecological study based on unusual nest structure and pollen analysis. Psyche 2009:268756

Sakagami SF (1982) Stingless bees. In: Hermann HR (ed) Social Insects III. Academic Press, New York, pp 361–423

Sakagami SF, Oniki Y (1963) Behavior studies of the stingless bees, with special reference to the oviposition process. I: *Melipona compressipes manaosensis* Schwarz. Journal of the Faculty of Science Hokkaido University Series VI, Zoology 15 (2):300–318

Sakagami SF, Zucchi R (1967) Behavior Studies of the Stingless Bees, with Special Reference to the Oviposition Process: VI. *Trigona (Tetragona) clavipes*. (With 6 Text-figures and 2 Tables). Journal of the Faculty of Science Hokkaido University Series VI Zoology 16 (2):292–313

Sakagami SF, Zucchi R (1974) Oviposition behavior of two dwarf stingless bees, *Hypotrigona (Leurotrigona) muelleri* and *H. (Trigonisca) duckei*, with notes on the temporal articulation of oviposition process in stingless bees Journal of the Faculty of Science Hokkaido University Series VI, Zoology 19 (2):361–421

Sakagami SF, Zucchi R, Araujo VdP (1977) Oviposition behavior of an aberrant african stingless bee *Meliponula bocandei*, with notes on the mechanism and evolution of oviposition behavior in stingless bees. Journal of the Faculty of Science Hokkaido University Series VI, Zoology 20 (4):647–690

Salmah S, Inoue T, Mardius P, Sakagami S (1987) Incubation period and post-emergence pigmentation in the Sumatran stingless bee, *Trigona (Trigonella) moorei*. Kontyu 55 (3):383–390

Salmah S, Inoue T, Sakagami S (1996) Incubation period and post-emergence pigmentation in the Sumatran stingless bee *Trigona (Heterotrigona) itama* (Apidae, Meliponinae). Japanese Journal of Entomology 64 (2):401–411

Santos DCJ (2013) Divisão de trabalho e sua relação com a dinâmica dos hidrobonetos cuticulares em *Melipona scutellaris* (Hymenoptera, Apidae, Meliponini). MSc Thesis, University of Sao Pãulo, Ribeirão Preto

Santos CF (2018) Cooperation and antagonism over time: a conflict faced by males of Tetragonisca angustula in nests. Insectes Sociaux 65:465–471

Scheiner R, Page RE, Erber J (2004) Sucrose responsiveness and behavioral plasticity in honey bees (*Apis mellifera*). Apidologie 35 (2):133–142

Schorkopf DLP (2016) Male meliponine bees (*Scaptotrigona aff. depilis*) produce alarm pheromones to which workers respond with fight and males with flight. Journal of Comparative Physiology A 202:667–678

Schorkopf DLP, Hrncir M, Mateus S, Zucchi R, Schmidt VM, Barth FG (2009) Mandibular gland secretions of meliponine worker bees: further evidence for their role in interspecific and intraspecific defence and aggression and against their role in food source signalling. Journal of Experimental Biology 212 (8):1153–1162

Schulz DJ, Sullivan JP, Robinson GE (2002) Juvenile hormone and octopamine in the regulation of division of labor in honey bee colonies. Hormones and Behavior 42 (2):222–231

Schwarz HF (1948) Stingless Bees (Meliponidae) of the Western Hemisphere. Bulletin of the American Museum of Natural History 90:1–546

Seeley TD (1982) Adaptive significance of the age polyethism schedule in honeybee colonies. Behavioral Ecology and Sociobiology 11 (4):287–293

Segers FHID, Menezes C, Vollet-Neto A, Lambert D, Grüter C (2015) Soldier production in a stingless bee depends on rearing location and nurse behaviour. Behavioral Ecology and Sociobiology 69:613–623

Segers FHID, von Zuben LG, Grüter C (2016) Local differences in parasitism and competition shape defensive investment in a polymorphic eusocial bee. Ecology 97:417–426

Seid MA, Traniello JFA (2006) Age-related repertoire expansion and division of labor in *Pheidole dentata* (Hymenoptera: Formicidae): a new perspective on temporal polyethism and behavioral plasticity in ants. Behavioral Ecology and Sociobiology 60:631–644

Sempo G, Detrain C (2004) Between-species differences of behavioural repertoire of castes in the ant genus *Pheidole*: a methodological artefact? Insectes Sociaux 51:48–54

Shorter JR, Tibbetts EA (2009) The effect of juvenile hormone on temporal polyethism in the paper wasp *Polistes dominulus*. Insectes Sociaux 56（1）:7–13

Silva FR（2008）Divisão de trabalho e processo de aprovisionamento e postura em *Trigona pallens* Fabricius, 1798（Hymenoptera, Apidae, Meliponini）. MSc Thesis, University of Sao Paulo, Brazil

Simões D, Bego LR（1979）Estudo da regulação social em *Nannotrigona*（*Scaptotrigona*）*postica* Latreille, em duas colónias（normal e com rainhas virgens）, com especial referência ao polietismo etário（Hym., Apidae, Meliponinae）. Boletim de Zoologia, Universidade de São Paulo 4:89–98

Simões D, Bego LR（1991）Division of labor, average life span and life table in *Nannotrigona*（*Scaptotrigona*）*postica* Latreille（Hymenoptera, Apidae, Meliponinae）. Naturalia 16:81–97

Simone-Finstrom M, Gardner J, Spivak M（2010）Tactile learning in resin foraging honeybees. Behavioral Ecology and Sociobiology 64:1609–1617

Slaa EJ（2006）Population dynamics of a stingless bee community in the seasonal dry lowlands of Costa Rica. Insectes Sociaux 53:70–79

Smith A（1776）The Wealth of Nations, book 1. Methuen, London

Sommeijer MJ（1984）Distribution of labour among workers of *Melipona favosa* F.: age-polyethism and worker oviposition. Insectes Sociaux 31:171–184

Sommeijer MJ, Bruijn LLMD, Van de Guchte C（1985）The Social Food-Flow Within the Colony of a Stingless Bee, *Melipona favosa*（F.）. Behaviour 92:39–58

Sommeijer MJ, De Rooy GA, Punt W, De Bruijn LLM（1983）A comparative study of foraging behavior and pollen resources of various stingless bees（*Hym., Meliponinae*）and honeybees（*Hym., Apinae*）in Trinidad, West-Indies. Apidologie 14:205–224

Spaethe J, Brockmann A, Halbig C, Tautz J（2007）Size determines antennal sensitivity and behavioral threshold to odors in bumblebee workers. Naturwissenschaften 94:733–739

Starr CK（1985）Enabling mechanisms in the origin of sociality in the Hymenoptera—the sting's the thing. Annals of the Entomological Society of America 78(6):836–840

Sullivan JP, Jassim O, Fahrbach SE, Robinson GE（2000）Juvenile hormone paces behavioral development in the adult worker honey bee. Hormones and Behavior 37（1）:1–14

Tenczar P, Lutz C, Rao VD, Goldenfeld N, Robinson GE (2014) Automated monitoring reveals extreme interindividual variation in plasticity in honeybee foraging activity levels. Animal Behaviour 95:41–48

Terada Y, Garofalo CA, Sakagami SF (1975) Age-survival curves for workers of two eusocial bees (*Apis mellifera* and *Plebeia droryana*) in a subtropical climate, with notes on worker polyethism in *P. droryana*. Journal of Apicultural Research 14 (3-4):161–170

Tofts C, Franks NR (1992) Doing the right thing: ants, honeybees and naked mole-rats. Trends in Ecology & Evolution 7:346–349

Tomé HVV, Rosi-Denadai CA, Pimenta JFN, Guedes RNC, Martins GF (2014) Age-mediated and environmentally mediated brain and behavior plasticity in the stingless bee *Melipona quadrifasciata anthidioides*. Apidologie 45:557–567

Ulrich Y, Saragosti J, Tokita CK, Tarnita CE, Kronauer DJC (2018) Fitness benefits and emergent division of labour at the onset of group living. Nature 560:635–638

van Benthem FDJ, Imperatriz-Fonseca VL, Velthuis HHW (1995) Biology of the stingless bee *Plebeia remota* (Holmberg): observations and evolutionary implications. Insectes Sociaux 42:71–87

van Veen J, Sommeijer MJ, Meeuwsen F (1997) Behaviour of drones in *Melipona* (Apidae, Meliponinae). Insectes Sociaux 44 (4):435–447

van Zweden JS, Grüter C, Jones SM, Ratnieks FLW (2011) Hovering guards of the stingless bee *Tetragonisca angustula* increase colony defensive perimeter as shown by intra- and inter-specific comparisons. Behavioral Ecology and Sociobiology 65:1277–1282

von Ihering H (1903) Biologie der stachellosen Honigbienen Brasiliens. Zoologische Jahrbücher Abteilung fuer Systematik Ökologie und Geographie der Tiere 19:179–287

Veiga JC, Leão KL, Coelho BW, Queiroz ACM de, Menezes C, Contrera FAL (2018) The life histories of the "Uruçu Amarela" males (*Melipona flavolineata*, Apidae, Meliponini). Sociobiology 65:780–783

Velez-Ruiz RI, Gonzales VH, Engel MS (2013) Observations on the urban ecology of the Neotropical stingless bee *Tetragonisca angustula* (Hymenoptera: Apidae: Meliponini). Journal of Melittology 1:1–8

Velthuis HH, Koedam D, Imperatriz-Fonseca VL (2005) The males of *Melipona* and other stingless bees, and their mothers. Apidologie 36 (2):169–185

Visscher K, Dukas R (1997) Survivorship of foraging honey bees. Insectes Sociaux 44:1–5

von Frisch K (1967) The dance language and orientation of bees. Harvard University Press, Cambridge

Waldschmidt A, Oliveira Campos L, De Marco P (1998) Behavioral plasticity of *Melipona quadrifasciata* (Hymenoptera: Meliponinae). Revista Brasileira de Biologia 58:25–31

West-Eberhard MJ (1975) The evolution of social behavior by kin selection. Quarterly Review of Biology 50:1–33

Wheeler DE (1991) The developmental basis of worker caste polymorphism in ants. The American Naturalist 138:1218–1238

Whitfield CW, Cziko A-M, Robinson GE (2003) Gene expression profiles in the brain predict behavior in individual honey bees. Science 302:296–299

Wille A (1966) Notes on two species of ground nesting stingless bees (*Trigona mirandula* and *T. buchwaldi*) from the pacific rain forest of Costa Rica. Revista de Biologia Tropical 14:251–277

Wilson EO (1953) The origin and evolution of polymorphism in ants. The Quarterly Review of Biology 28:136–156

Wilson EO (1971) The insect societies. Harvard University Press, Cambridge

Wilson EO (1980) Caste and division of labor in leaf-cutting ants (Hymenoptera: Formicidae: *Atta*). I. The overall pattern in *A. sexdens*. Behavioral Ecology and Sociobiology 7:143–156

Wilson EO (1984) The relation between caste ratios and division of labor in the ant genus *Pheidole* (Hymenoptera: Formicidae). Behavioral Ecology and Sociobiology 16:89–98

Winston ML (1987) The biology of the honey bee. Harvard University Press, Cambridge

Wittmann D (1985) Aerial defense of the nest by workers of the stingless bee *Trigona* (*Tetragonisca*) *angustula*. Behavioral Ecology and Sociobiology 16:111–114

Wittmann D, Ratke R, Zeil J, Lübke G, Francke W (1990) Robber bees (*Lestrimelitta limao*) and their host chemical and visual cues in nest defence by *Trigona* (*Tetragonisca*) *angustula* (Apidae: Meliponinae). Journal of Chemical Ecology 16:631–641

Wittwer B, Elgar MA (2018) Cryptic castes, social context and colony defence in a social bee, *Tetragonula carbonaria*. Ethology 124:617–622

Wolf TJ, Schmid-Hempel P (1989) Extra loads and foraging life span in honeybee workers. The Journal of Animal Ecology 58:943–954

Yerushalmi S, Bodenhaimer S, Bloch G (2006) Developmentally determined attenuation in circadian rhythms links chronobiology to social organization in bees. Journal of Experimental Biology 209:1044–1051

Źrałka J, Helmke C, Sotelo L, Koszkul W (2018) The discovery of a beehive and identification of apiaries among the ancient Maya. Latin American Antiquity 29:514–531

7　无刺蜂的天敌、威胁及蜂群防御

无刺蜂蜂巢中含有很多宝贵的资源，如蜂蜜、花粉、巢质、巢蜡、树脂、幼虫食物，还有蜂巢和无刺蜂个体（第 3 章）。因此其他动物想要获得这些资源也就不足为奇了。这反过来又会导致无刺蜂蜂群因其他动物的捕食而产生相关的大量死亡。例如乌干达每年就有 5 种无刺蜂蜂种近 12% 的蜂群死于其他动物的捕食（Kajobe and Roubik,2006），而哥斯达黎加的一个蜂群栖息地所有蜂群死亡事件中有 40% 都是由其他动物捕食造成的（Slaa,2006）。印度尼西亚的 *Tetragonula Minangkabau* 无刺蜂所有蜂群的 47% 死于动物捕食或寄生（Inoue et al., 1993）。而 Eltz 等（2002）则发现马来西亚的无刺蜂栖息地中动物捕食作用几乎不会引起无刺蜂蜂群死亡。

无刺蜂为了应对这些外部威胁进化出了许多令人值得注意的防御特征，但它们同时也丧失了最明显的防御性状：螯针（第 1 章）。因为螯针是蜂类的一种强大武器，所以无刺蜂进化过程中功能性螯针的退化看起来实在令人费解。然而由于无刺蜂祖先的体型较小，螯针的有效性可能那时就已经降低了（Melo,2020），目前还不清楚这种螯针是否是对抗当前无刺蜂的天敌的有效武器。例如，无刺蜂较之比蜜蜂面对的更多的是无脊椎动物天敌，而螯针攻击对有昆虫外骨骼的天敌来说效果较差（Schwarz,1948）[1]。蜜蜂在与同样大小的 *Melipona scutellaris* 无刺蜂工蜂互相攻击时经常能取得胜利（Schwarz,1948）。蜜蜂和一些胡蜂螫到天敌时，往往会导致天敌受到相当大的

---

[1] 也有案例表明螯针是应对无脊椎动物天敌的有效武器。例如，当蜜蜂被欧洲狼蜂（*Philanthus triangulum*）攻击时偶尔可以用螯针成功杀死这种攻击者（Tobias Engl）。

伤害，而这些螯击的蜂自己的结局就是死亡。因此螯击的代价高昂，而螯击的好处就取决于天敌的种类。此外，发育出螯针和生产毒液也会耗费大量的能量（Michener,1974）。因此，无刺蜂功能性螯针器官的退化表明拥有螯针的成本超过了拥有螯针的收益（Kerr and de Lello,1962）。

无刺蜂丧失的功能性螯针可以通过其他防御特征来弥补。不同蜂种的无刺蜂的防御策略也大不相同，且会随天敌种类而发生变化。工蜂体型较小的无刺蜂蜂种通常不会攻击大型脊椎动物，但是对节肢动物则会有进攻性反应（Johnson and Wiemer,1982; Roubik,1989）。无刺蜂蜂种则通常会以组织大量蜂进攻的方式攻击大型脊椎动物，它们会试图进入脊椎动物的鼻孔、耳朵、毛发，或者攻击脊椎动物的眼睛（Schwarz,1948; Kerr and de Lello,1962; Wille,1983; Roubik,2006; Shackleton et al., 2015）。Roubik（1983）测试了40个无刺蜂蜂种，发现其中大约有一半的蜂种对人类表现出攻击性。建造暴露型蜂巢的无刺蜂蜂种的攻击性尤其强烈（Michener,1974; Roubik,1983, 2006），而且最具攻击性的无刺蜂（特别是一些 *Trigona* 蜂种）经常会持续长时间地咬击它们的攻击对象，即使后者已经移动到离蜂巢几百米远以外的地方（Schwarz,1948; Shackleton et al., 2015）。建造暴露型蜂巢的无刺蜂蜂种攻击性增强的原因可能是它们更容易成为脊椎动物的捕食目标（Roubik,1983）。攻击性越强的蜂种咬击也越痛，其上颚内侧齿也更锋利（图 7.1）（Schwarz,1948; Shackelton et al., 2015）。Shackleton 等（2015）提出无刺蜂咬住不放的行为的致命性可能不亚于被蜜蜂蜇的致命性。

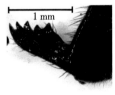  

Trigona hyalinata 疼痛感 = 5　　*Tetragona clavipes* 疼痛感 = 3　　*Melipona quadrifasciata* 疼痛感 = 不会咬击　　*Apis mellifera* 西方蜜蜂

图 7.1　无刺蜂和蜜蜂的上颚腺内侧

注：攻击性越强的蜂种咬击痛感也更强，齿状上颚也更强壮（Shackleton et al.,2015）。

不仅不同蜂种的攻击性存在差异，而且即便是同一蜂种攻击性也有差异（Roubik,1989）。例如，一些人认为 *Melipona fuliginos* 无刺蜂性情温和，而另一些人则认为这种无刺蜂十分好斗（Roubik,1989）。笔者曾经遇到过

Tetragonisca angustula 无刺蜂蜂群，尽管有传言它们对人类非常温驯，但那些蜂群却表现出非常强的攻击性。

7.1 无刺蜂的天敌

7.1.1 小型天敌

蚂蚁

蚂蚁是无刺蜂最重要的天敌之一（Schwarz,1948; Roubik,1989）。研究者观察到新热带区的 *Azteca*、*Camponotus*、*Eciton* 和 *Pheidole* 蚂蚁会入侵无刺蜂蜂巢并杀死里面的无刺蜂（Roubik,1989）因此，巢穴入口经常出现与蚂蚁打斗的痕迹（图 7.2 A），而且无刺蜂的一些防御特征似乎是专门针对蚂蚁入侵者的（第 7.3 节）。人工饲养的 *Melipona* 属无刺蜂蜂群特别容易受到蚂蚁攻击，因为养蜂的过程常常会造成蜂巢出现小缝隙和空洞，蚂蚁则可以通过这些孔隙进入蜂群中（Schwarz,1948）。*Atta* 切叶蚁群偶尔还用挖出的泥土覆盖在树基处的无刺蜂蜂巢上，从而杀死无刺蜂（个人观察）。诸如亚洲 *Oecophylla Smaragdina* 等蚂蚁则经常在花朵周围猎杀落单的无刺蜂（图 7.2 B），而新热带 *Ectatomma tuberculatum* 蚂蚁则会在无刺蜂蜂巢入口处伏击无刺蜂守卫蜂（如 *Tetragonisca angustula* 无刺蜂中的飞悬守卫蜂和站岗守卫蜂，第 6 章）(Ostwald et al., 2018)。

图 7.2 对于无刺蜂来说，蚂蚁和蚤蝇是主要危险

注：A 图中黑色的 *Camponotus* 蚂蚁尸体覆盖了一个巴西 *Tetragonisca angustula* 无刺蜂的蜂巢入口。图中可以看到一只蚂蚁正在向蜂巢入口爬去（蓝色箭头）。同时图中也可以观察到一只蚤蝇（黄色箭头）正在试图进入蜂巢。B 图所示为 *Oecophylla smaragdina* 编织蚁经常在花丛中捕食无刺蜂，图中 3 只编织蚁正在拉扯一只在斯里兰卡的花丛中捕获的 *Tetragonula* 无刺蜂采集蜂（摄影：C. Grüter）。

蚤蝇

蚤蝇幼虫（如 *Pseudohypocera*）是无刺蜂蜂群的主要害虫，特别是弱蜂群或被养蜂人频繁操作的蜂群（Nogueira-Neto,1997; Sommeijer,1999; Hernández and Gutiérrez,2001）。成年苍蝇会试图通过蜂巢入口处进入蜂巢（图 7.2 A），它们好像是被蜂群气味所吸引而来的（Roubik,1989）。它们会在蜂巢内生产大量的幼虫，这些幼虫就以巢内的蜂蜜、花粉和未成熟幼蜂为食（图 7.3 A）（Nogueira-Neto,1997）。一些蚤蝇还会以成年采集蜂和雄蜂集群中的雄蜂为目标（Simões et al., 1980; Brown,1997; Sommeijer et al., 2004; Santos et al., 2014）。Brown（1997）收集 *Cephalotrigona capitata* 无刺蜂雄蜂时发现近 50% 的 *Cephalotrigona capitata* 无刺蜂雄蜂被 *Apocephalus apivorus* 蚤蝇寄生了。这表明蚤蝇寄生可能是这个蜂种雄蜂集群中的重大威胁（Brown,1997）。目前尚不清楚其他蜂种中是否也普遍存在相似水平的寄生情况。新热带区 *Melaloncha* 属蚤蝇（约 170 种）是以无刺蜂蜂巢入口附近的工蜂为目标。雌蝇会将卵通过工蜂外骨骼板之间的膜区注射到工蜂体内。不断成长的蚤蝇幼虫会不断蚕食宿主，并且在化蛹前杀死工蜂宿主（Simões et al.,1980; Brown,2016）。Simóes 等（1980）的一项研究中检测的 *Scaptorigona* 无刺蜂工蜂中有高达 37% 的工蜂都被蚤蝇寄生了。

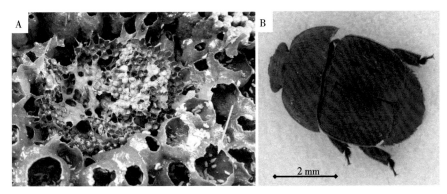

图 7.3 一个被蚤蝇幼虫侵害的 *Melipona scutellaris* 无刺蜂蜂群（图 A，黄色箭头）（摄影：Diego Vedovato）和小型蜂巢甲虫 *Aethina tumida*（图 B，Al Toufailia et al.,2017）

其他节肢动物

尽管无刺蜂天敌的相关报告中白蚁出现的频率极低，但已有研究观察到白蚁会袭击并杀死小型 *Trigonisca atomaria* 无刺蜂和弱势的 *Melipona fasciata* 蜂群（Roubik,1989）。胡蜂偶尔也会进入无刺蜂蜂巢来窃取资源

（如 *Polybia*），或者它们会猎捕在花丛中采集的无刺蜂或雄蜂集群中的雄蜂（Camargo and Posey, 1990; Rasmussen, 2008; Koedam et al., 2009）。*Scaptorigona Postica* 无刺蜂的每个蜂巢前每天都会有多达 50 只雄蜂被单只 *Trachypus boharti* 胡蜂所捕食（Koedam et al., 2009），这表明这些胡蜂是无刺蜂雄蜂集群中的一个重大威胁。澳洲和亚洲的 *Bembix* 沙胡蜂会猎捕 *Tetragonula* 无刺蜂工蜂来喂养它们的幼虫（Schwarz, 1948; Rasmussen, 2008）。*Syntretus trionaphagus* 茧蜂则是澳洲 *Tetragonula Cararia* 无刺蜂的一种寄生蜂（Gloag et al., 2009）。胡蜂还会在无刺蜂蜂巢入口处的工蜂腹部下或花朵上产卵，随后胡蜂幼虫就会在宿主腹部发育。Camargo 和 Posey（1990）报道发现新热带区的熊蜂（*Bombus* sp.）会从无刺蜂那里窃取花粉。

对于群势较弱的无刺蜂蜂群来说，甲虫及其幼虫是一种重大威胁（第 3.5 节）。蜂箱小甲虫 *Aesina umida*（small hive beetle，SHB）（图 7.3B）是一种无刺蜂天敌物种，越来越多研究者关注到这种生物，它原产于非洲撒哈拉以南，而一些地区由于 SHB 的引入，对当地蜜蜂蜂群构成了新的威胁（Neumann and Elzen, 2004）。SHB 是以蜂群储存的食物、蜂幼虫和成年蜂为食，它可以在短时间内摧毁蜂群。这些甲虫也自然而然地出现在一些非洲无刺蜂蜂群中，而且现在在澳洲的无刺蜂（*Austroplebeia australis* 和 *Tetragonula carbonaria*）蜂群以及古巴的 *Melipona beecheii* 无刺蜂蜂群中也都发现了这种引入害虫（第 3 章）（Greco et al., 2010; Halcroft et al., 2011; Halcroft et al., 2013; Peña et al., 2014; Bobadoye et al., 2018）。最近研究者在巴西的蜜蜂蜂巢中也首次报道发现了 SHB 的存在（Al Toufailia et al., 2017）。而这附近的无刺蜂蜂群似乎还没有受到其影响（Al Toufailia et al., 2017），但人们应密切监测 SHB 的扩散传播和新热带区无刺蜂的受害情况。

蜘蛛，特别是跳蛛（跳蛛科）通常会在无刺蜂蜂巢入口附近猎捕工蜂（图 7.4）（Sakagami et al., 1983; Penney and Gabriel, 2009; Shackleton et al., 2019）。离蜂巢入口太近的蜘蛛经常会受到无刺蜂守卫蜂的攻击，这会导致蜘蛛暂时撤退（Schwarz, 1948, pers. obs.）。*Partamona helleri* 无刺蜂是一种建立"蛤蟆嘴"入口蜂巢的蜂种（第 3 章；图 3.1），如果跳蛛守着该蜂种蜂巢的"蛤蟆嘴"入口时，这种无刺蜂的采集蜂就会以飞快的速度进入巢穴，这样可能降低被等在巢口的跳蛛捕获的风险（Shackleton et al., 2019）。人们已知的无刺蜂捕食者还有蟹蛛（蟹蛛科）和园蛛（园蛛科）。还有一些捕食者会在蜂巢入口处或在花朵上猎捕无刺蜂，其中包括螳螂（Mantidae 科）、食虫虻（Asilidae

科）、猎蝽（Reduviidae 科）和捕食性螨类（*Amblyseius* 属）（Schwarz,1948; Roubik,1989; Rasmussen,2008）。猎蝽还会猎捕无刺蜂雄蜂集群中的雄蜂（Cortopassi-Laurino,2007）。*Plega hagenella* 螳蛉的幼虫可以在无刺蜂幼虫巢室内发育，它们会在化蛹之前杀死巢室内正在发育中的无刺蜂幼虫。目前，*Melipona* 属无刺蜂蜂群就已经受到了大量 *Plega hagenella* 幼虫的严重损害（Maia-Silva et al., 2013）。研究者在印度观察到切叶蜂（*Megachile*）会从 *Tetragonula Iridipennis* 无刺蜂的蜂巢入口管中窃取蜡和树脂（Schwarz,1948）。美洲大陆和东半球的蜡蛾幼虫（小蜡蛾 *Achroia Grisella*）则会掠食无刺蜂的蜡和花粉（Hockings,1883; Roubik,1989; Nogueira-Neto,1997; Cepeda-Aponte et al., 2002）。和许多其他天敌一样，蜡蛾幼虫似乎在群势较弱的蜂群中生长的更好。

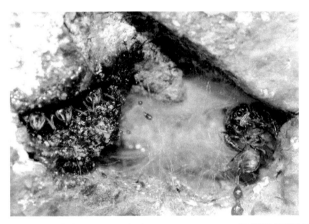

图 7.4　巴西的一只跳蛛（*Salticidae*）在 *Plebeia droryana* 无刺蜂蜂群的蜂巢入口处猎杀守卫蜂和采集蜂

注：*P. droryana* 无刺蜂则使用了小液滴的树脂来防御体型更小的天敌（摄影：C. Grüter）。

7.1.2　大型天敌

人类会掠取无刺蜂蜂蜜、破坏其栖息地、使用杀虫剂、引入非本土竞争者或者拆除城市地区的蜂巢，因而人类是无刺蜂生存的最大威胁（第 1 章）（Fowler,1979; Inoue et al., 1993; Antonini and Martins,2003; Cairns et al., 2005; Kajobe and Roubik,2006; Slaa,2006; Freitas et al., 2009; Venturieri,2009; Velez-Ruiz et al., 2013; Pioker-Hara et al., 2014; Barbosa et al., 2015）。人类根据狩猎传统掠取无刺蜂蜂蜜可能会破坏一个无刺蜂种群中的许多蜂巢。例如，Inoue 等（1993）发现在印度尼西亚一个受人类干扰的无刺蜂栖息地中，蜂蜜收割

是 *Tetragonula Minangkabau* 无刺蜂种群最重要的致死原因，所有蜂群死亡事件中有近 40% 都可归因于人类活动的干扰。而人类对无刺蜂群落最大的威胁则来自对环境的大规模改变，其中主要是人类将自然栖息地转变为城市或集约化耕种的土地。在一些严重受影响地区，无刺蜂栖息地丧失将导致蜂巢密度减少和蜂种消失（第 1 章，第 3 章）（Cairns et al., 2005; Slaa, 2006; Brosi, 2009; Freitas et al., 2009; Pioker-Hara et al., 2014）。

其他能够威胁无刺蜂蜂群的灵长类动物还有黑猩猩（*Pan troglodytes*）、大猩猩（*Gorilla Gorilla berengei*）、狒狒（*Papio anubis*）和其他一些猿猴物种（Schwarz, 1948; Kajobe and Roubik, 2006; Estienne et al., 2017 a）。乌干达的一个森林保护区里，被掠食的无刺蜂蜂巢中有 82% 是由人类和黑猩猩取食造成的（Kajobe and Roubik, 2006）。地面筑巢的无刺蜂蜂种受到的影响更加厉害。Estienne 等（2017 a）研究了加蓬地区地面筑巢的 *Plebeiella lendliana* 无刺蜂，发现黑猩猩在其他食物来源匮乏的旱季时偷盗其蜂蜜的行为更为频繁。它们为了避免挖掘坚硬的土壤则喜欢从生活在较软土地中的无刺蜂蜂巢中盗取蜂蜜。黑猩猩为了成功率更高，甚至会使用复杂的工具从地面和树上筑巢的无刺蜂蜂巢中取出蜂蜜（Boesch et al., 2009; Estienne et al., 2017 b）。

无刺蜂蜂产品的食用者中体型最大的可能是大象（*Loxodonta cyclotis*），它们偶尔会从土里挖出无刺蜂蜂群（Estienne et al., 2017 a），蜜獾（*Mellivora capensis*）也会在夜间打开无刺蜂蜂巢（Schwarz, 1948; Estienne et al., 2017 a）。东南亚马来熊（*Helarctos malayanus*）主要食用大蜜蜂的蜂蜜，但它们偶尔也会食用 *Tetragonilla collina* 等无刺蜂的蜂蜜（Schwarz, 1948; Eltz et al., 2002）。众所周知，新热带区的眼镜熊（*Tremarctos ornatus*）会破坏无刺蜂蜂巢（Roubik, 1989, 2006）。食蚁兽（*Tamandua* spp.）、猪鼻臭鼬（*Conepatus* spp.）和獾徐会打开无刺蜂蜂巢去吃里面的食物，这些无刺蜂蜂巢则多半是建在地下的。泰拉（*Eira barbara*）是一种和狗差不多大小的新热带鼬，它们喜欢从地里和在树洞里的蜂巢中偷取蜂蜜（Schwarz, 1948; Roubik, 1989, 2006）。它在瓜拉尼语中的意思就是"蜂蜜之王"（Schwarz, 1948）。另一种广为人知的喜欢无刺蜂蜂蜜的动物是埃拉（eyra）猫或小豹猫（*Herpailurus yagouaroundi*）。埃拉（eyra）这个名字来源于"ira"或"eira"，在图皮瓜拉尼语中是"蜂"的意思（Schwarz, 1948）。无刺蜂还有一些值得注意的哺乳动物天敌是金卡茹（kinkajou）或称"蜜熊"（*Potos flavus*，原尾狼科）、格里森（grison）或称南美狼獾（*Galictis* 属，鼬科）和郊狼（*Canis latrans*）

（Schwarz，1948）。有很多关于猴子和郊狼将尾巴浸入蜜罐中提取蜂蜜的轶事的记载（Schwarz，1948）。研究者发现壁虎和其他蜥蜴会在无刺蜂蜂巢入口处猎捕成年无刺蜂（Schwarz，1948；Grüter et al.，2011），而蟾蜍（蟾蜍科）如果停留在靠近蜂巢入口的地方就可以吃掉许多成年工蜂（Roubik，1989；Halcroft et al.，2013）。

啄木鸟（如 *Dryocopus lineatus*，啄木鸟科）偶尔也会吃无刺蜂。无刺蜂其他鸟类天敌还包括食蜂鸟（蜂虎科）、砍林鸟（裂雀科）、卷尾鸟（卷尾科）、鹟䴗鸟（鹟䴗科）、苍鹭（鹭科）、毕生鸟（霸鹟科）、鹟鸟（如 *Myiarchus yucatanasis*、霸鹟科）和雨燕（雨燕科）（Schwarz，1948；Rasmussen，2008）。澳洲的蜜雀（食蜜鸟科）偶尔会把蜜蜂加到它们的食物中。吸蜜鸟（吸蜜鸟科）是无刺蜂的一种特殊威胁。非洲吸蜜鸟 *Indicator indicator* 已经发展出一种与人类采蜜者的互惠互利的关系，这样造成了蜜蜂和无刺蜂的蜂巢（例如 *Hyptrigona araujoi*）的损坏（Spottiswoode et al.，2016）。吸蜜鸟会积极引导采蜜人到蜂巢，在采蜜人打开蜂巢并取出蜂蜜后，鸟儿会吃掉剩下的蜂蜡和幼虫（Schwarz，1948；Spottiswoode et al.，2016）。

7.1.3 疾病

无刺蜂较之西方蜜蜂受疾病的影响更小（Schwarz，1948；Nogueira-Neto，1997；Roubik，2006；Medina et al.，2009；Al Toufailia et al.，2016）。这表明无刺蜂拥有更有效的疾病管理策略，而且由于它们的生物学特性减少了疾病暴露。例如无刺蜂产卵后，工蜂就会立即对育虫巢室进行封盖（第5章），而蜜蜂则会不断地向巢室内进行食物供应，无刺蜂这种行为意味着工蜂和幼虫之间没有直接接触，这样可以减少无刺蜂幼虫与病原体的接触。另外，无刺蜂不会重复使用育虫巢室（第3章），这样也可以降低疾病风险。无刺蜂的食物储存在特定的食物罐中，食物罐则通常远离育虫区（第3章），这样降低了通过积攒资源而造成育虫巢房污染的可能性（Medina et al.，2009）。

蜂群减少疾病影响的重要行为策略就是卫生行为。蜜蜂会打开装有死亡的与受感染的幼虫和蛹的封盖巢室并将它们移走，这样可以降低疾病在蜂群内传播的风险（Rothenbuhler，1964；Spivak et al.，2003；Bigio et al.，2013）。人们对无刺蜂的卫生行为则知之甚少，但几个新热带无刺蜂蜂种的实验研究表明无刺蜂展现了高效的卫生行为（Medina et al.，2009；Nunes-Silva et al.，2009；Al Toufailia et al.，2016）。当在实验中冷冻杀死或用针杀死无刺蜂幼虫时，50%～99%的死亡幼虫会在48 h内被移出巢室（Medina et al.，2009；Nunes-

Silva et al., 2009; Al Toufailia et al., 2016; Jesus et al., 2017）。然而，卫生行为似乎存在相当大的种间和种内差异，这可能是由于工蜂嗅觉敏感性存在差异，并且移除死蛹的效率似乎低于去除死幼虫的效率（Jesus et al., 2017）。

我们对无刺蜂健康的相关知识仍然十分有限，可能有许多疾病尚未被发现。例如，Al Toufailia 等（2016）观察到一些 *Scaptotrigona depilis* 无刺蜂蜂群的后代中有相当一部分后代翅膀皱缩［蜜蜂中的残翅病毒（Deformed Wing Virus, DWV）的典型症状］，因此这些无刺蜂可能患有疾病。事实上，DWV 在阿根廷、巴西和墨西哥的其他无刺蜂蜂种中非常常见（Guzman-Novoa et al., 2015; Alvarez et al., 2018; Souza et al., 2019; Tapia-González et al., 2019; Guimarães-Cestaro et al., 2020）[2]。近期研究发现巴西的一个无刺蜂种群中 *Melipona scutellaris* 蜂群减少与蜜蜂中常见的急性蜜蜂麻痹病毒（Acute Bee Paralysis Virus, ABPV）有关（Ueira-Vieira et al., 2015）。在受感染地区的所有 10 个蜂群中都检测到了这种病毒，而健康无刺蜂蜂种的蜂群则没有携带这种病毒。在蜜蜂中，ABPV 的症状包括由于毛发脱落导致的无刺蜂外观颜色变深和变得更为锃亮、肢体运动颤抖和不能飞行（Miranda et al., 2010）。

早在很多年以前就有研究记录了发生在巴西南部夏末的 *Melipona quadrifasciata* 无刺蜂蜂群消失的现象（Díaz et al., 2017; Caesar et al., 2019）。由于这些蜂群的消失，现在在受影响地区几乎都没有野生 *M. quadrifasciata* 蜂群（Díaz et al., 2017）。那些受影响的无刺蜂所表现出的症状在不同地点有所不同，主要症状包括方向性障碍、颤抖、瘫痪和长喙行走，这些症状与携带了 ABPV 和以色列急性麻痹病毒（Israeli Acute Paralysis Virus, IAPV）的蜜蜂所表现出的症状相似（Díaz et al., 2017；Caesar et al., 2019）。Caesar 等（2019）在被感染的蜂群中鉴定出几种新的病毒，而 ABPV 和 IAPV 似乎与无刺蜂感染症状或蜂群消失无关。Caesar 等（2019）讨论提出了一种可能性，即已鉴定的病毒都不是造成这些蜂群消失的原因，但这些已鉴定的病毒在被未知病原体或寄生虫削弱的蜂群中会进行机会性自我复制。总而言之，越来越多的来自新热带区的证据发现已知的会感染蜜蜂的病毒也存在于无刺蜂蜂群中。然而，这些病毒对人工饲养的和野生的无刺蜂群的影响知之甚少，显然亟须

---

[2] 有趣的是，*M. subnitida* 无刺蜂蜂群主要被 C- 变体 DWV 感染，而相似地区的蜜蜂蜂群则主要被 A- 变体感染。这就提出了一个问题：无刺蜂中的 DWV 是否是蜜蜂病毒溢出的结果，或者说 DWV 是否可能是一种更为普遍存在的膜翅目病毒（Souza et al., 2019）。

这些病毒对无刺蜂群健康影响的深入研究。

最近研究者在澳洲的 *Tetragonula carbonaria* 无刺蜂蜂群中发现了第一例细菌性幼虫病，该病可能也感染了 *Austroplebeia australis* 无刺蜂（Shanks et al., 2017）。该病是被球形赖氨酸杆菌 *Lysinibacillus sphaericus* 感染所致，感染的蜂群表现出生长减弱，气味刺鼻发臭，幼虫颜色变深，工蜂嗜睡（Shanks et al., 2017）。欧洲幼虫腐臭病是一种众所周知的蜜蜂幼虫疾病，由 *Melissococcus puropoius* 细菌感染引起，研究者最近发现巴西的 *Melipona* 属无刺蜂蜂群也被其感染了（Teixeira et al., 2020）。被感染的无刺蜂蜂群出现了幼虫死亡，一些蜂群最终全部死亡。

当前人类饲养的蜜蜂和熊蜂会通过与其他蜂种共享资源或劫掠行为等方式将病原体和疾病传递给其他蜂种，这种疾病传播方式对无刺蜂来说逐渐构成了新的威胁（Genersch et al., 2006; Meeus et al., 2011; Fürst et al., 2014; Guzman-Novoa et al., 2015; Purkiss and Lach, 2019; Teixeira et al., 2020）。已经有研究检测到阿根廷（Porrini et al., 2017）、澳洲（Purkiss and Lach, 2019）和巴西（Guimarães-Cestaro et al., 2020; Nunes-Silva et al., 2016）等地的多种无刺蜂蜂群中存在一种常感染蜜蜂的微孢子虫 *Nosema ceranae*。澳洲的 *Tetragonula hockingsi* 无刺蜂工蜂只因为与感染蜜蜂造访过同一朵花就沾染了微孢子虫孢子。无刺蜂摄入的孢子在无刺蜂肠道中萌发，进入宿主细胞，2～4 d 内就会产生新的孢子。摄入了东方微孢子虫孢子的 *T.hockingsi* 无刺蜂死亡率远远高于未受感染的无刺蜂（Purkiss and Lach, 2019）。欧洲幼虫腐臭病会通过人类给无刺蜂喂食蜜蜂的蜂蜜和花粉（无刺蜂蜂农的一种常见做法）的过程中从蜜蜂传染给无刺蜂（Teixeira et al., 2020）。然而也不能排除某些疾病和病原体是无刺蜂的天然病原体的可能性（Alvarez et al., 2018）。

7.2 抢劫蜂和盗窃蜂

无刺蜂蜂群的主要威胁之一其实是其他无刺蜂蜂群（Nogueira-Neto, 1970; Laroca and Orth, 1984; Bego et al., 1991; Sakagami et al., 1993; Grüter et al., 2016），而且 Schwarz（1948）提出无刺蜂蜂群之间的斗争比无刺蜂与任何其他昆虫之间的斗争要多得多。然而，这也可能是因为观察者的观察偏差（蚂蚁入侵在夜间更频繁），也可能是因为研究者研究无刺蜂的活动通常是在蜂群密度较高的人工蜂箱中进行的。这种情况很可能会促进蜂群之间的战争[3]。尽管如此，野生无刺蜂蜂巢确实经常被其他无刺蜂攻击，这种情况并不奇怪，因为无刺蜂可以充分利用从其他无刺蜂蜂巢中掠夺到的所有资源。相应的，据人所知，许多无刺蜂蜂种也存在盗食共生或者说盗主寄生（Iyengar,2008; Breed et al., 2012）的情况：研究者已经观察到有30多个无刺蜂蜂种都有攻击其他蜂巢（包括蜜蜂巢）的行为（Grüter et al., 2016），这些蜂种要么是去窃取其他蜂巢的资源，要么是去霸占其他巢穴（分蜂到已经被其他蜂占领洞穴）（Gloag et al., 2008; Cunningham et al., 2014）[4]。而据报道大约50种无刺蜂是其他蜂攻击的受害者（Grüter et al., 2016），而且抢劫行为似乎是许多种无刺蜂蜂群生活的一部分。当无刺蜂争夺食物或巢址的竞争很激烈的时候，将篡夺其他蜂巢的行为作为一种分蜂策略可能是特别有益的（Foitzik and Heinze,1998; Quezada-Euán and González-Acereto,2002; Gloag et al., 2008; Rangel et al., 2010）。篡夺其他蜂巢是 *Tetragonula hockingsi* 和 *Tetragonisca angustula* 无刺蜂分蜂的一种常见策略，而且篡夺其他蜂巢的行为可能是 *Lestrimelitta* 无刺蜂盗蜂蜂群分蜂的唯一模式（第4章）（Schwarz,1948; Gloag et al., 2008; Cunningham et al., 2014; Grüter et al., 2016）。而对于入侵蜂来说，一个潜在风险就是可能会被受攻击蜂群传染上疾病和病原体，正如在蜜蜂中发现的那样（Lindström et al., 2008; Breed et al., 2012）。

　　偷取其他无刺蜂蜂群的食物是无刺蜂盗蜂与生俱来的生活方式。盗蜂主要是以新热带区的 *Lestrimelita* 属（约23种，第2章）（Gonzalez and Griswold, 2012）和非洲的 *Cleptorubula* 属（1种，*C. cubiceps*）（Eardley, 2004）无刺蜂为代表。这些蜂种是专性盗食寄生昆虫，这意味着它们不会访花去收集花蜜或花粉（Müller,1874; Friese,1931; Michener,1946; Sakagami and Laroca,1963; Roubik,1989）。遗憾的是，几乎没有任何关于 *Cleptotrigona cubiceps* 无刺蜂的生物学研究［仅在 Portugal-Araújo（1958）的研究中有关于其行为的稀少描述］。盗蜂工蜂的花粉篮已经退化或者完全消失了（图1.1）（Friese,1931；Parizotto,2010），但它们仍然可以利用后胫骨将筑巢材料或花粉从被袭击蜂群那里运回它们自己的蜂巢（von Zuben and Nunes,2014）。

---

［3］*Melipona* 无刺蜂人工饲养蜂群面临的风险可能不能完全代表野生蜂群面临的风险。
［4］有时候盗蜂不会进入蜂群中，但是会从其他蜂群巢入口管盗取巢质和树脂（Schwarz,1948）。非洲化蜜蜂也有类似的情况，笔者就经常观察到非洲化蜜蜂盗取无刺蜂蜂巢入口上和入口附近的树脂。

Lestrimelitta 无刺蜂盗蜂拥有强大的上颚，这使它们能够轻松杀死被攻击蜂群的工蜂（Nogueira Neto，1970）[5]。

令人费解的是 *Trichotrigona* 属无刺蜂（2个种）也可能是专性盗蜂。尽管研究者尚未观察到这个属无刺蜂的强盗行为，但是该属无刺蜂后腿上没有花粉梳（花粉耙），且该属无刺蜂蜂巢中也没有食物储存，这些迹象都表明这个属的无刺蜂是盗食寄生昆虫，它们可能是通过侵略 *Frieseomelitta* 属无刺蜂蜂群来存活的（第3.5.2节）（Camargo and Moure,1983; Camargo and Pedro,2007）。总而言之，所有无刺蜂中有4%～5%的无刺蜂是天生的盗食寄生蜂。

M. fuliginosa 无刺蜂是另一种经常攻击其他无刺蜂蜂巢的蜂种，它也是已知的体型最大的无刺蜂（Nogueira-Neto,1970; Camargo and Pedro,2008）[6]。*M. fuliginosa* 无刺蜂蜂群可以在数小时内完全摧毁一个群势较强的 *Melipona* 属无刺蜂蜂群。然而，*M. fuliginosa* 无刺蜂却不是天生的盗蜂，它们在特定的环境条件下也可能会劫掠其他蜂巢（Nogueira-Neto, 1970）。常在报道中作为侵略者的无刺蜂蜂种还有 OxyTribula 族和 *Tetragonisca angustula* 无刺蜂（Nogueira-Neto,1970; Roubik,1989; Grüter et al., 2016）。其中，*Tetragonisca angustula* 是侵略者这是出人意料的，因为 *Tetragonisca angustula* 无刺蜂的体型与被它攻击的蜂种相比简直就是个"小玩意儿"。

无刺蜂攻击方和防守方蜂群的死亡率会因蜂种和劫掠情况不同而有很大的差异。在许多涉及 *Lestrimelitta* 盗蜂的情况下，无刺蜂的劫掠行为并不会导致战斗，并且成年工蜂的死亡率接近于零（Nogueira-Neto,1970; Sakagami et al., 1993; Grüter et al., 2016）。然而盗蜂劫掠行为偶尔也会造成极大的损失（例如 *Scaptotrigona* 或 *Tetragonisca angustula* 无刺蜂蜂种参与其中的情况）（Johnson,1987; Nogueira-Neto,1970; Sakagami et al., 1993; Grüter et al., 2016）。在非洲 *Cleptotrigona cubiceps* 无刺蜂抢劫案中双方的死亡率取决于宿主蜂种。Portugal-Araujo（1958）研究发现，*Cleptotrigona cubiceps* 无刺蜂攻击 *Hypotrigona araujoi* 无刺蜂而导致的死亡率远高于攻击 *H. braunsi* 无

---

[5] 据报道 *Lestrimelitta* 盗蜂的蜂蜜是有毒的，这可能是因为其工蜂对蜂蜜进行了发酵或添加了分泌物的缘故（第1章）（Friese,1931; Sakagami and Laroca,1963）。

[6] *Melipona fuliginosa* 无刺蜂的工蜂体重约为125mg（Roubik,1989，表2.3）。*M.uliginosa* 属下另一个蜂种 *M. titania* 无刺蜂也是差不多大小。而相比起来，*Apis dorsata* 大蜜蜂的采集蜂的平均体重则为107～118 mg（Tan,2007）。

刺蜂。蜂群间大规模的战斗可能导致成百上千的工蜂死亡（Schwarz,1948; Nogueira-Neto,1970; Hubbell and Johnson,1977; Johnson,1987; Sakagami et al., 1993; Cunningham et al., 2014; Grüter et al., 2016）。例如，澳洲的 *Tetragonula carbonaria* 和 *Tetragonula hockingsi* 无刺蜂就会为了巢址进行大规模的种内和种间战斗，在此期间不管是进攻方还是防守方，蜂群中都有上千的工蜂死亡（Cunningham et al., 2014）。Schwarz（1948）记录了 *Melipona scutellaris* 无刺蜂两个蜂群之间的一场争斗，这次斗争导致了大约 3 000 只蜂死亡。一群 *Lestrimelita* 盗蜂对一个 *Trigona fulviventris* 蜂群的一场攻击中就有近 8 000 只蜂死亡（Johnson,1987）。这个盗蜂群尽管表面上赢得了这场战斗，但它们也失去了 1 162 只自己的蜂群成员。盗蜂群攻击一个好斗的 *Trigona* 属无刺蜂蜂群产生的巨大损失看似非常不合算，而 Johnson（1987）推测二者之间的斗争可能由于信息错误而逐步升级，因为 *T. fulviventris* 无刺蜂的报警信息素与柠檬醛类似，而柠檬醛是 *Lestrimelita* 盗蜂的公认的招募信息素（见下文）。

如果两个蜂群间斗争的胜利者可以赢得最终的战斗果实——一个新蜂巢及其中所有的食物，那么双方蜂群一定比率的死亡也就是合理的，事实上无刺蜂因其他无刺蜂蜂群导致蜂群死亡的情况是非常常见的（Müller,1874; Michener,1946; Schwarz,1948; Portugal-Araújo,1958; Sakagami and Laroca,1963; Nogueira-Neto,1970; Johnson,1987; Bego et al., 1991; Sakagami et al., 1993; Quezada-Euán and González-Acereto,2002; Pompeu and Silveira,2005; Cunningham et al., 2014; Mascena et al., 2017）。即使抢劫行为没有导致双方的成年蜂死亡，也有可能造成被抢劫蜂群的死亡，因为被抢劫蜂群通常会失去大部分食物储备，而抢劫蜂从其巢室中取出幼虫食物的同时也会杀死所有在巢室内的卵和正在发育的幼虫（Grüter et al., 2016）。因此，一个蜂群可能会在一场攻击后的数周或数月后灭亡。盗蜂偶尔也会成为被盗者，有研究观察到 *Tetragonisca angustula* 和 *Scaptotrigona pectoralis* 无刺蜂会攻击 *Lestrimelitt* 蜂巢并杀死 *Lestrimelitt* 蜂（Schwarz,1948; Sakagami et al., 1993; Nogueira-Neto, 1997）。

7.2.1 无刺蜂攻击的组织安排

我们对盗蜂攻击时蜂群的协调配合所涉及的招募和交流过程知之甚少。*Lestrimelitta* 盗蜂的相关研究表明，参与突袭的蜂数量从几十到上千只不等，但盗蜂突袭的最初阶段都是由一个或几个侦察蜂先确定好合适的进攻目标（Wittmann,1985; Sakagami et al., 1993）。在此之后不久，一大群盗蜂就会到达受害蜂群的蜂巢（Sakagami et al., 1993; Grüter et al., 2016; Mascena et al., 2017）。然后，盗蜂就开始守住被攻击蜂群蜂巢的入口，并且它们经常会建造自己的入口管（Sakagami and Laroca,1963; Nogueira-Neto,1970; Mascena et al., 2017）。在盗蜂招募更多同伴进攻的这一过程中，信息素好像起到了重要作用。已经有研究者推测柠檬醛就是一种起到招募盗蜂同伴作用的信息素（Blum et al., 1970; Wittmann et al., 1990; Sakagami et al., 1993; van Zweden et al., 2011; von Zuben et al., 2016），它是由盗蜂的上颚腺产生的一种柠檬味物质。此外，*Lestrimelitta limao* 盗蜂口唇腺释放的挥发物（主要是乙酸十六酯和乙酸9-十六酯）似乎也在 *Lestrimelitta limao* 的攻击过程中发挥了重要作用（von Zuben et al., 2016）。还有一些无刺蜂蜂种为了获取食物资源会使用唇腺化合物来招募蜂群采集（第 10 章）（Jarau et al., 2004, 2006, 2010; Schorkopf et al., 2007, 2009; Stangler et al., 2009），而 *Lestrimelita* 盗蜂则是在突袭过程中使用这种信息素来招募蜂群。"火蜂"是一种经常抢劫其他蜂群筑巢材料和食物储备的盗蜂（Roubik et al., 1987; Rinderer et al., 1988; Grüter et al., 2016），这种盗蜂的头部化合物在其进攻期间似乎也发挥了重要作用（Rinderer et al., 1988）。然而，目前尚不清楚这些化合物到底是招募蜂群的信息素还是宿主防御蜂的趋避剂。因为，当蜜蜂在受到 *Oxytrigona* 无刺蜂攻击时似乎会表现出一种方向迷失的反应，这一观察结果为前面提到的宿主防御蜂趋避剂这一功能提供了佐证（Rinderer et al., 1988）。

在 *Lestrimelitta* 盗蜂进攻其他蜂群期间，被进攻蜂群的返巢采集蜂通常不会进入蜂巢，而是在入口前飞悬或降落在附近的植被上（Michener,1946; Kerr,1951; Nogueira-Neto,1970; Sakagami et al., 1993）。这一现象引发了一种假说，即盗蜂能够产生破坏宿主蜂防御行为的信息素（"气味取代"假说）（Kerr,1951; Moure et al., 1958; Blum et al., 1970）。这种假说认为宿主蜂群的反应符合盗蜂的利益，而不符合宿主蜂群的利益。即便如此，研究者依然认为盗蜂上颚腺释放的柠檬醛才是导致宿主蜂群这样反应的原因（Blum et al., 1970），*Frieseomelitta varia* 无刺蜂的返蜂采集蜂也确实会对 *Lestrimelita* 盗蜂的唇腺化合物作出反应并停止进入自己蜂群（von Zuben et al., 2016）。唇部（酯类）和下颚部化合物（柠檬醛）各自在盗蜂突袭过程中发挥的功能仍不清楚。Nogueira Neto（1970）认为宿主蜂群对盗蜂信息素的行为反应是符合宿主蜂群的利益的，且有助于宿主蜂群避免更大的损失（"撤退信息"假说）。根据这一观点，盗蜂信息素是一种具有报警作用的利他信息素（即以有利于信息接收者的方式影响种间行为的化学信息素）(Wittmann,1985; Roubik,1989; Kärcher and Ratnieks,2009）。

盗蜂突袭行为的持续时间上有很大的差异（Sakagami and Laroca,1963; Nogueira-Neto,1970; Sakagami et al., 1993; Grüter et al., 2016）。它们会持续几个小时或者几天甚至几周不断进攻，直到搬走受害蜂群的大部分资源。一些蜂群会反复遭到盗蜂攻击，这些攻击可能是来自同一个盗蜂群，这意味着学习行为在盗蜂蜂群选择宿主蜂群时发挥了一定作用。Nogueira-Neto（1970）推测宿主蜂群也会从过去被突袭的经历中学习进而改变宿主蜂群守卫蜂的守卫防御行为。他提到了一个被 *Lestrimelitta limao* 攻击的 *Droryana* 无刺蜂群可能是通过一些适应过程学会了不攻击入侵盗蜂（Weaver et al., 1975）。Nogueira Neto（1970）也提出盗蜂好像更倾向于去攻击过去成功抢劫过的蜂群（Sakagami et al., 1993）。对于盗蜂和受害蜂群来说，在对峙过程中获取信息是十分有意义的，这将有助于双方在未来做出更有效的反应，但在无刺蜂劫掠过程中的学习行为尚无实验性的研究。

而当 *Tetragonula hockingsi* 无刺蜂试图侵占 *T. carbonaria* 无刺蜂蜂巢时，战斗时间从持续几个小时小规模冲突到持续数周的大规模攻击不等，并可能导致双方上千无刺蜂的伤亡（Cunningham et al., 2014）。造成这种战斗规模差异的原因尚不清楚，但有可能是蜂群会评估各自的实力，如果对手蜂群过于强大，入侵蜂群就会中止进攻。

7.2.2 影响无刺蜂蜂群易受袭击的因素

一些无刺蜂蜂种似乎比其他蜂种更容易受到盗蜂的攻击（Nogueira-Neto,1970; Sakagami et al., 1993; Roubik,1989; Quezada-Euán and González-Acereto,2002）。例如，非洲 *Cleptotrigona cubiceps* 无刺蜂攻击 *Hypotrigona braunsi* 无刺蜂的频率要远高于攻击同区域 *H. Araujoi* 无刺蜂的频率（Portugal-Araújo,1958），*Lestrimelitta* 无刺蜂则频繁地攻击 *Plebeia*、*Nannotriona* 和 *Scaptorigona* 属无刺蜂（Nogueira-Neto,1970; Laroca and Orth,1984; Roubik, 1989; Bego et al., 1991; Sakagami et al., 1993; Grüter et al., 2016）。这表明了盗蜂对寄主具有偏好性，但特定蜂群的易受袭性似乎也存在着区域差异。Emery's 法则指出，社会性寄生蜂往往与其宿主之间的亲缘关系较近（Emery,1909）。考虑到 *Lestrimelitta* 袭击的宿主蜂种范围很广（既包括了近亲蜂种如 *Nannotrigona* 属无刺蜂，也包括远亲蜂种如 *Melipona* 属无刺蜂），这种规律似乎并不适用于 *Lestrimelitta* 盗蜂。而有人提出 *Trichotrigola* 属无刺蜂

会专门盗取其近亲蜂种——*Frieseomelitta* 属无刺蜂的蜂群食物（见上文）。

无刺蜂的进攻方如何选择受害者目前仍不清楚，但有证据表明 *Lestrimelita* 盗蜂更喜欢攻击会生产高糖含量蜂蜜的蜂种（图 7.5）（Quezada-Euán and Gonzales-Acereto, 2002; Grüter et al., 2016）。有研究发现被入侵蜂巢中资源的能量值会影响盗蜂对其他无刺蜂（第 8 章，第 10 章）（Roubik, 1989; Biesmeijer and Ermers, 1999; Schmidt et al., 2006）和蜜蜂（例如，von Frisch, 1967; Seeley, 1995）的入侵行为，这与前面的发现相一致。而被入侵蜂群中储存食物的量是否也会影响盗蜂的攻击偏好还有待研究。而盗蜂掠夺其他蜂群的蜂巢时，蜂巢的容量和入口的尺寸可能在盗蜂攻击偏好中起到了重要作用。例如，洞巢型的 *Lestrimelita* 盗蜂蜂群不太会去掠夺比它体型小很多的蜂种的蜂巢，因为小体型蜂种使用的是小洞穴，*Lestrimelita*

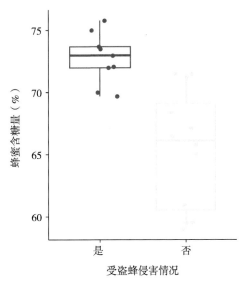

图 7.5　在同一地区栖息的受害和非受害无刺蜂蜂种的蜂蜜含糖量（t-value = 2.94, p = 0.008 4）（Grüter et al., 2016）

盗蜂也不会去掠夺地下的和半自建型蜂的蜂巢。

入侵无刺蜂会以比其体型小的蜂群为进攻目标，因为守卫这种小蜂群的工蜂数量更少，这可能会降低进攻方蜂群的死亡成本。如果蜂群斗争的结果遵循战斗的"平方定律"，那么攻击方和被攻击方的蜂群规模则尤为重要（Whitehouse and Jaffe, 1996; McGlynn, 2000）。依照这一定律，工蜂被杀死的速度与对手的数量成正比，而且拥有战斗蜂数量更多的一方蜂群将最终获胜。*Tetragonula hockingsi* 和 *T. carbonaria* 无刺蜂掠夺其他蜂巢的情况似乎就符合这种现象（Cunningham et al., 2014）。另外，如果某些蜂群因疾病或寄生虫而导致规模变小，那么喜欢这种小蜂群的入侵蜂入侵就有感染这些病原体的风险（Michener, 1974; Lindström et al., 2008; Breed et al., 2012）。此外，小蜂群可能拥有的资源也少，这种情况就促使进攻方选择攻击较大一些的蜂群（Pohl and Foitzik, 2011）。

蜂群的大小似乎确实会影响袭击的结果。例如，研究者已观察到 *Lestrimelitta limao* 盗蜂会破坏小型的 *Melipona rufiventris* 无刺蜂蜂群，而群势很强的 *M. rufiventris* 无刺蜂蜂群则能够成功地抵抗盗蜂的袭击（Pompeu and Silveira,2005）。同样，蜂群规模较小的 *Apis mellifera* 蜜蜂蜂群也不太能阻止盗蜂的入侵（Nogueira-Neto,1997）。然而，种水平蜂群规模的大小似乎并不能预测出盗蜂对其的袭击偏好（Grüter et al., 2016）。另一个重要因素可能是蜂本身的体型大小，这一因素确实已被证明会影响无刺蜂之间的战斗结果（Dworschak and Bluethgen,2010; Grüter et al., 2012）。然而，在物种水平上蜂体型大小似乎也并不能预测盗蜂是否会经常攻击某些特定蜂种的蜂群（Kerr 1951; Grüter et al., 2016），而且 *Lestrimelitta* 和 *Cleptotrigona* 属盗蜂的体型甚至比其经常攻击的蜂种的体型还要相对小一些。例如，单只 *Lestrimelitta limao* 无刺蜂体重约为 13 mg，但是它经常攻击比它大的 *Melipona* 属（单只体重 50～100 mg）无刺蜂（Grüter et al., 2016）。而 *Melipona fuliginosa* 无刺蜂在攻击其他蜂巢时可能会受益于其庞大的体型和力量（见上文）。

无刺蜂蜂群能否阻止一场攻击通常取决于其守卫蜂识别入侵者的能力。入侵者可能会尝试使用化学模拟、伪装或"化学弱识别化"来阻止守卫蜂的探查（Lenoir et al., 2001; Uboni et al., 2012; Grüter et al., 2018）。Quezada-Euán 等（2013）发现 *Lestrimelitta niitkib* 无刺蜂的化学表面特征与其首选的宿主蜂种十分接近，近似程度要高于其与其他宿主蜂种。然而，目前尚不清楚这是化学模拟的证据，还是盗蜂因为从其他蜂群掠夺资源而获得了某些化合物所导致的。大多数被盗蜂攻击蜂群的守卫蜂在面临盗蜂时会大幅改变自身的行为，要么表现出攻击性，要么表现出回避，这表明它们能够识别 *Lestrimelitta* 盗蜂（Nogueira-Neto,1970; Sakagami et al., 1993; van Zweden et al., 2011; Nunes et al., 2014）。

7.2.3　防御反应的不同

虽然有些无刺蜂蜂种在被盗蜂攻击时会采取强烈反击的防御措施，但其余无刺蜂蜂种则几乎没有表现出任何攻击性。研究者观察到其他动物分类群面临窃食寄生时也会出现这种"报复/容忍"二分法现象（Iyengar,2008）。不战斗的好处在于大多数成年工蜂和一些食品储存都可以幸免于难，而且所付出的代价也是相对可预测的。被攻击蜂群的工蜂（例如 *Aparatrigona*、*Nannotrigona* 或 *Plebeia*）可以在被攻击期间通过吸入液体食物将其储存在公共蜂群中来挽救部分资源。这样工蜂就可以用自己的身体临时储存食物并躲在角落里或者包壳下面直到盗蜂袭击结束（图7.6）（Noguiera-Neto,1970; Grüter et al., 2016）。

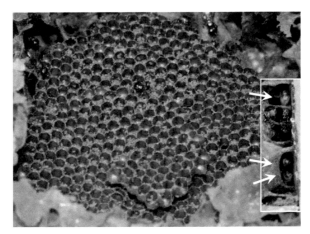

图7.6 在巴西贝伦附近的一个正在被 *Lestrimelitta maracaia* 盗蜂袭击的 *Aparatrigona* sp. 无刺蜂子脾

注：图中，盗蜂已经将子室中的幼虫食物全部搬走了。被攻击蜂群的工蜂则正躲在巢箱的角落里（白色箭头）。许多工蜂的腹部变长了，可能是因为它们将食物储存在蜜囊中（摄影：C. Grüter）。

如果被攻击的蜂群进行反击，那么代价将变得更加难以预测（Grüter et al., 2016）。如果反击的守卫蜂可以在早期阻止攻击，例如通过杀死盗蜂的侦察蜂，那么代价就比较低。然而如果守卫蜂无法阻止盗蜂的全面入侵，那么蜂群的死亡率会变得很高，因为蜂群会失去它们的食物储备、卵、幼虫和许多成年工蜂。

7.3　无刺蜂的防御策略

人们在无刺蜂身上发现了一些特别引人注意的行为特征和建筑特征，这显然是其应对其他动物攻击风险的一种应答反应，而且无刺蜂蜂群会将大量资源投入到蜂群防御中，要么是以守卫的形式，要么就是收集大量的树脂，或者是建筑防御性的巢结构。

7.3.1　避免被攻击的策略

无刺蜂蜂群在初始阶段就避免被攻击的是一种潜在的非常成功的策略。这种策略可以通过在天敌较少的栖息地筑巢、伪装、模仿其他动物或在能够提供一些保护的蜂群附近筑巢来实现（Kerr and de Lello,1962; Rech et al.,

2013；Grüter et al.，2018）。目前尚不清楚无刺蜂蜂群是否喜欢在更安全的环境中筑巢，但有证据表明无刺蜂的分蜂过程偶尔会中止（第 4 章），并且正在建立过程中蜂巢被干扰可能是造成无刺蜂分蜂突然中止的原因（个人观察）。因此，在侦察蜂分蜂的侦察阶段（第 4 章）就要评估新巢穴附近是否存在天敌。

地面筑巢型蜂种可能会试图在更难以到达的洞穴中筑巢。例如，如果巢腔与表面的入口孔之间的水平距离越大，则黑猩猩找到蜂巢的可能性就越低，因为距离越大使无刺蜂蜂巢的位置越难以被确定（Estienne et al.，2017 a）。无刺蜂的蜂巢伪装行为则包括蜂巢拥有一个非常小且不显眼的入口，或者蜂群的守卫蜂与入口周围环境颜色相匹配（第 6 章）（Grüter et al.，2017 a）。

有些无刺蜂居住在白蚁或蚂蚁巢穴附近（第 3 章），这样的好处在于攻击者不仅必须对抗守卫蜂，而且还可能需要面对有攻击性的蚂蚁或白蚁。一些养蜂人利用了蜂群攻击性差异的特点，将防御性较差的无刺蜂群放置在防御性较强的无刺蜂蜂群附近（Nogueira-Neto,1997；Rech et al.，2013）。Schwarz（1948）和 Kerr（1951）还提到了无刺蜂中出现贝氏拟态的几个案例，即无刺蜂的外观或行为与有刺的蜜蜂或胡蜂相似的案例。然而我们尚不明确这种相似之处是巧合还是确实是贝氏拟态的一种情况。目前研究者发现一些 *Frieseomelitta* 属（例如 *F. silvestrii*、*F. flavicornis* 或 *F. varia*）、*Tetragonilla* 和 *Tetrigona* 属无刺蜂的黑色翅膀尖端上有白色花纹，这类似于一些社会性胡蜂（例如 *Parachartergus fraternus*），这些胡蜂也是白尖黑翅的，这种情况是公认的贝氏拟态的一个案例（Kerr,1951；Kerr et al.，1967）。

7.3.2　蜂巢入口守卫策略

当入侵蜂在探测潜在的进攻目标蜂群时，目标蜂群的蜂巢守卫蜂通常会尝试通过攻击和通过释放警报信息素来招募更多防御蜂去阻止入侵蜂进入蜂巢（图 3.7）（Cruz - López et al.，2007；Gloag et al.，2008；Schorkopf et al.，2009；Nunes et al.，2014；Jernigan et al.，2018）。无刺蜂守卫蜂会非常专注于它们的守卫任务并很多天都负责担任守卫工作（相比之下，蜜蜂工蜂的守卫工作时长平均约为 1 d，Moore et al.，1987）。*Tetragonisca angustula* 无刺蜂的守卫蜂（士兵蜂，见下文）体型较大，它在蜂巢入口处出现的时长平均约为 5 d（Hammel et al.，2016），但有些无刺蜂守卫蜂则会守卫蜂巢入口长达近 3 周（第 6 章）（Grüter et al.，2011；Baudier et al.，2019）。蜂巢入口处执勤的守卫蜂数量一天中随时间变化而有所不同，清晨和傍晚时蜂巢入口处的守卫蜂数量最少（图 6.6）（Grüter et al.，2011；van Oystaeyen et al.，2013）。在夜间，无刺蜂守卫蜂通常在巢内（Schwarz,1948），但偶尔也可以在入口管内看到一些守卫蜂（Schwarz,1948；个人观察）。Michener（1946）报道提到一个 *Partamona testacea* 蜂群在夜晚攻击了干扰了蜂巢的人类，这个蜂群在夜间的攻击凶猛程度不亚于白天。一些无刺蜂守卫蜂会连续几天从早到晚都在入口处或入口附近的完全相同的位置上进行守卫（Zeil and Wittmann,1989; Kelber and Zeil,1997; Grüter et al., 2011）。

无刺蜂守卫蜂在大多数情况下很容易识别敌人，因为入侵者看起来和/或闻起来有所不同，而且守卫蜂可以根据视觉和化学特征进行识别（见下文）。守卫蜂针对潜在威胁做出的最初的攻击性反应通常是展开翅膀呈"V"形或振动翅膀（Johnson and Wiemer,1982）。下一个阶段就是咬击对手（Kerr and de Lello,1962; Michener,1974; Shorter and Rueppell, 2012; Shackleton et al., 2015）。通过咬住对手的翅膀底部是使有翼攻击者丧失移动能力的最为常见且有效的一种攻击策略（图 7.7 A）（Schwarz, 1948; Sakagami and Laroca, 1963; Grüter et al., 2012）。一些无刺蜂种群，特别是 *Oxytrigona* 无刺蜂，会分泌腐蚀性化学物质对抗脊椎动物（Michener,1974; Rinderer et al., 1988; Roubik,1989）。这些"火蜂"（巴西语为"cagafogo"[7]）向攻击者吐出混有二酮和其他化合物（可能是蛋白质类毒素）的甲酸，这会引起一种皮肤灼烧感，并导致皮肤起泡（Michener,1974; Roubik et al., 1987; Rinderer et al., 1988）。被 *Oxytrigona* 无刺蜂攻击过的人类通常会留下永久性的皮肤疤痕（Michener,1974）。与此相关的现象是，有人观察到"火蜂"具有异常大的上颚腺，占据其头部囊的大部分（Michener, 1974; Roubik et al., 1987）。*Oxytrigona* 无刺蜂的红色头部也很不寻常（图 7.7），这是无刺蜂中罕见的警告性颜色（警戒色）。

澳洲的 *Tetragonula* 无刺蜂蜂群会采用防御性蜂群来应对其他无刺蜂群的攻击（Gloag et al., 2008）。数百只工蜂蜂群聚集在蜂巢入口附近。蜂巢入口附近存在少量非同巢无刺蜂时，会触发这种群体反应（Gloag et al., 2008; Cunningham et al., 2014）。防御性蜂群还具有视觉上展示蜂群力量的功能，用以阻止对手蜂群发起全面攻击（Grüter et al., 2016）。一些 *Melipona* 属和 *Tetragonisca angustula* 无刺蜂蜂群在感应到 *Lestrimelitta* 盗蜂的化学信号后也

---

[7]"Cagafogo"的字面意思是"排火"，因此略有误导，因为该蜂种是从头腺中释放腐蚀性物质的，即它们是"吐火"。*O. tataira* 这个名字则是源自当地的"tata-ira""eira-tata"或"ei-tata"（图皮和瓜拉尼语中的"火蜂"）（Schwarz,1948）。

可以观察到类似的工蜂大量外流的反应（Wittmann et al., 1990; Segers et al., 2016; Jernigan et al., 2018; Campollo-Ovalle and Sánchez, 2018）。

图 7.7　无刺蜂的防御特征

注：A 显示一只 *Tetragonisca angustulade* 无刺蜂的守卫蜂咬住了一只 *Lestrimelitta limao* 盗蜂的翅膀根部，即使这只守卫蜂已死且头断了但是它头部的上颚仍紧锁在盗蜂身上使其无法飞行（摄影：C. Grüter）。B 显示 *Melipona flavolineata* 无刺蜂一个蜂群中的巢质球堆（摄影：Túlio Nunes）。C 显示正在被 *Oxytrigona tataira* 攻击的 *Tetragonisca angustula* 无刺蜂，*Tetragonisca angustula* 守卫蜂在其蜂巢入口附近放置了树脂液滴来阻退敌人（摄影：C. Grüter）。D 显示多种无刺蜂蜂群（图中为 *Nannotrigona testaceicornis* 无刺蜂）会在夜间使用柔软的巢质膜关闭其蜂巢入口。巢质膜上有小孔，可以进行空气交换（摄影：C. Grüter）。

7.3.3　警报信息素

目前尚不清楚无刺蜂是否普遍使用警报信息素（Roubik,1989; Campollo-Ovalle and Sánchez,2018），但毫无疑问的是许多无刺蜂蜂种在受到攻击时都会使用警报信息素。然而截至目前，研究者只鉴别出少数的警报物质。这些警报物质的一个共同的特征是它们主要在上颚腺中产生并且具有高度的挥发性（Blum et al., 1970; Luby et al., 1973; Weaver et al., 1975; Johnson and Wiemer,1982; Smith and Roubik,1983; Cruz-López et al., 2005; Cruz-López et al., 2007; Schorkopf et al., 2009; Alavez-Rosas et al., 2019）。Lindauer 和 Kerr（1958）提出无刺蜂将上颚腺的物质既用作警报信息素又用作采集活动中的追踪信息

素（"一个腺体—两个功能"假设），但最新的研究认为无刺蜂上颚腺化合物并未起到追踪信息素的作用（第 10 章）（Jarau et al., 2004; Schorkopf et al., 2009; Solórzano-Gordillo et al., 2018）。相反，上颚腺化合物主要起到了警报信息素的作用，而追踪信息素通常来自无刺蜂的唇腺（第 10 章）。

警报信息素可能有强烈的气味，人类闻到不同蜂种的这种气味会有好闻或不好闻等感受（例如，某些 Oxytrigona、Scaptotrigona 和 Trigona 属无刺蜂蜂种的警报信息素闻起来类似"腐臭黄油"或"蓝奶酪"气味，而一些其他蜂种的警报信息素则有令人愉悦的柠檬味或玫瑰味[8]）（Schwarz,1948）。Trigona fulviventris 无刺蜂的警报信息素的关键化合物可能是橙花醇（Johnson and Wiemer,1982）。柠檬醛这种具有令人愉悦的柠檬味的物质则在 Geotrigona subterranean 蜂群发挥了警报信息素的功能，并且以剂量依赖方式发挥作用（Blum et al., 1970; Schorkopf et al., 2009）。两种 Scaptotrigona 属无刺蜂和 Tetragonisca angustula 无刺蜂的警报信息素的重要组分是苯甲醛（Luby et al., 1973; Wittmann et al., 1990; Jernigan et al., 2018）。Melipona triplaridis 无刺蜂则采用强烈粪便气味的粪臭素作为警报信息素，这种粪臭素起到了一种威慑脊椎动物的作用（Smith and Roubik,1983; Wille,1983）。兵蚁袭击方也会释放粪臭素，因此粪臭素可能对军蚁攻击的小型敌人具有强大的威慑作用（Roubik,1989）。Melipona beecheii 工蜂暴露在一定浓度的香叶醇和乙酸金合欢酯会产生攻击性，这个反应浓度与该物种（Melipona beecheii）上颚提取物中发现的浓度相当（Cruz-Lopez et al., 2005）。无刺蜂上一些已鉴定的化合物也在其他社会性昆虫中作为警报物质发挥作用，还有些已鉴定的化合物是已知的驱虫剂，例如苯甲醛就是蚂蚁和蜜蜂的趋避剂（Luby et al., 1973; Schorkopf et al., 2009）。2-庚酮（闻起来像蓝奶酪）很可能是 Oxytrigona mediorufa 无刺蜂警报信息素的关键成分（Cruz-Lopez et al., 2007）。Melipona solani、Trigona spinipes 和 Scaptotrigona depilis 无刺蜂（可能还有许多其他蜂种）也使用 2-庚醇作为关键的警报化合物（Smith and Roubik,1983; Schorkopf et al., 2009; Leonhardt,2017; Alavez-Rosas et al., 2019）。一些系统发育距离较远的无刺蜂蜂种都使用 2-庚醇作为警报信息素，这表明 2-庚醇可能是无刺蜂普遍使用的一种警报信息素化合物。

---

[8] 无刺蜂的气味有时也会使蜂群更易受到攻击，因为澳洲和南美洲的土著人会利用无刺蜂这些气味来定位无刺蜂的巢穴（Schwarz,1948）。

无刺蜂的警报信息素不仅会释放到空气中，无刺蜂也经常会通过诸如咬或摩擦等方式主动将警报信息素转移到无刺蜂天敌的身体上，从而为正在攻击的同巢无刺蜂提供化学坐标（Roubik,1989）。一些无刺蜂蜂种的雄蜂集群或等在蜂巢入口处的雄蜂在受到攻击时也会释放警报信息素，蜂巢内的工蜂和其他雄蜂就会通过攻击（工蜂）或飞逃（雄蜂）对雄蜂释放出的警报信息素做出反应（Engels and Engels,1984; Sommeijer and De Bruijn,1995; Schorkopf,2016）。

7.3.4　无刺蜂蜂巢的入口封锁和树脂的使用

无刺蜂其他防御性行为反应还包括堵塞蜂巢入口，或者堵塞通往子区和食物罐的通道。*Melipona Seminigra* 无刺蜂工蜂会制作小泥球来阻挡入口（Kerr,1984），而 *Melipona flavolineata*、*M. fasciata* 和 *M. panamica* 无刺蜂则会制作巢脂球或硬化树脂球（图 7.7 B）来阻挡入口，这些行为主要发生在它们被 *Lestrimelitta* 袭击期间或当蜂群面临兼性盗蜂 *M. fuliginosa* 的攻击的时候（Roubik,1983, 2006; Nunes et al., 2014）。据观察，在蚂蚁攻击期间，*Lestrimelitta* 无刺蜂自己也会用蜡和树脂块挡住它们的蜂巢入口（Kerr and Lello,1962），而 *Trigona cilipes* 无刺蜂则主要使用树脂来阻挡入口（Kerr et al., 1967）。同样，非洲 *Apotrigona nebulata* 的守卫蜂在被蚂蚁攻击时会使用树脂球来缩窄它们的入口隧道（Darchen,1969）。据报道，*Hypotrigona braunsi* 工蜂在被 *Cleptotrigona* 盗蜂攻击时，会将蜂蜜倒在蜂巢入口管中来堵住它们的巢穴入口（Portugal-Araujo,1958）。进攻者随后就收走了这些蜂蜜，这些蜂蜜似乎成功阻止了攻击（Portugal-Araujo,1958）。一些无刺蜂的蜂巢入口较小，如 *Frieseomelitta* 属无刺蜂，守卫蜂甚至可以用头直接挡住它们的蜂巢入口（Schwarz,1948; Sakagami et al., 1993）。

研究者还普遍观察到，无刺蜂的守卫蜂会在它们的花粉筐或上颚上携带黏性物质（由不同比例的树脂和蜡混合而成），它们会将这些黏性物质涂到进攻者身上来使其丧失移动能力（Sakagami,1982; Roubik,2006; Lehmberg et al., 2008; Gastauer et al., 2011; Nunes et al., 2014）。以这种方式应用黏性物质可以有效地抵抗蚂蚁、蜂类、胡蜂和人类（Schwarz,1948; Camargo and Pedro,2003; Alves et al., 2018）。无刺蜂守卫蜂还会在入口孔附近堆放黏性液滴（图 7.7 C）（Roubik,2006; Alves et al., 2018）。一些澳洲和亚洲的 *Tetragonula* 属无刺蜂以及亚洲的 *Lophhotrigona canifrons* 无刺蜂则在此基础上更进一步，它们会在入口管道的外缘周围建造一个柔软而黏稠的环（Hockings,1883; Kerr and de

Lello,1962; Inoue et al., 1993), Schwarz（1948）将这种结构比作中世纪城堡的护城河。这一结构看起来对抵抗蚂蚁十分有效，例如常见的亚洲织布蚂蚁 *Oecophylla smaragdina*（Schwarz,1948; Inoue et al., 1993; Duangphakdee et al., 2009）。此外，树脂中存在的非挥发性化学物质具有驱虫作用，这可能可以阻止一些蚂蚁穿过树脂屏障（Wang et al., 2018）。有研究观察到 *Tetragonilla collina* 和 *Tetragonula melanocephala* 无刺蜂在受到蚂蚁攻击时，蜂群中的树脂采集蜂会急剧增加，这一现象进一步印证了树脂对于对抗蚂蚁的重要性（Leonhardt and Blüthgen,2009）。

无刺蜂的采集蜂和守卫蜂的不同身体部位上通常覆盖有黏性树脂（图 6.6 A）（Gastauer et al., 2011; Grüter et al., 2012），而且已知亚洲无刺蜂蜂种还会在其角质层上携带树脂衍生化合物（Leonhardt et al., 2009, 2011 a ,2015）。这些化合物的功能尚不为人所知，但它们可以降低无刺蜂对蚂蚁、蜘蛛或胡蜂等捕食者的吸引程度（Lehmberg et al., 2008; Drescher et al., 2014; Leonhardt et al., 2015）。例如，研究者检测了蚂蚁对几种亚洲无刺蜂蜂种的选择，与未清洗的蜂相比，蚂蚁更喜欢捕食用去除植物源性化合物的溶剂清洗过的蜂（Lehmberg et al., 2008）。

众所周知，树脂具有抗菌特性，蚂蚁和蜂都会使用树脂来抵御病原体（第 3 章）（Messer,1985; Christe et al., 2003; Chapuisat et al., 2007; Simone-Finstrom and Spivak,2010; Drescher et al., 2014; Brütsch et al., 2017）。一些无刺蜂蜂种还会使用树脂、巢质和泥浆来覆盖巢内不需要的寄生昆虫（第 3 章）。例如，澳洲的 *Austroplebeia australis* 和 *Tetragonula carbonaria* 无刺蜂会"木乃伊化"成年蜂粉小甲虫（SHB）（*Aethina tumida*）（Greco et al., 2010; Halcroft et al., 2011）。*Tetragonisca angustula* 无刺蜂甚至能将大型甲虫"木乃伊化"（Schwarz,1948; Nogueira-Neto,1997），并且经常可以看到蚂蚁"粘"在 *Tetragonisca angustula* 无刺蜂的蜂巢入口管上（图 7.2 A）。

一些无刺蜂蜂种会在晚上用一层柔软多孔的巢质封闭它们的蜂巢入口（图 7.7 D）（Michener,1946; Schwarz,1948; Kerr and de Lello,1962; Sakagami et al., 1973; Wille and Michener,1973; Roubik,1983; Roubik,2006; Grüter et al., 2011; Rasmussen and Gonzalez,2017）。在 Roubik（1983）观察的 40 个蜂种中有 17 个蜂种（43.5%）在夜间关闭了蜂巢入口。*Tetragonisca angusula* 无刺蜂蜂群中则只有大约一半的蜂群会关闭其蜂巢入口，而更多的蜂群即使在下雨天也仍然开放着它们的蜂巢入口（Grüter et al., 2011）。夜间温度可能是决定蜂群

是否以及如何在夜间关闭入口的另一个因素（Schwarz,1948）。笔者还观察到，*Austroplebeia australis* 无刺蜂的蜂巢入口在下雨天的早上关闭的时间更长，而在寒冷天气里蜂巢入口甚至会保持关闭状态好几天。这些观察结果反映出气候条件（可能通过它们对觅食活动的影响）和蜂群的通风需要决定了蜂群是否关闭蜂巢入口以及其关闭入口的时长。由于入口巢质通常非常柔软，建造这种入口巢质的主要目的可能是为了防止小型寄生虫（例如鬼蝇）、蚂蚁或昆虫的误入。*Lestrimelitta* 属和 *Scaura* 属无刺蜂的工蜂可以通过向内侧拉蜂巢入口处巢质的柔软边缘直到只剩下一个狭窄的间隙，从而可在几秒钟内就能关闭蜂巢入口（Roubik,1989; 个人观察）。

7.3.5 同巢与非同巢蜂的识别

有效的蜂群保护依赖于守卫蜂区分自己的同巢蜂和源自其他蜂巢的非同巢蜂的能力，这些非同巢蜂进入蜂巢可能是因为误入蜂群（漂流），或者是为了窃取资源（盗窃）或掠夺蜂巢，也可能是为了寄生繁殖（Roubik,1989; van Oystaeyen et al., 2013; Cunningham et al., 2014; Grüter et al., 2016，2018）。大多数研究者研究过的无刺蜂蜂种的工蜂都能够区分同巢蜂和非同巢蜂，并且会对非同巢蜂表现出攻击性（Kirchner and Friebe,1999; Breed and Page,1991; Suka and Inoue,1993; Inoue et al., 1999; Jungnickel et al., 2004; Buchwald and Breed,2005; Couvillon and Ratnieks,2008; Nunes et al., 2008; Kärcher and Ratnieks,2009; Leonhardt et al., 2010; Nascimento and Nascimento,2012; Wittwer and Elgar,2018; Bobadoye，2019）。在食物资源附近也会出现这种区别对待现象，例如无刺蜂采集蜂会在食物资源处要么避开要么攻击来自其他蜂群的蜂（Johnson and Hubbell,1974; Nagamitsu and Inoue,1997; Hrncir and Maia-Silva,2013）[9]。无刺蜂守卫蜂能够通过触角受体感知气味，从而帮助它们决定是允许一只蜂进入蜂巢，还是击退这只蜂。无刺蜂守卫蜂还可使用视觉特征来识别异种无刺蜂（Bowden et al., 1994）。

尽管我们对无刺蜂守卫蜂用来辨别同巢和非同巢蜂的特征化合物仍然知之甚少，但现有证据表明，无刺蜂角质层的碳氢化合物（cuticular hydrocarbons，CHCs）、脂肪酸和酯类（可能还有花的精油）在无刺蜂的辨别同巢蜂中发挥着重要作用（Jungnickel et al., 2004; Buchwald and Breed,2005; Nunes et al., 2008;

---

[9] Howard（1985）提到在树脂收集地来自不同蜂群的无刺蜂工蜂之间的攻击性尤为激烈。有趣的是，一些蜂群好像会持续地守着较大的树脂收集地。

Leonhardt et al., 2010; Nascimento and Nascimento, 2012; Martin et al., 2017）。这表明无刺蜂使用的识别同巢蜂的线索类别与蜜蜂使用的相似（Breed, 1998; Buchwald and Breed, 2005; van Zweden and D'Ettorre, 2010）。

大多数昆虫的身体表面都覆盖了 CHCs。这些 CHCs 形成了半流体的蜡质层，这对于昆虫的防水、防微生物感染、信号识别和传递都很重要（van Zweden and D'Ettorre, 2010; van Oystaeyen et al., 2014; Menzel et al., 2017）。研究者发现在无刺蜂的表皮中主要的碳氢化合物类别包括烷烃类、烯烃类和二烯烃类，这些碳氢化合物主要由 19～33 个碳原子的碳链构成（Breed, 1998; Abdalla et al., 2003; Jungnickel et al., 2004; Nunes et al., 2008, 2009 a, 2009 b; Leonhardt et al., 2009, 2011 a; van Zweden and D'Ettorre, 2010; Nascimento and Nascimento, 2012; Martin et al., 2017; Balbuena et al., 2018）。萜烯是一种在生物合成上与其他碳氢化合物有所不同的碳氢化合物，它在一些无刺蜂蜂种的角质层中普遍存在（Leonhardt et al., 2009, 2011 a, 2011 b, 2011 c, 2013; Balbuena et al., 2018; Kämper et al., 2019）。无刺蜂的 CHCs 有的是从头合成的，因此很可能含有一些可遗传的成分，一些则是通过食物、树脂、同巢蜂或巢结构等环境因子中获得的（Pianaro et al., 2007; Guerrieri et al., 2009; Leonhardt et al., 2009, 2011 a, 2013; van Zweden and D'Ettorre, 2010; Jones et al., 2012; Gutiérrez et al., 2016; Kämper et al., 2019）。对西方蜜蜂（*Apis mellifera*）来说，烯烃类似乎是比烷烃类更为重要的 CHCs 识别信号（Dani et al., 2005），而对于 *Trigona fulviventris* 无刺蜂而言，烯烃和正烷烃都会影响其攻击性（Buchwald and Breed, 2005）。

蜡脾是蜜蜂蜂群气味的主要来源（Breed et al., 1988; Breed, 1998; Downs and Ratnieks, 1999; D'Ettorre et al., 2006; Couvillon et al., 2007），而且交换蜂群间的蜡脾会引起蜂群对来自交换蜂群的非同巢蜂的接受度增加（D'Ettorre et al., 2006）。蜂蜡的主要成分是碳氢化合物（蜜蜂蜂蜡中碳氢化合物含量约为 15%，无刺蜂蜂蜡中碳氢化合物含量为 60%～90%）、酯类（蜜蜂蜂蜡中约 35%，无刺蜂蜂蜡中 6%～25%）和脂肪酸（在蜜蜂中约为 12%，在无刺蜂中为 2%～6%）（第 3 章）（Blomquist et al., 1985; Milborrow et al., 1987; Breed, 1998; Koedam et al., 2002; Pianaro et al., 2007）。这种成分组成使蜂蜡以及巢质也成为了无刺蜂识别因子的主要来源之一（Buchwald and Breed, 2005）。但无刺蜂巢质（第 3 章）是否调控了无刺蜂的群体气味仍不清楚。研究者发现，如果更换了 *Tetragonisca angustula* 无刺蜂的守卫蜂所站立的巢质入口管，在交换后 *Tetragonisca angustula* 无刺蜂的守卫蜂对入口管供体蜂群的非同巢蜂 24 h 内的接受度并未增加（Jones et al., 2012）。此外，几种亚洲无刺蜂蜂种的蜂巢入口管道材料（主要是巢质）的化学成分和蜂表皮化学成分谱差异很大（Leonhardt et al., 2011 a）。而 *Frieseomelitta varia* 无刺蜂工蜂在接触了另一个蜂群的巢质后，同巢蜂对其的攻击行为则更加剧烈（Nunes et al., 2011）[10]。这意味着巢质的气味因子也会影响无刺蜂的表皮物质谱。

无刺蜂为了区分同巢蜂和非同巢蜂，守卫蜂会将欲进入蜂群的蜂的气味（表皮气味谱或"标签"）与已有的被其记住的蜂群气味模板进行比较（van Zweden and D'Ettorre,2010）。守卫蜂这项工作的挑战性在于，即使是同巢蜂，其化学谱也各不相同，而非同巢蜂的身体上可能也会携带一些相同的气味因子（Jungnickel et al., 2004; Nunes et al., 2008, 2009 a, 2009 b; Leonhardt et al., 2009, 2011 b; Ferreira-Caliman et al., 2010; Martin et al., 2017; Balbuena et al., 2018）。特别是当同种蜂入侵蜂巢时，入侵蜂和同巢蜂的表皮气味因子的差异通常是数量上的差异而不是性质上的差异（Leonhardt et al., 2009; Nunes et al., 2009, 2009b; Nascimento and Nascimento,2012），而这种情况下避免错误辨别同巢蜂将变得更具挑战性。如果蜂群间的气味非常相似，那么一个蜂群的守卫蜂甚至会更频繁地接收来自另一个蜂群的工蜂（Nunes et al., 2008; Nascimento and Nascimento, 2012）[11]。例如，*Tetragonilla collina* 无刺蜂相邻的蜂群间的工蜂的表皮化学谱有很大的重叠，并且非同巢蜂之间也几乎不存在攻击性（Leonhardt et al., 2011 b）。

无刺蜂守卫蜂决定是否放行来巢蜂进入蜂巢时会犯两种错误。它一方面可能会错误地将非同巢蜂放行入巢（接收错误），一方面则可能会错误地拒绝同巢蜂入巢（拒绝错误）（van Zweden and D'Ettorre,2010）。这两种类型的错误可能是相关联的：蜂群选择低接收错误率就可能会增加拒绝错误率，因为这种策略下守卫蜂会变得更加严格（van Zweden and D'Ettorre,2010）。例如当蜂群面临被入侵的风险时，引发单只蜂接受或拒绝来巢蜂的行为阈值可以相对迅速

---

[10] 这个作者也探索了用于识别的气味谱是否受到采集来的食物的影响，但是他通过给不同的蜂群提供相同的食物发现并没有减弱这些蜂群对非同巢蜂的攻击性，这意味着食物的不同并不是导致蜂群拒绝非同巢蜂入巢的原因（Downs et al., 2001）。

[11] 守卫蜂对来巢蜂的反应通常不是二元的而是呈等级的，这种等级取决于蜂群气味模板和来巢蜂标记之间的错配程度（van Zweden and D'Ettorre,2010）。一些非同巢蜂想要进入蜂巢时可能只会被守卫蜂轻微攻击下或者被密切监察一下，而其他一些非同巢蜂则是一到蜂巢就被所有守卫蜂咬击或者推出蜂巢（在 *Tetragonisca angustula* 无刺蜂中观察到）。

地改变（Reeve,1989; Inoue et al., 1999; Downs and Ratnieks,2000; Couvillon et al., 2008 a）。

有趣的是，当无刺蜂非同巢蜂身上出现熟悉的气味因子时比当无刺蜂同巢蜂身上额外出现不熟悉的气味因子时守卫蜂攻击性的变化小（Couvillon and Ratnieks,2008; Ratnieks et al., 2011; Jones et al., 2012）。在非同巢蜂身上出现熟悉气味因子这种情况下，守卫蜂对非同巢蜂的攻击性仍然很高，而当同巢蜂身上额外出现不熟悉的气味因子时，守卫蜂对同巢蜂的攻击性会增加。这表明无刺蜂守卫蜂识别（或注意到的）是化学差异性而不是相似性（van Zweden and D'Ettorre,2010）。从机制上讲，这可能是因为守卫蜂触角感受器中受体神经元对熟悉的气味脱敏（Ozaki et al., 2005），也可能是因为无刺蜂守卫蜂的高级大脑中枢不断地在接收同巢蜂气味谱从而形成了习惯（van Zweden and D'Ettorre,2007）。而由于无刺蜂非同巢蜂（以及非同巢蜂气味因子处理过的同巢蜂）的气味谱与守卫蜂每天接触的蜂群气味谱不同，这些不熟悉的气味因子将引起守卫蜂触角感受器或更高的大脑中枢的应答反应。这种对无刺蜂守卫蜂行为的解释被归类为"非期望出现型"（"undesirable-present"，U-present）的识别模型（van Zweden and D'Ettorre,2010）。然而，无刺蜂 CHC 谱的改变并不都会增加守卫蜂的攻击性。Leonhardt 等（2010）在婆罗洲无刺蜂蜂种的气味谱中添加某些萜烯（特别是倍半萜烯）后发现，无刺蜂守卫蜂对这样处理后的蜂的攻击性会降低。作者认为这些萜烯具有"缓和"作用，并且可以解释为什么一些无刺蜂（例如 *Tetragonilla collina* 无刺蜂）蜂群经常在相邻的位置筑巢而互相不表现出攻击性。这种假设的缓和效应背后的机制仍有待深入研究。

不同无刺蜂蜂种（Jones et al., 2012）以及相同无刺蜂蜂种（Grüter et al., 2017 b）的不同蜂群的守卫蜂识别准确度（或攻击动机）存在着很大差异，有些蜂种会接受大多数非同巢蜂入巢（例如 *Melipona scutellaris*、*Tetragonilla collina* 无刺蜂）（Breed and Page,1991; Leonhardt et al., 2010），而有些无刺蜂蜂种则会近 100% 的拒绝非同巢蜂（例如 *Tetragonula minangkabau*、*Tetragonisca angustula* 无刺蜂）（Suka and Inoue,1993; Inoue et al., 1999; Kärcher and Ratnieks,2009; Jones et al., 2012）。这种种间和种内差异可以从以下方面来解释：①生态因素，例如蜂群被劫窃的风险差异会对应转化成不同的攻击水平；②蜂群间气味谱重叠程度存在种间差异；③诸如触角感受器的数量和灵敏度不同等原因造成的守卫蜂感知气味差别的能力存在不同（Downs and

Ratnieks,2000; Quezada-Euán et al., 2013; Grüter et al., 2017 b; Leonhardt,2017）。然而，由于不同的研究采用了不同的方法，所以蜂种间的识别准确度的差异很难阐释清除。这些研究方法在很多方面有所不同，例如有辨别力蜂的数量的呈现方式（有的研究用个数，有的研究则计算守卫蜂的群数）、依据分工分组还是依据蜂日龄分组（有的研究将采集蜂和守卫蜂作为有辨别力的蜂，有的研究则采用巢内的随机工蜂作为有辨别力的蜂）、观察的持续时间（从几秒到几小时不等）、评价蜂辨别力的环境（在蜂巢入口、蜂巢内部或在人造容器、盒子或管子中）、研究开展的季节，以及是否进行盲察等。这些不同的方法很可能会影响观察到蜂攻击性行为的可能性（Couvillon et al., 2013; van Wilgenburg and Elgar,2013）。例如，在蜂巢入口处，即在自然环境中进行研究时，*Tetragonisca angustula* 无刺蜂的守卫蜂只接受了 13.3% 的非同巢蜂，但在有盖培养皿中进行研究时，它们接受了 60%～70% 的非同巢蜂（Couvillon et al., 2013），而 *Tetragonula carbonaria* 无刺蜂的攻击行为则很大程度上取决于蜂巢气味的存在（Wittwer and Elgar,2018）。

无刺蜂对来自其他蜂群的新出房无刺蜂通常有攻击性，这种情况与蜜蜂相似（Bassindale,1955; Suka and Inoue,1993; Kirchner and Friebe,1999; Nunes et al., 2011; Inoue et al., 1999），这很大可能是因为新出房无刺蜂的表皮气味谱尚未发育完全（"空白板"假设）（Breed et al., 2004），而且 / 或者是因为无刺蜂没有使用非常小日龄的无刺蜂身上存在的化合物来进行同巢蜂识别（Nunes et al., 2011）。无刺蜂的表皮气味不仅会随着蜂日龄的增加而出现量上的增加，而且相对的气味因子的数量也会发生变化，并且通常与特定的生命阶段行为角色相关（Abdalla et al., 2003; Nunes et al., 2009, 2009 b, 2011; Ferreira-Caliman et al., 2010; Balbuena et al., 2018）。无刺蜂蜂群中的工蜂、蜂王和雄蜂的表皮化合物都有所不同（Abdalla et al., 2003; Nunes et al., 2009, 2009 b; Ferreira-Caliman et al., 2013），这使它们成了潜在的信息载体，例如它们可能带有蜂巢中是否有蜂王存在的信息（第 4 章）。

7.3.6 形态适应：大型守卫蜂和兵蜂

一些无刺蜂蜂种的护卫蜂体型更大（第 6 章）（Grüter et al., 2012; Grüter et al., 2017 a; Segers et al., 2015; Wittwer and Elgar,2018）。一些无刺蜂蜂种（*Tetragonisca angustula*、*T.fiebrigi*，可能还有 *Frieseomelitta longipe* 无刺蜂）的守卫蜂和采集蜂在体型大小方面大相径庭，而且这些守卫蜂形成了一个独特的兵蜂级型（或亚级型）（Grüter et al., 2012, 2017 a; Baudier et al.,

2019）。*Tetragonisca angustula* 无刺蜂的守卫蜂与采集蜂的体型差异最为明显（图 6.6，图 6.7）（Grüter et al., 2012; Segers et al., 2015, 2016; Hammel et al., 2016）。*Tetragonisca angustula* 无刺蜂守卫蜂的头部相对较小（即负体型变异），但后肢相对较大。*Tetragonisca angustula* 无刺蜂也是 *Lestrimelitta* 盗蜂的主要袭击目标（Nogueira-Neto, 1970; Bego et al., 1991; Sakagami et al., 1993; Grüter et al., 2016），而且有研究者猜测该蜂种的兵蜂级型就是其为了应对盗蜂的袭击而进化出来的（Grüter et al., 2012, 2017 a; Segers et al., 2016）。例如，*T. angustula* 无刺蜂的守卫蜂会对视觉上（黑色）或化学上（例如含有柠檬醛或盗蜂头部研磨物）与 *Lestrimelitta* 工蜂相似的物体会做出攻击性的反应（Wittmann,1985; Bowden et al., 1994; van Zweden et al., 2011; Segers et al., 2016; Campollo-Ovalle and Sánchez,2018）。如果将 *T. angustula* 无刺蜂蜂群暴露在盗蜂释放的挥发性气体中，这往往会引起数十只甚至数百只 *T. angustula* 无刺蜂离开蜂巢去与盗蜂搏斗（Wittmann,1985; Wittmann et al., 1990; Segers et al., 2016; Campollo-Ovalle and Sánchez,2018）。有趣的是，*T. angustula* 无刺蜂的守卫蜂的触角似乎比采集蜂的触角对柠檬醛（但不是花香）更为敏感（Balbuena and Farina, 2020）。此外，盗蜂越多的地区，*T. angustula* 无刺蜂蜂群中兵蜂的数量也就越多，而且研究者实验性将盗蜂头部研磨物暴露给 *T. angustula* 无刺蜂时，*T. angustula* 无刺蜂蜂群蜂巢入口处的兵蜂数量呈现出长时间的增加（Segers et al., 2016）。盗蜂可能不仅仅是 *T. angustula* 无刺蜂进化出兵蜂级型的驱动力，还是新热带区大多数无刺蜂进化出兵蜂级型的驱动力：*Lestrimelitta* 盗蜂袭击目标的无刺蜂蜂群中出现体型增大的守卫蜂的概率要远大于不是盗蜂袭击目标的无刺蜂蜂群（图 7.8）（Grüter et al., 2017 a）。

Tetragonisca angustula 无刺蜂

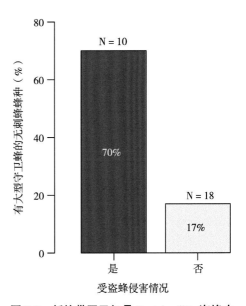

图 7.8 新热带区已知是 *Lestrimelitta* 盗蜂攻击目标的无刺蜂蜂种拥有大体型守卫蜂的可能性是非盗蜂攻击目标蜂种的 4 倍（Pagel's 法，*p* = 0.017，Grüter et al.,2017 a）

的兵蜂可以分为两类（第6章）。第一类主要在靠近巢入口处面对通向入口的飞行走廊不停飞悬（图6.4）（Zeil and Wittmann,1989; Kelber and Zeil,1997; van Zweden et al., 2011; Shackleton et al., 2018; Baudier et al., 2019; *Tetrigona apicalis* 是另一种有两类守卫蜂的蜂种，Burgett et al., 2020）。飞悬在空中无刺蜂守卫蜂经常会短暂地接近来巢蜂，这可能是为了更仔细地进行视觉侦察。飞悬守卫蜂会协调它们各自的位置以便在入口管的左右侧飞悬的守卫蜂数量相当（Shackleton et al., 2018）。与没有飞悬守卫蜂的蜂种相比，拥有飞悬守卫蜂增加了蜂群的防御范围，并且有助于及早发现蜂群的潜在威胁（van Zweden et al., 2011）。第二类的兵蜂则驻守在蜂巢入口或临近入口的地方（Wittmann,1985; Kärcher and Ratnieks,2009; Grüter et al., 2011; van Zweden et al., 2011）。无刺蜂兵蜂的体型大小是其防守表现的重要预测指标。当 *T. angustul* 无刺蜂与体型更大的 *Lestrimelitta limao* 盗蜂工蜂战斗时，其兵蜂的体型大小与盗蜂击败兵蜂所需的时间呈正相关（Grüter et al., 2012）。体型较大的守卫蜂似乎也更善于识别种内的非同巢蜂，这可能是因为体型越大守卫蜂触角也越大，触角上可以容纳的感觉器也就越多（Grüter et al., 2017 b; Wittwer and Elgar,2018; Month-Juris et al., 2020）。

7.3.7 建筑防御

无刺蜂蜂巢入口的建造对蜂群防御有重要影响。入口越狭窄则越容易防守，例如蜂群可以通过封锁入口来进行防御，但这样也限制了采集蜂的流通（Biesmeijer et al., 2007; Couvillon et al., 2008 b）。较大的入口则有利于采集，但也会使得入侵者更容易进入蜂巢。而蜂巢入口相对较大的蜂种可能是为了应对这种风险，通常有数量更多且攻击性更强的守卫蜂（Biesmeijer et al., 2007; Couvillon et al., 2008 b）。*Partamona* 属无刺蜂会建造"蛤蟆嘴"形入口（图3.1），该蜂种会通过依次修建两个入口来解决防御和采集效率之间的矛盾（Couvillon et al., 2008b）。*Partamona* 无刺蜂首先以极快的速度进入非常大的入口孔，然后从大入口处触顶反弹后落入较小的入口孔口（Shackleton et al., 2019）。

一些 *Partamona* 属和 *Ptilorigona* 属无刺蜂蜂种会在紧挨蜂巢入口处用泥土和树脂建造错综复杂的前厅结构（图3.6）（Michener,1974; Roubik,1989; Camargo and Pedro,2003）。有人猜测这种前厅能够分散和延缓袭击者，也能为大量的守卫蜂提供空间，还可以让食蚁兽望而却步（Michener,1974; Roubik,1989; Camargo and Pedro,2003）。更耐人寻味的是无刺蜂会在真巢附近建造所谓的假巢，假巢里面有空巢室、食物罐和包壳片（Roubik,1989; Camargo and Pedro,2003）。Camargo 和 Pedro（2003）观察到一个 *Lestrimelitta rufa* 蜂群在攻击一个 *Partamona vicina* 蜂群时，*Lestrimelitta rufa* 盗蜂仅占据了假巢，而 *Partamona vicina* 无刺蜂蜂群真正的育虫巢室则没有受到伤害。无刺蜂这种延迟进攻者攻击的策略可以为蜂群争取更多的时间，来封锁通往育虫室和食物室的通道，就像研究者在一次 *Partamona Ferreirai* 无刺蜂的骚乱中观察到的那样（Camargo and Pedro,2003）。因此，假巢或当作诱饵的蜂巢可能有助于蜂群在盗蜂群袭击时保护真正的育虫蜂巢。

参考文献

Abdalla FC, Jones GR, Morgan ED, da Cruz-Landim C（2003）Comparative study of the cuticular hydrocarbon composition of Melipona bicolor Lepeletier, 1836（Hymenoptera, Meliponini）workers and queens. Genetics and Molecular Research 2:191–199

Alavez-Rosas D, Sánchez-Guillén D, Malo EA, Cruz-López L（2019）（S）-2-Heptanol, the alarm pheromone of the stingless bee Melipona solani（Hymenoptera, Meliponini）. Apidologie 50:277–287

Al Toufailia H, Alves DA, Bená DC, Bento JM, Iwanicki NS, Cline AR, Ellis JD, Ratnieks FLW（2017）First record of small hive beetle, Aethina tumida Murray, in South America. Journal of Apicultural Research 56（1）:76–80

Al Toufailia H, Alves DA, Bento JM, Marchini LC, Ratnieks FLW（2016）Hygienic behaviour in Brazilian stingless bees. Biology Open 5（11）:1712–1718

Alvarez LJ, Reynaldi FJ, Ramello PJ, Garcia MLG, Sguazza GH, Abrahamovich AH, Lucia M（2018）Detection of honey bee viruses in Argentinian stingless bees（Hymenoptera: Apidae）. Insectes Sociaux 65:191–197

Alves A, Sendoya SF, Rech AR（2018）Fortress with Sticky Moats: the Functional Role of Small Particles around Tetragonisca angustula Latreille（Apidae: Hymenoptera）Nest Entrance. Sociobiology 65:330–332

Antonini Y, Martins RP（2003）The value of a tree species（Caryocar brasiliense）for a stingless bee Melipona quadrifasciata quadrifasciata. Journal of Insect Conservation 7（3）:167–174

Balbuena MS, Farina WM（2020）Chemosensory reception in the stingless bee

Tetragonisca angustula. Journal of Insect Physiology 125:104076

Balbuena MS, González A, Farina WM (2018) Characterization of cuticular hydrocarbons according to colony duties in the stingless bee Tetragonisca angustula. Apidologie 49:185–195

Barbosa WF, Smagghe G, Guedes RNC (2015) Pesticides and reduced-risk insecticides, native bees and pantropical stingless bees: pitfalls and perspectives. Pest Management Science 71(8):1049–1053

Bassindale R (1955) The biology of the Stingless Bee Trigona (Hypotrigona) gribodoi Magretti (Meliponidae). Proceedings of the Zoological Society of London 125(1):49–62

Baudier KM, Ostwald MM, Grüter C, Segers FHID, Roubik DW, Pavlic TP, Pratt SC, Fewell JH (2019) Changing of the guard: mixed specialization and flexibility in nest defense (Tetragonisca angustula). Behavioral Ecology 30:1041–1049

Bego LR, Zucchi R, Mateus S (1991) Notas sobre a estratégia alimentar: Cleptobiose de Lestrimelitta limao Smith (Hymenoptera, Apidae, Meliponinae). Naturalia 16:119–127

Biesmeijer JC, Ermers MCW (1999) Social foraging in stingless bees: how colonies of Melipona fasciata choose among nectar sources. Behavioral Ecology and Sociobiology 46:129–140

Biesmeijer JC, Slaa EJ, Koedam D (2007) How stingless bees solve traffic problems. Entomologische Berichten-Nederlandsche Entomologische Vereeniging 67(1/2):7–13

Bigio G, Schürch R, Ratnieks FLW (2013) Hygienic behavior in honey bees (Hymenoptera: Apidae): effects of brood, food, and time of the year. Journal of Economic Entomology 106(6):2280–2285

Blomquist GJ, Roubik DW, Buchmann SL (1985) Wax chemistry of two stingless bees of the Trigonisca group (Apididae: Meliponinae). Comparative Biochemistry and Physiology Part B: Comparative Biochemistry 82(1):137–142

Blum MS, Crewe RM, Kerr WE, Keith LH, Garrison AW, Walker MM (1970) Citral in stingless bees: isolation and functions in trail-laying and robbing. Journal of Insect Physiology 16:1637–1648

Bobadoye B (2019) Potential cues signalling nest mate recognition behaviour in

African meliponine bee species (Hymenoptera: Meliponini). Journal of Entomology and Zoology Studies 7:257–268

Bobadoye B, Fombong AT, Kiatoko N, Suresh R, Teal PEA, Salifu D, Torto B (2018) Behavioral responses of the small hive beetle, Aethina tumida, to odors of three meliponine bee species and honey bees, Apis mellifera scutellata. Entomol Exp Appl 166:528–534

Boesch C, Head J, Robbins MM (2009) Complex tool sets for honey extraction among chimpanzees in Loango National Park, Gabon. Journal of Human Evolution 56 (6):560–569

Breed MD (1998) Recognition pheromones of the honey bee. BioScience 48:463–470

Breed MD, Cook C, Krasnec MO (2012) Cleptobiosis in Social Insects. Psyche 2012:484765

Breed MD, Page RE (1991) Intra- and interspecific nestmate recognition in Melipona workers (Hymenoptera: Apidae). Journal of Insect Behavior 4:463–469

Breed MD, Perry S, Bjostad LB (2004) Testing the blank slate hypothesis: why honey bee colonies accept young bees. Insectes Sociaux 51 (1):12–16

Breed MD, Williams KR, Fewell JH (1988) Comb wax mediates the acquisition of nest-mate recognition cues in honey bees. Proceedings of the National Academy of Sciences of the United States of America 85:8766–8769

Brosi BJ (2009) The complex responses of social stingless bees (Apidae: Meliponini) to tropical deforestation. Forest Ecology and Management 258 (9):1830–1837

Brown BV (1997) Parasitic Phorid Flies: A Previously Unrecognized Cost to Aggregation Behavior of Male Stingless Bees. Biotropica 29 (3):370–372

Brown BV (2016) Two new bee-killing flies from Brazil (Insecta: Diptera: Phoridae: Melaloncha). Biodiversity Data Journal 4 (4):e7715

Brütsch T, Jaffuel G, Vallat A, Turlings TC, Chapuisat M (2017) Wood ants produce a potent antimicrobial agent by applying formic acid on tree collected resin. Ecology and Evolution 7 (7):2249–2254

Buchwald R, Breed MD (2005) Nestmate recognition cues in a stingless bee, Trigona fulviventris. Animal Behaviour 70:1331–1337

Burgett M, Sangjaroen P, Yavilat J, Chuttong B (2020) First report of hovering guard bees of the Paleotropical stingless bee Tetrigona apicalis (Hymenoptera: Apidae:

Meliponini). Apidologie 51:88–93

Caesar L, Cibulski SP, Canal CW, Blochtein B, Sattler A, Haag KL (2019) The virome of an endangered stingless bee suffering from annual mortality in southern Brazil. Journal of General Virology 100:1153–1164

Cairns CE, Villanueva-Gutiérrez R, Koptur S, Bray DB (2005) Bee populations, forest disturbance, and africanization in Mexico. Biotropica 37 (4):686–692

Camargo J, Moure J (1983) Trichotrigona, un novo genero de Meliponinae (Hymenoptera, Apidae) do Rio Negro, Amazonas Brasil. Acta Amazonica 13 (2):421–429

Camargo JM, Pedro SRM (2003) Neotropical Meliponini: the genus Partamona Schwarz, 1939 (Hymenoptera, Apidae, Apinae) -bionomy and biogeography. Revista Brasileira de Entomologia 47 (3):311–372

Camargo JM, Pedro SRM (2007) Notes on the bionomy of Trichotrigona extranea Camargo & Moure (Hymenoptera, Apidae, Meliponini). Revista Brasileira de Entomologia 51 (1):72–81

Camargo JM, Pedro SRM (2008) Revision of the species of Melipona of the fuliginosa group (Hymenoptera, Apoidea, Apidae, Meliponini). Revista Brasileira de Entomologia 52 (3):411–427

Camargo JM, Posey DA (1990) O conhecimento dos Kayapó sobre as abelhas sociais sem ferrão (Meliponinae, Apidae, Hymenoptera): notas adicionais. Boletim do Museu Paraense Emílio Goeldi, Zoologia 6:17–42

Campollo-Ovalle A, Sánchez D (2018) Temporal response of foragers and guards of two stingless bee species to cephalic compounds of the robber bee Lestrimelitta niitkib (Ayala) (Hymenoptera, Apidae). Neotropical Entomology 47:791–797

Cepeda-Aponte OI, Imperatriz-Fonseca VL, Velthuis HHW (2002) Lesser wax moth Achroia grisella: first report for stingless bees and new capture method. Journal of Apicultural Research 41 (3-4):107–108

Chapuisat M, Oppliger A, Magliano P, Christe P (2007) Wood ants use resin to protect themselves against pathogens. Proceedings of the Royal Society of London B: Biological Sciences 274 (1621):2013–2017

Christe P, Oppliger A, Bancala F, Castella G, Chapuisat M (2003) Evidence for collective medication in ants. Ecology Letters 6 (1):19–22

Cortopassi-Laurino M (2007) Drone congregations in Meliponini: what do they tell us? Bioscience Journal 23:153–160

Couvillon MJ, Caple JP, Endsor SL, Kärcher M, Russell TE, Storey DE, Ratnieks FLW (2007) Nest-mate recognition template of guard honeybees (Apis mellifera) is modified by wax comb transfer. Biology Letters 3:228–230

Couvillon MJ, Ratnieks FLW (2008) Odour transfer in stingless bee marmelada (Frieseomelitta varia) demonstrates that entrance guards use an "undesirable-absent" recognition system. Behavioral Ecology and Sociobiology 62:1099–1105

Couvillon MJ, Robinson EJ, Atkinson B, Child L, Dent KR, Ratnieks FLW (2008a) En garde: rapid shifts in honeybee, Apis mellifera, guarding behaviour are triggered by onslaught of conspecific intruders. Animal Behaviour 76(5):1653–1658

Couvillon MJ, Segers FHID, Cooper-Bowman R, Truslove G, Nascimento DL, Nascimento FS, Ratnieks FLW (2013) Context affects nestmate recognition errors in honey bees and stingless bees. Journal of Experimental Biology 216:3055–3061

Couvillon MJ, Wenseleers T, Imperatriz-Fonseca VL, Nogueira-Neto P, Ratnieks FLW (2008b) Comparative study in stingless bees (Meliponini) demonstrates that nest entrance size predicts traffic and defensivity. Journal of Evolutionary Biology 21:194–201

Cruz-López L, Malo EA, Morgan ED, Rincon M, Guzmán M, Rojas JC (2005) Mandibular gland secretion of Melipona beecheii: chemistry and behavior. Journal of Chemical Ecology 31(7):1621–1632

Cruz-López L, Aguilar S, Malo E, Rincón M, Guzman M, Rojas J (2007) Electroantennogram and behavioral responses of workers of the stingless bee Oxytrigona mediorufa to mandibular gland volatiles. Entomologia Experimentalis et Applicata 123(1):43–47

Cunningham JP, Hereward JP, Heard TA, De Barro PJ, West SA (2014) Bees at War: Interspecific Battles and Nest Usurpation in Stingless Bees. American Naturalist 184(6)

D'Ettorre P, Wenseleers T, Dawson J, Hutchinson S, Boswell T, Ratnieks FLW (2006) Wax combs mediate nestmate recognition by guard honeybees. Animal Behaviour 71(4):773–779

Dani FR, Jones GR, Corsi S, Beard R, Pradella D, Turillazzi S (2005) Nestmate recognition cues in the honey bee: differential importance of cuticular alkanes and alkenes. Chemical Senses 30(6):477–489

Darchen R (1969) Sur la biologie de Trigona (Apotrigona) nebulata komiensis Cock. I. Biologia Gabonica 5:151–183

Díaz S, de Souza Urbano S, Caesar L, Blochtein B, Sattler A, Zuge V, Haag KL (2017) Report on the microbiota of Melipona quadrifasciata affected by a recurrent disease. Journal of Invertebrate Pathology 143:35–39

Downs SG, Ratnieks FL (1999) Recognition of conspecifics by honeybee guards uses nonheritable cues acquired in the adult stage. Animal Behaviour 58 (3):643–648

Downs SG, Ratnieks FLW (2000) Adaptive shifts in honey bee (*Apis mellifera* L.) guarding behavior support predictions of the acceptance threshold model. Behavioral Ecology 11:326–333

Downs SG, Ratnieks FLW, Badcock NS, Mynott A (2001) Honeybee guards do not use food-derived odors to recognize non-nest mates: a test of the odor convergence hypothesis. Behavioral Ecology 12:47–50

Drescher N, Wallace HM, Katouli M, Massaro CF, Leonhardt SD (2014) Diversity matters: how bees benefit from different resin sources. Oecologia 176:943–953

Duangphakdee O, Koeniger N, Deowanish S, Hepburn HR, Wongsiri S (2009) Ant repellent resins of honeybees and stingless bees. Insectes Sociaux 56:333–339

Dworschak K, Bluethgen N (2010) Networks and dominance hierarchies: does interspecific aggression explain flower partitioning among stingless bees? Ecological Entomology 35 (2):216–225

Eardley C (2004) Taxonomic revision of the African stingless bees (Apoidea: Apidae: Apinae: Meliponini). African Plant Protection 10 (2):63–96

Eltz T, Brühl CA, van Der Kaars S, Linsenmair EK (2002) Determinants of stingless bee nest density in lowland dipterocarp forests of Sabah, Malaysia. Oecologia 131 (1):27–34

Emery C (1909) Über den Ursprung der dulotischen, parasitischen und myrmekophilen Ameisen. Biologisches Centralblatt 29:352–362

Engels E, Engels W (1984) Drohnen-Ansammlungen bei Nestern der stachellosen Biene Scaptotrigona postica. Apidologie 15 (3):315–328

Estienne V, Mundry R, Kühl HS, Boesch C (2017 a) Exploitation of underground bee nests by three sympatric consumers in Loango National Park, Gabon. Biotropica 49:101–109

Estienne V, Stephens C, Boesch C (2017 b) Extraction of honey from underground bee nests by central African chimpanzees (Pan troglodytes troglodytes) in Loango National Park, Gabon: Techniques and individual differences. American Journal of Primatology 79 (8):22672

Ferreira-Caliman M, Nascimento FS, Turatti I, Mateus S, Lopes N, Zucchi R (2010) The cuticular hydrocarbons profiles in the stingless bee Melipona marginata reflect task-related differences. Journal of Insect Physiology 56 (7):800–804

Fiebrig K (1908) Skizzen aus dem Leben einer Melipone aus Paraguay. Zeitschrift für wissenschaftliche Insektenbiologie 12:374–386

Foitzik S, Heinze J (1998) Nest site limitation and colony takeover in the ant Leptothorax nylanderi. Behavioral Ecology 9 (4):367–375

Fowler HG (1979) Responses by a stingless bee to a subtropical environment. Revista de Biologia Tropical 27:111–118

Freitas BM, Imperatriz-Fonseca VL, Medina LM, Kleinert A de MP, Galetto L, Nates-Parra G, Quezada-Euán JJG (2009) Diversity, threats and conservation of native bees in the Neotropics. Apidologie 40:332–346

Friese H (1931) Wie können Schmarotzerbienen aus Sammelbienen entstehen. Zoologische Jahrbücher Abteilung für Systematik, Geographie und Biologie der Tiere 62:1–14

Fürst M, McMahon DP, Osborne J, Paxton R, Brown M (2014) Disease associations between honeybees and bumblebees as a threat to wild pollinators. Nature 506 (7488):364

Gastauer M, Campos LAO, Wittmann D (2011) Handling sticky resin by stingless bees (Hymenoptera, Apidae). Revista Brasileira de Entomologia 55:234–240

Genersch E, Yue C, Fries I, de Miranda JR (2006) Detection of deformed wing virus, a honey bee viral pathogen, in bumble bees (Bombus terrestris and Bombus pascuorum) with wing deformities. Journal of Invertebrate Pathology 91 (1):61–63

Gloag R, Heard T, Beekman M, Oldroyd BP (2008) Nest defence in a stingless bee: What causes fighting swarms in Trigona carbonaria (Hymenoptera, Meliponini)? Insectes Sociaux 55 (4):387–391

Gloag R, Shaw SR, Burwell C (2009) A new species of Syntretus Foerster (Hymenoptera: Braconidae: Euphorinae), a parasitoid of the stingless bee Trigona carbonaria Smith (Hymenoptera: Apidae: Meliponinae). Austral Entomology 48 (1):8–14

Gonzalez VH, Griswold T (2012) New species and previously unknown males of neotropical cleptobiotic stingless bees (Hymenoptera, Apidae, Lestrimelitta). Caldasia 34 (1):227–245

Greco MK, Hoffmann D, Dollin A, Duncan M, Spooner-Hart R, Neumann P (2010) The alternative Pharaoh approach: stingless bees mummify beetle parasites alive. Naturwissenschaften 97 (3):319–323

Grüter C, Jongepier E, Foitzik S (2018) Insect societies fight back: the evolution of defensive traits against social parasites. Philosophical Transactions of the Royal Society B: Biological Sciences 373:20170200

Grüter C, Kärcher M, Ratnieks FLW (2011) The natural history of nest defence in a stingless bee, Tetragonisca angustula (Latreille) (Hymenoptera: Apidae), with two distinct types of entrance guards. Neotropical Entomology 40:55–61

Grüter C, Menezes C, Imperatriz-Fonseca VL, Ratnieks FLW (2012) A morphologically specialized soldier caste improves colony defence in a Neotropical eusocial bee. Proceedings of the National Academy of Sciences of the United States of America 109:1182–1186

Grüter C, Segers FHID, Menezes C, Vollet-Neto A, Falcon T, von Zuben L, Bitondi MMG, Nascimento FS, Almeida EAB (2017a) Repeated evolution of soldier sub-castes suggests parasitism drives social complexity in stingless bees. Nature Communications 8:4

Grüter C, Segers FHID, Santos LLG, Hammel B, Zimmermann U, Nascimento FS (2017b) Enemy recognition is linked to soldier size in a polymorphic stingless bee. Biology Letters 13:20170511

Grüter C, von Zuben L, Segers F, Cunningham J (2016) Warfare in stingless bees. Insectes Sociaux 63:223–236

Guerrieri FJ, Nehring V, Jorgensen CG, Nielsen J, Galizia CG, D'Ettorre P (2009) Ants recognize foes and not friends. Proceedings of the Royal Society of London Series B-Biological Sciences 276:2461–2468

Guimarães-Cestaro L, Martins MF, Martínez LC, Alves MLTMF, Guidugli-Lazzarini KR, Nocelli RCF, Malaspina O, Serrão JE, Teixeira ÉW (2020) Occurrence of virus, microsporidia, and pesticide residues in three species of stingless bees(Apidae: Meliponini) in the field. Science of Nature 107:16

Gutiérrez E, Ruiz D, Solís T, May-Itzá WJ, Moo-Valle H, Quezada-Euán JJG（2016）Does larval food affect cuticular profiles and recognition in eusocial bees? a test on Scaptotrigona gynes（Hymenoptera: Meliponini）. Behavioral Ecology and Sociobiology 70（6）:871–879

Guzman-Novoa E, Hamiduzzaman MM, Anguiano-Baez R, Correa-Benítez A, Castañeda-Cervantes E, Arnold NI（2015）First detection of honey bee viruses in stingless bees in North America. Journal of Apicultural Research 54:93–95

Halcroft M, Spooner-Hart R, Dollin LA（2013）Australian stingless bees. In: Vit P, Pedro SRM, Roubik DW（eds）Pot-Honey. Springer, New York, pp 35–72

Halcroft M, Spooner-Hart R, Neumann P（2011）Behavioral defense strategies of the stingless bee, Austroplebeia australis, against the small hive beetle, Aethina tumida. Insectes Sociaux 58（2）:245–253

Hammel B, Vollet-Neto A, Menezes C, Nascimento FS, Engels W, Grüter C（2016）Soldiers in a stingless bee: work rate and task repertoire suggest guards are an elite force. The American Naturalist 187:120–129

Hernández FO, Gutiérrez A（2001）Avoiding Pseudohypocera attacks（Diptera: Phoridae）during the artificial propagation of Melipona beecheii colonies（Hymenoptera: Apidae: Meliponini）. Folia Entomologica Mexicana 40:373–379

Hockings HJ（1883）Notes on two Australian species of Trigona. Transactions of the Entomological Society of London 2:149–157

Howard JJ（1985）Observations on Resin Collecting by Six Interacting Species of Stingless Bees（Apidae: Meliponinae）. Journal of the Kansas Entomological Society 58:337–345

Hrncir M, Maia-Silva C（2013）On the diversity of forging-related traits in stingless bees. In: Vit P, Pedro SRM, Roubik DW（eds）Pot-Honey: A legacy of stingless bees. Springer, New York, pp 201–215

Hubbell SP, Johnson LK（1977）Competition and nest spacing in a tropical stingless bee community. Ecology 58:949–963

Inoue T, Nakamura K, Salmah S, Abbas I（1993）Population dynamics of animals in unpredictably-changing tropical environments. Journal of Biosciences 18（4）:425–

455

Inoue T, Roubik DW, Suka T (1999) Nestmate recognition in the stingless bee Melipona panamica (Apidae, Meliponini). Insectes Sociaux 46 (3):208–218

Iyengar EV (2008) Kleptoparasitic interactions throughout the animal kingdom and a re-evaluation, based on participant mobility, of the conditions promoting the evolution of kleptoparasitism. Biological Journal of the Linnean Society 93 (4):745–762

Jarau S, Dambacher J, Twele R, Aguilar I, Francke W, Ayasse M (2010) The trail pheromone of a stingless bee, Trigona corvina (Hymenoptera, Apidae, Meliponini), varies between populations. Chemical Senses 35:593–601

Jarau S, Hrncir M, Zucchi R, Barth F (2004) A stingless bee uses labial gland secretions for scent trail communication (Trigona recursa Smith 1863). Journal of Comparative Physiology A 190 (3):233–239

Jarau S, Schulz CM, Hrncir M, Francke W, Zucchi R, Barth FG, Ayasse M (2006) Hexyl decanoate, the first trail pheromone compound identified in a stingless bee, Trigona recursa. Journal of Chemical Ecology 32 (7):1555–1564

Jernigan CM, Birgiolas J, McHugh C, Roubik DW, Wcislo WT, Smith BH (2018) Colony-level non-associative plasticity of alarm responses in the stingless honey bee, Tetragonisca angustula. Behavioral Ecology and Sociobiology 72 (3):58

Jesus JN, Chambó ED, da Silva Sodré G, de Oliveira NTE, de Carvalho CAL (2017) Hygienic behavior in Melipona quadrifasciata anthidioides (Apidae, Meliponini). Apidologie 48:504–512.

Johnson LK, Wiemer D (1982) Nerol: an alarm substance of the stingless bee, Trigona fulviventris (Hymenoptera: Apidae). Journal of Chemical Ecology 8(9):1167–1181

Johnson LK (1987) The pyrrhic victory of nest-robbing bees: did they use the wrong pheromone? Biotropica 19:188–189

Johnson LK, Hubbell SP (1974) Aggression and competition among stingless bees: field studies. Ecology 55 (1):120–127

Jones SM, van Zweden JS, Grüter C, Menezes C, Alves DA, Nunes-Silva P, Czaczkes TJ, Imperatriz-Fonseca VL, Ratnieks FLW (2012) The role of wax and resin in the nestmate recognition system of a stingless bee, Tetragonisca angustula. Behavioral Ecology and Sociobiology 66:1–12

Jungnickel H, da Costa AJS, Tentschert J, Patricio EFLRA, Imperatriz-Fonseca VL,

Drijfhout FP, Morgan ED (2004) Chemical basis for inter-colonial aggression in the stingless bee Scaptotrigona bipunctata (Hymenoptera: Apidae). Journal of Insect Physiology 50:761–766

Kajobe R, Roubik DW (2006) Honey-making bee colony abundance and predation by apes and humans in a Uganda forest reserve. Biotropica 38:210–218

Kämper W, Kaluza BF, Wallace H, Schmitt T, Leonhardt SD (2019) Habitats shape the cuticular chemical profiles of stingless bees. Chemoecology 29:125–133

Kärcher M, Ratnieks FLW (2009) Standing and hovering guards of the stingless bee Tetragonisca angustula complement each other in entrance guarding and intruder recognition. Journal of Apicultural Research 48:209–214

Kelber A, Zeil J (1997) Tetragonisca guard bees interpret expanding and contracting patterns as unintended displacement in space. Journal of Comparative Physiology A-Neuroethology Sensory Neural and Behavioral Physiology 181:257–265

Kerr WE (1951) Bases para o estudo da genética de populações dos Hymenoptera em geral e dos Apinae sociais em particular. Anais da Escola Superior de Agricultura Luiz de Queiroz 8:219–354

Kerr WE (1984) Virgilio de Portugal Brito Araújo (1919-1983). Acta Amazonica 13:327–328

Kerr WE, de Lello E (1962) Sting glands in stingless bees: a vestigial character (Hymenoptera: Apidae). Journal of the New York Entomological Society 70:190–214

Kerr WE, Sakagami SF, Zucchi R, Portugal-Araújo Vd, Camargo JMF (1967) Observações sobre a arquitetura dos ninhos e comportamento de algumas espécies de abelhas sem ferrão das vizinhanças de Manaus, Amazonas (Hymenoptera, Apoidea). In: Atas do Simpósio sobre a biota Amazônica. Conselho Nacional de Pesquisas Rio de Janeiro, pp 255–309

Kirchner WH, Friebe R (1999) Nestmate discrimination in the African stingless bee Hypotrigona gribodoi Magretti (Hymenoptera: Apidae). Apidologie 30(4):293–298

Koedam D, Jungnickel H, Tentschert J, Jones G, Morgan E (2002) Production of wax by virgin queens of the stingless bee Melipona bicolor (Apidae, Meliponinae). Insectes Sociaux 49(3):229–233

Koedam D, Slaa E, Biesmeijer J, Nogueira-Neto P (2009) Unsuccessful attacks dominate a drone-preying wasp's hunting performance near stingless bee nests.

Genetics and Molecular Research 8:690–702

Laroca S, Orth A (1984) Pilhagem de um ninho de Plebeia catamarcensis meridionalis por Lestrimelitta limao (Apidae, Meliponinae) em Itapiranga, SC, Sul do Brasil. Dusenia 14 (3):123–127

Lehmberg L, Dworschak K, Blüthgen N (2008) Defensive behavior and chemical deterrence against ants in the stingless bee genus Trigona (Apidae, Meliponini). Journal of Apicultural Research 47 (1):17–21

Lenoir A, d'Ettorre P, Errard C, Hefetz A (2001) Chemical ecology and social parasitism in ants. Annual Review of Entomology 46 (1):573–599

Leonhardt SD (2017) Chemical ecology of stingless bees. Journal of Chemical Ecology 43 (4):385–402

Leonhardt SD, Blüthgen N (2009) A sticky affair: resin collection by Bornean stingless bees. Biotropica 41 (6):730–736

Leonhardt SD, Blüthgen N, Schmitt T (2009) Smelling like resin: terpenoids account for species-specific cuticular profiles in Southeast-Asian stingless bees. Insectes Sociaux 56:157–170

Leonhardt SD, Blüthgen N, Schmitt T (2011 a) Chemical profiles of body surfaces and nests from six Bornean stingless bee species. Journal of Chemical Ecology 37:98–104

Leonhardt SD, Form S, Blüthgen N, Schmitt T, Feldhaar H (2011 b) Genetic relatedness and chemical profiles in an unusually peaceful eusocial bee. Journal of Chemical Ecology 37:1117–1126

Leonhardt SD, Jung L-M, Schmitt T, Blüthgen N (2010) Terpenoids tame aggressors: role of chemicals in stingless bee communal nesting. Behavioral Ecology and Sociobiology 64:1415–1423

Leonhardt SD, Rasmussen C, Schmitt T (2013) Genes versus environment: geography and phylogenetic relationships shape the chemical profiles of stingless bees on a global scale. Proceedings of the Royal Society of London Series B-Biological Sciences 280:20130680

Leonhardt SD, Schmitt T, Blüthgen N (2011 c) Tree resin composition, collection behavior and selective filters shape chemical profiles of tropical bees (Apidae: Meliponini). PLoS ONE 6:e23445

Leonhardt SD, Wallace HM, Blüthgen N, Wenzel F（2015）Potential role of environmentally derived cuticular compounds in stingless bees. Chemoecology 25:159–167

Lindauer M, Kerr WE（1958）Die gegenseitige Verständigung bei den stachellosen Bienen. Journal of Comparative Physiology A: Neuroethology, Sensory, Neural, and Behavioral Physiology 41（4）:405–434

Lindström A, Korpela S, Fries I（2008）Horizontal transmission of Paenibacillus larvae spores between honey bee（Apis mellifera）colonies through robbing. Apidologie 39（5）:515–522

Luby JM, Regnier FE, Clarke ET, Weaver EC, Weaver N（1973）Volatile cephalic substances of the stingless bees, Trigona mexicana and Trigona pectoralis. Journal of Insect Physiology 19（5）:1111–1127

Maia-Silva C, Hrncir M, Koedam D, Machado RJP, Imperatriz-Fonseca VL（2013）Out with the garbage: the parasitic strategy of the mantisfly Plega hagenella mass-infesting colonies of the eusocial bee Melipona subnitida in northeastern Brazil. Naturwissenschaften 100:101–105

Martin SJ, Shemilt S, Lima CBdS, de Carvalho CA（2017）Are isomeric alkenes used in species recognition among neo-tropical stingless bees（Melipona spp.）. Journal of Chemical Ecology 43（11-12）:1066–1072

Mascena VM, Nogueira DS, Silva CM, Freitas BM（2017）First record of the stingless bee Lestrimelitta rufa（Friese）（Hymenoptera: Apidae: Meliponini）in NE Brazil and its cleptobiotic behavior. Sociobiology 64（3）:359–362

McGlynn TP（2000）Do Lanchester's laws of combat describe competition in ants? Behavioral Ecology 11（6）:686–690

Medina L, Hart A, Ratnieks F（2009）Hygienic behavior in the stingless bees Melipona beecheii and Scaptotrigona pectoralis（Hymenoptera: Meliponini）. Genetics and Molecular Research 8（2）:571–576

Meeus I, Brown MJ, De Graaf DC, Smagghe G（2011）Effects of invasive parasites on bumble bee declines. Conservation Biology 25（4）:662–671

Menzel F, Blaimer BB, Schmitt T（2017）How do cuticular hydrocarbons evolve? Physiological constraints and climatic and biotic selection pressures act on a complex functional trait. Proceedings of the Royal Society B: Biological Sciences 284

（1850）:20161727

Messer AC（1985）Fresh dipterocarp resins gathered by megachilid bees inhibit growth of pollen-associated fungi. Biotropica 17:175–176

Michener CD（1946）Notes on the habits of some Panamanian stingless bees（Hymenoptera, Apidae）. Journal of the New York Entomological Society 54:179–197

Michener CD（1974）The Social Behavior of the Bees. Harvard University Press, Cambridge

Milborrow B, Kennedy J, Dollin A（1987）Composition of wax made by the Australian stingless bee Trigona australis. Australian Journal of Biological Sciences 40（1）:15–26

Miranda JR, Cordoni G, Budge G（2010）The acute bee paralysis virus–Kashmir bee virus–Israeli acute paralysis virus complex. Journal of Invertebrate Pathology 103:S30–S47

Month-Juris E, Ravaiano SV, Lopes DM, Salomão TMF, Martins GF（2020）Morphological assessment of the sensilla of the antennal flagellum in different castes of the stingless bee Tetragonisca fiebrigi. Journal of Zoology 310:110–125

Moore AJ, Breed MD, Moor MJ（1987）The guard honey bee: ontogeny and behavioural variability of workers performing a specialized task. Animal Behaviour 35:1159–1167

Moure JS, Nogueira-Neto P, Kerr WE（1958）Evolutionary problems among Meliponinae（Hymenoptera, Apidae）. Proceedings of the Xth International Congress of Entomology 2:481–493

Müller F（1874）The habits of various insects. Nature 10:102–103

Nagamitsu T, Inoue T（1997）Aggressive foraging of social bees as a mechanism of floral resource partitioning in an Asian tropical rainforest. Oecologia 110（3）:432–439

Nascimento D, Nascimento FS（2012）Acceptance threshold hypothesis is supported by chemical similarity of cuticular hydrocarbons in a stingless bee, Melipona asilvai. Journal of Chemical Ecology 38（11）:1432–1440

Neumann P, Elzen P（2004）The biology of the small hive beetle（Aethina tumida, Coleoptera: Nitidulidae）: Gaps in our knowledge of an invasive species. Apidologie 35（3）:229–247

Nogueira-Neto P（1970）Behavior problems related to the pillages made by some parasitic stingless bees（Meliponinae, Apidae）. In: Aronson LR（ed）Development and evolution of behavior: Essays in memory of TC Schneirla. W. H. Freeman San Francisco, California, pp 416–434

Nogueira-Neto P（1997）Vida e Criação de Abelhas Indígenas Sem Ferrão. Editora Nogueirapis, São Paulo

Nunes-Silva P, Imperatriz-Fonseca VL, Gonçalves LS（2009）Hygienic behavior of the stingless bee Plebeia remota（Holmberg, 1903）（Apidae, Meliponini）. Genetics and Molecular Research 8（2）:649–654

Nunes-Silva P, Piot N, Meeus I, Blochtein B, Smagghe G（2016）Absence of Leishmaniinae and Nosematidae in stingless bees. Scientific Reports 6:32547

Nunes TM, Mateus S, Turatti IC, Morgan ED, Zucchi R（2011）Nestmate recognition in the stingless bee Frieseomelitta varia（Hymenoptera, Apidae, Meliponini）: sources of chemical signals. Animal Behaviour 81:463–467

Nunes TM, Nascimento FS, Turatti IC, Lopes NP, Zucchi R（2008）Nestmate recognition in a stingless bee: does the similarity of chemical cues determine guard acceptance? Animal Behaviour 75:1165–1171

Nunes TM, Turatti IC, Lopes NP, Zucchi R（2009 a）Chemical signals in the stingless bee, Frieseomelitta varia, indicate caste, gender, age, and reproductive status. Journal of Chemical Ecology 35（10）:1172

Nunes TM, Turatti ICC, Mateus S, Nascimento FS, Lopes NP, Zucchi R（2009 b）Cuticular hydrocarbons in the stingless bee Schwarziana quadripunctata（Hymenoptera, Apidae, Meliponini）: differences between colonies, castes and age. Genetics and Molecular Research 8:589–595

Nunes TM, von Zuben LG, Costa L, Venturieri GC（2014）Defensive repertoire of the stingless bee Melipona flavolineata Friese（Hymenoptera: Apidae）. Sociobiology 61（4）:541–546

Ostwald MM, Ruzi SA, Baudier KM（2018）Ambush Predation of Stingless Bees（Tetragonisca angustula）by the Solitary-Foraging Ant Ectatomma tuberculatum. Journal of Insect Behavior 31:503–509

Ozaki M, Wada-Katsumata A, Fujikawa K, Iwasaki M, Yokohari F, Satoji Y, Nisimura T, Yamaoka R（2005）Ant nestmate and non-nestmate discrimination by a chemosensory sensillum. Science 309（5732）:311–314

Parizotto DR (2010) Morfologia externa da operária de Lestrimelitta ehrhardtti (Hymenoptera: Meliponini). Acta Biológica Colombiana 15 (2):131

Peña WL, Fonte Carballo L, Demedio Lorenzo J (2014) Reporte de Aethina tumida Murray (Coleoptera, Nitidulidae) en colonias de la abeja sin aguijón Melipona beecheii Bennett de Matanzas y Mayabeque. Revista de Salud Animal 36 (3):201–204

Penney D, Gabriel R (2009) Feeding behavior of trunk-living jumping spiders (Salticidae) in a coastal primary forest in the Gambia. Journal of Arachnology 37:113–115

Pianaro A, Flach A, Patricio EF, Nogueira-Neto P, Marsaioli AJ (2007) Chemical changes associated with the invasion of a Melipona scutellaris colony by Melipona rufiventris workers. Journal of Chemical Ecology 33 (5):971–984

Pioker-Hara FC, Drummond MS, Kleinert AMP (2014) The influence of the loss of Brazilian savanna vegetation on the occurrence of nests of stingless bees (Apidae: Meliponini). Sociobiology 61 (4):393–400

Pohl S, Foitzik S (2011) Slave-making ants prefer larger, better defended host colonies. Animal Behaviour 81 (1):61–68

Pompeu M, Silveira F (2005) Reaction of Melipona rufiventris Lepeletier to citral and against an attack by the cleptobiotic bee Lestrimelitta limao (Smith) (Hymenoptera: Apidae: Meliponina). Brazilian Journal of Biology 65 (1):189–191

Porrini MP, Porrini LP, Garrido PM, Porrini DP, Muller F, Nuñez LA, Alvarez L, Iriarte PF, Eguaras MJ (2017) Nosema ceranae in South American Native Stingless Bees and Social Wasp. Microbial Ecology 74 (4):761–764

Portugal-Araújo V (1958) A contribution to the bionomics of Lestrimelitta cubiceps (Hymenoptera, Apidae). Journal of the Kansas entomological Society 31:203–211

Purkiss T, Lach L (2019) Pathogen spillover from Apis mellifera to a stingless bee. Proceedings of the Royal Society London B 286:20191071

Quezada-Euán JJG, González-Acereto J (2002) Notes on the nest habits and host range of cleptobiotic Lestrimelitta niitkib (Ayala 1999) (Hymenoptera: Meliponini) from the Yucatan Peninsula, Mexico. Acta Zoológica Mexicana 86:245–249

Quezada-Euán JJG, Ramírez J, Eltz T, Pokorny T, Medina R, Monsreal R (2013) Does sensory deception matter in eusocial obligate food robber systems? A study of

Lestrimelitta and stingless bee hosts. Animal Behaviour 85:817–823

Rangel J, Griffin SR, Seeley TD (2010) Nest-Site Defense by Competing Honey Bee Swarms During House-Hunting. Ethology 116(7):608–618

Rasmussen C (2008) Catalog of the Indo-Malayan/Australasian stingless bees (Hymenoptera: Apidae: Meliponini). Zootaxa 1935:1–80

Rasmussen C, Gonzalez VH (2017) The neotropical stingless bee genus Nannotrigona Cockerell (Hymenoptera: Apidae: Meliponini): An illustrated key, notes on the types, and designation of lectotypes. Zootaxa 4299(2):191–220

Ratnieks FLW, Kärcher M, Firth V, Parks D, Richards A, Richards P, Helanterä H (2011) Acceptance by honey bee guards of non-nestmates is not increased by treatment with nestmate odours. Ethology 117:655–663

Rech AR, Schwade MA, Schwade MRM (2013) Abelhas-sem-ferrão amazônicas defendem meliponários contra saques de outras abelhas. Acta Amazonica 43:389–394

Reeve HK (1989) The evolution of conspecific acceptance thresholds. The American Naturalist 133:407–435

Rinderer TE, Blum MS, Fales HM, Bian Z, Jones TH, Buco SM, Lancaster VA, Danka RG, Howard DF (1988) Nest plundering allomones of the fire bee Trigona (Oxitrigona) mellicolor. Journal of Chemical Ecology 14:495–501

Rothenbuhler WC (1964) Behavior genetics of nest cleaning in honey bees. IV. Responses of F1 and backcross generations to disease-killed brood. American Zoologist 4(2):111–123

Roubik DW, Smith B, Carlson R (1987) Formic acid in caustic cephalic secretions of stingless bee, Oxytrigona (Hymenoptera: Apidae). Journal of Chemical Ecology 13(5):1079–1086

Roubik DW (1983) Nest and colony characteristics of stingless bees from Panama (Hymenoptera: Apidae). Journal of the Kansas Entomological Society 56:327–355

Roubik DW (1989) Ecology and Natural History of Tropical Bees. Cambridge University Press, New York

Roubik DW (2006) Stingless bee nesting biology. Apidologie 37:124–143

Sakagami SF, Inoue T, Yamane S, Salmah S (1983) Nest architecture and colony composition of the Sumatran stingless bee Trigona (Tetragonula) laeviceps. Kontyu 51(1):100–111

Sakagami SF, Roubik DW, Zucchi R (1993) Ethology of the robber stingless bee, Lestrimelitta limao (Hymenoptera: Apidae). Sociobiology 21:237–277

Sakagami SF (1982) Stingless bees. In: Hermann HR (ed) Social Insects III. Academic Press, New York, pp 361–423

Sakagami SF, Camilo C, Zucchi R (1973) Oviposition behavior of a Brazilian stingless bee, Plebeia (Friesella) schrottkyi, with some remarks on the behavioral evolution in stingless bees. Journal of the Faculty of Science Hokkaido University Series VI, Zoology 19 (1):163–189

Sakagami SF, Laroca S (1963) Additional observations on the habits of the cleptobiotic stingless bees, the genus Lestrimelitta Friese (Hymenoptera, Apoidea). Journal of the Faculty of Science Hokkaido University 15:319–339

Santos CF, Menezes C, Vollet-Neto A, Imperatriz-Fonseca VL (2014) Congregation sites and sleeping roost of male stingless bees (Hymenoptera: Apidae: Meliponini). Sociobiology 61 (1):115–118

Schmidt VM, Schorkopf DLP, Hrncir M, Zucchi R, Barth FG (2006) Collective foraging in a stingless bee: dependence on food profitability and sequence of discovery. Animal Behaviour 72:1309–1317

Schorkopf DLP (2016) Male meliponine bees (Scaptotrigona aff. depilis) produce alarm pheromones to which workers respond with fight and males with flight. Journal of Comparative Physiology A 202:667–678

Schorkopf DLP, Hrncir M, Mateus S, Zucchi R, Schmidt VM, Barth FG (2009) Mandibular gland secretions of meliponine worker bees: further evidence for their role in interspecific and intraspecific defence and aggression and against their role in food source signalling. Journal of Experimental Biology 212 (8):1153–1162

Schorkopf DLP, Jarau S, Francke W, Twele R, Zucchi R, Hrncir M, Schmidt VM, Ayasse M, Barth FG (2007) Spitting out information: Trigona bees deposit saliva to signal resource locations. Proceedings of the Royal Society of London Series B-Biological Sciences 274:895–898

Schwarz HF (1948) Stingless Bees (Meliponidae) of the Western Hemisphere. Bulletin of the American Museum of Natural History 90:1–546

Seeley TD (1995) The wisdom of the hive: The social physiology of honey bee colonies. Harward University Press, Cambridge

Segers FHID, Menezes C, Vollet-Neto A, Lambert D, Grüter C (2015) Soldier production in a stingless bee depends on rearing location and nurse behaviour. Behavioral Ecology and Sociobiology 69:613–623

Segers FHID, Von Zuben LG, Grüter C (2016) Local differences in parasitism and competition shape defensive investment in a polymorphic eusocial bee. Ecology 97:417–426

Shackleton K, Al Toufailia H, Balfour NJ, Nascimento FS, Alves DA, Ratnieks FLW (2015) Appetite for self-destruction: suicidal biting as a nest defense strategy in Trigona stingless bees. Behavioral Ecology and Sociobiology 69:273–281

Shackleton K, Alves DA, Ratnieks FLW (2018) Organization enhances collective vigilance in the hovering guards of Tetragonisca angustula bees. Behavioral Ecology 29:1105–1112

Shackleton K, Balfour NJ, Toufailia HA, Alves DA, Bento JM, Ratnieks FLW (2019) Unique nest entrance structure of Partamona helleri stingless bees leads to remarkable 'crash-landing' behaviour. Insectes Sociaux 66:471–477

Shanks JL, Haigh AM, Riegler M, Spooner-Hart RN (2017) First confirmed report of a bacterial brood disease in stingless bees. Journal of Invertebrate Pathology 144:7–10

Shorter J, Rueppell O (2012) A review on self-destructive defense behaviors in social insects. Insectes Sociaux 59(1):1–10

Simões D, Bego LR, Zucchi R, Sakagami SF (1980) Melaloncha sinistra Borgmeier, an endoparasitic phorid fly attacking Nannotrigona (Scaptotrigona) postica Latreille (Hymenoptera, Meliponinae). Revista Brasileira de Entomologia 24:137–142

Simone-Finstrom M, Spivak M (2010) Propolis and bee health: the natural history and significance of resin use by honey bees. Apidologie 41(3):295–311

Slaa EJ (2006) Population dynamics of a stingless bee community in the seasonal dry lowlands of Costa Rica. Insectes Sociaux 53:70–79

Smith B, Roubik D (1983) Mandibular glands of stingless bees (Hymenoptera: Apidae): Chemical analysis of their contents and biological function in two species of Melipona. Journal of Chemical Ecology 9(11):1465–1472

Solórzano-Gordillo E, Rojas JC, Cruz-López L, Sánchez D (2018) Associative learning of non-nestmate odor marks between colonies of the stingless bee Scaptotrigona mexicana Guérin (Apidae, Meliponini) during foraging. Insectes Sociaux 65:393–400

Sommeijer M, De Bruijn L (1995) Drone congregations apart from the nest in Melipona favosa. Insectes Sociaux 42 (2):123–127

Sommeijer MJ (1999) Beekeeping with stingless bees: a new type of hive. Bee World 80 (2):70–79

Sommeijer MJ, de Bruijn LLM, Meeuwsen FJAJ (2004) Behaviour of males, gynes and workers at drone congregation sites of the stingless bee Melipona favosa(Apidae: Meliponini). Entomologische Berichten 64:10–15

Souza FS, Kevill JL, Correia-Oliveira ME, de Carvalho CAL, Martin SJ (2019) Occurrence of deformed wing virus variants in the stingless bee Melipona subnitida and honey bee Apis mellifera populations in Brazil. Journal of General Virology 100:289–294

Spivak M, Masterman R, Ross R, Mesce KA (2003) Hygienic behavior in the honey bee (Apis mellifera L.) and the modulatory role of octopamine. Developmental Neurobiology 55 (3):341–354

Spottiswoode CN, Begg KS, Begg CM (2016) Reciprocal signaling in honeyguide-human mutualism. Science 353 (6297):387–389

Stangler ES, Jarau S, Hrncir M, Zucchi R, Ayasse M (2009) Identification of trail pheromone compounds from the labial glands of the stingless bee Geotrigona mombuca. Chemoecology 19 (1):13–19

Suka T, Inoue T(1993) Nestmate recognition of the stingless bee Trigona(Tetragonula) minangkabau (Apidae: Meliponinae). Journal of Ethology 11 (2):141–147

Tan NQ (2007) Biology of Apis dorsata in Vietnam. Apidologie 38:221–229

Tapia-González JM, Morfin N, Macías-Macías JO, De la Mora A, Tapia-Rivera JC, Ayala R, Contreras-Escareño F, Gashout HA, Guzman-Novoa E (2019) Evidence of presence and replication of honey bee viruses among wild bee pollinators in subtropical environments. Journal of Invertebrate Pathology 168:107256

Teixeira ÉW, Ferreira EA, Luz CFP da, Martins MF, Ramos TA, Lourenço AP (2020) European Foulbrood in stingless bees (Apidae: Meliponini) in Brazil: Old disease, renewed threat. Journal of Invertebrate Pathology 172:107357

Uboni A, Bagnères A-G, Christidès J-P, Lorenzi MC (2012) Cleptoparasites, social parasites and a common host: chemical insignificance for visiting host nests, chemical mimicry for living in. Journal of Insect Physiology 58 (9):1259–1264

Ueira-Vieira C, Almeida LO, De Almeida FC, Amaral IMR, Brandeburgo MAM, Bonetti AM (2015) Scientific note on the first molecular detection of the acute bee paralysis virus in Brazilian stingless bees. Apidologie 46(5):628–630

van Oystaeyen A, Alves DA, Oliveira RC, Nascimento DL, Nascimento FS, Billen J, Wenseleers T (2013) Sneaky queens in Melipona bees selectively detect and infiltrate queenless colonies. Animal Behaviour 86:603–609

van Oystaeyen A, Oliveira RC, Holman L, van Zweden JS, Romero C, Oi CA, D'Ettorre P, Khalesi M, Billen J, Wäckers F, Millar JG, Wenseleers R (2014) Conserved class of queen pheromones stops social insect workers from reproducing. Science 343:287–290

van Wilgenburg E, Elgar MA (2013) Confirmation bias in studies of nestmate recognition: a cautionary note for research into the behavior of animals. PLoS ONE 8:e53548

van Zweden JS, D'Ettorre P (2010) Nestmate recognition in social insects and the role of hydrocarbons. In: Blomquist GJ, Bagnères A-G (eds) Insect Hydrocarbons: Biology, Biochemistry, and Chemical Ecology. Cambridge University Press, Cambridge, pp 222–243

van Zweden JS, Grüter C, Jones SM, Ratnieks FLW (2011) Hovering guards of the stingless bee Tetragonisca angustula increase colony defensive perimeter as shown by intra- and inter-specific comparisons. Behavioral Ecology and Sociobiology 65:1277–1282

Velez-Ruiz RI, Gonzales VH, Engel MS (2013) Observations on the urban ecology of the Neotropical stingless bee Tetragonisca angustula (Hymenoptera: Apidae: Meliponini). Journal of Melittology 1:1–8

Venturieri G (2009) The impact of forest exploitation on Amazonian stingless bees (Apidae, Meliponini). Genetics and Molecular Research 8(2):684–689

von Frisch K (1967) The dance language and orientation of bees. Harvard University Press, Cambridge

von Zuben L, Schorkopf DLP, Elias L, Vaz A, Favaris A, Clososki G, Bento J, Nunes T (2016) Interspecific chemical communication in raids of the robber bee Lestrimelitta limao. Insectes Sociaux 63:339–347

von Zuben LG, Nunes TM (2014) A scientific note on the presence of functional tibia for pollen transportation in the robber bee Lestrimelitta limao Smith (Hymenoptera: Apidae: Meliponini). Sociobiology 61(4):570–572

Wang S, Wittwer B, Heard TA, Goodger JQD, Elgar MA (2018) Nonvolatile chemicals provide a nest defence mechanism for stingless bees Tetragonula carbonaria (Apidae, Meliponini). Ethology 124:633–640

Weaver N, Weaver EC, Clarke ET (1975) Reactions of five species of stingless bees to some volatile chemicals and to other species of bees. Journal of Insect Physiology 21 (3):479–494

Whitehouse MEA, Jaffe K (1996) Ant wars: combat strategies, territory and nest defence in the leaf-cutting ant Atta laevigata. Animal Behaviour 51 (6):1207–1217

Wille A (1983) Biology of the stingless bees. Annual Review of Entomology 28:41–64

Wille A, Michener C (1973) The nest architecture of stingless bees with special reference to those of Costa Rica. Revista de Biologia Tropical 21:9–278

Wittmann D (1985) Aerial defense of the nest by workers of the stingless bee Trigona (Tetragonisca) angustula. Behavioral Ecology and Sociobiology 16:111–114

Wittmann D, Ratke R, Zeil J, Lübke G, Francke W (1990) Robber bees (Lestrimelitta limao) and their host chemical and visual cues in nest defence by Trigona (Tetragonisca) angustula (Apidae: Meliponinae). Journal of Chemical Ecology 16:631–641

Wittwer B, Elgar MA (2018) Cryptic castes, social context and colony defence in a social bee, Tetragonula carbonaria. Ethology 124:617–622

Zeil J, Wittmann D (1989) Visually controlled station-keeping by hovering guard bees of Trigona (Tetragonisca) angustula (Apidae, Meliponinae). Journal of Comparative Physiology A 165 (5):711–718

8 无刺蜂的采集

无刺蜂为了饲喂幼虫、建造蜂巢结构以及保卫蜂群需要采集不同类型的资源。无刺蜂的采集对热带生态系统产生了重大影响，因为无刺蜂在收集花粉（幼虫的主要蛋白质来源和花蜜和碳水化合物来源）的过程中也实现了授粉。全世界有成千上万的植物物种可能都受益于无刺蜂授粉（第9章）。无刺蜂与蜜蜂在采集上有明显不同，无刺蜂采集的非花资源（如树脂材料、果汁、腐肉；见下文）也占了很大比例（Roubik,1989; Lorenzon and Matrangolo,2005）。

8.1 无刺蜂采集的物质

8.1.1 蛋白质

花粉

花粉是无刺蜂的主要蛋白质来源。一些花只会为无刺蜂提供花粉，其他一些花则会既提供花粉又提供花蜜（Vogel,1983; Roubik,1989）。然而，花粉不仅能为无刺蜂提供蛋白质（占干重的10%～60%）和氨基酸，花粉中还含有水分（含量通常约为20%，最高可达50%）、维生素（如维生素A、维生素B、维生素C、维生素D和E族维生素）、碳水化合物（含量高达干重的40%）、脂类（含量为1%～20%）、萜类、类固醇和类胡萝卜素（Solberg and Remedios,1980; Vogel,1983; Roubik,1989; Roulston and Cane,2000; Vossler et al., 2010; Vossler,2015）。类胡萝卜素、萜类和脂类物质通常存在于花粉的最外层，即花粉壁中（Roulston and Cane, 2000）。而在无刺蜂采集的花粉中发现的碳水化合物则是源自无刺蜂反刍到收集的花粉团里的花蜜或蜂蜜（见下文）（Solberg and Remedios,1980; Roulston and Cane,2000）。花朵的花粉有一种特有的气味，这种气味可以被蜂类学习并被随后用来定位相同类型的花朵（这方面的研究主要集中在蜜蜂和熊蜂上：von Frisch,1967; Vogel,1983; Dobson,1987; Dobson et al., 1996; Pernal and Currie,2002; Arenas and Farina,2012, 2014; Muth et al., 2016 a, 2016 b）。

无刺蜂采集的花粉的营养成分在蜂种内高度保守，但不同植物物种的花粉的营养成分则有很大的差异（Roulston et al., 2000; Roulston and Cane,2000）。无刺蜂采集蛋白质含量较高的花粉可能对其有益，例如，食用蛋白质含量较高的花粉能够促进无刺蜂的体型增大（Quezada-Euán et al., 2011）。目前还不清楚无刺蜂是否根据花粉的营养价值来区分不同类型的花粉。一般来说，无刺蜂似乎不太看重花粉本身的蛋白质含量（Roulston et al., 2000; Pernal and Currie,2002; Vossler,2015; Beekman et al., 2016; Zarchin et al., 2017），而且我们已经知道蜜蜂偶尔会采集花粉以外的非营养物质，如砖头、碳和锯末等（von Frisch,1967; Shaw,1990）。然而，蜜蜂表现出了对含有大量营养素的花粉的偏好，这些花粉中的大量营养素与蜂群过去采集物质的大量营养素是互补的（Hendriksma and Shafir,2016; Zarchin et al., 2017），而且有证据表明花粉来源的多样性越高越能增强蜜蜂健康（Alaux et al., 2010; Pasquale et al., 2013; Dolezal et al., 2019）。目前尚不清楚无刺蜂蜂群是否也会通过获得多样化的花粉饮食来平衡其营养需求。

无刺蜂的前基跗节和长吻可帮助其从花朵的花粉囊中取下花粉（Michener et al., 1978）。大体型的 *Melipona* 属无刺蜂可以在抓住花粉囊时快速且重复地收缩其飞行肌肉来提取花粉。这种行为使花粉从花朵的花粉囊中释放出来，并伴随着出现无刺蜂特有的嗡嗡声，这就是为什么人们称其为嗡嗡采集或嗡嗡授粉的原因（Wille,1963; Sommeijer et al., 1983; Roubik,1989; Nunes-Silva et al., 2013; Vallejo-Marín,2019）。也存在一些无刺蜂（主要是 *Trigona* 属无刺蜂）为了获得花粉（或花蜜）而破坏花朵的花粉囊或其他花朵结构的情况，无刺蜂采用这种所谓的花粉（或花蜜）抢夺行为时通常不为植物授粉（Willmer and Corbet,1981; Roubik,1982 b; Renner,1983; Roubik et al., 1985; Roubik,1989; Tezuka and Maeta,1995; Murphy and Breed,2008 a; Rego et al., 2018）。无刺蜂进行花粉（或花蜜）抢夺行为后形成的洞而后可以被其他蜂种或蜂鸟使用（Roubik,1989；表 2.5）。*Trigona* 属无刺蜂能这样抢夺花粉和花蜜行为的一个有利形态特征是它们的上颚十分强壮且具有齿结构（图7.1）（Shackleton et al., 2015）。这种可以有效应对天敌的上颚也是穿透花粉囊的理想工具。无刺蜂这种花粉或花蜜抢夺行为因为阻碍了植物的有效授粉者——蜂鸟等对其进行授粉，进而减少了植物果实的形成，从而影响了植物的继代繁衍（McDade and Kinsman,1980; Roubik,1982 b; Murphy and Breed,2008 a）。Rego 等（2018）提出 *Trigona fulviventris* 无刺蜂花粉抢夺行为对大西洋雨林中 *Eriocnema fulva* 造成了极大的负面影响，再加上 *Eriocnema fulva* 的栖息地也在不断丧失，因此使 *Eriocnema fulva* 濒临灭绝。

无刺蜂主要是用前足和中足将身体不同部位的花粉粒转移到花粉筐中的（关于无刺蜂梳理行为的详细描述见 Michener et al., 1978），无刺蜂在花粉筐中将花粉粒与自己反刍的液体混合在一起。将花粉转移到花粉筐的过程多发生在无刺蜂飞悬的时候，因为那样无刺蜂就可以腾出中足和后足来进行花粉粒转移（Schwarz,1948; Wille,1962; Roubik,1982 a; Baumgartner and Roubik,1989; Camargo and Roubik,1991; Noll et al., 1996; Breed et al., 1999）。为了实现这一目的，无刺蜂的前足以及后足和中足的跗节和胫节上都有刷状或梳状结构（Roubik,1989）。而到了蜂巢内部，携带花粉的蜂只需要移动中足，花粉包就会掉进花粉罐中（Roubik,1989）。

动物蛋白

目前研究者已经发现一些无刺蜂蜂种会采集动物（如蜥蜴、青蛙、鸟类和各种大小的哺乳动物等）尸体上的肉（Schwarz,1948; Wille,1962; Roubik,1982 a; Baumgartner and Roubik,1989; Camargo and Roubik,1991; Noll et al., 1996; Breed et al., 1999; Mateus and Noll,2004）。例如，Baumgartner 和 Roubik（1989）曾在秘鲁不同的无刺蜂栖息地放置了死鱼和牛肝诱饵，而后他们发现约 30 种无刺蜂的采集蜂访问了这些诱饵。这种采肉无刺蜂中有些可能是用肉类作为筑巢材料，有些是将肉类作为蜂群的盐分来源，但有些采肉无刺蜂则已经完全不再使用花粉蛋白，转而永久地将从动物尸体或被遗弃的昆虫幼虫采集来的蛋白作为蜂群的蛋白来源，至少 *Trigona* 属的 3 个亲缘关系很近的蜂种（*Trigona crassipes*、*T. hypogea* 和 *T. necrophaga*）就是这种情况（Roubik,1982 a; Camargo and Roubik,1991; Noll,1997; Mateus and Noll,2004）。无刺蜂使用动物蛋白的一个好处可能是动物蛋白比花粉粒中的蛋白质更容易被消化（Cruz-Landim and Serrao,1994）。

食肉无刺蜂通常不会采集已经死了一段时间的或被苍蝇幼虫侵袭过的动

物尸体（Roubik,1982a；Roubik,1989）。*Trigona* 属无刺蜂中 3 种食肉性的蜂种的工蜂的花粉筐都已经退化了（图 8.1）（这些蜂种仍然会使用花粉筐来采集树脂，但不会使用该结构来采集花粉），它们的口器上也缺少典型的帮助采集蜂从花朵花粉囊中提取花粉的毛发（Roubik,1982 a, 1989; Camargo and Roubik,1991）。肉食性无刺蜂的幼虫食物中腺体分泌物占了很大一部分，其腺体分泌物中的蛋白质含量约为 20%。其蜂巢内的储物罐中储存的也不是花粉，而是这些富含蛋白质的分泌物（Roubik,1982 a）。*Trigona hypogea* 无刺蜂采集蜂一般不去花丛中寻找花蜜，而是采集果汁或从花外蜜腺中采集富含糖分的液体（Noll et al., 1996）。这表明食肉性无刺蜂已经不属于授粉蜂的范畴（Camargo and Roubik,1991）。

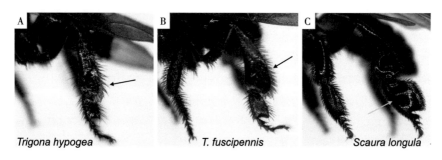

图 8.1 3 种无刺蜂的后足

注：新热带区专性食尸性的 *Trigona hypogea* 无刺蜂（A）和 *T. fuscipennis* 无刺蜂（B）的花粉采集蜂的后足。与 *T.fuscipennis* 无刺蜂相比，*T.hypogea* 无刺蜂采集蜂的花粉筐已经退化了（黑色箭头）。图 C 显示 *Scaura longula* 无刺蜂后足有囊状后基跗节结构（箭头），这种结构可能有助于该种无刺蜂采集蜂收集掉在地上或植被上的花粉（摄影：Katja Aleixo）。

3 种肉食性 *Trigona* 属无刺蜂（第 10 章）都具有有效的招募行为（Noll,1997; Jarau et al., 2003），这意味着蜂群可以霸占着腐肉，而且它们锋利的上颚齿使它们能在几个小时内把一个小尸体变成骨架。Roubik（1982 a）观察到无刺蜂会使用头部侧腺分泌的消化酶先消化腐肉。而且研究者已经观察到这些分泌物可以逼退蚂蚁和苍蝇（Noll,1997）。然后无刺蜂会用它们的蜜囊携带液化的食物回到蜂巢内，然后再将其反刍出来。研究者在无刺蜂的腺体分泌物和幼虫食物中发现了一些有益微生物（几种芽孢杆菌），他们认为这些微生物在代谢蛋白质和产生抗生素等方面具有重要作用（Gilliam et al., 1985; Camargo and Roubik,1991; Noll et al., 1996）。

真菌孢子

一些无刺蜂蜂种的采集蜂会食用真菌孢子（Roubik,1989; Oliveira and Morato,2000）。已经有研究者观察到亚马孙地区的几种 *Trigona* 属无刺蜂会舔食含有臭角真菌（*Phallaceae*）产孢体的孢子，有些无刺蜂还会把孢子放在花粉筐里带回蜂巢（Burr et al., 1996; Oliveira and Morato,2000）。Oliveira 和 Morato（2000）认为真菌孢子是 *Trigona crassipes* 无刺蜂的一种重要的蛋白质来源。还有研究者观察到亚洲的 *Tetragonilla collina* 无刺蜂会采集霉菌 *Rhizopus*（*Mucoraceae*）的孢子（Eltz et al., 2002）。总的来说，人们对真菌孢子是否是无刺蜂的重要蛋白质来源知之甚少，但相关研究发现一些 *T. collina* 无刺蜂蜂群会连续几天采集大量的霉菌孢子，这表明真菌孢子可能短时间内是这些无刺蜂蜂群食物的重要组成部分（Eltz et al., 2002）。而无刺蜂的幼虫发育是否受到真菌孢子食物的影响也还不清楚（Shaw,1990; Menezes et al., 2015）。研究者认为真菌孢子与许多类型的花粉相比，蛋白质含量略低，但如果在某些时期真菌孢子数量十分充足时，它们也可以成为能吸引无刺蜂的花粉替代品（Shaw,1990; Eltz et al., 2002）。

8.1.2 碳水化合物

花蜜

许多植物会分泌花蜜来吸引和回馈授粉者。花蜜是包括大多数无刺蜂在内的许多蜂类的主要碳水化合物来源。花蜜中主要糖类有 3 种，即一种二糖——蔗糖和两种单糖——果糖和葡萄糖（Vogel,1983; Chalcoff et al., 2017），其中蔗糖是大多数花蜜中最主要的糖类（Chalcoff et al., 2017）[1]。此外，花蜜中可能还有其他几种痕量的单糖、双糖和寡糖，这些糖包括了甘露糖（在较高的浓度下对蜜蜂和无刺蜂有害，Zucoloto and Penedo,1977）、戊糖、麦芽糖、蜜二糖以及棉籽糖。这些糖通常以特定的含量存在于特定植物物种的花蜜中。目前还不清楚无刺蜂是否偏好于某些含有特定的糖分比例的花蜜，但 Biesmeijer 等（1999a）发现 *Melipona beecheii* 和 *M. fasciata* 无刺蜂对蔗糖的偏好程度超过了葡萄糖和果糖，蜜蜂似乎也是这样（Değirmenci et al., 2018）。然而，蜜蜂的选择试验也表明花蜜中的糖分组成不如糖浓度重要（Vogel,1983）。花蜜还可能含有少量的酵母、碳酸、脂质、蛋白质、氨基酸、维生素（如维

---

[1] 请注意，这与无刺蜂的蜂蜜不同，在无刺蜂蜂蜜中只有少量的蔗糖（第 1 章），这表明在其蜂蜜生产过程中蔗糖被生物化学转化成了其他类型的糖。

生素C)、碱性金属(如钾)和生物碱如咖啡因、尼古丁等(Vogel,1983; Kretschmar and Baumann,1999; Singaravelan et al., 2005)。这些化合物中的一些可能是微生物活动产物、植物新陈代谢副产物，还可能是一些花朵的特定访花者留下的吸引物(Vogel,1983; Kim and Smith,2000; Afik et al., 2014; Couvillon et al., 2015)。例如，花蜜会因含有的咖啡因(存在于咖啡和柑橘的花蜜中)浓度不同而对采集蜂产生吸引作用或阻止访花的作用(Kretschmar and Baumann,1999; Singaravelan et al., 2005; Wright et al., 2013; Couvillon et al., 2015; Thomson et al., 2015)。植物可能会产含低浓度咖啡因的花蜜来提高授粉者的访花量(Couvillon et al., 2015)，或通过产含高浓度咖啡因的花蜜来阻止授粉者访花(Singaravelan et al., 2005)。目前我们对这些化合物对无刺蜂采集行为的影响知之甚少(Leonhardt,2017)，但有研究发现咖啡因并不会影响巴西 *Plebeia droryana* 无刺蜂的采集行为(Peng et al., 2019)。*P. droryana* 无刺蜂对咖啡因的响应缺失可能是其对咖啡因潜在有害作用产生了适应的结果(Couvillon et al., 2015)，也可能提示无刺蜂和蜜蜂在神经生理学上存在差异(Peng et al., 2019)。

研究者发现一些无刺蜂蜂种的采集行为是不受花蜜中氨基酸的影响的(Roubik et al., 1995; Gardener et al., 2003)，但是 *Melipona fuliginosa* 和 *Melipona panamica* 的采集蜂会拒绝采食含有丙氨酸、精氨酸、谷氨酸、甘氨酸、脯氨酸和丝氨酸的蔗糖溶液(Roubik et al., 1995)。当糖溶液中的钾含量接近或高于鳄梨(*Persea americana*)花蜜中自然水平的钾含量时，*Frieseomelitta nigra* 和 *Nannotrigona perilampoides* 对这种糖溶液的采食量也会降低(Afik et al., 2014)。

蜂类能感知糖是一种奖励，它们用触角或跗节触碰到糖时会诱发其吻部的反射性伸长(伸吻反射，proboscis extension response，PER)(Takeda,1961; Menzel,1999; Scheiner et al., 2004)。能够诱发一只蜂产生伸吻反射的糖浓度(蔗糖反射阈值)可以提供该蜂的味觉敏感度和对植物糖奖励的反应等相关信息。Mc Cabe等(2007)比较了巴西的两种新热带无刺蜂 *Melipona quadrifasciata* 和 *Scaptotrigona depilis* 与非洲化蜜蜂的糖反射情况，发现诱发这3个蜂种的伸吻反射的最低糖浓度平均都在3%～10%。研究者发现非洲的 *Meliponula bocandei* 和 *Axestotrigona ferruginea* 无刺蜂的蔗糖敏感性也与之相似。相反，大多数 *Melipona eburnea* 无刺蜂即使面对10%的蔗糖溶液也没有表现出PER(Amaya-Márquez et al., 2019)。一只蜂的蔗糖反射阈值(sucrose response threshold，SRT）在其一生中也并不是固定不变的，而是会随日龄、前期经验、花蜜的可利用度以及季节发生变化（蜜蜂：Lindauer,1948; Scheiner et al., 2003; Pankiw et al., 2004。无刺蜂：Mc Cabe et al., 2007）。

我们对无刺蜂感知糖奖励涉及的神经生物学基础知之甚少。对蜜蜂（和其他昆虫）而言，如辛胺和多巴胺等生物胺是其感知和学习糖奖励的重要媒介（Mercer and Menzel,1982; Scheiner et al., 2002; Perry and Barron,2013）。有证据表明，辛胺在无刺蜂感知糖奖励方面也起着类似的作用。*Melipona scutellaris* 无刺蜂采集蜂在摄取辛胺后对蔗糖表现出更高的敏感性（Mc Cabe et al., 2017），而 *Plebeia drorayna* 无刺蜂采集蜂在摄取辛胺后采集速度和招募强度都有所增加（Peng et al., 2020）。

大多数植物提供的花蜜的糖浓度都是高于蜂类的探测阈值的（Gu et al., 1977；Southwick et al., 1981；Vogel，1983；Roubik and Buchmann，1984；Roubik，1989；Chalcoff et al., 2006，2017）。例如，巴拿马地区无刺蜂采集的液体糖浓度从 9% 到 72% 不等（Roubik and Buchmann,1984; Roubik et al., 1986; Roubik et al., 1995），巴西无刺蜂采集的液体的糖浓度则从 0%（全是水）到 75% 不等（图 8.2）（I'Anson Price,2018），乌干达地区的无刺蜂采集的液体的糖浓度则在 9% ~ 67%（Kajobe,2007 a）。无刺蜂采集的液体的糖浓度会影响采集蜂在一次采集活动中采集的液体的总量[2]。一些无刺蜂蜂种的采集蜂在采集到较甜的食物时会将蜜囊填得更满一些（*Melipona triplaridis*），还有一些蜂种的采集蜂则是在遇到糖含量越高的食物时反而采集的量更少（*M. fuliginosa* 和 *Scaptotrigona depilis*）。甚至有一些无刺蜂蜂种采集蜂的蜜囊携糖量不受食物糖含量的影响（*Melipona micheneri* 和 *M. fasciata*）（Roubik and Buchmann,1984; Schmidt et al., 2006）。我们对无刺蜂这种有趣的现象仍不甚明了，但这可能与特定蜂种的招募行为有关：能够招募同巢蜂的采集蜂在找到优质食物源后，可能会半载而归以便更快地招募更多的同巢蜂到这个食物源（"信息传递假说"，Roces and Núñez,1993; Dornhaus et al., 2006; Bollazzi and Roces, 2011）。

---

［2］即使采集蜂的蜜囊可以容纳的液体量可高达采集蜂体重的 50%（Roubik and Buchmann,1984; Schmidt et al., 2006），但是事实上采集蜂通常是未填满蜜囊就返回蜂巢了（蜜蜂：Schmid-Hempel et al., 1985），这很可能是因为如果采集蜂在采集时填满蜜囊会增加它们的能量损耗和时间成本。

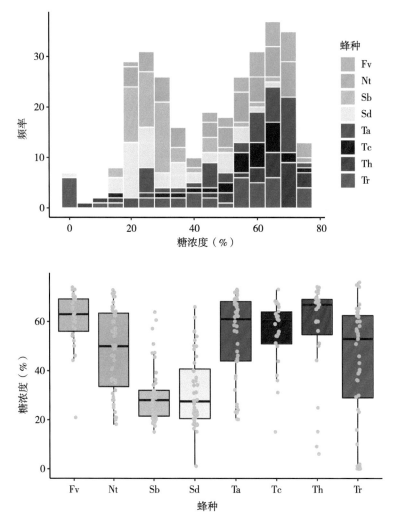

图 8.2　8 种无刺蜂采集蜂的蜜囊中获得的液体食物的糖浓度
（I'Anson Price，2018）

注：Fv = *Frieseomelitta varia*, Nt = *Nannotrigona testaceicornis*, Sb = *Scaptotrigona bipunctata*, Sd = *Scaptotrigona depilis*, Ta = *Tetragonisca angustula*, Tc = *Tetragona clavipes*, Th = *Trigona hypogea*, Tr = *Trigona recursa*.

研究者猜测无刺蜂采集蜂在其采集决策中会同时考虑蜂群的收益和支出（von Frisch,1967; Heinrich,1975; Schmid-Hempel et al., 1985; Roubik,1989;

Seeley,1995）[3]。而正如人们所猜想的那样，无刺蜂更偏好含糖量高的食物（Roubik and Buchmann,1984; Biesmeijer et al., 1999; Biesmeijer and Ermers,1999; León et al., 2015; Peng et al., 2019; Silva et al., 2019）：在其他条件相同的情况下，较甜的食物来源会增加无刺蜂的蜂群招募（第10章）、采集节奏（例如减少蜂巢停留时间），还会增加采集蜂返回该食物并积极捍卫该食物的概率（Johnson and Hubbell,1974; Biesmeijer and Ermers,1999; León et al., 2015; Schorkopf et al., 2016; Krausa et al., 2017; Peng et al., 2019）。然而，糖浓度过高的食物黏度也过高，致使其更难以被无刺蜂利用（von Frisch,1967; Vogel,1983; Roubik and Buchmann,1984; Schmidt et al., 2006; Kim et al., 2011）。鉴于这种物理阻隔的存在，因此对于无刺蜂来说存在一种最佳糖浓度范围，而这种最佳糖浓度低于天然花蜜中发现的最大糖浓度。包括无刺蜂和蜜蜂在内的大多数蜂类都是使用"黏滞插管"（将长吻插入花蜜）来饮蜜的。鉴于蜂类这种饮蜜方式，研究者认为适宜无刺蜂的最佳糖浓度是55%～60%，而且无刺蜂的体型大小与适宜无刺蜂的最佳糖浓度无关（Roubik and Buchmann,1984; Biesmeijer et al., 1999 a; Kim et al., 2011）。然而，无刺蜂的体型大小可能与其饮蜜速度呈正相关关系（Roubik，1989；表2.3）。例如，体型较大的 *Melipona fuliginosa* 无刺蜂（体重约125 mg）吸取27%～30%的蔗糖溶液的速度是2.6 μL/s，而体型小很多的 *Trigona pallens* 无刺蜂（体重约10 mg）的饮蜜速度则仅为0.3 μL/s（Roubik,1989）。蜂类还会使用技巧来摄取浓度更高甚至是干燥的糖类。例如，众所周知，蜜蜂会在干燥的花蜜中加入自己的唾液，然后吸食糖溶液（Simpson and Riedel,1964）。

尽管适宜无刺蜂的最佳糖浓度为55%～60%（Kim et al., 2011），但是不同无刺蜂蜂种采集的花蜜的糖含量却存在差异，一些蜂种会始终采集糖浓度很低的液体（图8.2）（Roubik et al., 1995; Biesmeijer et al., 1999; I' Anson Price,2018）。但是，*Melipona beecheii* 无刺蜂采集的花蜜浓度就比同域共生的 *M. fasciata* 无刺蜂采集的花蜜浓度更高，即使这两种无刺蜂的花蜜都是

---

[3] 研究者很少关注到无刺蜂采集蜂优先考虑哪个蜜源的问题。蜜蜂对蜜源区质量的评估则是基于净能量率[（收益-成本）/成本]而不是净能量摄入速率[（收益-成本）/时间]（Schmid-Hempel et al., 1985; Seeley,1986），这可能是因为采集蜂运输食物的代谢成本相对较高（Wolf and Schmid-Hempel, 1989）。其他可能影响首选蜜源抉择的因素还有蜜源地的天敌情况和竞争风险，或者是给同巢蜂的信息值（Núñez,1982; Nieh,2010）。采集蜂和同巢蜂间传递的信息量可能会影响蜂群在一个蜜源区中采集的持续时间。

从相同的植物上采集来的（Biesmeijer et al., 1999, 1999 b）。同样，Roubik 和 Buchmann（1984）发现 *Melipona compressipes* 和 *M. marginata* 无刺蜂采集的花蜜的糖含量也是高于 *M. fasciata* 和 *M. fuliginosa* 无刺蜂采集的花蜜的。巴西的 *Scaptotrigona bipunctata* 与 *S. depilis* 无刺蜂采集花蜜的糖浓度（20%～30%）就远低于它们的 6 个同域无刺蜂蜂种采集的花蜜的糖浓度（图 8.2）（I'Anson Price,2018）。*Scaptotrigona depilis* 无刺蜂也比较独特，它的招募行为并不太依赖于蔗糖浓度本身（第 10 章）（Schmidt et al., 2006）。这种现象进一步表明还有其他因素决定了食物源对无刺蜂的吸引力，这些因素包括食物源距离、花蜜流速、竞争情况或采集蜂形态限制（如蜂的体型大小、颜色或吻长度等方面的不同）（第 8.6 节）。

植物汁液和水果汁液

植物汁液或水果汁液是一些无刺蜂重要的碳水化合物来源，特别是在蜜粉资源匮乏的时期（图 8.3）（Wille,1962; Roubik,1989; Figueiredo,1996; Koch et al.,2011; Vijayakumar and Jeyaraaj,2016; Santos et al., 2019）。植物汁液不仅含有糖类，还含有氨基酸、蛋白质、维生素和矿物质（Roubik,1989）。无刺蜂采集的植物汁液里有蚜虫、介壳虫（*Coccidae* 和 *Pseudococcidae*）和角蝉等分泌的蜜露，有蜂直接从植物的开口处采集的（Roubik,1989; Camargo and Pedro,2002; Santos et al., 2019）。角蝉 *Aethalion reticulatum* 就会分泌蜜露，据发现它是与新热带区无刺蜂相关性最高的物种（Santos et al., 2019），*Trigona* 属无刺蜂会在角蝉若虫和成虫上走动，并且用触角刺激其腹部使其释放蜜露（Oda et al., 2009; Baronio et al., 2012; Santos et al., 2019）。无刺蜂会与蚂蚁竞争蜜露，并且会保护它们的蜜露供体以免被蚂蚁攻击（Oda et al., 2009; Koch et al., 2011; Barônio et al., 2012）。例如，在马达加斯加，*Liotrigona mahafalya* 无刺蜂和 *L. madecassa* 无刺蜂会与 *Monomorium destructor* 蚁和 *Paratrechina longicornis* 蚁争夺蛤蚧分泌的蜜露（第 8.6 节）（Koch et al., 2011）。

果汁和果肉是无刺蜂采集的另一种资源，这些资源有些源自仍在树上的水果，有些则是来自已经落地的成熟水果（图 8.3）（Roubik,1989; Peruquetti et al., 2010）。果汁与许多花蜜相比，含糖量较低（通常为 8%～25%）（White and Stiles,1985; Peruquetti et al., 2010; Shackleton et al., 2016），但无刺蜂通常采集果汁的速度更快，因此，对无刺蜂而言，果汁是它们的潜在的廉价资源（在能量和时间支出方面）（Shackleton et al., 2016）。

8　无刺蜂的采集　　　　　　　　　　　　　　　　　　　　　　　　　363

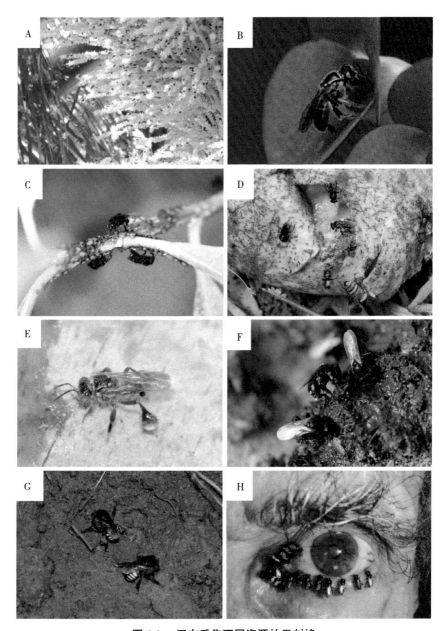

图 8.3.　正在采集不同资源的无刺蜂

注：A 为正在棕榈树上采集花粉的 *Trigona hyalinata* 无刺蜂。B 为正在采集 *Euphorbia milii* 花蜜的 *Plebeia droryana* 无刺蜂。C 为正在采集蜜露的 *Trigona* sp. 无刺蜂。D 为正在从一个掉落在地的杧果上采集果汁的 *Trigona spinipes* 无刺蜂。E 为正在从一株受伤的树上采集树脂的 *Frieseomelitta varia* 无刺蜂。F 为正在采集动物粪便的 *Trigona hyalinata* 无刺蜂。G 为正在采集泥浆的 *Melipona quadrifasciata* 无刺蜂。H 为正在采集人类眼泪的 *Lisotrigona furva* 无刺蜂 [照片 A–G 为 C. Grüter 拍摄于巴西圣保罗；照片 H 源自 Bänziger（2018）]。

8.1.3 树脂和其他黏性植物材料

植物树脂主要是由黏性的、不溶性的萜类化合物（单萜、二萜、三萜和倍半萜）组成，但也可能包含其他化合物，如没食子酸（Velikova et al., 2000; Patricio et al., 2002; Leonhardt et al., 2011）。一些花、水果和树木都会分泌液体状的树脂，通常是为了应对伤害和威慑食草昆虫（Roubik,1989）。无刺蜂，特别是 *Trigona* 属无刺蜂，偶尔会持续数天或数周啃咬植物的芽、嫩叶、花或树皮来刺激其分泌树脂（Schwarz,1948; Wille and Michener,1973; Howard,1985; Reyes-González and Zamudio,2020）。

一个无刺蜂蜂群通常采集多种植物的树脂，但似乎会对某些植物表现出偏好性（例如亚洲的 *Dipterocarpace*，美洲的 *Caesalpinioideae* 与 *Papilionoideae*），可能是因为这些植物的树脂具有抗菌活性［Leonhardt et al., 2009; Leonhardt and Blüthgen,2009; 关于无刺蜂采集树脂行为的详细描述，见 Bassindale（1955）与 Gastauer 等（2011）］。不同无刺蜂蜂种偏好的植物树脂的种类也有所不同（Roubik, 1989; Patricio et al., 2002）。例如一些 *Melipona* 属无刺蜂就偏好于 *Vismia* 果实产生的树脂（Roubik,1989）。如果树脂的混合物比单一类型的树脂能够更好地防御潜在的有害生物，那么无刺蜂从多种植物采集树脂的策略就可能是有益的（Drescher et al., 2014; Kämper et al., 2019）。

有些花种（如 Clusiaceae 和 Euphorbiaceae 科）甚至会从其花结构中分泌树脂来吸引授粉者（Armbruster,1984; Murphy and Breed,2008b）。在这种情况下，树脂已经取代花蜜和花粉成为花对授粉者的主要奖励[4]。花结构分泌树脂可能有一个优点就是这种树脂保持液体状态的时间更长（长达几周），因此无刺蜂采集和储存花朵树脂的持续时间可以更久，而从树伤采集来的树脂因为通常会在被分泌后几小时内变硬而采集储存时间要短些（Armbruster, 1984）。也有人观察到无刺蜂（例如 *Melipona, Scaptotrigona* 和 *Trigona*）会采集乳胶，即一种乳白色液体，通常是植物受伤后出现的，这种物质暴露在空气中会凝固（Absy and Kerr,1977; Roubik,1989; Pereira and Tannús-Neto,2009）。乳胶可能被无刺蜂用作筑巢材料，并且众人皆知它具有抗菌特性（Pereira and Tannús-Neto, 2009）。

---

[4] 研究者观察到一些 *Trigona* 属无刺蜂，特别是 *T. fulviventris* 无刺蜂甚至会强抢花朵树脂（Murphy and Breed,2008 b）。

8.1.4 其他资源

无刺蜂还会采集许多其他材料，如泥土、木浆、树叶、树皮、小石头、种子、果皮和粪便（图 8.3）（Schwarz,1948; Wille,1983; Baumgartner and Roubik,1989; Roubik,1989, 2006; Bastos Garcia et al., 1992）。无刺蜂主要用这些资源作为蜂巢的建筑材料（第 3 章）[5]。在城市地区的无刺蜂偶尔还会采集油漆和其他人造材料（Wille and Michener,1973）。目前研究者尚未全面研究无刺蜂对这些物质的采集。例如，泥浆是许多 *Melipona* 属无刺蜂采集的一种重要资源，但我们关于无刺蜂采集泥浆的自然历史和调控因素仍然知之甚少。Costa-Pereira（2014）的研究则是一个特例，他发现 *Melipona* 属无刺蜂采集蜂（图 8.3）喜欢不太潮湿的泥浆，这可能是因为潮湿的泥浆里的多余水分会增加重量。体型较大的采集蜂确实能够负载更多的资源（Costa-Pereira,2014），但目前还不知道采集泥浆的工蜂是否比采集其他类型资源的同巢蜂体型更大。

无刺蜂采集过程中会有意或无意地运输种子，这导致了无刺蜂介导的种子传播，澳洲的 *Corymbia torelliana* 桉树的种子就是这样传播的，澳洲无刺蜂在试图采集桉树种子囊上的树脂时也运输传播了桉树的种子（Wallace and Trueman,1995; Wallace and Lee,2010）。一个无刺蜂蜂群在一年中会采集到成千上万的桉树种子。然后无刺蜂工蜂通常会把种子搬出蜂巢并且丢弃在附近，这就导致 *C. torreliana* 树经常会在无刺蜂蜂巢附近生长（Wallace and Trueman,1995）。无刺蜂也会用采集来的种子作筑巢材料（第 3 章），而且当这些种子被用于建筑蜂巢入口时还可以发芽（Bastos Garcia et al., 1992）。

因为尿液、汗水或泥水这类资源中含有无刺蜂必需的矿物质和盐分，因此无刺蜂也会采集这类资源（Wille,1962; Roubik,1989; Bänziger et al., 2009; Karunaratne et al., 2017; Roubik,2018），而在蜂的生理过程中这些必需矿物质和盐分发挥了极为重要的营养作用。由于花蜜和花粉中只是含有少量的这类营养物质，因此无刺蜂经常积极采集这些更为不寻常的资源来补充它们的饮食。例如，当研究者为 *Trigona silvestriana* 无刺蜂提供了去离子水、矿化水和蔗糖溶液作为选择时，*Trigona silvestriana* 采集蜂采集的矿化水量几乎和蔗糖溶液一样多，但对去离子水却不屑一顾（Dorian and Bonoan,2016）。氯化钙（$CaCl_2$）、氯化钠（$NaCl$）和氯化钾（KCl）对于无刺蜂来说也特别有

---

[5] 无刺蜂在采集泥浆或巢质等材料过程中的行为动作顺序的详细描述参见 Michener 等（1978）。

吸引力。研究者观察到几种亚洲无刺蜂有采集鸟类、哺乳动物和爬行动物眼泪的情况（图 8.3），有人认为这说明无刺蜂可能对蛋白质以及盐类感兴趣（Bänziger et al., 2009; Karunaratne et al., 2017; Bänziger, 2018）。还有一些研究观察到一些新热带区无刺蜂蜂种（*Tetragona* 属、*Melipona* 属和几种 *Trigona* 属无刺蜂）会采集植物油（Roubik, 1989; Simpson et al., 1990; Lorenzon and Matrangolo, 2005; Ferreira et al., 2019）。无刺蜂采集油脂的生物功能仍然不清楚，但 Simpson 等（1990）猜测无刺蜂是将油用作筑巢材料，而不是为了油中含有的脂质（Roubik, 1989）。

8.2 无刺蜂采集的资源时空分布

热带花卉往往比温带花卉开花时间短（Primack, 1985; Roubik, 1989; Bawa, 1990）。例如，据估测热带干旱森林和热带雨林的平均开花期分别为 1.1 d 和 1.3 d，而且大多数会开花的物种的开花期不超过 3 d（然而也有开花几天甚至几周的热带花种）（Primack, 1985）。温带森林平均开花期则为 2.5～6.9 d，这取决于开花的时节（Primack, 1985）。这意味着热带无刺蜂采集蜂可能要比温带蜜蜂或熊蜂更频繁地寻找新的花卉资源，这种情况则很可能会凸显无刺蜂蜂群招募交流的重要性（Schürch and Grüter, 2014），并会影响无刺蜂学习记忆过程的动态性（Menzel, 1985）。

另一个重要的花卉物候特征则是开花物种的种内和种间的开花同步性。开花期短的植物经常同时开出大量的花（同步大量开花），而开花期较长的植物则是在较长时间内每天只开几朵花（稳态开花）（Rathcke and Lacey, 1985）。稳态开花多发现于林下植物，而同步大量开花现象则多发现于树木。研究者认为同步大量开花树木是许多无刺蜂蜂种最重要的食物来源（第 9 章）（Kleinert-Giovannini and Imperatriz-Fonseca, 1987; Wilms et al., 1996; Ramalho, 2004; Hrncir and Maia-Silva, 2013b; Roubik and Moreno Patiño, 2018），但开花灌木、附生植物、藤本植物、草本植物和小型树木也是无刺蜂食物来源的重要组成部分（Roubik and Moreno Patiño, 2018）。植物物种间开花不同步现象可能是源于植物物种对传粉者的竞争，也可能是为了防止种间花粉转移而产生的自然选择的结果。植物物种内的开花不同步可以通过促进植物间基因移动来影响基因流动（Roubik, 1989）。研究者发现在植物群落中既存在同步开花，也存在随机开花（Rathcke and Lacey, 1985）。

在特定日子里，花朵产生的花粉、花蜜和花香的数量（和质量）上表现出周期性（Butler,1945; Vogel,1983; Kajobe,2007 a）。无刺蜂会适应这些波动并会在糖浓度或花蜜流量最高的时期增加访花量（Kleber,1935; Butler,1945; Vogel,1983）。根据这一规律，无刺蜂就经常在清晨采集花粉，因为清晨是花粉这种资源可利用率最高的时间段（见下文）（Roubik,1989）。然而，应该注意的是，植物一天中产生花蜜量最高的时间段不一定与花蜜糖浓度最高的时间段相一致，而且这些还取决于开花是在阴面还是阳面（Biesmeijer et al., 1999 b）。

8.3　无刺蜂采集的专业化

无刺蜂的采集蜂个体常会表现出对某一种特定资源的偏好（Sommeijer et al., 1983; Biesmeijer and Tóth,1998; Hofstede and Sommeijer,2006; Leonhardt et al., 2007; Gomes et al., 2015; Gostinski et al., 2017; Mateus et al., 2019）。例如，Sommeijer 等（1983）追踪了 *Melipona favosa* 采集蜂，发现大多数采集蜂要么只采集花蜜要么只采集花粉，只有约 25% 的采集蜂会两种资源都采集。*Melipona beecheii* 无刺蜂中大约一半的采集蜂在其采集生涯中会只专门采集花粉或花蜜，剩下的另外一半则会采集一种以上的资源（Biesmeijer and Toth,1998）。在后一种情况中采集蜂往往在上午采集花粉，下午采集花蜜（也见下文）。在温室中观察 *Plebeia tobagoensis* 无刺蜂 3 d 后，发现在观察期内 71% 的采集蜂会专门采集一种资源（花粉、花蜜、水或树脂）（Hofstede and Sommeijer,2006）。有趣的是，在采集资源选择上，比较灵活的采集蜂个体总体上比专一采集一种资源的采集蜂效率更低：灵活采集蜂采集的花粉量更小，并且/或者它们需要更多的时间来采集蔗糖溶液，这一现象支持了个体专业化增加任务绩效的假设（Biesmeijer and Tóth,1998）。无刺蜂花粉采集蜂以及小部分树脂和泥浆采集蜂在离巢时会在蜜囊中携带浓缩花蜜（Leonhardt et al., 2007; Harano et al., 2020）[6]。无刺蜂这种携带花蜜出巢采集的行为可能有两个功能。首先它是无刺蜂飞行的能源，其次它可以作为黏附剂帮助采集蜂将花粉粒附着在花粉筐上（Leonhardt et al., 2007）。研究者发现 *Melipona marginata* 无刺蜂采集蜂中泥浆采集蜂表现出最明显的行为专业化，这可能是因为这种专业化减少了储存在食物罐的食物被采集来的泥浆中微生物污染的风险（Mateus et al., 2019）。蜜蜂中采集不同资源的采集蜂对糖奖励的敏感性不同

---

[6] 研究者未从离巢的 *Melipona subnitida* 无刺蜂花蜜采集蜂蜜囊中检测到任何花蜜，这提示它们是从离蜂巢很近的食物源收集花蜜的。

（Scheiner et al., 2004; Simone-Finstrom et al., 2010）。无刺蜂也存在类似差异，Balbuena 和 Farina（2020）最近首次为此提供了证据，他们发现 *Tetragonisca angustula* 无刺蜂的花粉采集蜂的蔗糖反应阈值（SRT）比非花粉采集蜂低，这与蜜蜂的花粉采集蜂和花蜜采集蜂的 SRT 差异情况一致。

无刺蜂采集花蜜或花粉的决策也取决于蜂群条件，如蜂群中储存的食物量。移走 *Melipona fasciculata* 无刺蜂蜂巢中的储存蜂蜜会导致蜂群持续几天增加对花蜜的采集量（Gostinski et al., 2017），而实验性地向 *M. subnitida* 无刺蜂蜂巢中添加花粉则会导致蜂群 24 h 减少对花粉的采集量（Maia-Silva et al., 2016）。这表明采集蜂能够获得蜂群内储存食物量的相关信息，并能相应地调整对资源采集行为。然而，研究者实验性地减少 *M. subnitida* 无刺蜂蜂巢的花粉储存量却没有即时影响 *M. subnitida* 无刺蜂对花粉的采集。Maia-Silva 等（2016）认为，对于采用渐饲幼虫食物的蜂种（无刺蜂）来说，花粉的暂时短缺产生的后果可能不那么严重，而对于采用零饲幼虫食物的蜂种（蜜蜂）而言，花粉暂时短缺产生的后果则更为严重。在后一种情况中突然缺乏花粉可能导致所有幼虫死亡，而在无刺蜂中缺乏花粉可能导致育虫的暂时中断，但对蜂巢中已经存在的幼虫则没有任何伤害。

8.4 无刺蜂的采集活动力

8.4.1 季节性采集活动

有些无刺蜂蜂种全年都在采集（图 8.4）[例如乌干达的 *Apotrigona nebulata*，Kajobe and Echazarreta（2005）；墨西哥的 *Melipona beecheii*，Di Trani and Villanueva-Gutiérrez（2018）；巴拿马的 *Partamona cupira*，Wolda and Roubik（1986）；澳大利亚的 *Tetragonula carbonaria*，Heard and Hendrikz（1993）]，而其他一些无刺蜂蜂种则会由于低温、资源匮乏、极端干旱或强降雨在某些月份减少或完全停止采集活动（第 5 章）（Roubik, 1982c; van Benthem et al., 1995; Ferreira Junior et al., 2010; Nascimento and Nascimento, 2012; Halcroft et al., 2013; Maia-Silva et al., 2015; Hrncir et al., 2019）。几乎所有的无刺蜂蜂种在其采集活动中都显示出一定程度的季节性。这种季节性直接影响了蜂巢内储存的资源量，因为当食物源越丰富时无刺蜂蜂群会储备越多资源（第 5 章）（Roubik, 1982 c; Nascimento and Nascimento, 2012; Maia-Silva et al., 2015; Aleixo et al., 2017; Hrncir et al., 2019）。例如，生活在巴西东北部的热带干旱森林 Caatinga 中的 *Melipona subnitida* 无刺蜂蜂群会经历长期的干旱，在干旱期间它们几乎完全不

会在采集和育子方面投入精力,直到采集条件突然改善它们才会解除这种锁定状态(Maia-Silva et al., 2015; Hrncir et al., 2019)[7]。*Austroplebeia australis* 无刺蜂也是如此,它们会在资源匮乏的时期几乎完全停止采集活动,这种情况在其生活的澳洲内陆的半干旱栖息地中并不少见(Halcroft et al., 2013)。除了采集的食物资源量,无刺蜂采集食物资源的质量也取决于季节:乌干达的 *Hypotrigona gribodoi* 和 *Axestotrigona ferruginea* 无刺蜂在旱季采集的花蜜的含糖量要高于在雨季采集的(Kajobe,2007 a; Campbell et al., 2019)。

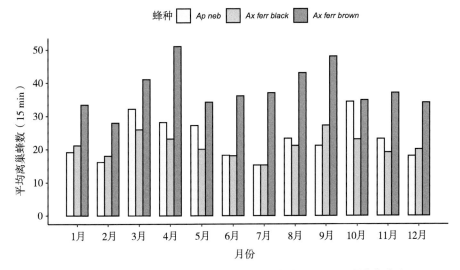

图 8.4 一年研究期间无刺蜂离巢蜂的平均数量(15 min 观察期内)

注:图中所示的无刺蜂蜂种是 *Axestotrigona ferruginea*(红褐色型)、*A. ferruginea*(黑色型)和 *Apotrigona nebulata* 无刺蜂(Kajobe and Echazarreta,2005)。

8.4.2 日常活动

无刺蜂一天中的采集活动分布也并不均匀,而是取决于光照条件、风、湿度、温度和蜂的体型大小(Fowler,1979; Imperatriz-Fonseca et al., 1985; Baumgartner and Roubik,1989; Roubik,1989; Heard and Hendrikz,1993; Hilário et al., 2000, 2001; Pick and Blochtein,2002; Teixeira and Campos,2005; Ferreira Junior et al.,2010; Sung et al.,2011; Hilário et al.,2012; Nascimento and Nascimento,2012; Keppner and Jarau,2016; Macías-Macías et al.,2017; Layek and Karmakar,2018)。与蜜蜂有些不同的是,一些无刺蜂在下雨时也会进行采集活

---

[7] 在 Caatinga,干旱甚至会持续数年,这迫使无刺蜂蜂群在较长的时间内保持最低育子水平,这种情况使得这个栖息地对于无刺蜂来说极具挑战性(Hrncir et al., 2019)。

动（Baumgartner and Roubik,1989; Kajobe and Echazarreta,2005; Keppner and Jarau, 2016）。这些外部因素可以通过影响无刺蜂飞行过程中的能量和水分损失等直接影响无刺蜂的飞行活动，也可以通过影响花生产花粉和花蜜而间接地影响无刺蜂的采集活动。

即使是在相同的环境下生活的无刺蜂，不同蜂种每日采集活力最高的时间段也存在很大差异（图 8.5）（Roubik,1983; Wolda and Roubik,1986; Roubik,1989; Keppner and Jarau,2016; I' Anson Price,2018）。一些无刺蜂蜂群表现出了明显的采集活动高峰期，而其他一些蜂群则在整个白天都有相对稳定的采集活动。无刺蜂（例如 *Tetragona clavipes*、*Trigona fuscipennis*、*T. recursa*、*Friesella schrottkyi*、*Plebeia droryana*）普遍会在 16—17 时达到傍晚的采集活动高峰（Roubik,1989）。这可能意味着在许多新热带环境中这个时间段的采集环境有所改善，也可能是傍晚时新手采集蜂的定向飞行活动增加造成的（Roubik,1989）[8]。研究者还发现 *Partamona helleri* 和 *P. orizabaensis* 无刺蜂的采集活动模式十分异常，它们在清晨和傍晚时分采集活动最多，而在一天的大部分时间里采集活动则相应减少（图 8.5）（Keppner and Jarau,2016; I' Anson Price,2018）。目前还不清楚为什么这些 *Partamona* 属无刺蜂表现出这种异常的活动模式。

具有大量招募行为的无刺蜂蜂种（第 10 章）则有所不同，主要表现在它们可以在相对短的采集活动暴发期内采集到相当大比例的食物（Roubik et al., 1986; Roubik,1989）。即使这种活动暴发期只持续 1 h，处于该时期的蜂群也能够在此期间采集到所需食物资源的一大部分（Roubik et al., 1986; Roubik,1989）。

新热带区无刺蜂上还存在一种普遍现象，即花粉和花蜜的采集时间有相对明显的时间分隔：无刺蜂主要在清晨采集花粉，而花蜜、树脂或泥浆的采集则是在中午前后和下午最密集（Sommeijer et al., 1983; Roubik and Buchmann,1984; Hilário et al., 2000; Pierrot and Schlindwein,2003; Nascimento and Nascimento,2012; Hilário et al., 2001）。无刺蜂开始采集花蜜的时间滞后可能是因为研究者观察到早上植物的花蜜含糖量较低，而后植物花蜜的含糖量不断增加（图 8.6）（Roubik and Buchmann,1984; Roubik et al., 1995; Kajobe,2007 a）。乌干达地区的无刺蜂采集花粉和花蜜则没有表现出这种的时间顺序，它们全天都会采集花粉（Kajobe and Echazarreta,2005）。

---

[8] 西方蜜蜂的幼蜂会在下午离开蜂巢进行排泄，而无刺蜂则是在巢内排泄的。

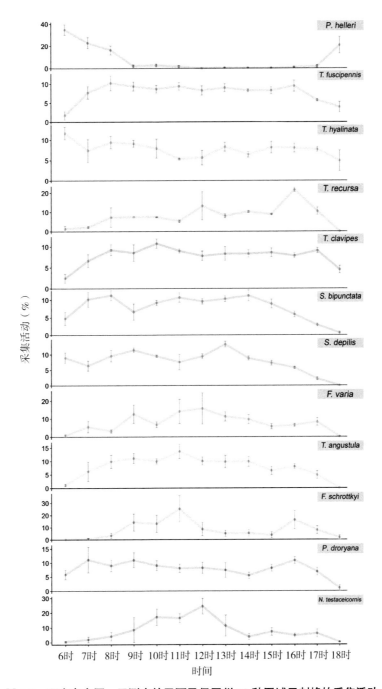

图 8.5 研究者在同一天测定的巴西圣保罗州 12 种同域无刺蜂的采集活动

注：图中数据是作者于夏末（雨季末期）的两天里收集的。数据显示为无刺蜂整日的采集活动的平均百分比 ± 标准误（I' Anson Price,2018）。

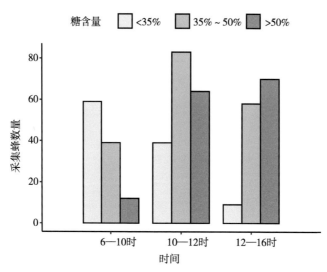

图 8.6 巴拿马地区 4 种 *Melipona* 无刺蜂采集液体的糖含量和采集活动时间段之间的关系（Roubik and Buchmann,1984 文中的 表 9）

蜜蜂形态的作用

体型越小的无刺蜂在飞行过程中温度降得越低，这是因为热对流相对较高（Pereboom and Biesmeijer,2003），而体型越大的无刺蜂白天进行采集活动时的过热风险也越高（Pereboom and Biesmeijer,2003；Hrncir and Maia-Silva,2013 a, 2013 b）。因此，温度和体型大小是决定无刺蜂采集活动力的关键因素（Fowler,1979; Roubik,1989; Heard and Hendrikz,1993; Corbet et al., 1995; Hilário et al., 2000, 2001; Teixeira and Campos,2005; Sung et al., 2011; Hrncir et al., 2019; Souza-Junior et al., 2020）。如 *Melipona* 属等大体型无刺蜂蜂种即使温度低至 11 ℃也能保持活力，而体型小得多的 *Friesella schrottkyi* 和 *Plebeia lucii* 无刺蜂则需要 21～22 ℃才能进行采集活动（Hilário et al., 2000; Teixeira and Campos,2005; Sung et al., 2011）。其他体型相对较小的蜂种，如 *Tetragonisca angustula*，*Tetragonula carbonaria* 和 *Plebeia pugnax* 无刺蜂在温度为 14～20 ℃时才能开始采集活动（Heard and Hendrikz,1993; Hilário et al., 2001; Malerbo-Souza and Halak, 2016）。

无刺蜂的体型大小也会影响其感觉器官的大小和敏感性，例如眼睛或触角的大小（Jander and Jander,2002; Spaethe et al., 2007; Streinzer et al., 2016; Grüter et al., 2017 b）。无刺蜂眼睛越小，光敏感性越低（Streinzer et al., 2016），这使得小体型蜂种在昏暗的光线条件下进行采集活动更加困难（Streinzer et al., 2016）。这种情况可能会促使不同蜂种在采集活动时长和花区选择方面产生差异（Willmer and Corbet,1981; Biesmeijer et al., 1999 a, 1999 b; Teixeira and Campos,2005）。诸如 *Melipona* 属等大体型的无刺蜂蜂种每天很早就会开始采集活动，有时甚至是在日出之前（Pereboom and Biesmeijer,2003; Teixeira and Campos,2005; Streinzer et al., 2016）。而像新热带的 *Friesella schrottkyi*[9] 等体型较小的无刺蜂则是在 8 时之前基本不活动（Pereboom and Biesmeijer,2003; I'Anson Price,2018）。

  无刺蜂的身体颜色是另一个影响无刺蜂采集活力和探索花区的形态特征（Hrncir and Maia-Silva,2013 b）。淡黄色的 *Melipona beecheii* 喜欢阳光充足的花区，而深色的 *M. fasciata* 则会选择较阴暗的花区（Biesmeijer et al. 1999 a,1999 b）。而这种情况也导致了这两种无刺蜂采集的花蜜的含糖量有所不同（第 8.1 节）。无刺蜂的形态学特性还可能影响无刺蜂的生物地理分布。着色浅的无刺蜂在开放和炎热的环境中具有优势，而着色较深的无刺蜂则可能更适宜在潮湿或多山地区进行栖息（Willmer and Corbet,1981; Pereboom and Biesmeijer,2003）。出现在高海拔地区（>1 500 m）的无刺蜂蜂种通常是深色的，而且其体型通常是中大型的（Pereboom and Biesmeijer,2003）。

8.5 无刺蜂的采集范围和采集旅程时长

  无刺蜂蜂群的采集范围会产生重要的群体生态后果，因为采集范围会影响蜂群之间的竞争和被访植物物种的种群遗传。此外，在一些受干扰的和零碎的栖息地中，无刺蜂蜂群的采集活动范围可能还会影响蜂群的采集成功率。

  研究者推测无刺蜂偏好于离蜂巢更近的食物源（León et al., 2015; Souza-Junior et al., 2020），因为这可以减少飞行成本，降低采集蜂暴露于如天敌或高温等环境的危险性。研究者多基于人工喂食器或归巢研究测定无刺蜂采集蜂的最大飞行距离（Michener,1974; Roubik and Aluja,1983; Van Nieuwstadt and Iraheta,1996; Silva et al., 2014; Campbell et al., 2019; Nunes-Silva et al., 2020）。对于归巢研究，研究者是通过在无刺蜂蜂巢中抓捕采集蜂并且给其做上标记，然后在离蜂巢的不同距离的位置释放这些采集蜂。如果采集蜂能够返回其蜂巢，则判定该采集蜂的释放距离在该蜂群的最大飞行范围内。返回蜂巢的采集蜂比例往往随着采集蜂的释放距离的增加而减少（Roubik and Aluja,1983;

---

[9] 由于 *F. schrottkyi* 无刺蜂每天开始采集的时间较晚（通常蜂群在 11 时之前几乎完全不出巢采集），*F. schrottkyi* 在巴西也被称为 "mirim-preguiça" 或 "懒蜂"。

van Nieuwstadt and Iraheta,1996; Nunes-Silva et al., 2020）。而对于基于人工喂食器的研究中，研究者会训练采集蜂习惯于从人工喂食器采集蔗糖溶液，然后将喂食器放在离蜂巢越来越远的距离，当采集蜂不再抵达喂食器时停止移动喂食器。喂食器所处位置与蜂巢间的距离就被认为是采集蜂的最大采集距离。van Nieuwstadt 和 Iraheta（1996）用 4 种无刺蜂比较了这两种方法，发现这两种方法测定的结果类似（Silva et al., 2014），但是采用标记、释放和重新捕获的方法测定出的最大采集距离会略偏大一些（约 +300 m）。

据估测无刺蜂采集蜂的最大采集距离在 120～4 000 m，亚洲 *Tetragonula iridipennis* 无刺蜂采集蜂的最大采集距离仅为 120 m，而美洲 *Melipona subnitida* 无刺蜂的最大采集活动距离则高达 4 000 m（表 8.1）。将 *Melipona fasciculata* 和 *M. subnitida* 采集蜂在离巢几千米外释放后，两种无刺蜂的部分采集蜂都能够找到回家的路（Silva et al., 2014; Nunes-Silva et al., 2020）。然而，将 *M. fasciculata* 采集蜂在离蜂巢 2 500m 以外的地方释放后，采集蜂则需要很长时间才能返回蜂巢，这表明它们迷路了。这一现象提示我们归巢实验中还必须记录采集蜂的回巢时长来阐释从归巢实验中获得的距离数据（Nunes-Silva et al., 2020）。图 8.7 中显示的数据指出无刺蜂采集蜂的飞行范围会随着采集蜂体型的增大而增加（van Nieuwstadt and Iraheta,1996; Araújo et al., 2004; Greenleaf et al., 2007）。

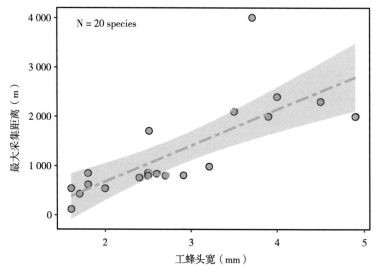

图 8.7　20 种无刺蜂的体型大小（头宽度）与估测最大采集距离之间的关系（线性回归，$R^2 = 0.57, p < 0.000\ 1$）（表 8.1）（灰色区域为 95% 置信区间）

8 无刺蜂的采集

表 8.1　无刺蜂的最大飞行范围、头宽和最大采集范围研究方法

无刺蜂蜂种	最大飞行范围	头宽	研究方法	参考文献
Cephalotrigona capitata	1 700 m	2.5 mm[a]	标记放归法	Roubik and Aluja（1983）
Melipona fasciculata	2 300 m	4.5 mm[b]	标记放归法	Araujo et al.（2004）；Nunes–Silva et al.（2019）
Melipona fasciata	2 400 m	4 mm[c]	标记放归法	Roubik and Aluja（1983）
Melipona fuliginosa	2 000 m	4.9 mm[c]	未提及	Wille（1983）
Melipona mandacaia	2 100 m	3.5 mm[c]	人工喂食器	Kuhn–Neto et al.（2009）
Melipona marginata	800 m	2.9 mm[d]	未提及	Wille（1983）
Melipona quadrifasciata	2 000 m	3.9 mm[d]	未提及	Araujo et al.（2004）
Melipona subnitida	4 000 m	3.7 mm[b]	两种方法	Silva et al. 2014（2014）
Nannotrigona perilampoides	620 m	1.8 mm[e]	两种方法	van Nieuwstadt and Iraheta（1996）
Partamona cupira	800 m	2.5 mm[e]	两种方法	van Nieuwstadt and Iraheta（1996）
Plebeia droryana	540 m	1.6 mm[b]	未提及	Araujo et al.（2004）
Plebeia mosquito	540 m	2 mm[f]	人工喂食器	Michener（1974）
Scaptotrigona postica	860 m	2.5 mm[d]	标记放归法	Campbell et al.（2019）
Tetragonula iridipennis	120 m	1.6 mm[g]	人工喂食器	Lindauer（1957）
Tetragonula minangkabau	430 m	1.7 mm[h]		Inoue et al.（1985）
Tetragonisca angustula	850 m	1.8 mm[b]	两种方法	van Nieuwstadt and Iraheta（1996）
Trigona amalthea	980 m	3.2 mm[a]	人工喂食器	Michener（1974）
Trigona corvina	760 m	2.4 mm[e]	两种方法	van Nieuwstadt and Iraheta（1996）
Trigona silvestriana	800 m	2.7 mm[a]	未提及	Wille（1983）
Trigona spinipes	840 m	2.6 mm[d]	人工喂食器	Michener（1974）

注：研究方法中提到的两种方法是指采用了人工喂食器训练法和标记放归法这两种方法。[a]Schwarz（1948），[b]Grüter et al.（2017a），[c] 由 Schwarz（1932）估算而出，[d] 依据 Kátia Aleixo 提供的照片，[e]van Nieuwstadt and Iraheta（1996），[f]von Ihering（1903），[g]Rasmussen（2013），[h]Sakagami and Inoue（1985）。

这两种用来估测无刺蜂采集活动范围的方法都有一定的局限性。首先，这两种方法都可能导致对无刺蜂最大采集范围的不准确估测，其次，这两种

方法提供的信息不能揭示蜂群中采集蜂普遍的采集距离（Breed et al., 1999; Greenleaf et al., 2007; Kuhn-Neto et al., 2009）。例如，喂食器实验中无刺蜂采集蜂是否继续访问某一特定距离的喂食器将取决于它们的采集动力，而这种动力反过来又可能取决于是否有其他的采集选择和实验所处季节。而归巢实验中，采集蜂归巢的成功率可能取决于采集蜂前期的采集经验。例如一只能够从北方 500 m 处返回的无刺蜂采集蜂很可能是因为它前几天已经去该位置采集过，而当这只采集蜂从 500 m 外的其他地点释放时就可能无法归巢。因此，无刺蜂实际的飞行范围可能超过归巢研究所提示的范围。然而，更重要的是，目前采用这两种方法估测无刺蜂最大采集距离的研究几乎都没有提供如下信息，即蜂群中大多数采集蜂在哪里采集食物，还有采集距离的分布是否取决于环境（Roubik, 1989; Campbell et al., 2019）。例如蜜蜂就能够在长达 12 km 的距离外访问人工喂食器，但其实大多数蜜蜂的采集活动则通常发生在 2.5 km 以内的范围（von Frisch, 1967; Steffan-Dewenter and Kuhn, 2003; Couvillon et al., 2014）。Breed 等（1999）使用诱饵来估测 *Trigona amalthea* 和 *T. corvina* 无刺蜂的采集距离。他们的数据表明这些无刺蜂大多数的采集活动发生在离蜂巢 200 m 范围以内，这比研究者估测出的最大采集范围（760～980 m）要小得多（表 8.1）。Roubik（1989）提出一个经验法则，即无刺蜂大部分的采集活动仅发生在其最大采集范围的 1/3～1/2 [10]。

蜂类的采集活动范围在很大程度上取决于环境和季节（蜜蜂研究参见：Visscher and Seeley, 1982; Waddington et al., 1994; Beekman and Ratnieks, 2000; Steffan-Dewenter and Kuhn, 2003; Couvillon et al., 2014）。例如，Campbell 等（2019）发现，巴西亚马孙河东部的 *Scaptotrigona postica* 无刺蜂采集蜂在干旱月份的采集活动距离要比在雨季的短。有趣的是，无刺蜂的采集活动距离与采集的糖浓度呈负相关。这表明诸如遇到亚马孙雨季等这些恶劣采集条件会迫使无刺蜂采集蜂飞到更远的地方去寻找食物（Campbell et al., 2019）。

无刺蜂大多数采集旅程的时长为几分钟到一个小时以内（Roubik, 1989; Harano et al., 2020）。I'Anson Price（2018）研究了圣保罗州 4 种小体型无刺蜂蜂种在雨季结束时的离巢时长，发现这些无刺蜂每次离开蜂巢的时长不超过 20 min，平均行程时长为 5～7 min（图 8.8）。*Partamona* sp. 无刺蜂采集

---

[10] 有研究者观察到食肉性 *Trigona hypogea* 无刺蜂在离它们蜂巢 800 m 的地方采集胡蜂幼虫（Noll, 1997），这个距离已经接近研究者推测出的这种大小的无刺蜂的最大采集距离。

蜂采集花粉的单程时长也与此类似（4～8 min）（Wille and Orozco,1975），*Melipona costaricensis* 无刺蜂的采集旅程的平均时长为 4 min（时长在 1～10 min）（De Bruijn and Sommeijer,1997）。这比温带蜜蜂的采集旅途时长要短得多，蜜蜂的采集旅程时长通常为 10～100 min（Park,1926; Butler et al.,1943; von Frisch,1967）。这些数据也进一步强调，尽管表 8.1 中无刺蜂采集活动范围之大令人惊奇，但实际上无刺蜂的采集蜂通常只在蜂巢附近探查食物。

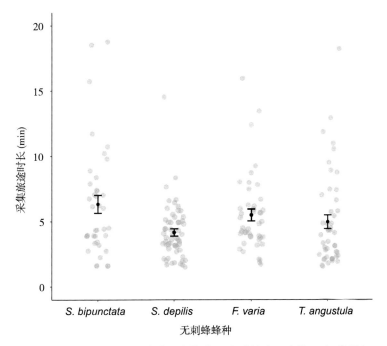

图 8.8 4 种无刺蜂采集蜂的采集旅途平均时长（平均值 ± 标准误）
（I'Anson Price,2018）

8.6 无刺蜂的资源竞争和资源分配

8.6.1 无刺蜂之间的竞争

许多热带栖息地是大量蜂群的家园，有时甚至有几十种无刺蜂在热带栖息地定居（第 3 章）。近期 Roubik（2018）报道了一个极端案例，他发现在厄瓜多尔亚马孙的 Yasuní 国家公园的一小片区域里（<10 km²）有约 100 种无刺蜂共存。由于无刺蜂这种共存情况以及无刺蜂的多元化（多元访花性）采集习性，这意味着不同无刺蜂蜂种的可用食物源可能会有所重叠，这就很可

能导致了无刺蜂蜂群间产生潜在采集竞争（Sommeijer et al., 1983; Imperatriz-Fonseca et al., 1989; Ramalho et al., 1989; Wilms et al., 1996; Eltz et al., 2001; Biesmeijer and Slaa, 2006; Myrtaceae, Solanaceae and Leguminosae（subfamily MimosoideaeTeixeira et al., 2007; Kajobe, 2007 b; Leonhardt and Blüthgen, 2009; Dworschak and Bluethgen, 2010; Lichtenberg et al., 2010, 2017; Hrncir and Maia-Silva, 2013 a, 2013 b）。总的来说，两个同域的无刺蜂蜂种之间的饮食生态位重叠［即两个物种共同使用资源，例如 Colwell 和 Futuyma（1971）］通常在 0.2～0.4（偶尔可能到 0.5 以上）（Wilms et al., 1996; Roubik and Moreno Patiño, 2018），但这个值也可能会随采集条件和季节而发生变化（Eltz et al., 2001）。与其他无刺蜂相比，新热带属 Melipona 无刺蜂的食物则相对独特，这可能是因为它们体型较大，因而它们能够利用特殊的花卉资源（第 9 章）（Ramalho et al., 1989）[11]。

研究者对不同热带栖息地的相关研究表明，在分类学上越相近的无刺蜂蜂种间饮食的重叠程度越大（Sommeijer et al., 1983; Ramalho et al., 1989; Wilms et al., 1996; Nagamitsu and Inoue, 1997; Biesmeijer and Slaa, 2006），同一蜂种的不同蜂群间饮食重叠现象最为明显（Sommeijer et al., 1983; Eltz et al., 2001; Roubik and Moreno Patiño, 2018）。因此，亲缘关系越近的蜂种之间采集竞争越强，这就可以解释为什么无刺蜂物种聚类中的属多样性往往大于通过物种数量预测出的属多样性，即聚类中的物种在系统发育上的距离似乎比预测的要远（Biesmeijer and Slaa, 2006）。

竞争意味着所有或部分参与到竞争中的无刺蜂都要付出代价，这些代价的付出会导致存活率下降或采集成功率下降。例如，无刺蜂们对食物源的竞争减少了采集蜂采集这些食物源所花费的时间以及采集蜂负载的花粉和花蜜量（Johnson and Hubbell, 1974）。竞争并不需要无刺蜂个体之间直接互斗。其形式还可能是，一方采集蜂已经开采某一资源后致使后来的采集蜂感受到的植物奖励减少。无刺蜂这种"开采竞争"好像十分常见，但不进行详细的实验很难对其进行量化（Balfour et al., 2015）。另外无刺蜂采集蜂之间也可能在面对资源时直接相斗从而阻止对方进行有效的采集，这就是所谓的"干扰性竞争"。这种竞争行为小到无刺蜂采集蜂呈现威胁性的身体

---

[11] 熊蜂与无刺蜂在食物源方面重叠最小，可能也是二者形态上的差异（例如体形大小）造成的（Biesmeijer and Slaa, 2006）。

姿势（张开上颚以及把翅膀展开呈"V"形），大到双方进行搏命性的战斗（Johnson and Hubbell,1974; Roubik,1989; Nagamitsu and Inoue,1997; Dworschak and Blüthgen,2010; Lichtenberg et al., 2010）。无刺蜂这种敌对行为的严重程度会随着食物源质量（Johnson and Hubbell,1974）以及敌对双方体型大小不对等程度（Dworschak and Blüthgen,2010）的增加而增加。在双方体型不对等的情况下，体型大无刺蜂偶尔会胜过体型较小的无刺蜂（Lichtenberg et al., 2010）。由于无刺蜂这种直接互斗行为十分显眼，因此大多数关于无刺蜂之间采集竞争的研究都涉及这种干扰性竞争。例如，Baumgartner 和 Roubik（1989）在研究无刺蜂采集人工饵料时观察到体型相对较小但能够迅速招募的 *Trigona fuscipennis* 无刺蜂往往最先抵达饵料，但随后体型较大的 *Cephalotrigona capitata* 无刺蜂就取代了 *Trigona fuscipennis* 无刺蜂占据了饵料。而第 3 种出现在饵料前的无刺蜂是非攻击性蜂种（*Scaptotrigona xanthotricha*）的采集蜂，这种无刺蜂只有在饵料上前两种更具攻击性的蜂种数量不多之后才会落到饵料上［新热带区的具有攻击性的采集蜂通常是 *Trigona* 属，例如 *T. amalthea*、*T. corvina*、*T. fuscipennis*、*T. hyalinata*、*T. sylvestriana*、*T. williana*，文献参考如 Johnson and Hubbell（1974）; Howard（1985）］。而研究者在研究亚洲无刺蜂间的竞争时发现，攻击性越强的无刺蜂到达人工食物源的时间越晚，这表明攻击性越低的蜂种更善于发现食物源（见下文），也可能表明攻击性较低的蜂种每天会更早去采集来避免直接的竞争（Nagamitsu and Inoue,1997）。应该注意的是，无刺蜂在天然食物源处的互斗行为要少于在人工饵料和树脂采集点处的互斗行为（Howard,1985; Leonhardt and Blüthgen,2009; Dworschak and Blüthgen,2010）。

无刺蜂生态位分割

鉴于无刺蜂竞争的可能性很大，那么这么多种不同的无刺蜂是如何能在同一环境中共存的？无刺蜂很多行为和形态上的特征影响着它们是否以及如何相互竞争。例如，竞争力较弱的蜂种可以改变它们的采集时间或食谱，以避免与其他物种竞争（Nagamitsu and Inoue,1997; Keppner and Jarau,2016; Lichtenberg et al., 2017）。因此，一个无刺蜂蜂种实际上最终占据的食物生态位将不同于其本质食物生态位，即该蜂种在完全没有竞争者的情况下会占据的食物生态位（Biesmeijer and Slaa,2006; Hrncir and Maia-Silva,2013 a,2013 b）。研究者观察到随着共生无刺蜂数量的增加，蜂种的生态位宽度会减少，这种现象就可以用竞争导致了无刺蜂蜂种的食谱调整来解释

（Biesmeijer and Slaa,2006）。根据 Johnson 和 Hubbell（1974）报道，小体型的无刺蜂专攻开花小且过度分散的植物，这也可能是其食物生态位转变的原因。

无刺蜂的形态特征会对其开采资源产生复杂的影响。如上所述，无刺蜂的体型大小和颜色会影响采集蜂出巢去采集的时间，这样就有助于资源的划分。无刺蜂吻部的长度是另一个会影响其花朵偏好性的形态特征（Corbet et al., 1995; Biesmeijer and Slaa,2006）。拥有长吻的蜂种可以触碰到花冠较深的花朵的花冠，因而这些无刺蜂可能获得更多种类的花蜜（Hrncir and Maia-Silva,2013 a）。

许多攻击性强、竞争激烈的无刺蜂蜂种有一个共同特征就是它们会大规模招募（Lichtenberg et al, 2010）。也就是说，这种无刺蜂会使用轨迹信息素来招募同巢蜂抵达采集目标（第 10 章）。大规模招募使蜂群可以独占食物源，排挤竞争对手（Hubbell and Johnson,1978; Hrncir,2009; Hrncir and Maia-Silva,2013 a）。然而，并不是所有会大规模招募的蜂种都具有相似的攻击性。*Scaptotrigona* 属、*Partamona* 属和一些 *Trigona* 属无刺蜂攻击性就不是太强，但这些蜂种会通过纯粹的数量优势来排挤其他竞争者（Biesmeijer and Slaa,2004; Lichtenberg et al., 2010; Hrncir and Maia-Silva,2013 a; Keppner and Jarau,2016）。而蜂群将大量采集蜂都分配去一处高质量资源采集就可能会降低蜂群发现新食物资源的能力（Johnson and Hubbell,1974, 1978）。Hubbell 和 Johnson（1987）曾在哥斯达黎加设置了由芳香蔗糖溶液构成的人工诱饵，他们发现独行采集的小体型 *Nannotrigona perilampoides*、*Plebeia frontalis* 和 *Trigonisca* sp. 无刺蜂[12]发现诱饵的速度比富有竞争优势的、大规模招募（或成群采集）的 *Scaptotrigona pectoralis* 和 *Trigona fuscipennis* 无刺蜂要快得多。无刺蜂这种探索资源—占领资源的权衡在蚂蚁中则是众所周知的（Lebrun and Feener,2007; Bertelsmeier et al., 2015），并且研究者认为这种探索资源—占领资源的权衡使得采集策略不同，但在采集资源上重叠的蜂种在同一栖息地中得以共存（Hubbell and Johnson,1978; Kneitel and Chase,2004; Lebrun and Feener,2007）。例如，如果富有竞争优势的大规模招募蜂种采集范围在大片集中开花区块，而低密度专一采集的蜂种则是在小块的且高度分散的植物区采集，那么富有竞争优势的大规模招募蜂种和低密度专一采集蜂种就可以在同一环境中共存（Hubbell and Johnson,1978; Biesmeijer and Slaa,2006）。有

---

[12] 在 Hubbell 和 Johnson（1978）文中写的是 *T. testaceicornis*、*T. frontalis* 和 *T. buyssoni*。然而哥斯达黎加既不存在 *N. testaceicornis* 也没有 *Trigonisca buyssoni*。笔者这里写的蜂种是依据 Camargo 和 Pedro（2013）一文中复原的无刺蜂分布信息而推断的。

研究者就曾观察到 *Trigona fulviventris* 和 *T. fuscipennis* 无刺蜂的采集蜂就同时在 *Cassia biflora* 灌木丛中采集，这种案例证实了上述情况（Johnson and Hubbell,1975）。会大规模招募采集的 *T. fuscipennis* 无刺蜂主要开采大型丛生且质量高的 *C. biflora* 植株丛，而相对独行采集的 *T. fulviventris* 无刺蜂采集蜂则会访问独立生长且间隔更远的 *C. biflora* 植株。这种模式可能是无刺蜂进化出的采集策略的差异与蜂种竞争能力的差异相结合而产生的结果（Johnson and Hubbell,1975; Lichtenberg et al., 2017）。对于无刺蜂的大规模招募采集的一个推测结果就是会大规模招募的蜂种的饮食谱可能比独行采集的蜂种的饮食谱要更窄一些（Lichtenberg et al., 2017）。

无刺蜂也可以通过占据不同森林层级来避免竞争，例如在垂直结构的热带森林中攻击性较低的无刺蜂会占据低层，这些蜂会在较小且分散的食物源中采集，而富有竞争优势的无刺蜂则在较高的层级采集，在那里同步大量开花型的树木会为其提供丛生的且回报率高的可供采集的资源。虽然 Nagamitsu 和 Inoue（1997）发现没有证据证实无刺蜂这种按森林层级的资源分配是基于蜂种的竞争优势性，但有证据显示不同的无刺蜂蜂种会始终在相应的高度层采集：研究者经常发现马来西亚的树木树冠上的花上有 *Tetragonula fuscobalteata* 无刺蜂存在，而 *T. melanocephala* 无刺蜂则更常被发现于树下花朵上（Nagamitsu et al., 1999）。同样，在巴拿马，在较高森林层较易发现 *Partamona cupira* 无刺蜂，而 *Trigona fulviventris* 无刺蜂则喜欢在靠近地面的地方采集（Roubik,1993；Ramalho,2004）。*Trigona corvina* 无刺蜂则是在相同环境中没有显示出高度的专一性。

在成片食物源处也会发生资源分割现象，因为具有竞争优势的无刺蜂采集过程中会遗落一些花粉在花瓣、叶子或地上，而一些竞争力较弱的无刺蜂采集蜂就会伺机采集这些遗落下来的花粉（Wille,1963; Roubik,1989; Pick and Blochtein,2002; Hrncir and Maia-Silva,2013 b）。这些捡拾其他无刺蜂的遗落物的无刺蜂通常体型较小，对它们而言这种采集策略也十分成功，且研究者发现一些 *Scaura* 属无刺蜂后基跗节呈囊状且带有长排刚毛，Roubik（2018）认为该结构是这种无刺蜂采集蜂采集落在表面的花粉的一种适应性性状特征（图1.1，图8.1 C）。这些例子都凸显出，多种无刺蜂之所以能在同一栖息地共存是因为它们在行为、形态、生理和生活方式上极具多样性（图8.9）。

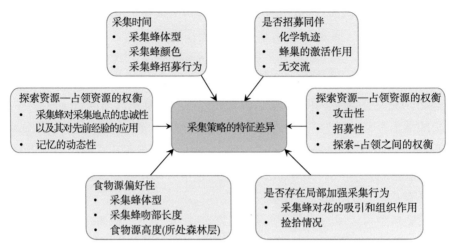

图 8.9 无刺蜂蜂种的行为和形态特征促使其采集策略形成

注：这些特征差异有利于蜂种间的生态位划分，也有利于无刺蜂蜂种在同一栖息地的共存。

8.6.2 无刺蜂与蜜蜂之间的竞争

蜜蜂是无刺蜂一个重要的潜在竞争类群。新热带区目前还引入了西方蜜蜂（主要是非洲化蜜蜂），因此西方蜜蜂是无刺蜂一个相对较新的竞争者。这意味着无刺蜂在特定群落中经过数百万年进化而成竞争的策略现在要在一个新的环境中接受考验，无刺蜂需要面对一个新的竞争者，这个竞争者食谱广泛、体型大且蜂群规模大，拥有复杂的招募通信系统（von Frisch, 1967; Roubik, 1989）。因此，非洲化蜜蜂可能会对新热带无刺蜂产生巨大压力，这已经毫无意外地引发了人们的担忧（Roubik,1978; Wilms et al., 1996）。

非洲化蜜蜂是 1957 年从非洲 *Apis mellifera scutellata* 分蜂群逃逸而来的，这些 *Apis mellifera scutellata* 蜂是由巴西遗传学家华威克柯尔（Warwick Kerr）在非洲南部收集并保存在圣保罗州的一个养蜂场的。这些非洲蜂王的子代蜂王迅速与诸如 17 世纪引入新热带区的 *A. m. ligustica* 等欧洲亚种蜂交配。由此产生的蜜蜂比它们的欧洲近亲更加高产且攻击性更强，也更能适应热带气候，因此这种蜜蜂迅速蔓延到整个南美洲的热带地区，并且在 20 世纪 80 年代初蔓延到中美洲。

无刺蜂和蜜蜂使用的食物资源有相当大部分的重叠（Koeniger and Vorwohl,1979; Sommeijer et al., 1983; Roubik,1978, 1983; Wilms et al., 1996; Wilms and Wiechers,1997; Biesmeijer and Slaa,2006; Kajobe,2007 b），至少一些无刺蜂的蜂群采集成功率似乎受到了引进的非洲化蜜蜂采集活动的负面影响

（Roubik,1978, 1983; Roubik et al., 1986）。例如，Roubik（1978）在法属圭亚那的一个栖息地，在非洲化蜜蜂首次出现后不到两年，就通过实验增加了当地非洲化蜜蜂的蜂群数量。他发现，非洲化蜜蜂的存在对无刺蜂采集一些植物物种产生了负面影响，但不影响无刺蜂采集其他植物物种。研究者还在巴拿马的低地森林中实验性地增加蜜蜂蜂群数量，结果发现他们所研究的 12 种无刺蜂蜂种中有 4 种的蜂群采集成功率降低了（Roubik et al., 1986）。特别是在高质量的食物资源地块，无刺蜂的采集活动受到了蜜蜂存在的影响（Roubik et al., 1986），这很可能是因为西方蜜蜂能够高效地招募同巢蜂抵达高质量食物地块（von Frisch,1967）。

蜜蜂会对新热带区无刺蜂采集产生负面影响，研究者在蜂天然食物资源处的观察结果为这一观点提供了进一步的证据。*Spondias mombin* 是一种热带植物，蜜蜂和无刺蜂都会访问这种热带植物。蜜蜂会每天更早地来采集这种植物的花粉，因此当 *Scaptotrigona tubiba* 无刺蜂来采集这种植物时蜜蜂已经将其采光了（Carneiro and Martins,2012; Menezes et al., 2007）。虽然蜜蜂与无刺蜂之间的竞争可能经常是"利用竞争"的情况（Roubik et al., 1986; Carneiro and Martins,2012），但是 Cairns 等（2005）已经观察到墨西哥蜜蜂会攻击性地赶走采食物资源的无刺蜂后自己去采集。Roubik（1980）在法属圭亚那的天然食物源和人工喂食器还观察到非洲化蜜蜂偶尔会攻击 *Melipona* 属无刺蜂。当食物源稀少且无刺蜂没有什么替代资源可供选择时，这种干扰性竞争可能会增加。因此，研究者推测在人类改变环境导致蜂类食物供应减少的环境中，蜜蜂蜂群对无刺蜂的负面影响会更大，或只在这种环境中才能够发现蜜蜂蜂群对无刺蜂的负面影响（Cairns et al., 2005）。

西方蜜蜂可能只影响新热带无刺蜂中的特定蜂种，而且只要蜜蜂的数量不是过多，一些好斗的大量招募型无刺蜂（例如一些 *Trigona*）也能够赶走蜜蜂占据食物资源（Roubik,1980; Cairns et al., 2005）。而其他一些无刺蜂蜂种还能够通过访问更小和更分散的食物资源来避免竞争。也有人认为，大型的同步大量开花型树木在缓解蜜蜂和无刺蜂之间的竞争（以及无刺蜂本身之间的竞争）方面发挥了重要作用，因为它们提供食物资源过于丰富以至于不会被任何一个蜂群所垄断，因此它们使得具有不同采集策略的蜂种能够互不干扰地进行采集（Wilms et al., 1996; Hrncir and Maia-Silva,2013 a）。

一些诸如 *Trigona pallens*、*Tetragona clavipes* 等无刺蜂能够通过分割食物资源来与蜜蜂一起采集（Roubik,1980），而且一项更长期的评估研究表明巴拿马和法属圭亚那地区入侵的蜜蜂总体而言对本地蜂的丰富度影响不大（Roubik,2000）。同样，一项在巴拿马 Barro Colorado 岛进行了 17 年的研究发现，蜜蜂的到来并不影响无刺蜂和其他本地蜂的丰富度（Roubik and Wolda,2001）。这个研究者认为该岛的面积相对较小（16 km^2），也就意味着蜜蜂群的密度仍然相对较低。总之，虽然有证据表明引进的蜜蜂会对一些无刺蜂的采集活动产生负面影响，但蜜蜂是否真的导致了新热带无刺蜂的减少还有待观察。评估新热带地区引进蜜蜂的生态影响是十分复杂的，因为蜜蜂通常与人类密切相关，因此蜜蜂在人类大幅改造的环境中更为常见（Stout and Morales,2009）。要了解蜜蜂对本地蜂的影响需要厘清是人类改变环境的影响还是蜜蜂活动的影响。

在非洲与亚洲这些蜜蜂本来就是本土蜂的地区，人们对蜜蜂与无刺蜂间的竞争互动所知更少（Koeniger and Vorwohl,1979）。在乌干达地区，蜜蜂和无刺蜂采集的资源中显示出相当大的重叠，这表明该地这两个群体之间可能存在相互竞争（Kajobe,2007 b）。然而，在喀麦隆进行的一项研究中发现蜜蜂和多元采集的 *Meliplebeia ogouensis* 无刺蜂在不同食物源处采集，这表明二者存在资源分割（Tropek et al., 2018）。研究者在斯里兰卡研究了人工食物源处蜜蜂和无刺蜂的相互竞争，发现小型的 *Tetragonula iridipennis* 无刺蜂似乎比小蜜蜂（*Apis florea*）和东方（*A. cerana*）蜜蜂更具有竞争优势（Koeniger and Vorwohl,1979），笔者认为 *T. iridipennis* 无刺蜂的飞行范围比那两种蜜蜂要小得多，因此这种无刺蜂会积极地捍卫在其较小领土内可以找到的食物源。无刺蜂与亚洲蜜蜂之间的竞争会随季节而有很大的变化，因为一些亚洲蜜蜂蜂种会进行季节性迁移，导致蜂群密度大幅度变化（Koeniger and Vorwohl,1979）。

8.6.3 无刺蜂与其他动物之间的竞争

研究者经常观察到无刺蜂采集蜜露或采集腐肉时与蚂蚁产生竞争，还观察到无刺蜂与苍蝇在腐肉处也产生了竞争（Roubik,1982 a, 1989; Baumgartner and Roubik,1989; Figueiredo,1996）。蚂蚁偶尔还会与无刺蜂竞争蜜源：*Ectatomma ruidum* 蚁就会积极阻止 *Trigona fulviventris* 无刺蜂采集 *Justicia aurea*，因而导致该种无刺蜂只能在清晨或傍晚这些没有蚂蚁出没的时间段进行采集（Willmer and Corbet,1981）。无刺蜂与苍蝇抢占腐肉时，无刺蜂会积极地赶走试图在动物尸体上产卵的苍蝇（如绿头苍蝇）（Roubik,1982 a; Baumgartner and Roubik,1989）。无刺蜂偶尔还会与蜂鸟争夺花蜜。例如巴拿马地区的 *Trigona ferricauda* 无刺蜂就会积极对抗 *Phaethornis supercilious* 蜂鸟

来捍卫 *Pavonia dasypetala* 的花朵（Roubik,1982 b）。

8.7 无刺蜂采集过程中的学习

工蜂从巢内工作过渡到采集工作之前通常会进行定向飞行，这可以让即将成为采集蜂的工蜂学习蜂巢周围的视觉环境（蜜蜂：Capaldi et al. ,2000）。随后采集蜂就可以使用路径整合和基于地标的学习实现在蜂巢和食物源之间的来回飞行（von Frisch,1967; Zeil and Wittmann,1993; Chittka et al., 1995; Capaldi et al., 2000）。蜂类导航学习的一个重要部分就是感知太阳的位置（"太阳罗盘"），或者当太阳不直接可见时感知日光的偏振光（von Frisch,1967；Rossel and Wehner,1986）。采集蜂通过这种日光感知与基于"视线流"估计出行距离和高度的能力相结合，能够计算出它们相对于蜂巢或资源的位置（Srinivasan et al., 2000; Hrncir et al., 2003; Eckles et al.,2012）。视线流是指在飞行过程中周围环境的图像在昆虫眼睛上移动的程度（Srinivasan et al., 2000; Hrncir et al., 2003）。除了少数研究以外（Zeil and Wittmann,1989, 1993; Hrncir et al., 2003）[13]，无刺蜂的导航仍然是一个几乎无人探索的研究领域。因此，地标、天际线特征、树冠形状或天体线索在无刺蜂导航中的相对重要性，我们知之甚少。

许多植物为了吸引蜂类来采花会产生香味，或者具有显眼的颜色和形状（Vogel,1983）。产生香味的时间通常与开花以及花粉和花蜜的产生时间相关（Vogel,1983），这使香味成为蜂类根据奖励进行关联学习的合适刺激物。花香通常由许多不同的化合物组成，这些化合物在特定植物物种中的组成比例相对稳定。这使蜂类能够区分植物种类，并学习到特定类型的植物所提供的奖励（Vogel,1983）。目前人们已经鉴定了 1 700 多种花的挥发性有机化合物，这些化合物主要是脂肪酸衍生物、苯类、苯丙类、萜类、含氮化合物和含硫化合物（Knudsen et al., 1993; Muhlemann et al., 2014）。

无刺蜂和人类一样也有三色视觉[14]，而且无刺蜂能非常有效地学习花的颜色和形状或者花产生花蜜或花粉的时间段信息（Beling,1929; von Frisch,1967;

---

[13] Zeil 和 Wittmann（1993）提出 *Tetragonisca angustula* 无刺蜂会使用蜂巢附近的地标而不是蜂巢本身来定位蜂巢。

[14] 目前研究者研究过的包括无刺蜂在内的许多蜂种都具有色觉，它们拥有对紫外线（UV）、蓝色（B）和绿色（G）敏感的光感器（三色觉）。到目前为止，研究者研究过的无刺蜂蜂种在对物体和颜色辨别能力方面要弱于蜜蜂和熊蜂，这可能是因为目前研究过的无刺蜂蜂种体型都相对较小（Spaethe et al., 2014; Streinzer et al., 2016; Dyer et al., 2016 b）。

Menzel et al., 1993; Slaa et al., 1998; Slaa et al., 2003）。例如，*Melipona fasciculata* 无刺蜂只经过 1 d 的训练就能学会在正确的时间点去查看人工食物源（Jesus et al., 2014）。关于无刺蜂这种学习时间地点的情况的研究仍相对较少（Breed et al., 2002; Murphy and Breed,2008a; Jesus et al., 2014），但这种学习能力对于无刺蜂开始采集活动的决策确立（Biesmeijer et al., 1998; Breed et al., 2002）和在竞争对手之前抵达食物源（Hrncir and Maia-Silva,2013 a,2013 b; Jesus et al., 2014）都很重要。

8.7.1　关联学习

无刺蜂学习花朵特征的一个关键过程就是关联学习。在关联学习中，无刺蜂要学会通过感知到的一种刺激（或一组刺激物，如食物源的颜色、气味或形状）预测一种奖励或惩罚［奖励或惩罚这两种刺激被称为非条件性刺激或 US（unconditioned stimuli）］。气味或颜色刺激［即条件性刺激或 CS（conditioned stimulus）］和非条件刺激之间的关联性取决于多种因素，包括采集蜂去食物源采集的次数、CS 和 US 之间的时间相关性、CS 的显著性（例如，由于无刺蜂的感官偏好等原因它们更容易学习某些特定的气味或颜色）以及 US 的强度（例如花蜜的糖含量）（Rescorla and Wagner,1972; Menzel et al., 1993; Menzel,1999）。

研究者已经对蜜蜂的关联学习进行了广泛的研究，这些研究有的是在实验室环境下采用伸吻反射模型（proboscis extension response，PRE）（最近也有测定 sting extension response, SER）进行研究，有的则是对自由飞行蜜蜂采用"Y"形迷宫的方法来进行研究（Bitterman et al., 1983; Menzel,1999; Giurfa,2007; Vergoz et al., 2007; Giurfa and Sandoz,2012）。在典型的 PER 条件反射中，人类观察者只要用蔗糖溶液触碰被束缚的蜜蜂的触角就会诱使蜜蜂伸吻（非条件性反射，UR）。然后，观察者给蜜蜂喂食少量的蔗糖溶液的同时，将蜜蜂暴露于一种气味中（Takeda,1961; Bitterman,1983; Menzel,1999; Giurfa,2007）。这就是条件反射的一种经典（或巴甫洛夫式的）形式。研究者还对自由飞行蜜蜂开展了学习实验，试验中涉及了操作性条件反射（operant conditioning），即通过这种反射蜜蜂能学会将自己的行为与奖励或惩罚联系起来（Giurfa, 2007）[15]。这项研究表明蜜蜂通过少量的学习，就可以建立条件刺激和非条件性刺激的关联，并形

---

[15] 在差异性条件反射中，研究者采用了两种条件刺激，一种条件刺激与奖励有关，另一种条件刺激则没有奖励或惩罚。两种条件刺激（如气味）以随机的顺序出现，蜜蜂必须学会对有奖励的条件刺激气味作出反应（显示 PER），而不会对另一种条件刺激气味作出反应（Bitterman et al., 1983）。

成终生记忆。例如，Menzel（1968）发现蜜蜂连续 3 次飞向蓝色或橙色的背景时获得蔗糖奖励后，这些蜜蜂在其余生都对奖励相关颜色有偏好性。蜜蜂通过一次嗅觉条件反射实验就能学会一种味道，而且一次学习通常就足以诱发蜜蜂的长期记忆，这种记忆会持续数天之久（Villar et al., 2020）。

而研究者首次尝试研究无刺蜂的经典嗅觉条件反射时却失败了：他们无法采用 PER 模型让 *Melipona scutellaris* 无刺蜂将气味与奖励关联起来［Abramson 等（1999）、Roselino 和 Hrncir（2012）和 Amaya-Márquez 等（2019）对 *M. eburnea* 和 *M. scutellaris* 无刺蜂的研究也得到类似结果］。而澳洲的 *Austroplebeia australis*、*Tetragonula carbonaria* 和 *T. hockingsi* 无刺蜂则能够在 PER 装置中学习气味（Frasnelli et al., 2011）[16]。Mc Cabe 等（2007）以及 Mc Cabe 和 Farina（2010）还研究了 *Melipona quadrifasciata*、*Scaptotrigona depilis* 和 *Tetragonisca angustula* 无刺蜂的经典伸吻条件反射，发现只有 *M. quadrifasciata* 无刺蜂能够将奖励与特定气味联系起来并展现出条件反射（伸吻反应）。*T. angustula* 无刺蜂采集蜂则只有在其蜂巢内感受过奖励气味而后进行 PER 条件反射实验才能学会区分奖励和非奖励的气味（Mc Cabe and Farina, 2010）。与非洲化蜜蜂相比，*M. quadrifasciata* 无刺蜂学习辨别有奖和无奖气味的能力（"差异性条件反射"）明显不足（McCabe et al., 2007）。*Meliponula bocandei* 和 *Axestotrigona ferruginea*（红棕色形态）这两种非洲无刺蜂则可以在 PER 模型中学会将气味与奖励联系起来并表现出差异性条件反射（Henske et al., 2015）。然而尽管 *A. ferruginea* 无刺蜂的气味辨别力高于 *M. bocandei* 无刺蜂，但是西方蜜蜂的气味辨别力仍然是最高的。

正如 Mc Cabe 等（2007）所讨论的，蜂种间的这些学习差异可能是由于采集行为和策略的差异造成的。例如，有些蜂种的食物比较专一（第 9 章），这些蜂种就不会经常经历必须学习新气味的情况，其关联学习能力差也就合情合理。此外一些蜂种的采集蜂在采集过程中可能较少依赖花香信息，而是更多地依赖视觉刺激或追踪信息素。例如 *Scaptotrigona depilis* 无刺蜂就会放置轨迹信息素来招募同巢蜂到食物源处（第 10 章）。因此在这个蜂种的采集过程中花香气味的关联学习可能就不太重要。

也有可能是方法上的原因导致一些无刺蜂蜂种中观察到的学习能力较

---

[16] 该作者还发现无刺蜂的左、右触角存在侧向化。当在右侧触角上呈现气味 1 h 后无刺蜂的记忆提取效果要好于左触角。而 5 h 后测试记忆提取效果时情况则相反（Frasnelli et al., 2011）。

差。研究者已经开发并优化了 PER 模型来研究蜜蜂的经典条件反射，但是这种学习模型可能不同样适用于无刺蜂（McCabe et al., 2007; Henske et al., 2015; Nocelli et al., 2018; Amaya-Márquez et al., 2019）。为了更好地研究无刺蜂在受控实验室条件下的关联学习，应该根据特定无刺蜂蜂种开发和优化 PER 模型。例如 Amaya-Márquez 等（2019）采用放在迷你笼中且能自由移动的采集蜂进行了 PER 模型研究并记录了学习效果，他发现这种方法测定出的学习能力要优于传统 PER 方法测出的结果（Nocelli et al., 2018）。

使用自由飞行蜂的学习模型似乎更适合研究无刺蜂的学习情况。一些研究检测了新热带无刺蜂在人工喂养站处对颜色、形状或图案与蔗糖奖励之间进行关联学习的能力。这些研究发现所有蜂种的采集蜂都能迅速学会区分有奖励和无奖励的颜色、形状和图案（图 8.10）（Pessotti and Lé'Sénéchal,1981; Menzel et al., 1989; Villa and Weiss,1990; Moreno et al., 2012; Sánchez and Vandame,2012; Balamurali et al., 2018; Aguiar et al., 2020; Koethe et al., 2020）。研究者发现 *Frieseomelitta varia* 无刺蜂学习过程较缓慢，而 *Melipona* 属无刺蜂则能高效地学习（Pessotti and Lé'Sénéchal,1981）。Mc Cabe 和 Farina（2010）还有研究了自由飞行的 *Tetragonisca angustula* 无刺蜂采集蜂的嗅觉学习情况，它们同样发现采集蜂会更倾向于它们前期采集过程中感受过的与奖励有关的气味。

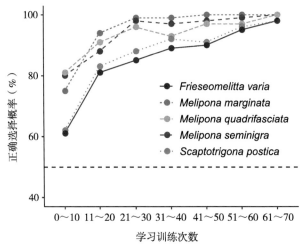

图 8.10　不同无刺蜂蜂种的自由飞翔蜂的学习成绩

注：研究者训练自由飞翔蜂在两种人工喂食器上采集食物，这两种喂食器有不同的背景色并提供了两种食物。研究者先在蓝色或黄色背景色的食物器给每种蜂的 10 只蜂奖励食物（60% 蔗糖溶液），然后在研究者记录了每只蜂 70 次训练后的着陆决策（蓝色与黄色）（Pessotti and Lé'Sénéchal,1981）。黑色虚线代表 50%（或随机）的选择概率。

无刺蜂采集蜂也能够学会将奖励与同类采集过程中在食物源上留下的印迹（如角质碳氢化合物）联系起来。研究者研究 *Melipona scutellaris* 和 *Scaptotrigona mexicana* 无刺蜂时发现，如果印迹是与奖励相关的，那采集蜂就喜欢在附着了这种印迹的食物源上着陆，而一旦印迹和空的喂食器或奖励减少的食物源关联起来后，采集蜂就会拒绝访问这些有印迹的喂食器（第10章）(Sánchez et al., 2008; Roselino et al., 2016)。这些发现也支持了上段中的解释，即一些无刺蜂在 PER 装置中无法形成学习关联是由于方法学因素，而不是无刺蜂将食物源特征和奖励进行关联学习的能力差。

当没有明显的奖励时，无刺蜂关联学习的作用还不是很清楚，例如在无刺蜂采集泥土、树脂、粪便或纸浆时就没有明显的奖励。有可能采集这些材料无刺蜂将这些材料中某些组分当作了奖励（如这些资源中的水）。对这些非主流资源的采集蜂的奖励感知和学习过程的研究，在未来也是一个有趣的研究方向。

蜂对食物气味的关联学习不仅发生在采集活动过程中，也可能当它们在巢内交哺分享食物时或者当它们对储存食物进行取样尝试时就已经发生了关联学习（在蜜蜂中：Farina et al., 2005; Grüter et al., 2006; Arenas et al., 2007, 2008; Balbuena et al., 2012)。研究者发现当 *Melipona quadrifasciata* 无刺蜂在蜂群内闻到了有香味的食物后再进行差异性 PER 条件反射试验，这些无刺蜂就会对在蜂群内循环的食物中存在过的气味表现出更好的辨别能力（Mc Cabe and Farina,2009)。研究数据还表明，被招募到香味食物处的同巢蜂需要有该香味的嗅觉记忆，而这种嗅觉记忆可能是巢内蜂与招募蜂通过交哺等互动时产生的（Mc Cabe and Farina,2009; Mc Cabe et al., 2015)。相反，*Tetragonisca angustula* 无刺蜂即使是在蜂群内接触过一些香味也并没有显示出对有这些香味的食物源的偏好（Mc Cabe and Farina,2010)。有趣的是，无刺蜂在巢内学习时，气味不一定要与奖励配对。Reichle 等（2010）发现 *Scaptotrigona pectoralis* 无刺蜂会优先选择带有其蜂巢空气中存在过的气味的人工喂食器。研究者在受控实验室环境中，进一步探讨了气味暴露对 *Melipona scutellari* 无刺蜂采食决策的影响（Roselino and Hrncir,2012)，他们发现受控蜂如果反复暴露于一种无奖励的有气味的空气，后来就会被有这种气味的食物源吸引。这表明，熟悉的巢内气味可能是驱动无刺蜂食物源偏好的一个重要因素（Reichle et al., 2011)。

8.7.2　无刺蜂对采集地点的忠诚性

采集蜂对食物源特征的学习将有助于其返回可利用的食物源（Biesmeijer

and Ermers,1999; Jesus et al., 2014; Peng et al., 2019）。众所周知，蜜蜂会连续数天或数周返回同一地点进行采集（Ribbands,1949; Al Toufailia et al., 2013），而无刺蜂好像也存在这种长时间的采集地点忠诚性（Biesmeijer et al., 1998; Biesmeijer and Ermers,1999; Hrncir and Maia-Silva,2013 b）。无刺蜂对采集地点的忠诚性在重要性上可能不如温带蜜蜂，因为热带花卉的寿命通常比较短（第 8.2 节）。因此，无刺蜂采集蜂可能要比蜜蜂采集蜂更频繁地寻找新的食物源。对于一些会从动物尸体上采集动物蛋白的无刺蜂而言，这种情况甚至更为明显。而采集泥浆、蜜露或植物树脂的采集蜂却更可能会长期访问同一地点（Leonhardt and Blüthgen,2009）。事实上，无刺蜂经常连续几天甚至几周回去访问同一个树脂采集点（Schwarz,1948; Leonhardt and Blüthgen,2009）。

一些有经验的采集蜂可能会由于下雨或夜间等原因而暂时不出采集，但是这些采集蜂一旦尝到回巢蜂带回食物样本有熟悉的味道，或者闻到其他蜂身上有它熟悉的气味（第 10 章）后，就会得到目标食物类型再次有奖励的信息，这些有经验的采集蜂就会恢复采集（Biesmeijer et al. ,1998; Biesmeijer and Ermers,1999; Hrncir and Maia-Silva,2013 b）。例如 Biesmeijer 等（1998）研究 *Melipona beecheii* 和 *M. fasciata* 无刺蜂的采集决策时发现，开始、继续或停止采集活动的决策主要取决于以往的经验和与回巢采集蜂的互动。而新手采集蜂则更依赖于成功的同巢采集蜂提供的信息。

8.7.3 无刺蜂的访花恒定性

授粉者采集花种恒定是指授粉者在一次采集旅程中只访问一种花（Grant,1950; Free,1970; Waser,1986; Chittka et al., 1999; Grüter and Ratnieks,2011），亚里士多德在公元前 340 年就已经提出过授粉者访花恒定性。这可能是花的呈簇性、授粉者对花类型的学习和对特定花特征具有先天偏好的结果（Marden and Waddington,1981; Chittka et al., 1997; Hill et al., 2001; Grüter and Ratnieks,2011; Dyer et al., 2016 a; Pangestika et al., 2017）。无刺蜂作为一个群体似乎没有对特定颜色有明确的先天偏好。例如无刺蜂偏好的蜂绿或蜂蓝等首选颜色似乎在不同的蜂种、背景颜色或实验设置中又有些不同（Dyer et al., 2016a; Koethe et al., 2016, 2018; Balamurali et al., 2018）。授粉者的访花恒定性对植物来说十分重要，因为这有利于授粉（Waser,1986; Goulson,1999），但访花恒定性是否也符合无刺蜂的最佳利益或者是否是无刺蜂学习和信息处理限制的结果仍存在争议（Waser,1986; Chittka et al., 1999; Goulson,2000; Hill et al., 2001; Grüter et al., 2011; Grüter and Ratnieks,2011; Ishii and Masuda,2014）。"学习限制"假说认为传粉者具有访花恒定性现象是因为其神经系统的限制。例如，学习如何在一种新花型上采集可能会扰乱关于在以前访问过的花型上采集的记忆。与此相关的观点是，授粉者需要具有访花恒定性以便授粉者可以为特定花型建立一个"搜索图像"，这有助于它们在这种花型上有效地采食（Waser,1986; Chittka et al., 1999; Goulson,2000; Grüter and Ratnieks,2011; Ishii and Masuda,2014）。另一个观点是，访花恒定性是一种适应性策略，是采集蜂平衡访问当前花型的好处和获取替代花型选择的信息的成本的结果（Chittka et al., 1999; Grüter and Ratnieks,2011）。不同的假说不一定是相互排斥的，鉴于授粉者拥有的不同花型选择的信息比较有限，"学习限制"的观点和"访花恒定性是一种适应性策略（'昂贵的信息假说'）"假说都有经验现象的支持（Chittka et al., 1999; Grüter and Ratnieks,2011; Ishii and Masuda,2014）。

无刺蜂似乎也是遵从访花恒定性的（Ramalho et al., 1994; Slaa et al., 1998; Slaa et al., 2003; Pangestika et al., 2017; Layek and Karmakar,2018; Cholis et al., 2020）。例如，Ramalho等（1994）观察了9个新热带区蜂种的采集蜂，发现97%的采集蜂在一次采集旅途中只访问一个花种。不同蜂种的访花恒定性程度上似乎存在差异，在相似的实验设置中无刺蜂访花恒定性程度似乎低于蜜蜂（但与中蜂相当）（Slaa et al., 1998）。研究者曾提出了多种因素来解释为什么不同蜂种都或多或少地具有访花恒定性，这些因素包括竞争、蜂群大小、蜂的体型、招募方法、食物位、栖息地多样性和稳定性等（Marden and Waddington,1981; Slaa et al., 1998, 2003; Grüter and Ratnieks,2011）。而由于不同研究采用了不同的实验装置，因而想要有意义的对比不同蜂种的访花恒定性极具挑战性（Slaa, 1998, 2003）。

参考文献

Abramson CI, Aquino IS, Stone SM（1999）Failure to find proboscis conditioning in one-day old Africanized honey bees（*Apis mellifera* L.）and in adult Uruçu honey bees（*Melipona scutellaris*）. International Journal of Comparative Psychology 12:242–262

Absy ML, Kerr WE（1977）Algumas plantas visitadas para obtenção de pólen por operárias de *Melipona seminigra merrillae* em Manaus. Acta Amazonica 7:309–315

Afik O, Delaplane KS, Shafir S, Moo-Valle H, Quezada-Euán JJG（2014）Nectar minerals as regulators of flower visitation in stingless bees and nectar hoarding wasps. Journal of Chemical Ecology 40:476–483

Aguiar JMRBV, Giurfa M, Sazima M（2020）A cognitive analysis of deceptive pollination: associative mechanisms underlying pollinators' choices in non-rewarding colour polymorphic scenarios. Scientific Reports 10:9476

Al Toufailia HM, Grüter C, Ratnieks FLW（2013）Persistence to unrewarding feeding locations by honeybee foragers（*Apis mellifera*）: the effects of experience, resource profitability and season. Ethology 119:1096–1106

Alaux C, Ducloz F, Crauser D, Le Conte Y（2010）Diet effects on honeybee immunocompetence. Biology Letters 6:562–565

Aleixo KP, Menezes C, Imperatriz Fonseca VL, Silva CI（2017）Seasonal availability of floral resources and ambient temperature shape stingless bee foraging behavior（*Scaptotrigona* aff. *depilis*）. Apidologie 48:117–127

Amaya-Márquez M, Tusso S, Hernández J, Jiménez JD, Wells H, I. Abramson C（2019）Olfactory Learning in the Stingless Bee *Melipona eburnea* Friese（Apidae: Meliponini）. Insects 10:412

Araújo ED, Costa M, Chaud-Netto J, Fowler HG（2004）Body size and flight distance in stingless bees（Hymenoptera: Meliponini）: Inference of flight range and possible ecological implications. Brazilian Journal of Biology 64:563–568

Arenas A, Farina WM（2012）Learned olfactory cues affect pollen-foraging preferences in honeybees, *Apis mellifera*. Animal Behaviour 83:1023–1033

Arenas A, Farina WM（2014）Bias to pollen odors is affected by early exposure and foraging experience. Journal of Insect Physiology 66:28–36

Arenas A, Fernández VM, Farina WM（2007）Floral odor learning within the hive affects honeybees' foraging decisions. Naturwissenschaften 94:218–222

Arenas A, Fernández VM, Farina WM（2008）Floral scents experienced within the colony affect long-term foraging preferences in honeybees. Apidologie 39:714–722

Armbruster WS（1984）The Role of Resin in Angiosperm Pollination: Ecological and Chemical Considerations. American Journal of Botany 71:1149–1160

Balamurali GS, Nicholls E, Somanathan H, Hempel de Ibarra N（2018）A comparative analysis of colour preferences in temperate and tropical social bees. The Science of Nature 105:8

Balbuena MS, Farina WM (2020) Chemosensory reception in the stingless bee *Tetragonisca angustula*. Journal of Insect Physiology 125:104076

Balbuena MS, Molinas J, Farina WM (2012) Honeybee recruitment to scented food sources: correlations between in-hive social interactions and foraging decisions. Behavioral Ecology and Sociobiology 66:445–452

Balfour NJ, Gandy S, Ratnieks FLW (2015) Exploitative competition alters bee foraging and flower choice. Behavioral Ecology and Sociobiology 69:1731–1738

Bänziger H, Boongird S, Sukumalanand P, Bänziger S (2009) Bees (Hymenoptera: Apidae) that drink human tears. Journal of the Kansas Entomological Society 82:135–151

Barônio G, Pires ACV, Aoki C (2012) *Trigona branneri* (Hymenoptera: Apidae) as a Collector of Honeydew from *Aethalion reticulatum* (Hemiptera: Aethalionidae) on *Bauhinia forficatai* (Fabaceae: Caesalpinoideae) in a Brazilian Savanna. Sociobiology 59:407–414

Bassindale R (1955) The biology of the Stingless Bee *Trigona* (*Hypotrigona*) *gribodoi* Magretti (Meliponidae). Proceedings of the Zoological Society of London 125:49–62

Baumgartner DL, Roubik DW (1989) Ecology of necrophilous and filth-gathering stingless bees (Apidae: Meliponinae) of Peru. Journal of the Kansas Entomological Society 62:11–22

Bawa KS (1990) Plant-pollinator interactions in Tropical rain forests. Annual Review of Ecology and Systematics 21:399–422

Beekman M, Ratnieks FLW (2000) Long-range foraging by the honey-bee, *Apis mellifera* L. Functional Ecology 14:490–496

Beekman M, Preece K, Schaerf TM (2016) Dancing for their supper: Do honeybees adjust their recruitment dance in response to the protein content of pollen? Insectes Sociaux 63:117–126

Beling I (1929) Über das Zeitgedächtnis der Bienen. Zeitschrift für vergleichende Physiologie 9:338

Bertelsmeier C, Avril A, Blight O, Jourdan H, Courchamp F (2015) Discovery-dominance trade-off among widespread invasive ant species. Ecology and Evolution 5:2673–2683

Biesmeijer JC, Ermers MCW (1999) Social foraging in stingless bees: how colonies of *Melipona fasciata* choose among nectar sources. Behavioral Ecology and Sociobiology 46:129–140

Biesmeijer JC, Richter JAP, Smeets MAJP, Sommeijer MJ (1999 a) Niche differentiation in nectar-collecting stingless bees: the influence of morphology, floral choice and interference competition. Ecological Entomology 24:380–388

Biesmeijer JC, Slaa EJ (2006) The structure of eusocial bee assemblages in Brazil. Apidologie 37:240–258

Biesmeijer JC, Smeets MJAP, Richter JAP, Sommeijer MJ (1999 b) Nectar foraging by stingless bees in Costa Rica: botanical and climatological influences on sugar concentration of nectar collected by *Melipona*. Apidologie 30:43–55

Biesmeijer JC, Tóth E (1998) Individual foraging, activity level and longevity in the stingless bee *Melipona beecheii* in Costa Rica (Hymenoptera, Apidae, Meliponinae). Insectes Sociaux 45:427–443

Biesmeijer JC, van Nieuwstadt MGL, Lukacs S, Sommeijer MJ (1998) The role of internal and external information in foraging decisions of *Melipona* workers (Hymenoptera : Meliponinae). Behavioral Ecology and Sociobiology 42:107–116

Bitterman ME, Menzel R, Fietz A, Schafer S (1983) Classical-conditioning of proboscis extension in honeybees (*Apis mellifera*). Journal of Comparative Psychology 97:107–119

Bollazzi M, Roces F (2011) Information needs at the beginning of foraging: grass-cutting ants trade off load size for a faster return to the nest. PLoS ONE 6:e17667

Breed MD, McGlynn TP, Sanctuary MD, Stocker EM, Cruz R (1999) Distribution and abundance of colonies of selected meliponine species in a Costa Rican tropical wet forest. Journal of Tropical Ecology 15:765–777

Breed MD, Stocker EM, Baumgartner LK, Vargas SA (2002) Time-place learning and the ecology of recruitment in a stingless bee, *Trigona amalthea* (Hymenoptera, Apidae). Apidologie 33:251–258

Burr B, Barthlott W, Westerkamp C (1996) *Staheliomyces* (*Phallales*) visited by *Trigona* (Apidae): melittophily in spore dispersal of an Amazonian stinkhorn? Journal of Tropical Ecology 12:441–445

Butler CG (1945) The influence of various physical and biological factors of the environment on honeybee activity - an examination of the relationship between activity and nectar concentration and abundance. Journal of Experimental Biology 21:5–12

Butler CG, Jeffree EP, Kalmus H (1943) The behaviour of a population of honeybees on an artificial and on a natural crop. Journal of Experimental Biology 20:65–73

Cairns CE, Villanueva-Gutiérrez R, Koptur S, Bray DB (2005) Bee populations, forest disturbance, and africanization in Mexico. Biotropica 37:686–692

Camargo JM, Roubik DW (1991) Systematics and bionomics of the apoid obligate necrophages: the *Trigona hypogea* group (Hymenoptera: Apidae; Meliponinae). Biological Journal of the Linnean Society 44:13–39

Camargo JM, Pedro SR (2002) Mutualistic association between a tiny Amazonian stingless bee and a wax-producing scale insect. Biotropica 34:446–451

Camargo JMF, Pedro SRM (2013) Meliponini Lepeletier, 1836. In: Moure JS, Urban D, Melo GAR (eds) Catalogue of Bees (Hymenoptera, Apoidea) in the Neotropical Region - online version. Available from http://www.moure.cria.org.br/catalogue

Campbell AJ, Gomes RLC, da Silva KC, Contrera FAL (2019) Temporal variation in homing ability of the neotropical stingless bee *Scaptotrigona* aff. *postica* (Hymenoptera: Apidae: Meliponini). Apidologie 50:720–732

Capaldi EA, Smith AD, Osborne JL, Fahrbach SE, Farris SM, Reynolds DR, Edwards AS, Martin A, Robinson GE, Riley JR (2000) Ontogeny of orientation flight in the honeybee revealed by harmonic radar. Nature 403:537–540

Carneiro LT, Martins CF (2012) Africanized honey bees pollinate and preempt the pollen of *Spondias mombin* (Anacardiaceae) flowers. Apidologie 43:474–486

Chalcoff VR, Aizen MA, Galetto L (2006) Nectar Concentration and Composition of 26 Species from the Temperate Forest of South America. Annals of Botany 97:413–421

Chalcoff VR, Gleiser G, Ezcurra C, Aizen MA (2017) Pollinator type and secondarily climate are related to nectar sugar composition across the angiosperms. Evolution and Ecology 31:585–602

Chittka L, Kunze J, Shipman C, Buchmann SL (1995) The significance of landmarks for path integration in homing honeybee foragers. Naturwissenschaften 82:341–343

Chittka L, Gumbert A, Kunze J (1997) Foraging dynamics of bumble bees: correlates of movements within and between plant species. Behavioral Ecology 8:239–249

Chittka L, Thomson JD, Waser NM (1999) Flower constancy, insect psychology, and plant evolution. Naturwissenschaften 86:361–377

Cholis MN, Alpionita R, Prawasti TS, Atmowidi T (2020) Pollen Load and Flower Constancy of Stingless Bees *Tetragonula laeviceps* (Smith) and *Heterotrigona itama* (Cockerell) (Apidae: Meliponinae). Atlantis Press 8:285–289

Colwell RK, Futuyma DJ (1971) On the Measurement of Niche Breadth and Overlap. Ecology 52:567–576

Corbet SA, Saville NM, Fussell M, Prŷs-Jones OE, Unwin DM (1995) The Competition Box: A Graphical Aid to Forecasting Pollinator Performance. Journal of Applied Ecology 32:707–719

Costa-Pereira R (2014) Removal of clay by stingless bees: load size and moisture selection. Anais da Academia Brasileira de Ciências 86:1287–1294

Couvillon MJ, Al Toufailia H, Butterfield TM, Schrell F, Ratnieks FLW, Schürch R (2015) Caffeinated Forage Tricks Honeybees into Increasing Foraging and Recruitment Behaviors. Current Biology 25:2815–2818

Couvillon MJ, Schürch R, Ratnieks FLW (2014) Waggle Dance Distances as Integrative Indicators of Seasonal Foraging Challenges. PLoS ONE 9:e93495

Cruz-Landim C, Serrao JE (1994) The evolutive significance of pollen use as protein resource by trigonini bees (Hymenoptera, Apidae, Meliponinae). Journal of Advanced Zoology 15:1–5

De Bruijn LLM, Sommeijer MJ (1997) Colony foraging in different species of stingless bees (Apidae, Meliponinae) and the regulation of individual nectar foraging. Insectes Sociaux 44:35–47

Değirmenci L, Thamm M, Scheiner R (2018) Responses to sugar and sugar receptor gene expression in different social roles of the honeybee (*Apis mellifera*). Journal of Insect Physiology 106:65–70

Di Trani JC, Villanueva-Gutiérrez R (2018) Annual Foraging Patterns of the Maya Bee *Melipona beecheii* (Bennett, 1831) in Quintana Roo, Mexico. In: Vit P, Pedro SRM, Roubik DW (eds) Pot-Pollen in Stingless Bee Melittology. Springer International Publishing, Cham, pp 131–138

Dobson HEM (1987) Role of flower and pollen aromas in host-plant recognition by solitary bees. Oecologia 72:618–623

Dobson HEM, Groth I, Bergström G (1996) Pollen advertisement: chemical contrasts between whole-flower and pollen odors. American Journal of Botany 83:877–885

Dolezal AG, Carrillo-Tripp J, Judd TM, Miller WA, Bonning BC, Toth AL (2019) Interacting stressors matter: diet quality and virus infection in honeybee health. Royal Society Open Science 6:181803

Dorian NN, Bonoan RE (2016) Salt Foraging of Stingless Bees at La Selva Biological Station, Costa Rica. Bee World 93:61–63

Dornhaus A, Collins EJ, Dechaume-Moncharmont FX, Houston AI, Franks NR, McNamara JM (2006) Paying for information: partial loads in central place foragers. Behavioral Ecology and Sociobiology 61:151–161

Drescher N, Wallace HM, Katouli M, Massaro CF, Leonhardt SD (2014) Diversity matters: how bees benefit from different resin sources. Oecologia 176:943–953

Dworschak K, Bluethgen N (2010) Networks and dominance hierarchies: does interspecific aggression explain flower partitioning among stingless bees? Ecological Entomology 35:216–225

Dyer AG, Boyd-Gerny S, Shrestha M, Lunau K, Garcia JE, Koethe S, Wong BBM (2016a) Innate colour preferences of the Australian native stingless bee *Tetragonula carbonaria* Sm. Journal of Comparative Physiology A 202:603–613

Dyer AG, Streinzer M, Garcia J (2016b) Flower detection and acuity of the Australian native stingless bee *Tetragonula carbonaria* Sm. Journal of Comparative Physiology A 202:629–639

Eckles MA, Roubik DW, Nieh JC (2012) A stingless bee can use visual odometry to estimate both height and distance. Journal of Experimental Biology 215:3155–3160

Eltz T, Brühl CA, Görke C (2002) Collection of mold (*Rhizopus* sp.) spores in lieu of pollen by the stingless bee *Trigona collina*. Insectes Sociaux 49:28–30

Eltz T, Brühl CA, van der Kaars S, Chey VK, Linsenmair KE (2001) Pollen foraging and resource partitioning of stingless bees in relation to flowering dynamics in a Southeast Asian tropical rainforest. Insectes Sociaux 48:273–279

Farina WM, Grüter C, Diaz PC (2005) Social learning of floral odours within the honeybee hive. Proceedings of the Royal Society of London Series B 272:1923–1928

Ferreira Junior NT, Blochtein B, Moraes JF (2010) Seasonal flight and resource collection patterns of colonies of the stingless bee *Melipona bicolor schencki* Gribodo (Apidae, Meliponini) in an Araucaria forest area in southern Brazil. Revista Brasileira de Entomologia 54:630–636

Ferreira NP, Chiavelli LUR, Savaris CR, Oliveira SM, Lucca DL, Milaneze-Gutierre MA, Faria RT, Pomini AM (2019) Chemical study of the flowers of the orchid *Oncidium baueri* Lindley and their visiting bees *Trigona spinipes* Fabricius. Biochemical Systematics and Ecology 86:103918

Figueiredo RA (1996) Interactions between stingless meliponine bees, honeydew-producing homopterans, ants and figs in a Cerrado area. Naturalia 21:159–164

Fowler HG (1979) Responses by a stingless bee to a subtropical environment. Revista de Biologia Tropical 27:111–118

Frasnelli E, Vallortigara G, Rogers LJ (2011) Origins of brain asymmetry: Lateralization of odour memory recall in primitive Australian stingless bees. Behavioural Brain Research 224:121–127

Free JB (1970) The flower constancy of bumblebees. Journal of Animal Ecology 39:395–402

Garcia MV, de Olivera ML, de Olivera Campos LA (1992) Use of Seeds of *Coussapoa asperifolia magnifolia* (Cecropiaceae) by Stingless Bees in the Central Amazonian Forest (Hymenoptera: Apidae: Meliponinae). Entomologia Generalis 17:255–258

Gardener MC, Rowe RJ, Gillman MP (2003) Tropical Bees (*Trigona hockingsi*) Show No Preference for Nectar with Amino Acids. Biotropica 35:119–125

Gastauer M, Campos LAO, Wittmann D (2011) Handling sticky resin by stingless bees (Hymenoptera, Apidae). Revista Brasileira de Entomologia 55:234–240

Gilliam M, Buchmann SL, Lorenz BJ, Roubik DW (1985) Microbiology of the Larval Provisions of the Stingless Bee, *Trigona hypogea*, an Obligate Necrophage. Biotropica 17:28–31

Giurfa M (2007) Behavioral and neural analysis of associative learning in the honeybee: a taste from the magic well. Journal of Comparative Physiology A 193:801–824

Giurfa M, Sandoz JC (2012) Invertebrate learning and memory: fifty years of olfactory conditioning of the proboscis extension response in honeybees. Learning & Memory 19:54–66

Gomes RLC, Menezes C, Contrera FAL (2015) Worker longevity in an Amazonian *Melipona* (Apidae, Meliponini) species: effects of season and age at foraging onset. Apidologie 46:133–143

Gostinski LF, Albuquerque PMC de, Contrera FAL (2017) Effect of honey harvest on the activities of *Melipona* (*Melikerria*) *fasciculata* Smith, 1854 workers. Journal of Apicultural Research 56:319–327

Goulson D (1999) Foraging strategies of insects for gathering nectar and pollen, and implications for plant ecology and evolution. Perspectives in Plant Ecology, Evolution and Systematics 2:185–209

Goulson D (2000) Are insects flower constant because they use search images to find flowers? Oikos 88:547–552

Grant V (1950) The flower constancy of bees. The Botanical Review 16:379–398

Greenleaf SS, Williams NM, Winfree R, Kremen C (2007) Bee foraging ranges and their relationship to body size. Oecologia 153:589–596

Grüter C, Acosta LE, Farina WM (2006) Propagation of olfactory information within the honeybee hive. Behavioral Ecology and Sociobiology 60:707–715

Grüter C, Moore H, Firmin N, Helanterä H, Ratnieks FLW (2011) Flower constancy in honey bee foragers (*Apis mellifera*) depends on ecologically realistic rewards. Journal of Experimental Biology 214:1397–1402

Grüter C, Ratnieks FLW (2011) Flower constancy in insect pollinators: Adaptive foraging behaviour or cognitive limitation? Communicative and Integrative Biology 4:6

Grüter C, Segers FHID, Menezes C, Vollet-Neto A, Falcon T, von Zuben LG, Bitondi MMG, Nascimento FS, Almeida EAB (2017a) Repeated evolution of soldier sub-castes suggests parasitism drives social complexity in stingless bees. Nature Communications 8:4

Grüter C, Segers FHID, Santos LLG, Hammel B, Zimmermann U, Nascimento FS (2017b) Enemy recognition is linked to soldier size in a polymorphic stingless bee. Biology Letters 13:20170511

Gut LJ, Schlising RA, Stopher CE (1977) Nectar-sugar concentrations and flower visitors in the Western Great Basin. The Great Basin Naturalist 37:523–529

Halcroft M, Haigh AM, Spooner-Hart R (2013) Ontogenic time and worker longevity in the Australian stingless bee, *Austroplebeia australis*. Insectes Sociaux 60:259-264

Harano K, Maia-Silva C, Hrncir M (2020) Adjustment of fuel loads in stingless bees (*Melipona subnitida*). Journal of Comparative Physiology A 206:85–94

Heard TA, Hendrikz JK (1993) Factors Influencing Flight Activity of Colonies of the Stingless Bee *Trigona carbonaria* (Hymenoptera, Apidae). Australian Journal of Zoology 41:343–353

Heinrich B (1975) Energetics of pollination. Annual Review of Ecology and Systematics 6:139–170

Hendriksma HP, Shafir S (2016) Honey bee foragers balance colony nutritional deficiencies. Behavioral Ecology and Sociobiology 70:509–517

Henske J, Krausa K, Hager FA, Nkoba K, Kirchner WH (2015) Olfactory associative learning in two African stingless bee species (*Meliponula ferruginea* and *M. bocandei*, Meliponini). Insectes Sociaux 62:507–516

Hilário SD, Imperatriz-Fonseca VL, Kleinert AMP (2000) Flight activity and colony strength in the stingless bee *Melipona bicolor bicolor* (Apidae, Meliponinae). Revista Brasileira de Biologia 60:299–306

Hilário SD, Imperatriz-Fonseca VL, Kleinert AMP (2001) Responses to climatic factors by foragers of *Plebeia pugnax* Moure (in litt.) (Apidae, Meliponinae). Revista Brasileira de Biologia 61:191–196

Hilário SD, Ribeiro M de F, Imperatriz-Fonseca VL (2012) Can climate shape flight activity patterns of *Plebeia remota* Hymenoptera, Apidae)? Iheringia, Série Zoologia 102:269–276

Hill PSM, Hollis J, Wells H (2001) Foraging decisions in nectarivores: unexpected interactions between flower constancy and energetic rewards. Animal Behaviour 62:729–737

Hofstede FE, Sommeijer MJ (2006) Effect of food availability on individual foraging specialisation in the stingless bee *Plebeia tobagoensis* (Hymenoptera, Meliponini). Apidologie 37:387

Howard JJ (1985) Observations on Resin Collecting by Six Interacting Species of Stingless Bees (Apidae: Meliponinae). Journal of the Kansas Entomological Society 58:337–345

Hrncir M (2009) Mobilizing the Foraging Force Mechanical Signals in Stingless

Bee Recruitment. In: Jarau S, Hrncir M (eds) Food Exploitation by Social Insects: Ecological, Behavioral, and Theoretical Approaches. CRC Press, Taylor & Francis Group, Boca Raton, FL

Hrncir M, Jarau S, Zucchi R, Barth FG (2003) A stingless bee (*Melipona seminigra*) uses optic flow to estimate flight distances. Journal of Comparative Physiology A 189:761–768

Hrncir M, Maia-Silva C (2013 a) On the diversity of forging-related traits in stingless bees. In: Vit P, Pedro SRM, Roubik DW (eds) Pot-Honey: A legacy of stingless bees. Springer, New York, pp 201–215

Hrncir M, Maia-Silva C (2013 b) The fast versus the furious - on competition, morphological foraging traits, and foraging strategies in stingless bees. In: Vit P, Roubik DW (eds) Stingless bees process honey and pollen in cerumen pots, Chapter 13. Universidad de Los Andes, Mérida

Hrncir M, Maia-Silva C, da Silva Teixeira-Souza VH, Imperatriz-Fonseca VL (2019) Stingless bees and their adaptations to extreme environments. Journal of Comparative Physiology A 205:415–426

Hubbell SP, Johnson LK (1978) Comparative foraging behavior of six stingless bee species exploiting a standardized resource. Ecology 59:1123–1136

I'Anson Price R (2018) The adaptive significance of communication and learning in bees. PhD Thesis, University of Lausanne, Switzerland

Imperatriz-Fonseca VL, Kleinert-Giovannini A, Pires JT (1985) Climate variation influence on the flight activity of *Plebeia remota* Holmberg (Hymenoptera, Apidae, Meliponinae). Revista Brasileira de Entomologia 29:427–434

Imperatriz-Fonseca VL, Kleinert-Giovannini A, Ramalho M (1989) Pollen Harvest by Eusocial Bees in a Non-Natural Community in Brazil. Journal of Tropical Ecology 5:239–242

Ishii HS, Masuda H (2014) Effect of flower visual angle on flower constancy: a test of the search image hypothesis. Behavioral Ecology 25:933–944

Jander U, Jander R (2002) Allometry and resolution of bee eyes (Apoidea). Arthropod Structure & Development 30:179–193

Jarau S, Hrncir M, Schmidt VM, Zucchi R, Barth FG (2003) Effectiveness of recruitment behavior in stingless bees (Apidae, Meliponini). Insectes Sociaux

50:365-374

Jesus TNCS, Venturieri GC, Contrera FAL (2014) Time-place learning in the bee *Melipona fasciculata* (Apidae, Meliponini). Apidologie 45:257-265

Johnson LK, Hubbell SP (1974) Aggression and competition among stingless bees: field studies. Ecology 55:120-127

Johnson LK, Hubbell SP (1975) Contrasting Foraging Strategies and Coexistence of Two Bee Species on a Single Resource. Ecology 56:1398-1406

Kajobe R (2007 a) Botanical sources and sugar concentration of the nectar collected by two stingless bee species in a tropical African rain forest. Apidologie 38:110-121

Kajobe R (2007 b) Pollen foraging by *Apis mellifera* and stingless bees *Meliponula bocandei* and *Meliponula nebulata* in Bwindi Impenetrable National Park, Uganda. African Journal of Ecology 45:265-274

Kajobe R, Echazarreta CM (2005) Temporal resource partitioning and climatological influences on colony flight and foraging of stingless bees (Apidae; Meliponini) in Ugandan tropical forests. African Journal of Ecology 43:267-275

Kämper W, Kaluza BF, Wallace H, Schmitt T, Leonhardt SD (2019) Habitats shape the cuticular chemical profiles of stingless bees. Chemoecology 29:125-133

Karunaratne W, Edirisinghe JP, Engel MS (2017) First record of a tear-drinking stingless bee *Lisotrigona cacciae* (Nurse) (Hymenoptera: Apidae: Meliponini), from the central hills of Sri Lanka. Journal of the National Science Foundation of Sri Lanka 45:79-81

Keppner EM, Jarau S (2016) Influence of climatic factors on the flight activity of the stingless bee *Partamona orizabaensis* and its competition behavior at food sources. Journal of Comparative Physiology A 202:691-699

Kim W, Gilet T, Bush JWM (2011) Optimal concentrations in nectar feeding. Proceedings of the National Academy of Sciences of the United States of America 108:16618-16621

Kim YS, Smith BH (2000) Effect of an amino acid on feeding preferences and learning behavior in the honey bee, *Apis mellifera*. Journal of Insect Physiology 46:793-801

Kleber E (1935) Hat das Zeitgedächtnis der Bienen biologische Bedeutung? Journal of Comparative Physiology A 22:221-262

Kleinert-Giovannini A, Imperatriz-Fonseca VL (1987) Aspects of the trophic niche of *Melipona marginata marginata* Lepeletier (Apidae, Meliponinae). Apidologie 18:69–100

Kneitel JM, Chase JM (2004) Trade-offs in community ecology: linking spatial scales and species coexistence. Ecology Letters 7:69–80

Knudsen JT, Tollsten L, Bergstrom LG (1993) Floral scents - A checklist of volatile compounds isolated by headspace techniques. Phytochemistry 33:253–280

Koch H, Corcoran C, Jonker M (2011) Honeydew collecting in Malagasy stingless bees (Hymenoptera: Apidae: Meliponini) and observations on competition with invasive ants. African Entomology 19:36–41

Koeniger N, Vorwohl G (1979) Competition for Food Among Four Sympatric Species of Apini in Sri Lanka (*Apis dorsata, Apis cerana, Apis florea* and *Trigona iridipennis*). Journal of Apicultural Research 18:95–109

Koethe S, Bossems J, Dyer AG, Lunau K (2016) Colour is more than hue: preferences for compiled colour traits in the stingless bees *Melipona mondury* and *M. quadrifasciata*. Journal of Comparative Physiology A 202:615–627

Koethe S, Banysch S, Alves-dos-Santos I, Lunau K (2018) Spectral purity, intensity and dominant wavelength: Disparate colour preferences of two Brazilian stingless bee species. PLOS ONE 13:e0204663

Koethe S, Fischbach V, Banysch S, Reinartz L, Hrncir M, Lunau K (2020) A Comparative Study of Food Source Selection in Stingless Bees and Honeybees: Scent Marks, Location, or Color. Frontiers in Plant Sciences 11:516

Krausa K, Hager FA, Kirchner WH (2017) The effect of food profitability on foraging behaviors and vibrational signals in the African stingless bee *Plebeina hildebrandti*. Insectes Sociaux 64:567–578

Kretschmar JA, Baumann TW (1999) Caffeine in Citrus flowers. Phytochemistry 52:19–23

Kuhn-Neto B, Contrera FAL, Castro MS, Nieh JC (2009) Long distance foraging and recruitment by a stingless bee, *Melipona mandacaia*. Apidologie 40:472–480

Layek U, Karmakar P (2018) Nesting characteristics, floral resources, and foraging activity of *Trigona iridipennis* Smith in Bankura district of West Bengal, India. Insectes Sociaux 65:117–132

Lebrun EG, Feener DH (2007) When trade-offs interact: balance of terror enforces dominance discovery trade-off in a local ant assemblage. Journal of Animal Ecology 76:58–64

León A, Arias-Castro C, Rodríguez-Mendiola MA, Meza-Gordillo R, Gutiérrez-Miceli FA, Nieh JC (2015) Colony foraging allocation is finely tuned to food distance and sweetness even close to a bee colony. Entomologia Experimentalis et Applicata 155:47–53

Leonhardt SD (2017) Chemical ecology of stingless bees. Journal of Chemical Ecology 43:385–402

Leonhardt SD, Blüthgen N (2009) A sticky affair: resin collection by Bornean stingless bees. Biotropica 41:730–736

Leonhardt SD, Blüthgen N, Schmitt T (2009) Smelling like resin: terpenoids account for species-specific cuticular profiles in Southeast-Asian stingless bees. Insectes Sociaux 56:157–170

Leonhardt SD, Dworschak K, Eltz T, Blüthgen N (2007) Foraging loads of stingless bees and utilisation of stored nectar for pollen harvesting. Apidologie 38:125–135

Leonhardt SD, Schmitt T, Blüthgen N (2011) Tree resin composition, collection behavior and selective filters shape chemical profiles of tropical bees (Apidae: Meliponini). PLoS ONE 6:e23445

Lichtenberg EM, Imperatriz-Fonseca VL, Nieh JC (2010) Behavioral suites mediate group-level foraging dynamics in communities of tropical stingless bees. Insectes Sociaux 57:105–113

Lichtenberg EM, Mendenhall CD, Brosi B (2017) Foraging traits modulate stingless bee community disassembly under forest loss. Journal of Animal Ecology 86:1404–1416

Lindauer M (1948) Über die Einwirkung von Duft- und Geschmacksstoffen sowie anderer Faktoren auf die Tänze der Bienen. Zeitschrift für vergleichende Physiologie 31:348–412

Lorenzon MCA, Matrangolo CAR (2005) Foraging on some nonfloral resources by stingless bees (Hymenoptera, Meliponini) in a Caatinga region. Brazilian Journal of Biology 65:291–298

Macías-Macías JO, Tapia-Gonzalez JM, Contreras-Escareño F (2017) Foraging behavior, environmental parameters and nests development of *Melipona colimana*

Ayala (Hymenoptera: Meliponini) in temperate climate of Jalisco, México. Brazilian Journal of Biology 77:383–387

Maia-Silva C, Hrncir M, da Silva CI, Imperatriz-Fonseca VL (2015) Survival strategies of stingless bees (*Melipona subnitida*) in an unpredictable environment, the Brazilian tropical dry forest. Apidologie 46:631–643

Maia-Silva C, Hrncir M, Imperatriz-Fonseca VL, Schorkopf DLP (2016) Stingless bees (*Melipona subnitida*) adjust brood production rather than foraging activity in response to changes in pollen stores. Journal of Comparative Physiology A 202:723–732

Malerbo-Souza DT, Halak AL (2016) Flight activity of stingless bees *Tetragonisca angustula* (Hymenoptera: Apidae: Meliponini) during in the year. Entomotropica 31:319–331

Marden JH, Waddington KD (1981) Floral choices by honey bees in relation to the relative distances to flowers. Physiological Entomology 6:431–435

Mateus S, Ferreira-Caliman MJ, Menezes C, Grüter C (2019) Beyond temporal-polyethism: division of labor in the eusocial bee *Melipona marginata*. Insectes Sociaux 66:317–328

Mateus S, Noll FB (2004) Predatory behavior in a necrophagous bee *Trigona hypogea* (Hymenoptera; Apidae, Meliponini). Naturwissenschaften 91:94–96

Mc Cabe SI, Farina WM (2009) Odor information transfer in the stingless bee *Melipona quadrifasciata*: effect of in-hive experiences on classical conditioning of proboscis extension. Journal of Comparative Physiology A 195:113–122

Mc Cabe SI, Farina WM (2010) Olfactory learning in the stingless bee *Tetragonisca angustula* (Hymenoptera, Apidae, Meliponini). Journal of Comparative Physiology A 196:481–490

Mc Cabe SI, Hartfelder K, Santana WC, Farina WM (2007) Odor discrimination in classical conditioning of proboscis extension in two stingless bee species in comparison to Africanized honeybees. Journal of Comparative Physiology A 193:1089–1099

Mc Cabe SI, Hrncir M, Farina WM (2015) Vibrating donor-partners during trophallaxis modulate associative learning ability of food receivers in the stingless bee *Melipona quadrifasciata*. Learning and Motivation 50:11–21

McDade LA, Kinsman S (1980) The Impact of Floral Parasitism in Two Neotropical Hummingbird-Pollinated Plant Species. Evolution 34:944–958

Menezes C, Silva CI da, Singer RB, Kerr WE (2007) Competição entre abelhas durante forrageamento em *Schefflera arboricola* (Hayata). Bioscience Journal 23:63–69

Menezes C, Vollet-Neto A, Marsaioli AJ, Zampieri D, Fontoura IC, Luchessi AD, Imperatriz-Fonseca VL (2015) A Brazilian social bee must cultivate fungus to survive. Current Biology 25:2851–2855

Menzel R (1999) Memory dynamics in the honeybee. Journal of Comparative Physiology A 185:323–340

Menzel R (1968) Das Gedächtnis der Honigbiene für Spektralfarben. I. Kurzzeitiges und langzeitiges Behalten. Zeitschrift für vergleichende Physiologie 60:82–102

Menzel R (1985) Learning in honey bees in an ecological and behavioral context. In: Hölldobler B, Lindauer M (eds) Experimental Behavioral Ecology. G. Fischer Verlag, Stuttgart, pp 55–74

Menzel R, Greggers U, Hammer M (1993) Functional organization of appetitive learning and memory in a generalist pollinator, the honey bee. In: Papaj D, Lewis AC (eds) Insect learning: Ecological and evolutionary perspectives. Chapman & Hall, New York, pp 79–125

Menzel R, Ventura DF, Werner A, Joaquim LCM, Backhaus W (1989) Spectral sensitivity of single photoreceptors and color vision in the stingless bee, *Melipona quadrifasciata*. Journal of Comparative Physiology A 166:151–164

Mercer AR, Menzel R (1982) The effects of biogenic amines on conditioned and unconditioned responses to olfactory stimuli in the honeybee *Apis mellifera*. Journal of Comparative Physiology A 145:363–368

Michener CD (1974) The Social Behavior of the Bees. Harvard University Press, Cambridge, MA

Michener CD, Michener CD, Winston ML, Jander R (1978) Pollen manipulation and related activities and structures in bees of the family Apidae. The University of Kansas Science Bulletin 51:575–601

Moreno AM, Souza DG, Reinhard J (2012) A comparative study of relational learning capacity in honeybees (*Apis mellifera*) and stingless bees (*Melipona rufiventris*). PLoS ONE 7:e51467

Muhlemann JK, Klempien A, Dudareva N (2014) Floral volatiles: from biosynthesis to function. Plant, Cell and Environment 37:1936–1949

Murphy CM, Breed MD (2008 a) Time-place learning in a Neotropical stingless bee, *Trigona fulviventris* Guérin (Hymenoptera: Apidae). Journal of the Kansas Entomological Society 81:73–76

Murphy CM, Breed MD (2008 b) Nectar and Resin Robbing in Stingless Bees. American Entomologist 54:36–44

Muth F, Francis JS, Leonard AS (2016 a) Bees use the taste of pollen to determine which flowers to visit. Biology Letters 12:20160356

Muth F, Papaj DR, Leonard AS (2016 b) Bees remember flowers for more than one reason: pollen mediates associative learning. Animal Behaviour 111:93–100

Nagamitsu T, Inoue T (1997) Aggressive foraging of social bees as a mechanism of floral resource partitioning in an Asian tropical rainforest. Oecologia 110:432–439

Nagamitsu T, Momose K, Inoue T, Roubik DW (1999) Preference in flower visits and partitioning in pollen diets of stingless bees in an Asian tropical rain forest. Researches on Population Ecology 41:195–202

Nascimento DL, Nascimento FS (2012) Extreme effects of season on the foraging activities and colony productivity of a stingless bee (*Melipona asilvai* Moure, 1971) in Northeast Brazil. Psyche: A Journal of Entomology 2012:267361

Nieh JC (2010) A negative feedback signal that is triggered by peril curbs honey bee recruitment. Current Biology 20:1–6

Nocelli RCF, Lourenço CT, Denardi-Gheller SE, Malaspina O, Pereira AM, Nominato FC (2018) New method to Proboscis Extension Reflex to the assessment of gustatory responses for stingless bees. Revista Ciência, Tecnologia & Ambiente 7:69–76

Noll FB (1997) Foraging behavior on carcasses in the nectrophagic bee *Trigona hypogea* (Hymenoptera: Apidae). Journal of Insect Behavior 10:463–467

Noll FB, Zucchi R, Jorge JA, Mateus S (1996) Food Collection and Maturation in the Necrophagous Stingless Bee, *Trigona hypogea* (Hymenoptera: Meliponinae). Journal of the Kansas Entomological Society 69:287–293

Nunes-Silva P, Hrncir M, Silva CI, Roldão YS, Imperatriz-Fonseca VL (2013) Stingless bees, *Melipona fasciculata*, as efficient pollinators of eggplant (*Solanum melongena*) in greenhouses. Apidologie 44:537–546

Nunes-Silva P, Costa L, Campbell AJ, Arruda H, Contrera FAL, Teixeira JSG, Gomes RLC, Pessin G, Pereira DS, de Souza P, Imperatriz-Fonseca VL (2020) Radiofrequency identification (RFID) reveals long-distance flight and homing abilities of the stingless bee *Melipona fasciculata*. Apidologie 51:240–253

Núñez JA (1982) Honeybee Foraging Strategies at a Food Source in Relation to its Distance from the Hive and the Rate of Sugar Flow. Journal of Apicultural Research 21:139–150

Oda FH, Aoki C, Oda TM, Silva RA da, Felismino MF (2009) Interação entre abelha *Trigona hyalinata* (Lepeletier, 1836) (Hymenoptera: Apidae) e *Aethalion reticulatum* Linnaeus, 1767 (Hemiptera: Aethalionidae) em *Clitoria fairchildiana* Howard (Papilionoideae). Entomo Brasilis 2:58–60

Oliveira ML, Morato EF (2000) Stingless bees (Hymenoptera, Meliponini) feeding on stinkhorn spores (Fungi, *Phallales*): robbery or dispersal? Revista Brasileira de Zoologia 17:881–884

Pangestika NW, Atmowidi T, Kahono S (2017) Pollen Load and Flower Constancy of Three Species of Stingless Bees (Hymenoptera, Apidae, Meliponinae). Tropical Life Sciences Research 28:179–187

Pankiw T, Nelson M, Page RE, Fondrk MK (2004) The communal crop: modulation of sucrose response thresholds of pre-foraging honey bees with incoming nectar quality. Behavioral Ecology and Sociobiology 55:286–292

Park WO (1926) Water-carriers versus nectar-carriers. Journal of Economic Entomology 19:656–664

Pasquale GD, Salignon M, Conte YL, Belzunces LP, Decourtye A, Kretzschmar A, Suchail S, Brunet J-L, Alaux C (2013) Influence of Pollen Nutrition on Honey Bee Health: Do Pollen Quality and Diversity Matter? PLOS ONE 8:e72016

Patricio EFLRA, Cruz-López L, Maile R, Tentschert J, Jones GR, Morgan ED (2002) The propolis of stingless bees: terpenes from the tibia of three *Frieseomelitta* species. Journal of Insect Physiology 48:249–254

Peng T, Segers FHID, Nascimento F, Grüter C (2019) Resource profitability, but not caffeine, affects individual and collective foraging in the stingless bee *Plebeia droryana*. Journal of Experimental Biology 222: jeb195503

Peng T, Schroeder M, Grüter C (2020) Octopamine increases individual and collective foraging in a neotropical stingless bee. Biology Letters 16:20200238

Pereboom JJM, Biesmeijer JC (2003) Thermal constraints for stingless bee foragers: the importance of body size and coloration. Oecologia 137:42–50

Pereira CD, Tannús-Neto J (2009) Observations on stingless bees (Hymenoptera: meliponini) collecting of latex in *Mammy mammea americana* (1.) Jacq. (Clusiaceae), Manaus, state of Amazonas, Brazil. Biosci J 25:133–135

Pernal SF, Currie RW (2002) Discrimination and preferences for pollen-based cues by foraging honeybees, *Apis mellifera* L. Animal Behaviour 63:369–390

Perry CJ, Barron AB (2013) Neural mechanisms of reward in insects. Annual Review of Entomology 58:543–562

Peruquetti RC, Costa L da SM da, Silva VS da, Drumond PM (2010) Frugivory by a stingless bee (Hymenoptera: Apidae). Neotropical Entomology 39:1051–1052

Pessotti I, Lé'Sénéchal AM (1981) Aprendizagem em abelhas. I - Discriminação simples em onze espécies. Acta Amazonica 11:653–658

Pick RA, Blochtein B (2002) Atividades de coleta e origem floral do pólen armazenado em colônias de *Plebeia saiqui* (Holmberg) (Hymenoptera, Apidae, Meliponinae) no sul do Brasil. Revista Brasileira de Zoologia 19:289–300

Pierrot LM, Schlindwein C (2003) Variation in daily flight activity and foraging patterns in colonies of uruçu - *Melipona scutellaris* Latreille (Apidae, Meliponini). Revista Brasileira de Zoologia 20:565–571

Primack RB (1985) Longevity of individual flowers. Annual Review of Ecology and Systematics 16:15–37

Quezada-Euán JJG, López-Velasco A, Pérez-Balam J, Moo-Valle H, Velazquez-Madrazo A, Paxton RJ (2011) Body size differs in workers produced across time and is associated with variation in the quantity and composition of larval food in *Nannotrigona perilampoides* (Hymenoptera, Meliponini). Insectes Sociaux 58:31–38

Ramalho M (2004) Stingless bees and mass flowering trees in the canopy of Atlantic Forest: a tight relationship. Acta Botanica Brasilica 18:37–47

Ramalho M, Giannini TC, Malagodi-Braga KS, Imperatriz-Fonseca VL (1994) Pollen Harvest by Stingless Bee Foragers (Hymenoptera, Apidae, Meliponinae). Grana

33:239–244

Ramalho M, Kleinert-Giovanni A, Imperatriz-Fonseca VL (1989) Utilization of floral resources by species of *Melipona* (Apidae, Meliponinae): floral preferences. Apidologie 20:185–195

Rasmussen C (2013) Stingless bees (Hymenoptera: Apidae: Meliponini) of the Indian subcontinent: Diversity, taxonomy and current status of knowledge. Zootaxa 3647:401–428

Rathcke B, Lacey EP (1985) Phenological patterns of terrestrial plants. Annual Review of Ecology and Systematics 16:179–214

Rego JO, Oliveira R, Jacobi CM, Schlindwein C (2018) Constant flower damage caused by a common stingless bee puts survival of a threatened buzz-pollinated species at risk. Apidologie 49:276–286

Reichle C, Aguilar I, Ayasse M, Jarau S (2011) Stingless bees (*Scaptotrigona pectoralis*) learn foreign trail pheromones and use them to find food. Journal of Comparative Physiology A 197:243–249

Reichle C, Jarau S, Aguilar I, Ayasse M (2010) Recruits of the stingless bee *Scaptotrigona pectoralis* learn food odors from the nest atmosphere. Naturwissenschaften 97:519–524

Renner S (1983) The Widespread Occurrence of Anther Destruction by *Trigona* Bees in Melastomataceae. Biotropica 15:251–256

Rescorla RA, Wagner AR (1972) A theory of Pavlovian conditioning: variations of the effectiveness of reinforcement and nonreinforcement. In: Black AH, Prokasy WT (eds) Classical Conditioning. II. Current Research and Theory. Appleton-Century-Crofts, New York, pp 64–99

Reyes-González A, Zamudio F (2020) Competition interactions among stingless bees (Apidae: Meliponini) for *Croton yucatanensis* Lundell resins. Int J Trop Journal Tnsect Sci (in press)

Ribbands CR (1949) The foraging method of individual honey-bees. Journal of Animal Ecology 18:47–66

Roces F, Núñez JA (1993) Information about food quality influences load-size selection in recruited leaf-cutting ants. Animal Behaviour 45:135–143

Roselino AC, Hrncir M (2012) Repeated unrewarded scent exposure influences the food choice of stingless bee foragers, *Melipona scutellaris*. Animal Behaviour 83:755–762

Roselino AC, Rodrigues AV, Hrncir M (2016) Stingless bees (*Melipona scutellaris*) learn to associate footprint cues at food sources with a specific reward context. Journal of Comparative Physiology A 202:657–666

Rossel S, Wehner R (1986) Polarization vision in bees. Nature 323:128–131

Roubik DW (1989) Ecology and Natural History of Tropical Bees. Cambridge University Press, New York

Roubik DW (1978) Competitive interactions between neotropical pollinators and africanized honey bees. Science 201:1030–1032

Roubik DW (1980) Foraging Behavior of Competing Africanized Honeybees and Stingless Bees. Ecology 61:836–845

Roubik DW (1982a) Obligate necrophagy in a social bee. Science 217:1059–1060

Roubik DW (1982b) The Ecological Impact of Nectar-Robbing Bees and Pollinating Hummingbirds on a Tropical Shrub. Ecology 63:354–360

Roubik DW (1982c) Seasonality in colony food storage, brood production and adult survivorship: studies of *Melipona* in tropical forest (Hymenoptera: Apidae). Journal of the Kansas Entomological Society 55:789–800

Roubik DW (1983) Experimental community studies: time-series tests of competition between African and neotropical bees. Ecology 971–978

Roubik DW (1993) Tropical pollinators in the canopy and understory: Field data and theory for stratum "preferences." Journal of Insect Behavior 6:659–673

Roubik DW (2000) Pollination System Stability in Tropical America. Conservation Biology 14:1235–1236

Roubik DW (2006) Stingless bee nesting biology. Apidologie 37:124–143

Roubik DW (2018) 100 Species of Meliponines (Apidae: Meliponini) in a Parcel of Western Amazonian Forest at Yasuní Biosphere Reserve, Ecuador. In: Vit P, Pedro SRM, Roubik DW (eds) Pot-Pollen in Stingless Bee Melittology. Springer International Publishing, Cham, pp 189–206

Roubik DW, Aluja M (1983) Flight Ranges of Melipona and Trigona in Tropical Forest. Journal of the Kansas Entomological Society 56:217–222

Roubik DW, Buchmann SL (1984) Nectar selection by *Melipona* and *Apis mellifera*

(Hymenoptera: Apidae) and the ecology of nectar intake by bee colonies in a tropical forest. Oecologia 61:1–10

Roubik DW, Holbrook NM, Parra GV (1985) Roles of nectar robbers in reproduction of the tropical treelet *Quassia amara* (Simaroubaceae). Oecologia 66:161–167

Roubik DW, Moreno JE, Vergara C, Wittmann D (1986) Sporadic food competition with the African honey bee: projected impact on neotropical social bees. Journal of Tropical Ecology 2:97–111

Roubik DW, Moreno Patiño JE (2018) The Stingless Honey Bees (Apidae, Apinae: Meliponini) in Panama and Pollination Ecology from Pollen Analysis. In: Vit P, Pedro SRM, Roubik DW (eds) Pot-Pollen in Stingless Bee Melittology. Springer International Publishing, Cham, pp 47–66

Roubik DW, Wolda H (2001) Do competing honey bees matter? Dynamics and abundance of native bees before and after honey bee invasion. Population Ecology 43:53–62

Roubik DW, Yanega D, S MA, Buchmann SL, Inouye DW (1995) On optimal nectar foraging by some tropical bees (Hymenoptera: Apidae). Apidologie 26:197–211

Roulston TH, Cane JH (2000) Pollen nutritional content and digestibility for animals. Plant Systematics and Evolution 222:187–209

Roulston TH, Cane JH, Buchmann SL (2000) What Governs Protein Content of Pollen: Pollinator Preferences, Pollen–Pistil Interactions, or Phylogeny? Ecological Monographs 70:617–643

Sakagami S, Inoue T (1985) Taxonomic notes on three bicolorous *Tetragonula* stingless bees in Southeast Asia. Kontyu 53:174–189

Sánchez D, Nieh JC, Vandame R (2008) Experience-based interpretation of visual and chemical information at food sources in the stingless bee *Scaptotrigona mexicana*. Animal Behaviour 76:407–414

Sánchez D, Vandame R (2012) Color and Shape Discrimination in the Stingless Bee *Scaptotrigona mexicana* Guérin (Hymenoptera, Apidae). Neotropical Entomology 41:171–177

Santos CF, Halinski R, de Souza Dos Santos PD, Almeida EAB, Blochtein B (2019) Looking beyond the flowers: associations of stingless bees with sap-sucking insects. Naturwissenschaften 106:12

Scheiner R, Barnert M, Erber J (2003) Variation in water and sucrose responsiveness during the foraging season affects proboscis extension learning in honey bees. Apidologie 34:67–72

Scheiner R, Page RE, Erber J (2004) Sucrose responsiveness and behavioral plasticity in honey bees (*Apis mellifera*). Apidologie 35:133–142

Scheiner R, Plückhahn S, Öney B, Blenau W, Erber J (2002) Behavioural pharmacology of octopamine, tyramine and dopamine in honey bees. Behavioural Brain Research 136:545–553

Schmid-Hempel P, Kacelnik A, Houston AI (1985) Honeybees maximize efficiency by not filling their crop. Behavioral Ecology and Sociobiology 17:61–66

Schmidt VM, Zucchi R, Barth FG (2006) Recruitment in a scent trail laying stingless bee (*Scaptotrigona* aff. *depilis*): Changes with reduction but not with increase of the energy gain. Apidologie 37:487–500

Schorkopf DLP, de Sá Filho GF, Maia-Silva C, Schorkopf M, Hrncir M, Barth FG (2016) Nectar profitability, not empty honey stores, stimulate recruitment and foraging in *Melipona scutellaris* (Apidae, Meliponini). Journal of Comparative Physiology A 202:709–722

Schürch R, Grüter C (2014) Dancing bees improve colony foraging success as long-term benefits outweigh short-term costs. PLoS ONE 9:e104660

Schwarz HF (1948) Stingless Bees (Meliponidae) of the Western Hemisphere. Bulletin of the American Museum of Natural History 90:1–546

Schwarz HF (1932) The genus *Melipona*: the type genus of the Meliponidae or stingless bees. Bulletin of the American Museum of Natural History 63:231–460

Seeley TD (1995) The wisdom of the hive: The social physiology of honey bee colonies. Harward University Press, Cambridge, MA

Seeley TD (1986) Social foraging by honeybees - How colonies allocate foragers among patches of flowers. Behavioral Ecology and Sociobiology 19:343–354

Shackleton K, Al Toufailia H, Balfour NJ, Nascimento FS, Alves DA, Ratnieks FLW (2015) Appetite for self-destruction: suicidal biting as a nest defense strategy in *Trigona* stingless bees. Behavioral Ecology and Sociobiology 69:273–281

Shackleton K, Balfour NJ, Al Toufailia H, Gaioski R, de Matos Barbosa M, Silva CA de S, Bento JMS, Alves DA, Ratnieks FLW (2016) Quality versus quantity: Foraging decisions in the honeybee (*Apis mellifera scutellata*) feeding on wildflower nectar and fruit juice. Ecology and Evolution 6:7156–7165

Shaw DE (1990) The incidental collection of fungal spores by bees and the collection of spores in lieu of pollen. Bee World 71:158–176

Silva AG, Pinto RS, Contrera FAL, Albuquerque PMC, Rêgo MMC (2014) Foraging distance of *Melipona subnitida* Ducke (Hymenoptera: Apidae). Sociobiology 61:494–501

Silva JG, Meneses HM, Freitas BM, Silva JG da, Meneses HM, Freitas BM (2019) Foraging behavior of the small-sized stingless bee *Plebeia* aff. *flavocincta*. Revista Ciência Agronômica 50:484–492

Simone-Finstrom M, Gardner J, Spivak M (2010) Tactile learning in resin foraging honeybees. Behavioral Ecology and Sociobiology 64:1609–1617

Simpson BB, Neff JL, Dieringer G (1990) The production of floral oils by *Monttea* (Scrophulariaceae) and the function of tarsal pads in *Centris* bees. Plant Systematics and Evolution 173:209–222

Simpson J, Riedel IBM (1964) Discharge and manipulation of labial gland secretion by workers of *Apis mellifera* (L.) (Hymenoptera: Apidae). Proceedings of the Royal Entomological Society of London Series A, General Entomology 39:76–82

Singaravelan N, Nee'man G, Inbar M, Izhaki I (2005) Feeding responses of free-flying honeybees to secondary compounds mimicking floral nectars. Journal of Chemical Ecology 31:2791–2804

Slaa EJ, Cevaal A, Sommeijer MJ (1998) Floral constancy in *Trigona* stingless bees foraging on artificial flower patches: a comparative study. Journal of Apicultural Research 37:191–198

Slaa EJ, Tack AJM, Sommeijer MJ (2003) The effect of intrinsic and extrinsic factors on flower constancy in stingless bees. Apidologie 34:457–468

Solberg Y, Remedios G (1980) Chemical composition of pure and bee-collected pollen. Meldinger fra Norges Landbrukshgskole 59:1–12

Sommeijer MJ, De Rooy GA, Punt W, De Bruijn LLM (1983) A comparative study of foraging behavior and pollen resources of various stingless bees (*Hym., Meliponinae*) and honeybees (*Hym., Apinae*) in Trinidad, West-Indies. Apidologie 14:205–224

Southwick EE, Loper GM, Sadwick SE (1981) Nectar Production, Composition, Energetics and Pollinator Attractiveness in Spring Flowers of Western New York. American Journal of Botany 68:994–1002

Souza-Junior JBF, Teixeira-Souza VH da S, Oliveira-Souza A, de Oliveira PF, de Queiroz JPAF, Hrncir M (2020) Increasing thermal stress with flight distance in stingless bees (*Melipona subnitida*) in the Brazilian tropical dry forest: Implications for constraint on foraging range. Journal of Insect Physiology 123:104056

Spaethe J, Brockmann A, Halbig C, Tautz J (2007) Size determines antennal sensitivity and behavioral threshold to odors in bumblebee workers. Naturwissenschaften 94:733–739

Spaethe J, Streinzer M, Eckert J, May S, Dyer AG (2014) Behavioural evidence of colour vision in free flying stingless bees. Journal of Comparative Physiology A 200:485–496

Srinivasan MV, Zhang SW, Altwein M, Tautz J (2000) Honeybee navigation: Nature and calibration of the "odometer." Science 287:851–853

Steffan-Dewenter I, Kuhn A (2003) Honeybee foraging in differentially structured landscapes. Proceedings of the Royal Society of London Series B-Biological Sciences 270:569–575

Streinzer M, Huber W, Spaethe J (2016) Body size limits dim-light foraging activity in stingless bees (Apidae: Meliponini). Journal of Comparative Physiology A 202:643–655

Sung IH, Yamane S, Lu SS, Ho KK (2011) Climatological influences on the flight activity of stingless bees (*Lepidotrigona hoozana*) and honeybees (*Apis cerana*) in Taiwan (Hymenoptera, Apidae). Sociobiology 58:835–850

Takeda K (1961) Classical conditioned response in the honey bee. Journal of Insect Physiology 6:168–179

Teixeira AF, Oliveira FF de, Viana BF (2007) Utilization of floral resources by bees of the genus *Frieseomelitta* von Ihering (Hymenoptera: Apidae). Neotropical Entomology 36:675–684

Teixeira LV, Campos F de NM (2005) Início da atividade de vôo em abelhas sem ferrão (Hymenoptera, Apidae): influência do tamanho da abelha e da temperatura ambiente. Zoociências 7:195–202

Tezuka T, Maeta Y (1995) Pollen robbing behaviors observed in two species of introduced stingless bees (Hymenoptera, Apidae). Japanese Journal of Entomology 63:759–762

Thomson JD, Draguleasa MA, Tan MG (2015) Flowers with caffeinated nectar receive more pollination. Arthropod-Plant Interactions 9:1–7

Tropek R, Padysakova E, Padysakova E, Janecek S, Janecek S (2018) Floral Resources Partitioning by Two Co-occurring Eusocial Bees in an Afromontane Landscape. Sociobiology 65:527–530

Vallejo-Marín M (2019) Buzz pollination: studying bee vibrations on flowers. New Phytologist 224:1068–1074

van Benthem FDJ, Imperatriz-Fonseca VL, Velthuis HHW (1995) Biology of the stingless bee *Plebeia remota* (Holmberg): observations and evolutionary implications. Insectes Sociaux 42:71–87

van Nieuwstadt MGL, Iraheta CR (1996) Relation between size and foraging range in stingless bees (Apidae, Meliponinae). Apidologie 27:219–228

Velikova M, Bankova V, Marcucci MC, Tsvetkova I, Kujumgiev A (2000) Chemical composition and biological activity of propolis from Brazilian meliponinae. Z Naturforsch, C, Journal of Bioscience 55:785–789

Vergoz V, Roussel E, Sandoz JC, Giurfa M (2007) Aversive learning in honeybees revealed by the olfactory conditioning of the sting extension reflex. PLoS ONE 2:e288

Vijayakumar K, Jeyaraaj R (2016) Floral Sources for Stingless Bees (*Tetragonula iridipennis*) in Nellithurai Village, Tamilnadu, India. Ambient Science 3:69–74

Villa JD, Weiss MR (1990) Observations on the use of visual and olfactory cues by *Trigona* spp. foragers. Apidologie 21:541–545

Villar ME, Marchal P, Viola H, Giurfa M (2020) Redefining Single-Trial Memories in the Honeybee. Cell Reports 30:2603–2613.e3

Visscher PK, Seeley TD (1982) Foraging strategy of honeybee colonies in a temperate deciduous forest. Ecology 63:1790–1801

Vogel S (1983) Ecophysiology of zoophilic pollination. In: Lange OL, Nobel PS, Osmond CB, Ziegier H (eds) Physiological plant ecology III. Springer, Berlin Heidelberg New York, pp 559–624

von Frisch K (1967) The dance language and orientation of bees. Harvard University

Press, Cambridge, MA

von Ihering H (1903) Biologie der stachellosen Honigbienen Brasiliens. Zoologische Jahrbücher Abteilung für Systematik Ökologie und Geographie der Tiere 19:179–287

Vossler FG (2015) Broad Protein Spectrum in Stored Pollen of Three Stingless Bees from the Chaco Dry Forest in South America (Hymenoptera, Apidae, Meliponini) and Its Ecological Implications. Psyche J Entomol 2015:659538

Vossler FG, Tellería MC, Cunningham M (2010) Floral resources foraged by *Geotrigona argentina* (Apidae, Meliponini) in the Argentine Dry Chaco forest. Grana 49:142–153

Waddington KD, Visscher PK, Herbert TJ, Richter MR (1994) Comparisons of forager distributions from matched honey-bee colonies in suburban environments. Behavioral Ecology and Sociobiology 35:423–429

Wallace HM, Lee DJ (2010) Resin-foraging by colonies of *Trigona sapiens* and *T. hockingsi* (Hymenoptera: Apidae, Meliponini) and consequent seed dispersal of *Corymbia torelliana* (Myrtaceae). Apidologie 41:428–435

Wallace HM, Trueman SJ (1995) Dispersal of *Eucalyptus torelliana* seeds by the resin-collecting stingless bee, *Trigona carbonaria*. Oecologia 104:12–16

Waser NM (1986) Flower constancy: definition, cause, and measurement. The American Naturalist 127:593–603

White DW, Stiles EW (1985) The Use of Refractometry to Estimate Nutrient Rewards in Vertebrate-Dispersed Fruits. Ecology 66:303–307

Wille A (1983) Biology of the stingless bees. Annual Review of Entomology 28:41–64

Wille A (1963) Behavioral adaptations of bees for pollen collecting from *Cassia* flowers. Revista de Biologia Tropical 11:205–210

Wille A (1962) A technique for collecting stingless bees under jungle conditions. Insectes Sociaux 9:291–293

Wille A, Michener CD (1973) The nest architecture of stingless bees with special reference to those of Costa Rica. Revista de Biologia Tropical 21:9–278

Wille A, Orozco E (1975) Observations on the founding of a new colony by *Trigona cupira* (Hymenoptera: Apidae) in Costa Rica. Revista de Biologia Tropical 22:253–287

Willmer PG, Corbet SA (1981) Temporal and microclimatic partitioning of the floral resources of *Justicia aurea* amongst a concourse of pollen vectors and nectar robbers. Oecologia 51:67–78

Wilms W, Imperatriz-Fonseca VL, Engels W (1996) Resource partitioning between highly eusocial bees and possible impact of the introduced Africanized honey bee on native stingless bees in the Brazilian Atlantic rainforest. Studies on Neotropical Fauna and Environment 31:137–151

Wilms W, Wiechers B (1997) Floral resource partitioning between native *Melipona* bees and the introduced Africanized honey bee in the Brazilian Atlantic rain forest. Apidologie 28:339–355

Wolda H, Roubik DW (1986) Nocturnal Bee Abundance and Seasonal Bee Activity in a Panamanian Forest. Ecology 67:426–433

Wolf TJ, Schmid-Hempel P (1989) Extra loads and foraging life span in honeybee workers. The Journal of Animal Ecology 58:943–954

Wright GA, Baker DD, Palmer MJ, Stabler D, Mustard JA, Power EF, Borland AM, Stevenson PC (2013) Caffeine in floral nectar enhances a pollinator's memory of reward. Science 339:1202–1204

Zarchin S, Dag A, Salomon M, Hendriksma HP, Shafir S (2017) Honey bees dance faster for pollen that complements colony essential fatty acid deficiency. Behavioral Ecology and Sociobiology 71:172

Zeil J, Wittmann D (1989) Visually controlled station-keeping by hovering guard bees of *Trigona* (*Tetragonisca*) *angustula* (Apidae, Meliponinae). Journal of Comparative Physiology A 165:711–718

Zeil J, Wittmann D (1993) Landmark orientation during the approach to the nest in the stingless bee *Trigona* (*Tetragonisca*) *angustula* (Apidae, Meliponinae). Insectes Sociaux 40:381–389

Zucoloto FS, Penedo MCT (1977) Physiological effects of mannose in *Nannotrigona* (*Scaptotrigona*) *postica* (Hymenoptera, Apoidea, Meliponinae). Boletim de Zoologia, Universidade de São Paulo 2:129–134

9 无刺蜂授粉

热带的开花植物有超过 90% 都是通过动物授粉的（Ollerton et al., 2011）。不管是在新大陆还是旧大陆蜂类都是热带地区最重要的授粉昆虫（Bawa,1990; Momose et al.,1998; Corlett,2004; Klein et al.,2007; Michener,2007; Giannini et al.,2015; Ollerton,2017），其中无刺蜂则是在多种环境中比较突出的访花种群（图 9.1）。例如，Ramalho（2004）记录了巴西大西洋雨林中的访花蜂，发现尽管所有访花蜂蜂种中无刺蜂只占 7%，但是却占到所有访花蜂总数的 70%（Wilms et al.,1996）。研究者在哥斯达黎加的一个栖息地观察到正在采集的蜂中有 50% 是无刺蜂，而在当地记录到的蜂种中无刺蜂仅占 16%（Brosi et al.,2008）。在墨西哥其他栖息地上也出现类似的情况，正在采花的蜂中有 52% 都是无刺蜂。Gutiérrez-Chacón 等（2018）在哥伦比亚安第斯山脉的森林边缘和牧场对访花蜂进行了采样，发现 49% 的访花蜂都是 *Meliponini* 属无刺蜂（11% 是意大利蜜蜂），其中无刺蜂蜂种占到所有访花蜂蜂种的 17.5%（图 9.1）。这 5 项研究记录到的所有访花蜂个体中有一半都是无刺蜂，蜜蜂则只占 18%，其他蜂种占到 29%。在亚洲访花蜂中无刺蜂也同样重要：马来西亚的一个低湿雨林中所有开花植物中约有 25% 是完全或部分由无刺蜂授粉的（11% 则是由亚洲蜜蜂属蜂种授粉的）（Momose et al.,1998; Sakai et al., 1999）。该地区所有授粉类群中，无刺蜂是访问植物种类数量最多的授粉蜂类群（Momose et al.,1998）。

图 9.1　热带美洲不同地区在开花植物上或食物诱饵上观察到的无刺蜂、西方蜜蜂（*Apis mellifera*）和其他蜂的丰度

注：Wilms 等（1996）发现在巴西大西洋热带雨林中无刺蜂占所有访花者的 44%，而 Ramalho（2004）发现无刺蜂占所有访花者的 70%。在哥伦比亚安第斯山脉的森林边缘和牧场中研究者观察到的所有访花蜂中 Meliponini 无刺蜂占 49%（Gutiérrez-Chacon et al.,2018），而在哥伦比亚山区栖息地研究者在开花植物上观察到的访花蜂中 46% 是无刺蜂（Cely-Santos and Philpott,2019）。在哥斯达黎加和墨西哥研究者取样得到的采集蜂种有 50% 是无刺蜂（Cairns et al.,2005; Brosi et al.,2008）。

在热带环境中 *Meliponine* 属无刺蜂十分重要，这主要有多方面的原因，例如，它们绝对数量庞大、形态多样、采集策略多样、采集习性多元（多元性采集），而且它们采集过程中遵循访花恒定性（第 8 章）。依据研究者报道的蜂巢密度和蜂群大小（第 1 章，第 3 章），提示在热带栖息地每平方千米内通常有 100 万只以上的无刺蜂。*Meliponine* 无刺蜂形态和行为的多样性意味着它们可以更为广泛地从开花植物中收集花粉和花蜜。然而，并不是所有的访花行为都会帮植物授粉（Heard,1999）。无刺蜂中的花蜜和花粉强盗蜂就是这种情况的典型案例，而且访问大型花朵的小型无刺蜂采集蜂甚至可以在不接触花柱头的情况下采到花粉（"食花粉者"）。这样导致那些本可以被有效授粉者采集的花粉被食花粉无刺蜂取走，因此无刺蜂偶尔也会对植物适应性产生负面影响（Roubik,1989）。即使无刺蜂正常访花也经常因为花柱头不再接受授粉等原因不能成功授粉（Roubik,1989），因此要确定无刺蜂授粉的有效性亟须更为详细地研究。

此外，无刺蜂蜂群中尽管大部分的采集工作是由雌性工蜂完成的（第6章），但是 Boongird 和 Michener（2010）观察到一些亚洲蜂种（*Tetrigona apicalis*、*Tetragonilla collina*、*Tetragonula fuscobalteata* 和 *T. pagdeni*）雄蜂的后胫节上也会携带花粉或树脂。这些蜂种的雄蜂的胫节大小形状与同蜂种工蜂相似（"等足"）。而与之亲缘关系较近的蜂种雄蜂则是"不等足"后胫节（不同于工蜂的），研究者也从未发现这些雄蜂会携带花粉。想要确认这种现象是否普遍存在并探讨雄蜂采集花粉的潜在意义，进行定量观察是十分必要的（Boongird and Michener,2010）。

目前已有大量研究通过调研访花蜂、鉴定采花归来的采集蜂上取下的花粉类型或者通过记录贮藏在花粉罐或蜜罐中的花粉类型等方法分析了无刺蜂的食物（表9.1）。花粉最终会在蜜罐中存在是因为花蜜采集蜂吸取花蜜时吸入了花中脱落出的花粉粒，或者是因为花粉粒会附着在花蜜采集蜂的身体上被带回蜂巢而后最终被放入食物罐（Maia-Silva et al.,2018）。这些研究发现无刺蜂蜂群会从多种多样的植物物种中收集的花粉和花蜜（图9.2）。例如，有一项研究检测了玻利维亚、巴拿马和秘鲁地区的 *Tetragonisca angustula* 无刺蜂蜂蜜样本，发现 *Tetragonisca angustula* 无刺蜂总计访采了175种不同的植物（Roubik and Patiño,2013, 2018）。墨西哥的 *Scaptotrigona hellwegeri* 无刺蜂的采集蜂则在12个月里采集了来自两个地点57个科165个种的植物的花粉（Quiroz-García et al.,2011）。印度西孟加拉邦的 *Tetragonula iridipennis* 无刺蜂采集蜂携带的花粉粒及其蜂巢中的花粉罐中有来自49个科117个种的不同类型植物的花粉（Layek and Karmakar,2018; Roopa et al.,2017），而哥斯达黎加低地地区的 *Trigona fulviventris* 无刺蜂则访采了95种植物（Roubik,1989）。这些研究和其他一些研究表明，一些无刺蜂蜂群（如亚洲 *Tetragonula iridipennis* 或美洲 *Plebeia droryana* 和 *Trigona spinipes*）较之于生活在同一地区的蜜蜂蜂群会从多种多样的植物中收集食物（Koeniger and Vorwohl,1979; Roubik,1989; Wilms et al.,1996; Lorenzon et al.,2003）。然而许多其他蜂种较之其蜜蜂族近亲食谱却相对较窄一些（Sommeijer et al.,1983; Imperatriz-Fonseca et al.,1989; Wilms et al.,1996）。

表 9.1　根据花粉和蜂蜜样品或访花蜂确定的无刺蜂访问的植物物种数量

无刺蜂蜂种	植物种属数	采样期	样品类型	采样地点	参考文献
Frieseomelitta varia	77	12 个月	携粉采集蜂	São Paulo state, Brazil	Aleixo et al.（2013）
Melipona beecheii	68	240 个月	花粉罐	Yucatán, Mexico	Villanueva–Gutiérrez et al.（2018）
Melipona bicolor	57	30 个月	访花蜂	Atlantic rain forest, SP, Brazil	Wilms et al.（1996）
Melipona eburnea	92	12 个月	蜜罐和花粉罐	Cundinamarca, Colombia	Obregon and Nates–Parra（2014）
Melipona marginata	43	30 个月	访花蜂	Atlantic rain forest, SP, Brazil	Wilms et al.（1996）
Melipona quadrifasciata	31	30 个月	访花蜂	Atlantic rain forest, SP, Brazil	Wilms et al.（1996）
Melipona rufiventris	51	12 个月	花蜜采集蜂	Amazonas state, Brazil	Absy et al.（1980）
Melipona seminigra	48	12 个月	花蜜采集蜂	Amazonas state, Brazil	Absy et al.（1980）
Melipona subnitida	19	12 个月	采集蜂和食物罐	Caatinga dry forest, RN, Brazil	Maia–Silva et al.（2018）
Paratrigona subnuda	95	30 个月	访花蜂	Atlantic rain forest, SP, Brazil	Wilms et al.（1996）
Partamona helleri	77	30 个月	访花蜂	Atlantic rain forest, SP, Brazil	Wilms et al.（1996）
Plebeia droryana	103	30 个月	访花蜂	Atlantic rain forest, SP, Brazil	Wilms et al.（1996）
Plebeia remota	97	12 个月	储蜜	São Paulo state, Brazil	Ramalho et al.（1985）
Scaptotrigona depilis	66	12 个月	储藏花粉	São Paulo state, Brazil	Aleixo et al.（2017）
Scaptotrigona bipunctata	51	30 个月	访花蜂	São Paulo state, Brazil	Wilms et al.（1996）
Scaptotrigona fulvicutis	97	12 个月	携粉采集蜂	Amazonas state, Brazil	Marques–Souza et al.（2007）
Scaptotrigona hellwegeri	165	12 个月	储藏花粉	State of Jalisco, Mexico	Quiroz–García et al.（2011）

续表

无刺蜂蜂种	植物种属数	采样期	样品类型	采样地点	参考文献
Schwarziana quadripunctata	61	30 个月	访花蜂	Atlantic rain forest, SP, Brazil	Wilms et al.（1996）
Tetragonisca angustula	73	17 个月	储蜜	Pará state, Brazil	Novaise and Absy（2015）
Trigona spinipes	89	30 个月	访花蜂	Atlantic rain forest, SP, Brazil	Wilms et al.（1996）
Trigona williana	56	12 个月	携带花粉	Amazonas state, Brazil	Marques-Souza et al.（1996）

注：此表中只包括新热带蜂种，且这些研究的采样期至少为 12 个月。

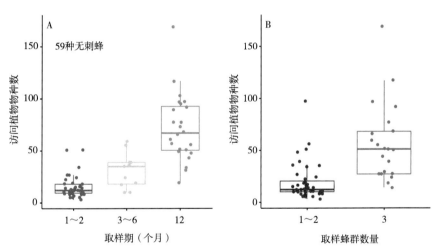

图 9.2　不同取样期（A）、取样的蜂群数量（B）情况下，研究者从无刺蜂采集蜂和蜂巢储藏食物中鉴定出的花粉类型（植物种类）数量

注：这些研究对 11 种无刺蜂中每个蜂种都进行了多次研究，因此每个蜂种多个数据点。取样期和取样蜂群数量对研究者鉴定出的花粉类型数量均有显著正影响（双预测因子泊松 GLM：取样期，z 值 = 24.4，$p<0.000\,1$；取样蜂群数量，z 值 = 12.1，$p<0.000\,1$）。图中数据源自 Koeniger and Vorwohl（1979）；Absy et al.（1980, 1984）；Ramalho et al.（1985, 1989）；Imperatriz-Fonseca et al.（1989）；Wilms et al.（1996）；Marques-Souza et al.（1996, 2006）；Nagamitsu and Inoue（2002）；Vossler et al.（2010）；Quiroz-Garcia et al.（2011）；Rech and Absy（2011）；Obregon and Nates-Parra（2014）；Novaise and Absy（2015）；Aleixo et al.（2013, 2017）；Ferreira and Absy（2018）；Ghazi et al.（2018）；Layek and Karmakar（2018）；Maia-Silva et al.（2018）；Saravia-Nava et al.（2018）；Ramirez-Arriaga et al.（2018）；Villanueva-Gutierrez et al.（2018）（Excel 文档可以由此网址下载：www.socialinsect-research.com/book.php）。

尽管研究者可以在无刺蜂样本中发现多种花粉类型，但研究者在单个无刺蜂蜂群中发现的花粉大多数来自10种以下的不同植物（Ramalho et al.,1985; Kleinert-Giovannini and Imperatriz-Fonseca,1987; Ramalho,1990; Teixeira et al., 2007; Ramírez-Arriaga and Martínez-Hernández,2007; Vossler et al.,2010; Obregon and Nates-Parra,2014; Roubik and Moreno Patiño,2018; Villanueva-Gutiérrez et al.,2018; Vossler,2018）。这种情况也是合情合理的，这可能是因为在特定时间里无刺蜂蜂群对特定的植物物种是有偏好的（Wilms and Wiechers,1997; Maia-Silva et al.,2015）。

由于不同研究中收集花粉粒样品的方法不同，因此往往难以比较这些研究所报道的无刺蜂访问植物种类数量。在确定蜂群生态位宽度中有两个因素起到了重要作用：被取样的蜂群数量和取样时间段（图9.2）。分析59种不同无刺蜂的数据发现那些采样时间小于等于2个月的研究中发现的无刺蜂采集花粉类型平均不到20种，而采样时间大于等于一年的研究发现的无刺蜂采集花粉类型平均约为70种（图9.2）。同样，只对1～2个蜂群采样的研究发现的无刺蜂采集花粉类型平均少于20种，而对3个群以上的蜂群采样的研究发现的无刺蜂采集花粉类型平均为57种（图9.2）。这是因为相同蜂种和位置的蜂群在食谱上存在一些不同（Rech and Absy,2011; Ramírez-Arriaga et al.,2018; Saravia-Nava et al.,2018）。从表9.1所示的数据中可以看出一个趋势：8种*Melipona*属无刺蜂采集的花粉类型种类数（51.1±22.4）少于13种非*Melipona*无刺蜂采集的花粉类型种类数（85.2±29.3）。这表明，*Melipona*属蜂群的食谱范围可能比非*Melipona*属无刺蜂要更窄，食性更专一（*Melipona eburnea*无刺蜂参见, Obregon and Nates-Parra,2014）。

考虑到单个无刺蜂会从许多植物上收集花粉，人们可能会问：所有无刺蜂加起来到底总共访问过多少开花植物呢？关于无刺蜂花访花行为的信息仍然非常有限，特别是关于亚洲和非洲的无刺蜂种访花的信息，但我们可以利用新热带无刺蜂种的现有数据来进行初步估计。笔者采用了21种新热带区无刺蜂的研究数据，这些研究都进行了至少12个月的采样来评估无刺蜂在其栖息地访问的植物科、属和种的数量（表9.1）。21种无刺蜂总计访问了919种开花植物，这些开花植物隶属113科，407～477个属[1]。而无刺蜂访问频率最高的植物科

---

[1] 这些开花植物的确切种属数量取决于不同研究中研究者如何界定那些未识别种属。在不同的研究和地点报告中将未识别种属（例如Fabaceae中的属）定义为相同种属或不同种属。为了准确，笔者在此假定了在不同地区进行的不同研究中报告的未识别种属为独特的种属。还应注意的是，由于分类学家偶尔会给同一个物种起不同的名字，所以此处的物种数量会略低一些（Joppa et al., 2011）。

为 Fabaceae（豆科）、Euphorbiaceae（大戟科）、Asteraceae（菊科）和 Myrtaceae（桃金娘科），这里的 21 种无刺蜂中至少有 19 种访问过这些科的植物（图 9.3）（Absy et al.,2018）。无刺蜂访问频率最高的植物属为 Croton（大戟属）、Eugenia（桃金娘属）和 Miconia（野牡丹属），这里的 21 种无刺蜂中至少有 15 种访问过这些属的植物（图 9.3）。目前研究者已经从大量无刺蜂物种获得了样本，并从中发现了一些无刺蜂的访花种属数，那么可以用稀释法和外推法从中估算出无刺蜂总共可能访问的植物类群数量（Chao et al.,2014）。例如，如果我们目前在哥斯达黎加收集到 60 种无刺蜂的样本，而这个蜂种数接近当地已发现的无刺蜂蜂种数量（Aguilar et al.,2013），那么预计我们会在样本中发现的花粉就会是来自约 140 个科、850 个属、2 000 个种植物的花粉。图 9.4 所示曲线的斜率和误差范围表明，想要可靠地预测目前已知的 426 种新热带无刺蜂所访问的新热带植物种属数，就必须采集更多种的无刺蜂的花粉样本。而由于外推法外推的物种范围很大时就不再可靠，因此，需谨慎看待此处这些外推数值（Chao et al.,2014）。这些数值可能大大低估了无刺蜂蜂种的生态位宽度，因为这些数值仅源自每个无刺蜂种的小部分栖息地中的小部分蜂群。

另一种预测无刺蜂访问植物种属数量的方法是测定特定环境下开花植物中被无刺蜂访问的比例。Roubik（1989）估测当地 15% ～ 20% 的被子植物区系已被无刺蜂访问过，这个比例略低于马来西亚栖息地报告的 25%（Sakai et al.,1999）。这表明全世界有 3 万～ 5 万种热带植物被无刺蜂访问[2]。

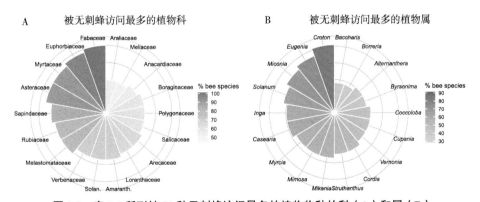

图 9.3 表 9.1 所列被 21 种无刺蜂访问最多的植物物种的科（A）和属（B）

注：21 种无刺蜂访问最多的都是豆科植物，其次是大戟科等（Solan. = Solanaceae, Amaranth. = Amaranthaceae）。21 种无刺蜂中有 18 种都访问过 Croton 这种植物（Excel 文档可以由此网址下载：www.socialinsect-research.com/book.php）。

---

[2] 全世界有 300 000 ～ 400 000 种开花植物（Ollerton et al.,2011; Joppa et al., 2011; Ollerton,2017），其中有一半以上都是热带植物。

无刺蜂的饮食多样性不仅取决于其访问植物分类群的总数，还取决于样本中某一特定分类群的丰度（Chao et al.,2014）。Simpson 多样性既考虑了分类群的数量，也考虑了它们在采集样本中的丰度。当样品中有优势科、属和种时，Simpson 多样性就会下降。Shannon 多样性是另一种常用的多样性度量方法，它考虑了被访问植物类群的数量和它们的丰度，但对优势植物类群的权重略低于 Simpson 多样性。图 9.4 中 Simpson 和 Shannon 多样性表明，新热带区无刺蜂作为一个群体并不是均匀地访问开花植物，而是更多地依赖于某些植物类群和物种。这可能并不奇怪，因为我们已经看到一些植物科和属特别受无刺蜂喜爱（图 9.3）。

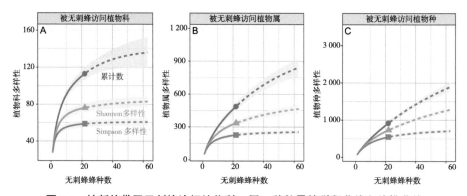

图 9.4　被新热带区无刺蜂访问植物科、属、种数量的稀释曲线和外推曲线

注：花粉鉴定则是基于巴西、墨西哥的不同位置以及阿根廷一处进行的研究。这里的分析使用了 R 3.6 软件（R Development Core Team，2013）和 iNEXT 软件包（Excel 文档可以由此网址下载：www.socialinsect-research.com/book.php）。

9.1　作物授粉

由于许多作物依赖或受益于蜂授粉，因此当前人们对于全球的蜂授粉充满担忧（Klein et al.,2007; Potts et al.,2010, 2016; Garibaldi et al.,2013）[3]。据估计仅在热带地区就栽种了大约 1 300 种植物，然而我们基本没有这些植物授粉需求以及哪种蜂为其授粉的相关信息（Roubik,1995）。众所周知，无刺蜂会访问许多热带作物（Roubik,1995; Heard,1999; Giannini et al., 2015, 2020）。例如，已经有研究者发现玛瑙斯市（巴西亚马孙州）市场最常见的商业化植物中有 60%（38 种商业化植物中的 23 种）被无刺蜂访花，因为研究

---

[3] Gallai 等（2009）估算全世界作物授粉的总价值约为 1 500 亿欧元，其中蔬菜和水果各占约 500 亿欧元。

者在无刺蜂蜂巢中发现了这些植物的花粉［Rabelo（2012）被 Absy 等（2018）引用］。特别是在无刺蜂完整自然栖息地附近种植的作物都受益于无刺蜂授粉（Slaa et al.,2006）。例如，种植在森林地区附近的泰国果园中的红毛丹（*Nephelium lappaceum*）的无刺蜂访花率和坐果率显著较高（Sritongchuay et al., 2016）。当前人类已经成功实现了在人工蜂箱养殖多种无刺蜂蜂群（主要是在洞巢型无刺蜂蜂种）（Nogueira-Neto,1997），这进一步促进了商业授粉，而且人类可以人工饲养多种蜂种来进行温室授粉（表9.2）［Heard,1999; Slaa et al.,2006; Greco et al.,2011; Nunes-Silva et al.,2013; Kishan et al.,2017; Azmi et al., 2019; Meléndez Ramírez et al., 2018; 更多关于温室中应用无刺蜂的详细讨论参见 Slaa 等（2006）和 Meléndez Ramírez 等（2018）］。例如，已经有人在温室中养殖 *Melipona fasiculata* 无刺蜂蜂群，这增加了温室中茄子（*Solanum melongena*）的果实重量（图9.5）。由于无刺蜂蜂种在体型、蜂群大小、采集策略和传粉行为（包括 *Melipona* 属无刺蜂的蜂鸣采集行为）方面存在差异，因此无刺蜂可以为多种作物授粉，这些作物中甚至包括一些不能用蜜蜂授粉的作物（Slaa et al.,2006; Meléndez Ramírez et al., 2018）。尽管无刺蜂在作物授粉方面很有潜力，但与蜜蜂相比，它也有缺点。人们常常无法获得大量的无刺蜂蜂群，而且无刺蜂蜂群的生长和繁殖速度相对缓慢（第1章，第4章）（Heard, 1999; Slaa et al., 2006）。

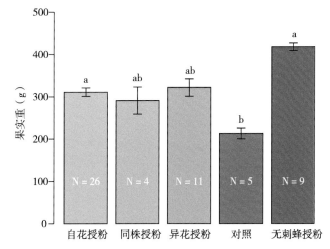

图9.5 研究者在温室内采用自花授粉（用同一朵花的花粉人工授粉）、同株授粉（用同一株植物上的其他花的花粉人工授粉）、异花授粉（用其他植株的花粉人工授粉）、不授粉（对照）以及使用 *Melipona fasiculata* 无刺蜂授粉

注：图中展示了5种方法产生的茄子果实的重量（平均值和SE）。柱状图中柱上字母不同表示两组之间存在显著差异，$p<0.05$（Nunes-Silva et al.,2013）。

表 9.2 实质上受益于无刺蜂授粉的 20 种作物

通用名	物种名	所属科	授粉无刺蜂属	参考文献
胭脂树	*Bixa orellana*	Bixaceae	*Melipona*	Heard（1999）
茄子	*Solanum melongena*	Solanaceae	*Melipona*	Nunes-Silva et al.（2013）
牛油果	*Persea americana*	Lauraceae	*Nannotrigona, Trigona*	Heard（1999）；Slaa et al.（2006）
卡姆果	*Myrciaria dubia*	Myrtaceae	*Melipona, Scaptotrigona, etc.*	Heard（1999）
杨桃	*Averrhoa carambola*	Oxalidaceae	*Trigona*	Heard（1999）
佛手瓜	*Choko Sechium edule*	Cucurbitaceae	*Trigona, Partamona*	Heard（1999）
椰子	*Cocos nucifera*	Arecaceae	various genera	Heard（1999）
咖啡	*Coffea arabica*	Rubiaceae	*Lepidotrigona, Trigona*	Heard（1999）；Slaa et al.（2006）
咖啡	*Coffea canephora*	Rubiaceae	*Lepidotrigona, Trigona*	Slaa et al.（2006）
黄瓜*	*Cucumis sativus*	Cucurbitaceae	*Nannotrigona, Scaptotrigona*	Heard（1999）；Slaa et al.（2006）
大花可可	*Theobroma grandiflorum*	Sterculiaceae	*Trigona*	Heard（1999）
澳洲坚果	*Macadamia integrifolia*	Proteaceae	*Trigona*	Heard（1999）
杧果	*Mangifera indica*	Anacardiaceae	*Trigona*	Heard（1999）
亚马逊葡萄	*Pourouma cecropiaefolia*	Moraceae	*Oxitrigona, Trigona*	Heard（1999）
粉萼鼠尾草*	*Salvia farinaceae*	Lamiaceae	*Nannotrigona, Tetragonisca*	Slaa et al.（2006）
红毛丹*	*Nephelium lappaceum*	Sapindaeae	*Scaptotrigona*	Heard（1999）；Slaa et al.（2006）

续表

通用名	物种名	所属科	授粉无刺蜂属	参考文献
哈密瓜*	*Cucumis melo*	Cucurbitaceae	*Heterotrigona*	Azmi et al.（2019）
草莓*	*Fragaria* sp.	Rosaceae	various genera	Heard（2019）；Slaa et al.（2006）
辣椒*	*Capsicum annuum* varieties	Solanaceae	*Austroplebeia, Melipona, Tetragonula*	Heard（1999）；Slaa et al.（2006）；Putra et al.（2017）
番茄*	*Lycopersicon esculentum*	Solanaceae	*Melipona, Nannotrigona*	Slaa et al.（2006）

注：*表示温室授粉。

无刺蜂授粉的重要性取决于作物，而对许多作物来说，无刺蜂授粉的重要性仍然不清楚（Heard,1999; Slaa et al.,2006）。以杧果（*Mangifera indica*）为例，无刺蜂不仅是其最频繁的访花者之一（Heard,1999），也是其最有效的传粉者之一（Anderson et al.,1982）。另一种非常依赖无刺蜂授粉的作物是胭脂树或者红木。在墨西哥、加勒比地区和中南美洲地区，人们将红木的种子作为食用色素和调味料用于传统药物和菜肴中。红木花特征特别适合蜂鸣传粉，这就是为什么蜂鸣传粉的 *Melipona beecheii* 无刺蜂比 *Apis mellifera* 蜜蜂更能有效地为其授粉（Caro et al.,2017）。相反，尽管无刺蜂（例如 *Trigona spinipes* 无刺蜂）会访问鸡蛋果或百香果的花，但观察结果表明无刺蜂的这些访花行为对其没有益处，甚至有些无刺蜂（如好斗的 *Trigona*）会阻止更有效的传粉者（如 *Xylocopa* 木蜂），这种情况下无刺蜂的访花行为就对该植物产生了负面影响（Sazima and Sazima,1989）。表 9.2 列出了 20 种实质上受益于无刺蜂授粉的作物。也存在许多其他作物可能或偶尔受益于无刺蜂授粉（表 9.3 列出了 75 种作物）[4]。例如，Roopa 等（2017）发现在印度 *Tetragonula iridipennis* 无刺蜂访问过许多具有药用价值的作物或植物，但还需要更多的研究来评估无刺蜂在这些植物授粉中的重要性。

表 9.3　75 种偶尔或者潜在由无刺蜂授粉的作物

（Heard,1999;Slaa,et al., 2006;Giannini et al., 2015）

通用名	物种名	所属科	参考文献
巴西莓	*Euterpe oleracea*	Arecaceae	Campbell et al.（2018）
鸡翅木	*Vouacapoua americana*	Fabaceae	Giannini et al.（2015）
西印度樱桃	*Malpighia punicifolia*	Malpighiaceae	Heard（1999）
西非荔枝果	*Blighia sapida*	Sapindaeae	Heard（1999）
苹果	*Malus domestica*	Rosaceae	Viana et al.（2014）
酒果椰属	*Oenocarpus distichus*	Arecaceae	Giannini et al.（2015）
酒果椰属	*Oenocarpus mapora*	Arecaceae	Giannini et al.（2015）
红叶麻风树	*Jatropha gossypifolia*	Euphorbiaceae	Heard（1999）

---

[4] 人们经常说在墨西哥和中美洲 *Melipona* 无刺蜂是香草兰花的天然授粉者。然而，几乎没有证据表明 *Melipona* 无刺蜂会为香草花授粉。相反，*Eulaema* 兰花蜂更可能是 *Vanilla* 香草植物的有效授粉者（Roubik, 1995；Heard, 1999；Lubinsky et al., 2006）。

续表

通用名	物种名	所属科	参考文献
苦瓜	*Momordica charantia*	Cucurbitaceae	Heard（1999）
巴西栗	*Bertholletia excelsa*	Lecythidaceae	Giannini et al.（2015）
面包树	*Artocarpus altilis*	Moraceae	Heard 1999
欧洲油菜	*Brassica napus*	Brassicaceae	Giannini et al.（2015）
绿豆蔻	*Elettaria cardamomum*	Zingiberaceae	Heard（1999）
胡萝卜	*Daucus carota* Cultivar Brasilia	Apiaceae	Giannini et al.（2015）
木薯	*Manihot esculenta*	Euphorbiaceae	Giannini et al.（2015）
蓖麻油	*Ricinus communis*	Euphorbiaceae	Heard（1999）
柑橘	*Citrus* spp.	Rutaceae	Heard（1999）
芫荽	*Coriander sativum*	Apiaceae	Heard（1999）
棉花	*Gossypium* spp.	Malvaceae	Giannini et al.（2015）
茴香	*Foeniculum vulgare*	Apiaceae	Heard（1999）
芸薹	*Brassica campestris*	Brassicaceae	Heard（1999）
瓜拉纳	*Paullinia cupana*	Sapindaeae	Heard（1999）
番石榴	*Psidium guajava*	Myrtaceae	Heard（1999）
黄槟榔青	*Spondias mombin*	Anacardiaceae	Heard（1999）
印加豆	*Inga edulis*	Mimosoideae	Heard（1999）
印度枣	*Zizyphus mauritiana*	Rhamnaceae	Heard（1999）
美人蕉	*Canna indica*	Cannaceae	Heard（1999）
铺地木蓝	*Indigofera endocaphylla*	Papilionoideae	Heard（1999）
波罗蜜	*Artocarpus heterophyllus*	Moraceae	Heard（1999）
乌墨	*Syzygium cumini*	Myrtaceae	Heard（1999）
麻风树	*Jatropha curcas*	Euphorbiaceae	Giannini et al.（2015）
荔宝果	*Myrciaria cauliflora*	Myrtaceae	Heard（1999）
木棉	*Ceiba pentandra*	Bombacaceae	Heard（1999）

续表

通用名	物种名	所属科	参考文献
银合欢	*Leucaena leucocephala*	Mimosoideae	Heard（1999）
荔枝	*Litchi chinensis*	Sapindaeae	Heard（1999）
龙眼	*Euphoria longan*	Sapindaeae	Heard（1999）
枇杷	*Eriobotrya japonica*	Rosaceae	Heard（1999）
丝瓜	*Luffa acutangula*	Cucurbitaceae	Heard（1999）
莲玉蕊	*Gustavia superba*	Lecythidaeae	Heard（1999）
藤黄科植物	*Allanblackia stuhlmannii*	Clusiaceae	Mrema and Nyundo（2016）
龟背竹	*Monstera deliciosa*	Araceae	Heard（1999）
树莓	*Rubus* sp.	Rosaceae	Giannini et al.（2015）
金虎尾科植物	*Byrsonima chrysophylla*	Malpighiaceae	Giannini et al.（2015）
厚叶金匙木	*Byrsonima crassifolia*	Malpighiaceae	Heard（1999）
小葵子	*Guizotia abyssinica*	Asteraceae	Heard（1999）
咖啡黄葵	*Abelmoschus esculentus*	Malvaceae	Giannini et al.（2015）
洋葱	*Allium cepa*	Alliaceae	Heard（1999）
橙子	*Citrus* spp.	Rutaceae	Giannini et al.（2015）
巴拿马草	*Carludovica palmata*	Cyclanthaceae	Heard（1999）
桃	*Prunus persica*	Rosaceae	Heard（1999）
棕榈果	*Bactris gasipeas*	Arecaceae	Heard（1999）
梨	*Pyrus communis*	Rosaceae	Heard（1999）
木豆	*Cajanus cajan*	Papilionoideae	Heard（1999）
牛睛果	*Talisia esculenta*	Sapindaeae	Giannini et al.（2015）
石榴	*Punica granatum*	Lythraceae	Giannini et al.（2015）
欧洲李	*Prunus domestica*	Rosaceae	Heard（1999）
南瓜	*Cucurbita moschata*	Cucurbitaceae	Slaa et al.（2006）
萝卜	*Raphanus sativus*	Cruciferae	Slaa et al.（2006）
省藤属	*Calamus* spp.	Arecaceae	Heard（1999）

续表

通用名	物种名	所属科	参考文献
马六甲蒲桃	*Syzigium malaccense*	Myrtaceae	Giannini et al.（2015）
蒲桃	*Syzygium jambos*	Myrtaceae	Heard（1999）
橡胶树	*Hevea brasiliensis*	Euphorbiaceae	Heard（1999）
西谷椰	*Metroxylon sagu*	Arecaceae	Heard（1999）
砂仁	*Amomum villosum*	Zingiberaceae	Heard（1999）
芝麻	*Sesamum indicum*	Pedaliaceae	Heard（1999）
剑麻	*Agave sisalana*	Agavaceae	Heard（1999）
无患子属植物	*Sapindus emarginatus*	Sapindaeae	Heard（1999）
西葫芦	*Cucurbita pepo*	Cucurbitaceae	Heard（1999）
圭亚那笔花豆	*Stylosanthes guianensis*	Papilionoideae	Heard（1999）
向日葵	*Helianthus annuus*	Asteraceae	Heard（1999）
酸豆	*Tamarindus indica*	Caesalpinioideae	Heard（1999）
槟榔青属植物	*Spondias tuberosa*	Anacardiaceae	Giannini et al.（2015）
西瓜	*Citrullus lanatus*	Cucurbitaceae	Heard（1999）
黄麻	*Corchorus capsularis*	Tiliaceae	Heard（1999）
野辣椒	*Capsicum frutescens*	Solanaceae	Giannini et al.（2015）

参考文献

Absy ML, Bezerra EB, Kerr WE（1980）Plantas nectaríferas utilizadas por duas espécies de *Melipona* da Amazônia. Acta Amazonica 10:271–282

Absy ML, Camargo JMF, Kerr WE, Miranda IPA（1984）Espécies de plantas visitadas por Meliponinae（Hymenoptera; Apoidea）, para coleta de pólen na região do médio Amazonas. Revista Brasileira de Biologia 44:227–237

Absy ML, Rech AR, Ferreira MG（2018）Pollen Collected by Stingless Bees: A Contribution to Understanding Amazonian Biodiversity. In: Vit P, Pedro SRM, Roubik DW（eds）Pot-Pollen in Stingless Bee Melittology. Springer International Publishing, Cham, pp 29–46

Aguilar I, Herrera E, Zamora G (2013) Stingless Bees of Costa Rica. In: Vit P, Pedro SRM, Roubik DW (eds) Pot-Honey: A legacy of stingless bees. Springer, New York, NY, pp 113–124

Aleixo KP, De Faria LB, Garófalo CA, Imperatriz-Fonseca VL, Da Silva CI (2013) Pollen collected and foraging activities of *Frieseomelitta varia* (Lepeletier) (Hymenoptera: Apidae) in an urban landscape. Sociobiology 60:266–276

Aleixo KP, Menezes C, Imperatriz Fonseca VL, da Silva CI (2017) Seasonal availability of floral resources and ambient temperature shape stingless bee foraging behavior (*Scaptotrigona* aff. *depilis*). Apidologie 48:117–127

Anderson DL, Sedgley M, Short JRT, Allwood AJ (1982) Insect pollination of mango in northern Australia. Australian Journla of Agricultural Research 33:541–548

Antonelli A, Zizka A, Silvestro D, Scharn R, Cascales-Miñana B, Bacon CD (2015) An engine for global plant diversity: highest evolutionary turnover and emigration in the American tropics. Frontiers in Genetics 6:130

Azmi WA, Wan Sembok WZ, Yusuf N, Mohd. Hatta MF, Salleh AF, Hamzah M a. H, Ramli SN (2019) Effects of Pollination by the Indo-Malaya Stingless Bee (Hymenoptera: Apidae) on the Quality of Greenhouse-Produced Rockmelon. Journal of Economic Entomology 112:20–24

Bawa KS (1990) Plant-pollinator interactions in Tropical rain forests. Annual Review of Ecology and Systematics 21:399–422

Brosi BJ, Daily GC, Shih TM, Oviedo F, Durán G (2008) The effects of forest fragmentation on bee communities in tropical countryside: Bee communities and tropical forest fragmentation. Journal of Applied Ecology 45:773–783

Cairns CE, Villanueva-Gutiérrez R, Koptur S, Bray DB (2005) Bee populations, forest disturbance, and africanization in Mexico. Biotropica 37:686–692

Campbell AJ, Carvalheiro LG, Maués MM, Jaffé R, Giannini TC, Freitas MAB, Coelho BWT, Menezes C (2018) Anthropogenic disturbance of tropical forests threatens pollination services to açaí palm in the Amazon river delta. Journal of Applied Ecology 55:1725–1736

Caro A, Moo-Valle H, Alfaro R, Quezada-Euán JJG (2017) Pollination services of Africanized honey bees and native *Melipona beecheii* to buzz-pollinated annatto(*Bixa orellana* L.) in the neotropics. Agricultural and Forest Entomology 19:274–280

Cely-Santos M, Philpott SM（2019）Local and landscape habitat influences on bee diversity in agricultural landscapes in Anolaima, Colombia. Journal of Insect Conservation 23:133–146

Chao A, Gotelli NJ, Hsieh TC, Sander EL, Ma KH, Colwell RK, Ellison AM（2014）Rarefaction and extrapolation with Hill numbers: a framework for sampling and estimation in species diversity studies. Ecological Monographs 84:45–67

Corlett RT（2004）Flower visitors and pollination in the Oriental（Indomalayan）Region. Biological Reviews 79:497–532

Ferreira MG, Absy ML（2018）Pollen niche of *Melipona*（*Melikerria*）*interrupta*（Apidae: Meliponini）bred in a meliponary in a terra-firme forest in the central Amazon. Palynology 42:199–209

Gallai N, Salles J-M, Settele J, Vaissière BE（2009）Economic valuation of the vulnerability of world agriculture confronted with pollinator decline. Ecological Economics 68:810–821

Garibaldi LA, Steffan-Dewenter I, Winfree R, Aizen MA, Bommarco R et al.（2013）Wild Pollinators Enhance Fruit Set of Crops Regardless of Honey Bee Abundance. Science 339:1608–1611

Ghazi R, Zulqurnain NS, Azmi WA（2018）Melittopalynological Studies of Stingless Bees from the East Coast of Peninsular Malaysia. In: Vit P, Pedro SRM, Roubik DW（eds）Pot-Pollen in Stingless Bee Melittology. Springer International Publishing, Cham, pp 77–88

Giannini TC, Boff S, Cordeiro GD, Cartolano EA, Veiga AK, Imperatriz-Fonseca VL, Saraiva AM（2015）Crop pollinators in Brazil: a review of reported interactions. Apidologie 46:209–223

Giannini TC, Alves DA, Alves R, Cordeiro GD, Campbell AJ, Awade M, Bento JMS, Saraiva AM, Imperatriz-Fonseca VL（2020）Unveiling the contribution of bee pollinators to Brazilian crops with implications for bee management. Apidologie

Greco MK, Spooner-Hart RN, Beattie AGAC, Barchia I, Holford P（2011）Australian stingless bees improve greenhouse *Capsicum* production. Journal of Apicultural Research 50:102–115

Gutiérrez-Chacón C, Dormann CF, Klein A-M（2018）Forest-edge associated bees benefit from the proportion of tropical forest regardless of its edge length. Biological

Conservation 220:149–160

Heard TA (1999) The role of stingless bees in crop pollination. Annual Review of Entomology 44:183–206

Imperatriz-Fonseca VL, Kleinert-Giovannini A, Ramalho M (1989) Pollen Harvest by Eusocial Bees in a Non-Natural Community in Brazil. Journal of Tropical Ecology 5:239–242

Joppa LN, Roberts, D. L., Pimm, S. L. (2011) How many species of flowering plants are there? Proceedings of the Royal Society B: Biological Sciences 278:554–559

Kishan TM, Srinivasan MR, Rajashree V, Thakur RK (2017) Stingless bee *Tetragonula iridipennis* Smith for pollination of greenhouse cucumber. Journal of Entomology and Zoology Studies 5:1729–1733

Klein AM, Vaissiere BE, Cane JH, Steffan-Dewenter I, Cunningham SA, Kremen C, Tscharntke T (2007) Importance of pollinators in changing landscapes for world crops. Proceedings of the Royal Society B: Biological Sciences 274:303–313

Kleinert-Giovannini A, Imperatriz-Fonseca VL (1987) Aspects of the trophic niche of *Melipona marginata marginata* Lepeletier (Apidae, Meliponinae). Apidologie 18:69–100

Koeniger N, Vorwohl G (1979) Competition for Food Among Four Sympatric Species of Apini in Sri Lanka (*Apis dorsata, Apis cerana, Apis florea* and *Trigona iridipennis*). Journal of Apicultural Research 18:95–109

Layek U, Karmakar P (2018) Nesting characteristics, floral resources, and foraging activity of *Trigona iridipennis* Smith in Bankura district of West Bengal, India. Insectes Sociaux 65:117–132

Lorenzon MCA, Matrangolo CAR, Schoereder JH (2003) Flora visitada pelas abelhas eussociais (Hymenoptera, Apidae) na Serra da Capivara, em caatinga do Sul do Piauí. Neotropical Entomology 32:27–36

Lubinsky P, Dam M van, Dam A van (2006) Pollination of Vanilla and evolution in Orchidaceae. Orchids 75:926–929

Maia-Silva C, Hrncir M, da Silva CI, Imperatriz-Fonseca VL (2015) Survival strategies of stingless bees (*Melipona subnitida*) in an unpredictable environment, the Brazilian tropical dry forest. Apidologie 46:631–643

Maia-Silva C, Limão AAC, Hrncir M, da Silva Pereira J, Imperatriz-Fonseca VL (2018)

The Contribution of Palynological Surveys to Stingless Bee Conservation: A Case Study with *Melipona subnitida*. In: Vit P, Pedro SRM, Roubik DW (eds) Pot-Pollen in Stingless Bee Melittology. Springer International Publishing, Cham, pp 89–101

Marques-Souza AC, Absy ML, Kerr WE (2007) Pollen harvest features of the Central Amazonian bee *Scaptotrigona fulvicutis* Moure 1964 (Apidae: Meliponinae), in Brazil. Acta Botanica Brasilica 21:11–20

Marques-Souza AC, Moura CO, Nelson BW (1996) Pollen collected by *Trigona williana* (Hymenoptera: Apidae) in central Amazonia. Revista de Biologia Tropical 44:567–573

Meléndez Ramírez V, Ayala R, Delfín González H (2018) Crop Pollination by Stingless Bees. In: Vit P, Pedro SRM, Roubik DW (eds) Pot-Pollen in Stingless Bee Melittology. Springer International Publishing, Cham, pp 139–153

Michener CD (2007) The bees of the world, 2nd edn. The Johns Hopkins University Press, Baltimore

Momose K, Yumoto T, Nagamitsu T, Kato M, Nagamasu H, Sakai S, Harrison RD, Itioka T, Hamid AA, Inoue T (1998) Pollination biology in a lowland dipterocarp forest in Sarawak, Malaysia. I. Characteristics of the plant-pollinator community in a lowland dipterocarp forest. American Journal of Botany 85:1477–1501

Mrema IA, Nyundo BA (2016) Pollinators of *Allanblackia stuhlmannii* (Engl.), Mkani fat an endemic tree in eastern usambara mountains, Tanzania. International Journal of Pure & Applied Bioscience 4:61–67

Nagamitsu T, Inoue T (2002) Foraging activity and pollen diets of subterranean stingless bee colonies in response to general flowering in Sarawak, Malaysia. Apidologie 33:303–314

Novais JS de, Absy ML (2015) Melissopalynological records of honeys from *Tetragonisca angustula* (Latreille, 1811) in the Lower Amazon, Brazil: pollen spectra and concentration. Journal of Apicultural Research 54:11–29

Nunes-Silva P, Hrncir M, da Silva CI, Roldão YS, Imperatriz-Fonseca VL (2013) Stingless bees, *Melipona fasciculata*, as efficient pollinators of eggplant (*Solanum melongena*) in greenhouses. Apidologie 44:537–546

Obregon D, Nates-Parra G (2014) Floral Preference of *Melipona eburnea* Friese (Hymenoptera: Apidae) in a Colombian Andean Region. Neotropical Entomology

43:53–60

Ollerton J (2017) Pollinator Diversity: Distribution, Ecological Function, and Conservation. Annual Review of Ecology, Evolution and Systematics 48:353–376

Ollerton J, Winfree R, Tarrant S (2011) How many flowering plants are pollinated by animals? Oikos 120:321–326

Potts SG, Biesmeijer JC, Kremen C, Neumann P, Schweiger O, Kunin WE (2010) Global pollinator declines: trends, impacts and drivers. Trends in Ecology & Evolution 25:345–353

Potts SG, Imperatriz-Fonseca VL, Ngo HT, Aizen MA, Biesmeijer JC, Breeze TD, Dicks LV, Garibaldi LA, Hill R, Settele J (2016) Safeguarding pollinators and their values to human well-being. Nature 540:220–229

Putra DP, Salmah S, Swasti E (2017) Daily Flight Activity of *Trigona laeviceps* and *T. minangkabau* in Red Pepper (*Capsicum annuum* L.) Plantations in Low and High Lands of West Sumatra. International Journal of Applied Environmental Sciences 12:1497–1507

Quiroz-García DL, Arreguín-Sánchez M de la L, Fernández-Nava R, Martínez-Hernández E (2011) Patrones estacionales de utilización de recursos florales por *Scaptotrigona hellwegeri* en la Estación de Biología Chamela, Jalisco, México. Polibotánica:89–119

R Development Core Team R (2013) R: A language and environment for statistical computing. R Foundation for Statistical Computing, Vienna

Ramalho M (2004) Stingless bees and mass flowering trees in the canopy of Atlantic Forest: a tight relationship. Acta Botanica Brasilica 18:37–47

Ramalho M (1990) Foraging by Stingless Bees of the Genus, *Scaptotrigona* (Apidae, Meliponinae). Journal of Apicultural Research 29:61–67

Ramalho M, Imperatriz-Fonseca VL, Kleinekt-Giovannini A, Cortopassi-Laurino M (1985) Exploitation of floral resources by *Plebeia remota* Holmberg (Apidae, Meliponinae). Apidologie 16:307–330

Ramalho M, Kleinert-Giovanni A, Imperatriz-Fonseca VL (1989) Utilization of floral resources by species of *Melipona* (Apidae, Meliponinae): floral preferences. Apidologie 20:185–195

Ramírez-Arriaga E, Martínez-Hernández E (2007) Melitopalynological

Characterization of *Scaptotrigona mexicana* Guérin (Apidae: Meliponini) and *Apis mellifera* L. (Apidae: Apini) Honey Samples in Northern Puebla State, Mexico. Journal of the Kansas Entomological Society 80:377–391

Ramírez-Arriaga E, Pacheco-Palomo KG, Moguel-Ordoñez YB, Zepeda García Moreno R, Godínez-García LM (2018) Angiosperm Resources for Stingless Bees (Apidae, Meliponini): A Pot-Pollen Melittopalynological Study in the Gulf of Mexico. In: Vit P, Pedro SRM, Roubik DW (eds) Pot-Pollen in Stingless Bee Melittology. Springer International Publishing, Cham, pp 111–130

Rech AR, Absy ML (2011) Pollen storages in nests of bees of the genera *Partamona*, *Scaura* and *Trigona* (Hymenoptera, Apidae). Revista Brasileira de Entomologia 55:361–372

Roopa AN, Eswarappa G, M. Sajjanar S, Gowda G (2017) Study on Identification of Pasturage Sources of Stingless Bee (*Trigona iridipennis* Smith.). International Journal of Current Microbiology and Applied Sciences 6:938–943

Roubik DW (1989) Ecology and Natural History of Tropical Bees. Cambridge University Press, New York

Roubik DW (1995) Pollination of cultivated plants in the tropics. FAO Bulletin of Agricultural Services 118:1–194

Roubik DW, Moreno Patiño JE (2018) The Stingless Honey Bees (Apidae, Apinae: Meliponini) in Panama and Pollination Ecology from Pollen Analysis. In: Vit P, Pedro SRM, Roubik DW (eds) Pot-Pollen in Stingless Bee Melittology. Springer International Publishing, Cham, pp 47–66

Roubik DW, Patiño JEM (2013) How to Be a Bee-Botanist Using Pollen Spectra. In: Vit P, Pedro SRM, Roubik D (eds) Pot-Honey: A legacy of stingless bees. Springer New York, New York, NY, pp 295–314

Sakai S, Momose K, Yumoto T, Nagamitsu T, Nagamasu H, Hamid AA, Nakashizuka T (1999) Plant Reproductive Phenology over Four Years Including an Episode of General Flowering in a Lowland Dipterocarp Forest, Sarawak, Malaysia. American Journal of Botany 86:1414–1436

Saravia-Nava A, Niemeyer HM, Pinto CF (2018) Pollen Types Used by the Native Stingless Bee, *Tetragonisca angustula* (Latreille), in an Amazon-Chiquitano Transitional Forest of Bolivia. Neotropical Entomology 47:798–807

Sazima I, Sazima M (1989) Mamangavas e irapuás (Apoidea): visitas, interações e consequências para polinização do maracujá (Passifloraceae). Revista Brasileira de Entomologia 33:109–118

Slaa EJ, Chaves LAS, Malagodi-Braga KS, Hofstede FE (2006) Stingless bees in applied pollination: practice and perspectives. Apidologie 37:293–315

Sommeijer MJ, De Rooy GA, Punt W, De Bruijn LLM (1983) A comparative study of foraging behavior and pollen resources of various stingless bees (*Hym., Meliponinae*) and honeybees (*Hym., Apinae*) in Trinidad, West-Indies. Apidologie 14:205–224

Sritongchuay T, Kremen C, Bumrungsri S (2016) Effects of forest and cave proximity on fruit set of tree crops in tropical orchards in Southern Thailand. Journal of Tropical Ecology 32:269–279

Teixeira AF, Oliveira FF de, Viana BF (2007) Utilization of floral resources by bees of the genus *Frieseomelitta* von Ihering (Hymenoptera: Apidae). Neotropical Entomology 36:675–684

Viana BF, Coutinho JG da E, Garibaldi LA, Castagnino GLB, Gramacho KP, Silva FO (2014) Stingless bees further improve apple pollination and production. Journal of Pollination Ecology 14:261–269

Villanueva-Gutiérrez R, Roubik DW, Colli-Ucán W, Tuz-Novelo M (2018) The Value of Plants for the Mayan Stingless Honey Bee *Melipona beecheii* (Apidae: Meliponini): A Pollen-Based Study in the Yucatán Peninsula, Mexico. In: Vit P, Pedro SRM, Roubik DW (eds) Pot-Pollen in Stingless Bee Melittology. Springer International Publishing, Cham, pp 67–76

Vossler FG (2018) Are Stingless Bees a Broadly Polylectic Group? An Empirical Study of the Adjustments Required for an Improved Assessment of Pollen Diet in Bees. In: Vit P, Pedro SRM, Roubik DW (eds) Pot-Pollen in Stingless Bee Melittology. Springer International Publishing, Cham, pp 17–28

Vossler FG, Tellería MC, Cunningham M (2010) Floral resources foraged by *Geotrigona argentina* (Apidae, Meliponini) in the Argentine Dry Chaco forest. Grana 49:142–153

Wilms W, Imperatriz-Fonseca VL, Engels W (1996) Resource partitioning between highly eusocial bees and possible impact of the introduced Africanized honey bee on native stingless bees in the Brazilian Atlantic rainforest. Studies on Neotropical Fauna and Environment 31:137–151

Wilms W, Wiechers B (1997) Floral resource partitioning between native *Melipona* bees and the introduced Africanized honey bee in the Brazilian Atlantic rain forest. Apidologie 28:339–355

10　无刺蜂采集活动中的招募和交流

对于一只无刺蜂来说，找到好的食物资源是一件很困难的事情，尤其是这只无刺蜂孤身在外的时候。但是，如果有同伴告诉它如何以及在哪里能找到好的食物资源，这项任务将被大大简化。而当蜂群分蜂时大量无刺蜂需要找到从旧巢到新巢的路，这时候来自同巢蜂的信息就更加重要了（第4章）。如果蜂群内的无刺蜂之间没有某种形式的交流的帮助，它们是如何做到这些的？由于尚无研究者深入研究无刺蜂分蜂过程中的交流和招募行为（第4章），本章重点关注了无刺蜂的采集行为（Nieh,2004; Barth et al.,2008; Hrncir,2009; Jarau,2009; Hrncir and Barth,2014; 关于无刺蜂交流的各个方面的综述参见 Leonhardt,2017）。研究者将社会性昆虫招募行为定义为将同伴带到需要其工作的某个地方的交流过程（Wilson,1971; Hölldobler and Wilson,1990）[1]。这种交流过程涉及一些在定位上具有非特异性的社会性信息，例如当一只侦察蜂激励同伴寻找一类特定食物或者特定位置上的食物，并引导同伴到这只侦察蜂找到食物的地方的过程中就有这种社会性信息交流。

Lindauer（1956, 1957）及 Lindauer 和 Kerr（1958, 1960）最早研究了无刺蜂的招募交流。无刺蜂招募交流的开拓性实验是 Lindauer 和 Kerr（1958,1960）在巴西开展的，他们研究了采集蜂发现食物源后是否能够把同伴带到这一食物源上。他们和大多数后续研究采用的基本方法是由 Karl von Frisch 最初研究蜜蜂采集活动时提出的（von Frisch,1919, 1923, 1967）。这种方法需要训练

---

[1] 这并不是意味着蜂群总是受益于招募交流。招募也可能不利于蜂群继代，并可能会增加死亡率和能量或时间损耗（Johnson,1987; Dechaume-Moncharmont et al.,2005; I' Anson Price et al.,2019）。

采集蜂从人工喂食器中采集蔗糖溶液。这种训练程序通常是先在蜂巢入口旁边（几厘米）放置高质量的蔗糖溶液（>1 mol/L），蜂巢内一些采集蜂离开蜂巢时会找到这个蔗糖溶液，并把蔗糖溶液运回蜂巢内后，再继续返回该蔗糖溶液处。当训练成的从人工喂食器中采集的采集蜂数量满足特定实验需求时，研究者再将有采集蜂正在其上采集的喂食器移到距离原位置几米外的新位置。这样采集蜂就会学习这个新位置并能够返回到这个新位置。研究者可以重复这个过程直至喂养器达到理想距离。当采集蜂正在喂食器上采集蔗糖溶液时，研究者通常会用涂画或数字牌给其做上标记（图 10.1）。这个训练过程看起来很简单，但也具有一些挑战性［参见 Nieh（2004）的讨论部分］。例如，训练无刺蜂某些蜂种的采集蜂十分困难，研究者有时也不可能在一年中持续很长时间来训练无刺蜂（例如天然食物来源丰富时），还有非研究者关注区域的同种无刺蜂蜂巢的采集蜂也可能会发现这些人工喂养器（观察者有时会观察到喂养器上打斗的无刺蜂，从而注意到这种情况）。另一个潜在问题是采集蜂可能会丢失自己的标记而被研究者当作是新来的被招募蜂，但实际上这只采集蜂是一个有经验的招募蜂。

图 10.1　在巴西里贝朗普雷图的圣保罗大学校园内拍摄的无刺蜂

注：图中 *Plebeia droryana* 无刺蜂采集蜂正在从一个人工喂食器中收集蔗糖溶液。其中一个采集蜂被用丙烯颜料做了标记（摄影：C. Grüter）。

在 Lindauer 和 Kerr（1958,1960）实验中，他们训练每种无刺蜂的 5 只采集蜂从蜂巢飞往到人工喂食器，喂食器中放置的是 2 mol/L 蔗糖溶液，采集时间为 1 h。他们总共用这种方法检测了 10 个无刺蜂种属，并与西方蜜蜂

（*Apis mellifera*）进行了比较。他们将喂食器放在离蜂巢 10～180 m 不等的位置（图 10.2）（Lindauer and Kerr,1958, 1960）[2]。但是对于 *Frieseomelitta silvestrii* 和 *Tetragonisca angustula* 无刺蜂来说，即使蔗糖溶液放在离蜂巢很近的地方，它们也只能招募少量同伴抵达喂食器（图 10.2）。*Cephalotrigona capitata*、*Geotrigona mombuca*、*Scaptotrigona postica* 和 *Trigona spinipes* 无刺蜂的招募能力最强。这些蜂种招募的同伴数量与蜜蜂相当，甚至更多，因此研究者将这几种无刺蜂认定为大规模招募型蜂种。Lindauer 和 Kerr（1958, 1960）在人工蜂箱内研究无刺蜂采集蜂行为时发现无刺蜂没有像蜜蜂一样的舞蹈行为，但是他们注意到当无刺蜂采集蜂从人工喂食器带着食物返回蜂巢时普遍会表现出一种兴奋状态。

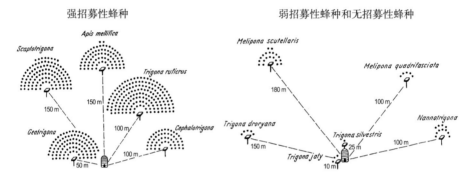

图 10.2　该图为 Lindauer 和 Kerr（1958）的原图

注：图中显示了 2 mol/L 蔗糖喂食器处不同无刺蜂蜂种和西方蜜蜂的 10 只工蜂在 1 h 内的招募成功率。每个点代表一个无标记的新被招募蜂。图中已经变更了部分蜂种的科学命名，*Trigona ruficrus* = *T. spinipes*，*Trigona droryana* = *Plebeia droryana*，*T. jaty* = *Tetragonisca angustula*，*T. silvestris* = *Frieseomelitta silvestrii*。

10.1　基于蜂巢的招募交流

10.1.1　"之"字形或推挤式前进

Lindauer 和 Kerr 观察到回巢的无刺蜂采集蜂会表现出这种兴奋状态，这些采集蜂会以"之"字形或"推挤"的方式前进，在此过程中它们冲撞着

---

[2] 作者承认，这些喂食距离的差异在一定程度上阻碍了他们建立一个标准化的实验方法来进行蜂种间的比较，但是他们这样进行实验的原因是因为这些蜂种所处地形并不总是允许所有蜂种有相近的喂食距离，而且不可能对所有蜂种进行相同距离的训练。训练 *Frieseomelitta silvestrii* 和 *Tetragonisca angustula* 无刺蜂尤其具有挑战性（图 10.2）。还应该注意的是研究者检测的蜂种在体型和蜂群大小上有很大的不同（第 1 章），这可能会影响它偏好的采集距离（第 8 章）和被招募蜂群的大小。

途中遇到的其他蜂（Lindauer and Kerr,1958, 1960; Hrncir et al.,2000）[3]。无刺蜂采集蜂通常在靠近蜂巢入口的地方会表现出这些行进方式（Nieh,1998; Hrncir,2009），在这种行进方式中的采集蜂会不规则地向各个方向行进，偶尔还会做半圆形的急转弯直至面对蜂巢入口时停止转弯（Lindauer and Kerr,1960; Nieh,1998; Hrncir,2009）。在自然环境下，无刺蜂采集蜂在狭窄的入口结构处会表现出这些行为，而这种狭窄入口结构几乎没有"舞蹈"的空间（这不像蜜蜂的舞蹈行为，蜜蜂的舞蹈通常发生在平面上）。Nieh（1998）测试了 *Melipona panamica* 无刺蜂采集蜂是否可以在这些行进方式中提供方向性信息，但是未发现这种情况。无刺蜂采集蜂在将食物转移给同伴而进行交哺行为时，会偶尔打断这种"之"字形或"推挤式"行进（图10.3）。新热带区无刺蜂种和亚洲无刺蜂种中普遍存在这种行为模式（Lindauer,1956; Lindauer and Kerr,1958, 1960; Hrncir et al.,2000），因此这种行为模式可能是所有无刺蜂都有的。"推挤式"前进在激活巢内采集蜂中发挥了重要作用。无刺蜂的归巢采集蜂在蜂巢停留期间与其他采集蜂的碰撞数量就预示了被招募到食物源的采集蜂数量（Hrncir et al.,2000），而在 *Melipona seminigra* 无刺蜂中未激活的采集蜂与招募蜂发生碰撞后自己就会开始碰撞其他工蜂（Hrncir,2009）。

活化的无刺蜂采集蜂是否会进行兴奋性行进方式部分取决于食物源的收益性和食物的通用性（Lindauer and Kerr,1958; Schmidt et al.,2008; Schmidt et al., 2006）。目前我们尚不知无刺蜂采集蜂是如何评估食物源的收益性，但这似乎取决于多种因素，如糖浓度、食物源距离、收集食物所需时间、是否存在其他蜂以及是否有前期的采集经验（第8章）（Hrncir,2009）。此外，季节甚至是时段都会影响无刺蜂对一个食物源的吸引力和是否值得招募同伴的判定（Hrncir,2009）。这其中一个原因就是无刺蜂工蜂对糖奖励的反应能力不是固定的[蜜蜂：Lindauer（1948）; Seeley（1995）; Martinez and Farina（2008）][4]。对于蜜蜂而言，当采集蜂能在巢内快速且大量地找到空载的同伴时，它们更能招募到其他工蜂（Lindauer,1948; Seeley,1986; Anderson and Ratnieks,1999; Farina,2000; De Marco,2006; Grüter and Farina,2009 a）。而蜂巢内是否有可供使用的空载工蜂也取决于采集条件，因为采集条件影响对空载工蜂的需求量（Seeley,1995）。然而招募等待时间和接收食物的工蜂数量是否

---

[3] 继代的熊蜂采集蜂也表现出相似的警戒性行进方式（Dornhaus and Chittka, 1999）。

[4] 例如，如果一个蜜蜂蜂群收集了大量非常甜的花蜜，这个蜂群中的工蜂的蔗糖反应阈值就会提高，它们就会变得更加苛刻（Pankiw et al., 2004;Martinez and Farina,2008）。

影响无刺蜂招募新成员的动机，这仍有待研究（Hrncir,2009）。

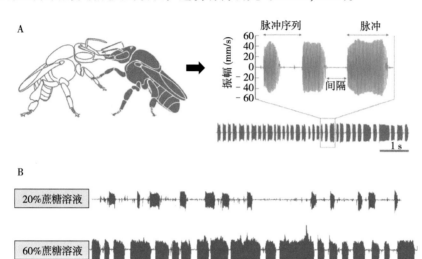

图 10.3　无刺蜂工蜂的交哺行为

注：A 显示两只 *Melipona* 属无刺蜂工蜂间的交哺行为。采集蜂通常在交哺过程中产生胸部的振动脉冲。图中左侧的工蜂（打开上颚）给右侧的工蜂（伸出长吻）提供了食物（Hrncir,2009）。B 显示研究者用激光多普勒比特计记录了 *Melipona seminigra* 采集蜂从 20% 或 60% 蔗糖溶液的食物源处返回后产生的胸部振动脉冲的情况（Hrncir et al.,2006）。

10.1.2　胸部振动

在无刺蜂推挤前进过程中，采集蜂还会用其胸肌产生不同持续时间的振动脉冲，人们听到的嗡嗡声就是这种振动脉冲（图 10.3）（Esch et al., 1965; Esch,1967; Nieh,1998; Hrncir et al.,2000, 2006, 2008 a, 2008 b; Aguilar and Briceño,2002; Nieh et al.,2003; Hrncir and Barth,2014; Krausa et al.,2017）[5]。这些脉冲的强度和频率会随蜂种而异，频率为 200～600 Hz/s（Lindauer and Kerr,1958, 1960; Aguilar and Briceño,2002; Hrncir and Barth,2014; Krausa et al., 2017）。

无刺蜂工蜂通常是在交哺（食物转移）时产生振动脉冲，这可能是为了进一步刺激振动接收工蜂开始采集活动（Hrncir et al.,2006, 2008 b; Hrncir and Barth,2014; Krausa et al.,2017 b）。有趣的是，*Melipona quadrifasciata* 无刺蜂的巢内工蜂中那些在接收食物时感受到更多振动的工蜂在嗅觉条件反射过程中也表现出了更好的学习能力（Mc Cabe et al.,2015）。这种现象支持了如下

---

[5] 无刺蜂采集蜂偶尔会在食物源处也产生嗡嗡声（Lindauer and Kerr,1960; Nieh,1998 b），但目前还不知道在食物源处的嗡嗡声是否有信号功能。

假说，即振动能够改变无刺蜂的机动状态及其对采集活动相关刺激的反应性（Mc Cabe et al.,2015）。无刺蜂的巢内同伴如何感知这些振动脉冲尚不清楚，但是交哺期的直接接触、基片振动和空气传播的声音（空气颗粒运动）都是信息传递的潜在途径（Hrncir et al.,2008 b）。虽然，无刺蜂的基片振动可以传递给在工蜂体长长度几倍距离外的工蜂，但是振动脉冲短程传送的主要形式是直接接触和空气颗粒振动（Hrncir and Barth,2014）。鉴于昆虫的听觉器官是昆虫振动感知的重要器官（Barth et al.,2008），无刺蜂的所有足部胫部下方都有这个器官，因此这也可能是无刺蜂接收振动的一个潜在器官，但无刺蜂的其他振动探测器官，如触角的琼氏器，可能也参与其中（Hrncir et al.,2006, 2008 b）。

当食物源的收益较低，无刺蜂工蜂可能不会产生胸部振动，"推挤式"前进行为也是如此（Lindauer and Kerr,1958; Nieh et al.,2003; Hrncir,2009; Hrncir and Barth, 2014）[6]。例如，当为 *Melipona scutellaris* 和 *M. quadrifasciata* 无刺蜂的采集蜂提供 0.75 mol/L 蔗糖溶液时它们不会产生振动声，而蔗糖溶液为 1.5 mol/L 时则可以听到采集蜂产生振动声（Jarau et al.,2000）。而且，无刺蜂采集蜂产生的声音脉冲幅度、脉冲时长以及声音脉冲总时长都与食物源的质量呈正相关（图 10.3）（Aguilar and Briceño,2002; Nieh et al.,2003; Hrncir et al., 2006; Schmidt et al.,2008; Krausa et al.,2017 b）。某些无刺蜂蜂种的声音脉冲之间的时间即脉冲间隔时间会随着食物源收益的增加而减少（Aguilar and Briceño,2002; Nieh et al.,2003; Hrncir et al.,2006; Schmidt et al.,2008）。而食物源的距离对振动声的产生具有负面作用，可能是因为距离越远的食物源开采成本越高，因此收益就越低（Aguilar and Briceño,2002; Hrncir,2009）。这样产生的结果就是，食物源收益越低就会导致被招募至食物源的新采集蜂就越少（Nieh et al.,2003; Schorkopf et al.,2016）。有趣的是，如果食物源有香味，那么无刺蜂交哺期的振动脉冲持续时间会更长且更频繁（Mc Cabe et al.,2015）。这表明无刺蜂认为香味越浓重的食物源更有利可图［Mc Cabe 等（2015）；蜜蜂中也有相似规律，参见 Kaschef（1957）］，这可能因为招募蜂更容易定位这种食物源，因此可以节约时间成本。

对于 *Melipona scutellaris* 无刺蜂来说，尽管仅仅是脉冲振动就足以促使采集蜂离开蜂巢去寻找同伴所宣传的食物源（Lindauer and Kerr,1958, 1960;

---

[6] Nieh 和 Sánchez（2005）发现，*Melipona panamica* 的胸部温度与食物源的收益性呈正相关。这种现象在社会性昆虫中似乎相当普遍（Nieh and Sánchez,2005）。这是否与 *M. panamica* 的声音产生有关尚不清楚。

Hrncir et al.,2000），但是当曲折推挤式前行与脉冲振动结合时似乎比单独的振动脉冲更能成功激活采集蜂离巢（Lindauer and Kerr,1958, 1960）。研究者观察到非洲 *Axestotrigona ferruginea* 无刺蜂有一种有趣的集体行为。在这种集体行为中，一个顺利采蜜的采集蜂发出的嗡嗡声似乎具有传染性，巢中的其他蜂会很快接收到这一信息（Kerr et al.,1963; Esch et al.,1965）。不久之后，该蜂群中大量的采集蜂会离开蜂巢去寻找这个食物源。

振动脉冲是否提供食物源距离的信息？

有证据表明 *Melipona bicolor*、*M. quadrifasciata*、*M. mandacaia*、*M. panamica* 和 *M. seminigra* 无刺蜂在蜂巢停留期间产生的声音脉冲的持续时间与食物源距离有关（图 10.4）（Kerr et al.,1963; Esch et al.,1965; Nieh and Roubik,1998 a; Aguilar and Briceño,2002; Nieh et al.,2003 b）而 Hrncir 等（2000）发现对于 *Melipona quadrifasciata* 和 *M. scutellaris* 无刺蜂，脉冲持续时间与食物源距离之间并不存在正相关关系。同样地，*M. costaricensis* 无刺蜂的脉冲持续时间与食物源距离也没有相关性（Aguilar and Briceño,2002）。但是当食物源离地面越近时，*Melipona panamica* 无刺蜂在卸载食物过程中产生的声音脉冲持续时间会越长（Nieh and Roubik,1998）。

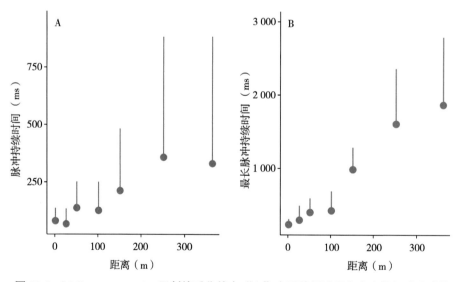

图 10.4 *Melipona panamica* 无刺蜂采集蜂在"之"字形前行过程中产生的振动脉冲的平均时长（A）和最长时长（B）

注：点代表平均值，线代表标准差（Nieh and Roubik,1998）。

这些相关性就引出了这样一个问题：无刺蜂是否可以通过声音脉冲来给同伴传达食物源的距离（和高度）信息，从而使同伴随后能用这些信息来定位食物源？然而，尽管有研究者尝试去证明被招募蜂会应用这些信息，但是这个问题目前仍然没有定论（Nieh and Roubik,1998; Hrncir et al.,2000; Nieh,2004），脉冲持续时间和食物源距离之间的正相关关系也可能只是生理过程的副产物，并没有任何交流功能（Esch,1967）。实际上，对于某一给定距离的食物源，无刺蜂产生的脉冲持续时间的平均值和最大值之间的差距是相当大的（图10.4）（Nieh and Roubik,1998; Hrncir et al.,2000; Aguilar and Briceño,2002; Schmidt et al.,2008; Hrncir and Barth,2014），这将使所有由声音和振动信号编码的空间信息非常不精确（Hrncir et al.,2000; Barth et al.,2008）。被招募采集蜂会尝试平衡这些声音脉冲以获得更准确的空间信息。对于蜜蜂而言，蜂舞追随蜂就采用了平均多次摆动轨迹的策略来从不同的摆动轨迹中获得更准确的信息（von Frisch and Jander,1957; Tanner and Visscher,2008）。然而，如果食物源的质量也影响无刺蜂产生的脉冲持续时间，那么这种策略很可能会失败，一些无刺蜂蜂种就出现了这种情况（Aguilar and Briceño,2002; Hrncir and Barth,2014）。然而，被招募蜂也可能会通过其他方式来区分低质量或高质量的食物源的对应声音，因此它们仍然能够利用声音来确定食物源的大概距离（Nieh,2004）。此外，还有一些其他非 *Melipona* 属无刺蜂蜂种产生的声音脉冲的持续时间与食物源距离无关（Esch et al.,1965）。因此，无刺蜂产生的声音脉冲和食物源距离间呈正相关也许不是个普遍现象。但是这种现象仍然是十分有趣并且值得进一步研究，因为它可以代表一种能够进化成诸如蜜蜂摇摆舞这种复杂信号的"原始状态"的例子。

10.1.3 食物气味的社会性学习

携带着花蜜或果汁返回蜂巢的无刺蜂采集蜂不仅在交哺时将食物转移给其他巢内蜂（第6章）（Hart and Ratnieks,2006），而且还会分发食物小样给潜在的被招募蜂品尝（Hrncir,2009）。这使食物小样的接收蜂可以评估食物源的质量，并有可能会学习食物源的气味（第8章）（Lindauer and Kerr,1958, 1960; Jarau,2009）。西方蜜蜂就可以通过交哺行为接收食物小样建立稳定的嗅觉联想记忆（Farina et al., 2005; Gil and De Marco,2005; Grüter et al., 2006, 2009）。无刺蜂蜂巢内，活跃采集蜂和同伴间的交哺行为通常发生在靠近巢内入口的地方，但是当采集活动很旺盛时也能在巢穴入口外面观察到这种现象（Kerr,1994; Krausa et al., 2017 b）。接收食物小样是无刺蜂招募活动的重要部分。*Melipona quadrifasciata* 和 *M. seminigra* 无刺蜂蜂群内的能够成功定位到招募蜂宣传的食物源的所有采集蜂中都收到了来自招募蜂的至少一个食物小样（Hrncir,2009）。值得注意的是，食物源中蔗糖含量越高，接收到食物小样的被招募蜂也就越多（图 10.5）[*Plebeina armata* 无刺蜂参见 Krausa 等（2017 b）]。这表明，高质量食物源的信息会更广泛地在蜂群同伴间传播。而一些无刺蜂蜂种没有或者只有在较弱的位置特异性招募行为，这些蜂种的采集蜂可以利用这种信息来寻找散发相同气味的食物源（Nieh et al., 2000; Aguilar et al., 2005; Jarau,2009）。研究者已证明新热带蜂种 *Plebeia tica* 和亚洲 *Tetragonula iridipennis* 无刺蜂在招募过程中的确会学习食物气味：在与成功采集的采集蜂互动后，离巢的被招募蜂会对散发着与招募蜂带回气味相同的食物源表现出强烈偏好（Lindauer,1956, 1957; Aguilar et al., 2005）。对于 *Tetragonula carbonaria* 无刺蜂而言，如果食物源是有气味的且食物源喂食器位于蜂巢的逆风处，它们的招募就会更容易成功（Nieh et al., 2000）。

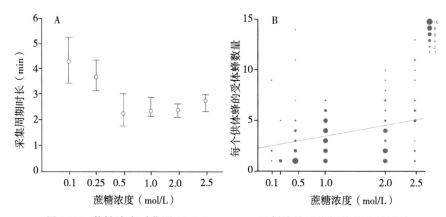

图 10.5　蔗糖浓度对非洲 *Plebeina armata* 无刺蜂的采集相关行为的影响

注：A 显示采集周期的时长随着蔗糖浓度的增加而降低。B 显示蜂群中接受采集蜂传递食物小样的受体工蜂的数量随着糖浓度的增加而增加（Krausa et al.,2017 b）。

10.1.4　有采集经验的采集蜂的重新激活

蜂巢内交流的一个重要作用是重新激活有采集经验的采集蜂。重新激活指的是一个暂时不活跃的经验丰富的采集蜂当遇到社会性提示或信号的刺激时，会去探查它以前发现过食物源的位置（von Frisch,1967; Biesmeijer and de Vries,2001; Grüter and Farina,2009 b; Hrncir and Maia-Silva,2013）。当无刺蜂蜂群的侦察蜂和巡视蜂已经发现了可用食物源，它们就最有可能会去担任重新激活有采集经验的采集蜂的任务。"侦察蜂"是一种会主动寻找新的食物源而不会等待招募蜂提供信息的工蜂，而"巡视蜂"则是经验丰富的采集蜂，它们偶尔会去检查过去获得过食物但是已经暂时没有食物的采集地点，而这些采集地点后来没有食物可能由于不利的天气条件或时段（von Frisch,1967; Biesmeijer and de Vries,2001; Al Toufailia et al.,2013; Hrncir and Maia-Silva,2013; Jesus et al.,2014）。图 10.6 概括了无刺蜂主要的采集策略和状况。

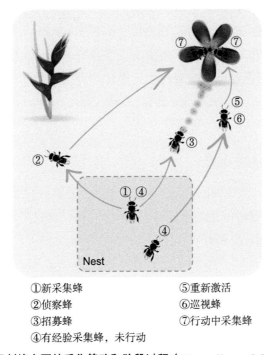

①新采集蜂　　　　　　　　⑤重新激活
②侦察蜂　　　　　　　　　⑥巡视蜂
③招募蜂　　　　　　　　　⑦行动中采集蜂
④有经验采集蜂，未行动

图 10.6　无刺蜂主要的采集策略和阶段过程（Biesmeijer and de Vries,2001; Grüter and Farina,2009 b; Hrncir and Maia-Silva et al.,2013）

注：首先，侦察蜂② 在没有招募蜂指示的情况下寻找食物源。然而，它们可能会依据幼时在巢内学到的食物气味信息去寻找食物源。新被招募蜂③利用社会信息去寻找食物源的位置，如信息素轨迹（点状痕迹）。一些新被招募蜂可能会因没有追踪到信息素轨迹等"错误"而发现其他新食物源。暂时没有被征用的采集蜂④也有各种选择来继续采集工作，包括探查它们已知的食物源地点⑤或被社会互动或熟悉食物源的线索重新激活⑥。被征用的采集蜂⑦则会依据食物源的质量和蜂种的不同选择招募同伴到这个食物源或单独去开采这个食物源（源自 Rodolfo Guimarães 对 *Scaptotrigona bipunctata* 的图解）。

对于蜜蜂而言，触发有采集经验的采集蜂重新激活，通常是由于这些采集蜂感知到巢内有熟悉的食物气味（如返回工蜂的身体上出现的气味或食物中存在的气味）或者它们感知了跳舞蜜蜂身上的角质层烃类化合物，还可能是被蜂舞本身触发（von Frisch,1923；Johnson,1967；Reinhard et al.,2004；Grüter et al.,2008；Grüter and Farina,2009 b；Gilley et al.,2012）。而无刺蜂只要感知到熟悉的食物气味似乎就足以被重新激活（Biesmeijer et al.,1998）。Lindauer 和 Kerr（1958）认为，*Melipona scutellaris* 无刺蜂的重新激活甚至不需要工蜂之间的直接身体接触，而只需要在一段距离外感知到同伴的胸部振动脉冲，就可能足以被重新激活（Hrncir and Barth,2014）。在这一蜂种蜂群中回放招募蜂产生的声音就足以重新激活有经验的采集蜂，但前提是食物源位于附近区域（Esch et al.,1965；Esch,1967）。另外，当蜂群内工蜂间有身体接触和推挤式前进行为时，就能更有效地重新激活有经验的采集蜂。重新激活是一个关键的过程，因为当觅食条件改善时，它能让蜂群迅速激活其经验丰富的采集蜂力量（von Frisch,1967；Sánchez et al.,2004；Granovskiy et al.,2012；Hrncir and Maia-Silva,2013）。

10.2　位置特异性招募

鉴于一些无刺蜂蜂种的采集蜂能够很高效地将同伴招募到食物源（图 10.1），很明显这些蜂种的采集蜂可以为它们的同伴提供关于食物源位置的信息（Lindauer and Kerr,1958，1960；Hubbell and Johnson,1977，1978）。例如，在 Lindauer 和 Kerr（1958，1960）[7] 进行的实验中，*Melipona scutellaris* 无刺蜂却不能引导其同伴到达特定位置的食物源（图 10.7 A），而 *Scaptotrigona postica* 无刺蜂的采集蜂在招募同伴时能提供具体的地点信息（图 10.7 B，图 10.7 C）（表 10.1）。特别是 *Scaptotrigona postica* 无刺蜂，有 42 只被招募采集蜂顺利到达了一个只有少数标记采集蜂访问过的喂食器，而同一时期在等距、无采集蜂访问过的对照组喂食器上则只捕获到 2 只采集蜂（图 10.7 B）。*S. postica* 采集蜂招募者看起来确实能够提供食物源的距离和方向信息，因为研究者发现在一个距离较远（150 m）的但有被标记蜂访问过的喂食器上观察到的采集蜂数量远高于一个不是同一方向的对照喂食器上的采集蜂数量，且这个对照喂食器距离蜂群只有 10 m 远（图 10.7 C）。如前所述，*Melipona panamica* 不仅能够引导同伴朝着特定的方向和距离寻找食物，而且似乎还能提供食物源垂直高度的信息（Nieh and Roubik,1995）。

---

[7] Lindauer 与 Kerr（1958，1960）提到他们测定的 *Melipona* 蜂种在招募中确实表现出一些方向性，并提出这可能是用引导飞行的方法实现的。然而，他们的研究中所展示的数据，例如他们文中的表 3，并没有为这个说法提供令人信服的证据。

10 无刺蜂采集活动中的招募和交流

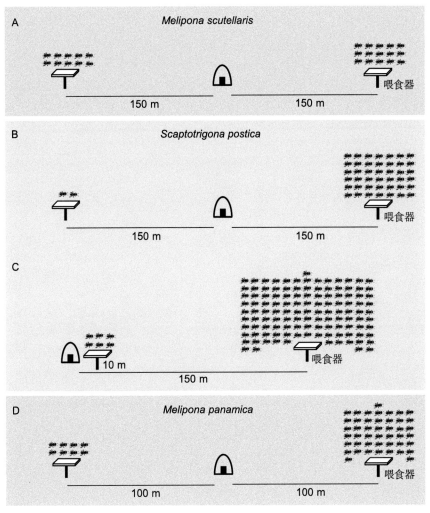

图 10.7 *Melipona scutellaris*（A）、*Scaptotrigona postica*（B、C）和 *M. panamica*（D）的招募成功率

注：研究者会同时使用两个喂食器，但只在一个喂食器中提供食物。图中蜂型图标数量表示新被招募来的采集蜂的数量（Lindauer and Kerr,1958; Nieh and Roubik,1995; 有修改）。

表 10.1　无刺蜂不同蜂种是否招募同伴至食物源的信息列表

无刺蜂蜂种	区域	气味特性是否确认	招募地点是否确认	备注及参考文献
Apotrigona nebulata	AT	否	否	Darchen（1969）

续表

无刺蜂蜂种	区域	气味特性是否确认	招募地点是否确认	备注及参考文献
Axestotrigona ferruginea	AT	否	否	Esch et al.（1965）
Tetragonula carbonaria	IM/AA	否	部分	仅有方位 Nieh et al.（2000）；Nieh（2004）
Tetragonula iridipennis	IM/AA	否	否	Lindauer（1956）
Cephaiotrigona capitata	NE	是	是	Lindauer and Kerr（1960）
Frieseonelitta silvestrii	NE	否	否	Lindauer and Kerr（1958,1960）
Frieseomnelitta varia	NE	否	否	Esch et al.（1965）：Jarau et al.（2003）
Geotrigona mombuca	NE	是	是	Lindauer and Kerr（1958,1960）；Stangler et al.（2009）
Lestrinelitta limao	NE	否	是	第7章
Melipona bicolor	NE	否	是	Nieh et al.（2003b）
Melipona mandacaia	NE	否	是	Nieh et al.（2003b）
Melipona pamanicat	NE	否	是	Nieh and Roubik（1995）
Melipona quadrifasciata	NE	否	部分	仅有方位 Jarau et al.（2000）
Melipona rufvenris	NE	否	是	Kerr and Rocha（1988）
Melipona scutellaris	NE	否	部分	仅有方位 Jarau et al.（2000）
Namnotrigona testaceicormis	NE	否	否	Lindauer and Kerr（1958, 1960）
Partamona helleri	NE	否	否	Esch et al.（1965）
Partanona orizabaensis	NE	否	是	Flaig et al.（2016）
Plebeia droryana	NE	否	部分	仅有方位 Lindauer and Kerr（1958, 1960）；Peng et al., in press
Plebeia tica	NE	否	部分	仅有方位 Aguilar et al.（2005）
Scaptotrigona bipunctata	NE	是	是	Kerr et al.（1963）
Scaptotrigona mexicamat	NE	是	是	Sanchez et al.（2004,2007）
Scaptotrigona pectoralis	NE	是	是	Hubbell and Johnson（1977）
Scaptotrigona postica	NE	是	是	Lindauer and Kerr（1958,1960）

续表

无刺蜂蜂种	区域	气味特性是否确认	招募地点是否确认	备注及参考文献
Scaptotrigona depilis	NE	是	是	Schorkopf et al.（2011）
Scaptotrigona xamthotricha	NE	是	是	Kerr et al.（1963）
Tetragona clavipes	NE	否	否	Jarau et al.（2003）
Tetragonisca amgustula	NE	否	部分	仅有方位 Lindauer and Kerr（1958,1960）；Aguilar et al.（2005）
Trigona amalthea	NE	是	是	Kerr et al.（1963）
Tyigona corvina	NE	可能	是	Aguilar et al.（2005）
Trigona fulviventris	NE	是	是	Johnson（1987），cited in Schmitt et al.（2003）
Trigona fuscipennis	NE	是	是	Hubbell and Johnson（1977）
Trigona hyalinata	NE	是	是	Nieh et al.（2003）
Trigona hypogea group	NE	是	是	Roubik 1982）；Nol1（1997）；Jarau et al.（2003）
Trigona recursa	NE	是	是	Jarau et al.（2003）
Trigona silvestriana	NE	是	是	Hubbell and Johnson（1977）
Trigona spinipes	NE	是	是	Lindauer and Kerr（1958,1960）：Nieh et al.（2004）

注：表中标注了确定会使用气味轨迹的蜂种。AT= 非洲热带区；IM/AA =印尼－马来半岛/澳大拉西亚；NE =新热带区。

与 Lindauer 和 Kerr（1958, 1960）进行的初步实验不同的是，最近更多关于 Melipona scutellaris 无刺蜂研究表明，这一蜂种的采集蜂能给同伴提供一些空间信息。然而，其同伴好像只能获取关于方向的信息，而不能获得关于食物源距离的信息（Jarau et al.,2000）[8]。这与一些其他蜂种相似，这些蜂种的采集蜂能够将同伴募集到特定方向的食物源，而不能提供食物源准确的距离信息（表10.1）（Jarau et al.,2000; Nieh et al., 2000; Aguilar et al., 2005; Peng et al., 出版

---

[8] 对于 M. Quadrifasciata 无刺蜂而言，在食物源相对靠近蜂巢时（<30 m），方向信息才是有效的（Jarau et al.,2000）。

中）。例如，*Plebeia tica* 新被招募采集蜂大多会降落在朝着侦察蜂所指示方向上离蜂巢最近的喂食器上（Aguilar et al.,2005）。因此，这些蜂种的被招募采集蜂好像只会朝着食物源的方向飞行，然后飞向气味与蜂巢内侦察蜂报告的气味相似的食物源处。然而，这些物种的采集蜂是如何将同伴指往食物源大致方向的，这仍然是一个谜。

10.2.1 信息素轨迹

当 Lindauer 和 Kerr（1958，1960）研究 *Scaptotrigona postica* 无刺蜂时观察到该蜂种采集蜂离开食物源后每隔几米就会落在植被、石头或泥土上，留下气味作为标记。他们在一个案例中观察到，采集蜂在距离蜂巢 50 m 的食物源与蜂巢之间留下了 32 个气味标记点（图 10.8）。留下气味标记的采集蜂通常不会进入蜂巢，而是在距蜂巢入口几米外的位置留下气味标记后就立即返回食物源。有趣的是，研究者从未观察到 *S. postica* 采集蜂在靠近蜂巢的位置留下气味标记，离蜂巢最近的气味标记点也离蜂巢有几米远的距离。这种行为可能有助于防止多条气味路径在蜂巢入口附近形成交叉，并防止迷惑新被招募蜂而产生时间损失（Lindauer and Kerr,1958, 1960）。只有在食物源距离蜂巢 10～20 m 开外时，无刺蜂才会进行气味标记（不同蜂种进行气味标记所需的最小距离存在差异，*Geotrigona mombuca* 无刺蜂仅为 3 m，而 *Trigona spinipes* 无刺蜂则需 35 m）（Lindauer and Kerr,1960; Kerr et al., 1963）。无刺蜂采集蜂留下的第一个气味标记通常在食物源上或离食物源很近的地方，且采集蜂留气味标记的频率随着离食物源距离的增加而降低（Lindauer and Kerr,1960; Kerr et al., 1963; Noll,1997; Nieh et al., 2003 a; Stangler et al., 2009; Schorkopf et al., 2011）。不同无刺蜂留下的气味标记之间的长度也大相径庭：研究发现 *Trigona amalthea* 无刺蜂的铺设的气味标记点间距离长达 900 m（Kerr,1960，此处与 *Trigona trinidadensis* 类似），而 *T. hyalinata* 和 *T. Spinipes* 无刺蜂的气味标记点间距离则相对较短。

T. hyalinata 和 *T. Spinipes* 无刺蜂的采集蜂留下气味标记的频率在食物源处最高，在距离食物源 30 m 处达到最大频率后逐渐减少，即使食物源在离蜂巢 150 m 开外也是如此（Nieh et al., 2003 a, 2004; Stangler et al., 2009）。这种现象表明，气味标记点分布具有极性，能够使新招募来的采集蜂确定食物源的方向和距离（Kerr et al., 1963; Nieh et al., 2003 a, 2004）。然而，这也引发了一个思考：新成员是如何找到这种踪迹信息素的起点的？

10 无刺蜂采集活动中的招募和交流

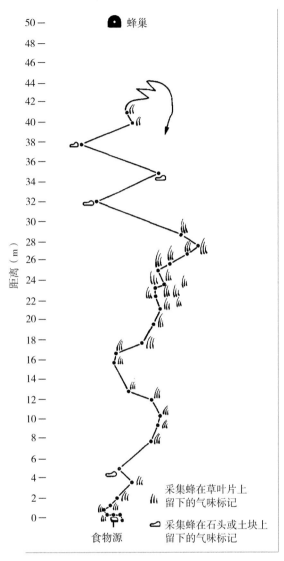

图10.8 *Scaptotrigona postica* 无刺蜂采集蜂从食物源返回巢穴时留下的气味标记（图中小点状标记）

注：这些气味标记大多数是留在草叶、石头或土块上。采集蜂没有进入蜂群就又折返回食物源了。最后一个气味标记是留在离其蜂巢几米远的地方。

Scaptotrigona postica 无刺蜂的采集蜂从食物源飞向蜂巢时会留下气味标记，而从蜂巢飞向食物源时则不会留下气味标记，而且这种气味痕迹的效果只能持续 15 min（Lindauer and Kerr,1958, 1960）。其他蜂种的气味标记的行

为学活性有的可持续 14 min（*Scaptotrigona bipunctata*），有的为 9 ~ 20 min
（*Trigona spinipes*, Kerr et al.,1963; Nieh et al.,2004），有的是 15 min（*T. hyalinata*,
Nieh et al.,2003 a）。这表明无刺蜂的轨迹信息素具有高度挥发性且活性短暂
（Jarau,2009）。例如，Kerr 等（1981）报告称，*T. spinipes* 释放的信息素能在
1 m 外被人类闻到。这种高挥发性有助于采集蜂从很远的地方发现信息素标
记，但也可能吸引来竞争对手（见下文）。

无刺蜂轨迹信息素的特性和分泌腺

在无刺蜂采集蜂留下气味标记过程中，采集蜂会用口部在物种特定间
隔（*Cephalalotrigona capitata* 的间隔是 5 m，*Scaptotrigona postica* 的间隔是
1 ~ 2 m）的标记基质上摩擦［打开下颌骨并且伸出长吻；Jarau 等（2004 b）；
Nieh 等（2004 a，2004 b）；Jarau（2009）］，这表明气味标记上的化学物质是
由头部腺体（即头腺）分泌的（Lindauer and Kerr,1958, 1960）[9]。也有一些其
他会进行气味标记的无刺蜂蜂种（如 *Trigona hypogea*）是用后腿和腹部摩擦
标记基质的（Noll, 1997）。早期研究提出了一个假设认为目前存在争议的
头腺其实就是上颚腺（Lindauer and Kerr,1958, 1960; Blum et al.,1970; Nieh
et al.,2003 a）。然而，最近的许多研究已经证明，无刺蜂的轨迹信息素是
来源于其唇腺并且是通过伸出的长吻留在标记点上（图 10.9）［*Geotrigona
mombuca*: Stangler 等（2009）；*Scaptotrigona pectoralis*: Jarau（2009）；
Trigona corvina: Jarau 等（2010）；*T. hyalinata*: Lichtenberg 等（2011）；*T.
recursa*: Jarau 等（2004, 2006）；*T. spinipes*: Schorkopf 等（2007）；综述于
Jarau（2009）］。而无刺蜂的上颚腺分泌物主要起到警戒信息素的作用（第
7 章）。*T. Spinipes* 无刺蜂的招募信息素由单一化合物——辛酯 – 辛酸酯构
成（Schorkopf et al.,2007）。*T. hyalinata* 无刺蜂的轨迹信息素的主要成分也
是由这种化合物与其他酯类化合物结合而成的（Lichtenberg et al.,2011）。*T.
Recursa* 无刺蜂的轨迹信息素组成则更复杂，但其主要成分也是一种酯类
（乙基癸酸盐，占所有挥发物的 72%）（图 10.9）（Jarau et al.,2006）。而对于
Geotrigona mombuca 无刺蜂来说，轨迹信息素中的法尼酯 – 丁酸酯才是其中
的关键酯类组分（Stangler et al.,2009）。

---

[9] 这些间隔也意味着，宽度大于 10 m 的河流等水体可能会严重阻碍无刺蜂的招募，从而阻碍大量
招募型蜂种的有效开采（Kerr,1960; Schorkopf et al.,2011）。

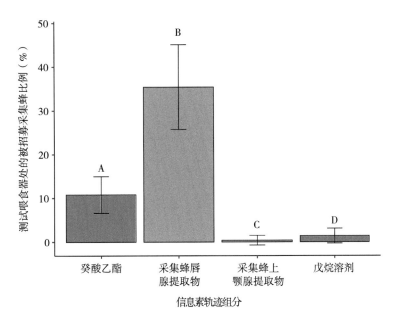

图 10.9 *Trigona recursa* 无刺蜂抵达测试喂食器的新被招募蜂的比例

注：该测试喂食器位于 10 m 外的人工气味轨迹的末端，研究中使用的人工气味分别是用癸酸乙酯（见正文）、采集蜂唇腺提取物、采集蜂上颚腺提取物、戊烷溶剂制成（Jarau et al.,2006 有修改）。由于无刺蜂铺设的天然气味轨迹是指向测试喂食器 10 m 外的竞争喂食器，人工气味轨迹与天然气味轨迹会竞争采集蜂［详见 Jarau 等（2006）文中的图 1］，因而这就解释了为什么测试喂食器的新被招募蜂百分比总是低于 50%。

一些蜂种的不同蜂群的轨迹信息素在计量成分组成上表现出相当大的差异，较之于同种其他蜂群的信息素，一个蜂群的采集蜂更倾向于尾随自己蜂群同伴留下的信息素混合物（Jarau et al.,2010; Reichle et al.,2011; John et al.,2012）。这有助于避免蜂群之间对食物源的竞争或激烈的打斗（第 10.3 节）。*Scaptotrigona pectoralis* 和 *S. subobscuripennis* 无刺蜂的采集蜂还能学习自己巢内同伴的轨迹追踪信息素的化学组成（Reichle et al.,2011, 2013）在异种蜂群中饲养的无刺蜂就能学会对异种同伴的轨迹追踪信息素做出反应（Reichle et al.,2013）。无刺蜂能够学习信息素特征的现象是十分令人惊奇的，因为研究者普遍认为群居昆虫对信息素的反应是天生的（Wyatt,2010; Grüter and Czaczkes,2019），但如果信息素的化学成分不是固定的，而是根据内部（基因结构，如蜂王替换）和外部（食物来源和气候条件的变化）条件的改变而发生变化，这种现象可能是有益的（Reichle et al., 2011）。*Scaptotrigona* 无刺蜂的信息素组成是否会随着季节等因素而变化仍有待研究。

食物源特征（即食物源的距离、质量和数量）是否以及如何影响无刺蜂形成的气味轨迹，目前还不太清楚。但 Schmidt 等（2006 a）发现 *Trigona recursa* 无刺蜂针对 40% 蔗糖溶液留下的气味轨迹招募到的采集蜂数量要高于 20% 蔗糖溶液。这表明，如果食物源的收益性越高，无刺蜂采集蜂留下轨迹标记的强度越大，但这一点需要进一步的研究来证实（Schmidt et al.,2006 a）。而对于 *Scaptotrigona depilis* 无刺蜂，蔗糖溶液的浓度本身似乎不影响招募（Schmidt et al., 2006 b），这意味着它不影响无刺蜂留下轨迹信息素标记点的频率。可是，如果食物源的糖浓度从 40% 降低到 20%，那么被招募到达人工食物源的新被招募蜂就会明显减少。因此，有采集经验的采集蜂可能会基于与其他食物源的对比或过往经验来决定是否前往招募蜂指引的食物源。

因为无刺蜂的轨迹信息素可以吸引大量同伴，这些同伴在招募过程中会产生强烈的正反馈。例如，*Scaptotrigona depilis* 无刺蜂采集蜂能在 4 h 内招募到 1 500 多个同伴到达同一个喂食器（Jarau et al.,2003）（图 10.7 C）。这样有一个缺点，那就是蜂群无法转向后来出现的其他更好的食物源（蚂蚁参见：Beckers et al.,1990; Camazine et al.,2001; Czaczkes et al.,2016）。*Trigona recursa* 无刺蜂是一种会大规模招募的蜂种，研究者至今仅研究了这种无刺蜂在蜂群水平上分配采集蜂的灵活性，研究者发现这种无刺蜂如果先发现了 20% 蔗糖溶液后才发现 40% 蔗糖溶液，蜂群会继续把更多的采集蜂派往 20% 蔗糖溶液处，原因很有可能是由于大量采集蜂持续留下信息素，因而指向劣质食物源处的轨迹信息素仍然能强烈吸引采集蜂（Schmidt et al.,2006a; Noll,1997）。

蜜蜂采用摇尾舞传递的是矢量信息，而无刺蜂在植物上留下气味标记的一个优势是进一步增加了三维环境的特异性。这在热带森林中可能特别有用，因为热带森林的无刺蜂食物源通常存在于不同的垂直层中（第 8 章）（Nieh and Roubik,1995; Nieh,2004）。例如，针对一个位于水塔上的食物源，*Scaptotrigona postica* 无刺蜂能够比 *Apis mellifera* 蜜蜂招募到更多的同伴（Lindauer and Kerr,1958, 1960）。

信息素轨迹和蜂群规模

研究者已经证实一些无刺蜂蜂种的信息素轨迹能够引导采集蜂抵达食物源处，但也有许多蜂种在采集活动中好像并没有用到轨迹信息素（表

10.1）[10]。气味轨迹对于无刺蜂蜂群来说是否是高效益的需要取决于蜂群规模等因素。由于气味轨迹会不断挥发，因此蜂群需要派遣大量工蜂去进行信息素标记，从而保持轨迹存在（Beekman et al.,2001; Beekman and Dussutour,2009）。此外，蜂群规模越大越能够垄断食物源，这也使蜂群招募的效益越高（Aguilar et al., 2005）。研究者也相应发现会用信息素轨迹的无刺蜂蜂种（表10.1）的蜂群规模比不用信息素轨迹蜂种的蜂群规模更大（图10.10）。

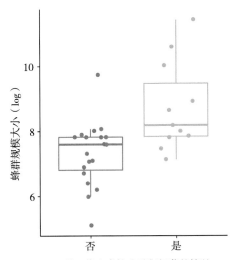

图10.10　已经确认会使用信息素轨迹和尚未确认会使用信息素轨迹的无刺蜂蜂种的蜂群规模（自然对数转换值）

注：蜂群大小数据取自第1章的表1.3。研究统计中纳入了物种间系统发育关系［基于Rasmussen和Cameron（2010）提供的系统发育关系和年表的进化树］并基于Ornstein–Uhlenbeck模型方法（Paradis,2011）的广义最小二乘法模型（GLS），研究比较是否基于信息素轨迹进行招募的蜂种时发现二者间的蜂群大小规模上存在显著差异（t-value =2.1, p=0.045）（表格文件和进化树下载网址：www.socialinsect-research.com/book.php）。

10.2.2　无刺蜂在食物源上留下的化学标记

许多无刺蜂蜂种的采集蜂，甚至是那些不会释放信息素的蜂种，通常都

---

[10] 基于信息素标记的招募行为在无刺蜂分蜂过程中更为常见（第4章）（Roubik, 1989）。这与蚂蚁相似，它们在寻找居所期间会通过"前后跑"等方式交流新巢址的位置，但在觅食时则不使用"前后跑"（Franklin,2014; Grüter et al.,2018）。无刺蜂之间的交流是否有利于蜂群继代取决于食物源的时空分布和竞争情况（Sherman and Visscher,2002; Dornhaus and Chittka,2004; I'Anson Price and Grüter,2015）。

会于采集期间在食物源上或在其附近留下气味标记（Kerr,1994; Nieh,1998 b; Aguilar and Sommeijer,2001; Schmidt et al.,2003, 2005; Hrncir et al.,2004; Jarau et al., 2004 a; Boogert et al.,2006; Contrera and Nieh,2007; Alavez-Rosas et al., 2017; Koethe et al.,2020; Jarau,2009）。这些化学"指示牌"不仅可以帮助同伴找到食物源，还可以使做标记的工蜂本身找到并返回这个采集点（Boogert et al., 2006）。这种化学化合物有的是工蜂主动刻意释放的信息素以此来传递食物源位置的信息（Kerr,1994; Schmidt et al.,2003; Jarau et al., 2004 a），有的则是工蜂采集过程中被动留下的脚印［例如，*Nannotrigona testaceicornis*，Schmidt 等（2005）］。在后一种被动留下脚印的情况中，脚印中的化学物质仅仅是信息线索而不是前往信号（Seeley,1995; Danchin et al.,2004; Dall et al.,2005; Schmidt et al.,2005; Jarau,2009）[11]。*Melipona seminigra* 无刺蜂在食物源上留下的这种气味标记具有吸引力的范围约为 1 m，其作用时间可持续 2 h 左右（Hrncir et al.,2004）。*Melipona panamica* 无刺蜂在食物源上留下的这种气味标记具有吸引力的范围为几米远（Nieh,1998 b），而 *Scaptotrigona depilis* 无刺蜂在食物源上留下的这种气味标记可以被 20 m 以外的其他采集蜂感知到（Schmidt et al.,2003）。

　　研究者已观察到某些 *Melipona* 属蜂种，如 *Melipona favosa* 和 *M. Mandacaia* 无刺蜂，采集蜂被吸引到人工喂食器上后会在该人工喂食器上和喂食器附近留下其肛门或腹部液体物（Aguilar and Sommeijer,2001; Nieh et al.,2003 c）。采集蜂采集的喂食器中蔗糖溶液浓度越低（即蜂群收益率低），采集蜂在喂食器上留下的腹部液体越多，表明这些液体是采集蜂释放的多余液体（Nieh,1998 b; Nieh et al.,2003 b）。目前尚不清楚采集蜂在访问自然食物源而不是人工喂食器时是否也会产生腹部液滴。Nieh 等（2003 c）则已证实 *M. Mandacaia* 无刺蜂确实会用腹部气味标记食物源。研究者还观察到一些 *Melipona* 属无刺蜂蜂种的采集蜂会在喂食器附近的植被上摩擦其口部（Aguilar and Sommeijer,2001; Kerr,1994），这种行为是采集蜂在自我清洁还是确实具有交流功能尚不清楚。

　　无刺蜂在食物源上留下的有吸引性的化学标记的来源需要进一步研究，而由于 *M. Favosa* 和 *M. Mandacaia* 无刺蜂的腹部液滴可能是用来标记食物源

---

[11] 研究者认为只有涉及信号的信息传递（即由自然选择形成的传递信息的特征）才是一种交流形式（Seeley,1995; Danchin et al.,2004; Dall et al.,2005）。根据这一定义，蜂群使用的任何信息线索，例如在采集过程中被动留下的作为食物印记的碳氢化合物（第 7 章）不能算作一种交流。

的，因此在研究其他 *Melipona* 属无刺蜂蜂种的这一问题时，已将这两种无刺蜂排除在外（Nieh,1998 b; Hrncir et al.,2004）。*Melipona solani* 无刺蜂摄取喂食器中的食物时会在喂食器上留下了烃类化合物，这种化合物更有可能是其被动留下的一种脚印。此外，这种无刺蜂的采集蜂会用一种化学混合物标记质量更高的食物源，该化学混合物的主要成分之一是由唇腺释放的油酸甲酯（Alavez–Rosas et al.,2017）。

研究者对无刺蜂在食物源上留下的气味标记的来源进行了详细研究，他们指出 *Melipona seminigra* 无刺蜂的跗分节提取物与其在食物源上留下的气味一样可以吸引同伴（Hrncir et al.,2004; Jarau et al.,2004 a）。Hrncir 等（2004）用指甲油覆盖 *Melipona seminigra* 无刺蜂采集蜂的跗趾和跗端节后发现这些采集蜂就不再能吸引同伴去它造访过的喂食器（Hrncir et al.,2004）。研究者对这些气味标记物进行了组织化学分析，发现其主要由烃类化合物组成［主要是五烷、七烷，以及相应的 7-（Z）- 五烯和 7-（Z）- 七烯］。它们来源于无刺蜂的足部腱腺（爪牵肌腱的腺上皮），并在足根部释放（Jarau et al.,2004 a）。

驱避型气味标记

工蜂在食物源上留下的气味标记也可能是趋避剂，这可以帮助工蜂避开自己最近采集过的或已凋谢的花朵［蜜蜂参见：Giurfa 和 Núñez（1992）；熊蜂参见：Stout 等（1998）］。研究发现，*Trigona fulviventris* 无刺蜂采 *Priva mexicana* 花时总能避开之前采集过的花朵，无论这些花是否能提供奖励，这表明这些无刺蜂在食物源上留下的气味标记是一种趋避剂（Goulson et al., 2001）。而与之相反的是，*Trigona fulviventris* 无刺蜂在 *Crotalaria cajanifolia* 花上留下的气味标记却不会阻止其采集蜂掠取该花种的花蜜。这可能是因为 *Crotalaria cajanifolia* 植物的花结构更简单，花蜜更容易获取，这可以减少了无刺蜂处理花朵的时间，因而减少了采集到无蜜花所造成的损耗（Goulson et al., 2001）。这表明在不同环境背景下，无刺蜂在食物源上留下的气味标记的作用不同。无刺蜂在食物源上留下的气味标记是引诱剂还是趋避剂需要取决于一只采集蜂的采集经验。例如，当 *Scaptotrigona mexicana* 采集蜂在前期在气味标记食物源上得到过奖励，那它们后期就会被这种气味标记吸引（Sanchez et al., 2008; Solórzano–Gordillo et al.,2018）。而如果它们前期遇到的气味标记食物源没有奖励，那么采集蜂后期就会避开这种气味（第10.4 节）（Saleh and Chittka,2006; Sanchez et al.,2008; Jarau,2009; Slaa and Hughes,2009）。研究者发现 *Melipona scutellaris* 无刺蜂也有相似的学习效应：当该蜂种采集蜂感受到气味标记食物源的奖励减少后,这些采集蜂后续就会倾向于选择未采集过的食物源,而采集蜂感受到气味标记食物源的奖励增加后,后续则会倾向于选择有气味标记的喂食器(Roselino et al.,2016)。另外,气味标记可能具有剂量依赖性。对于 *M. seminigra* 无刺蜂而言,当食物源上的气味标记的量相当于一个食物源处有 40 只左右采集蜂时,其标记是起吸引作用的,但当食物源上的气味标记的量过高时则具有驱避作用(Hrncir et al., 2004; Jarau et al.,2004 a)。在自然情况下,高浓度的气味标记可能表示这是一个反复被采集过的食物源,因而这个食物源可能已经被耗尽了。采集蜂是否使用气味标记还取决于食物源有无强烈气味。当食物源无气味时,气味标记似乎更为重要(Hrncir et al.,2004; Schmidt et al.,2005)。

10.2.3 无刺蜂的引导飞行(引路)行为

Lindauer 和 Kerr(1958)观察到,*Scaptotrigona postica* 无刺蜂的被招募蜂通常与做气味标记的采集蜂一起到达食物源处。他们认为这是因为新被招募蜂使用了其他信息,比如招募蜂的视觉引导或招募蜂飞行过程中释放的化学物质。*Trigona corvina* 无刺蜂中也有类似的发现,其新被招募蜂通常会在招募蜂着陆之后才降落在喂食器上(第 10.4 节)(Aguilar et al.,2005)。Lindauer 和 Kerr(1958, 1960)推测,许多无刺蜂的表皮呈现黑色且有光泽,这会使它们在视觉上很显眼,因而使其同伴在飞行中更容易跟随它们。

许多 *Melipona* 属无刺蜂蜂种,如 *Melipona panamica*、*M. quadrifasciata*、*M. rufiventris* 以及 *M. scutellaris* 能够招募同伴抵达特定位置的食物源,而且在招募过程中它们没有任何使用了气味轨迹的迹象(Nieh and Roubik,1995; Hrncir et al.,2000; Jarau et al.,2000; Jarau et al.,2003)[12]。例如,Hrncir 等(2000)观察到,*M. quadrifasciata* 和 *M. scutellaris* 采集蜂返回蜂巢的飞行中,约 99.5% 都完全没有降落在植物上过。当 Nieh 和 Roubik(1995)训练 *M. panamica* 无刺蜂采集蜂到距离蜂巢 100 m 的一个食物源采集时,尽管没有明显的气味标记,但是几乎所有的新被招募蜂都到达了这一喂食器,而不是飞至相对蜂巢方向相反的一个等距对照的喂食器上(图 10.7 D)。*M. panamica* 无刺蜂招募的位置特异性与使用有轨迹信息素的蜂种招募的位置特异性相当。尽管可以用引导飞行行为来解释这种位置精确性,但是 Nieh 和 Roubik(1995, 1998 b)

---

[12] Kerr 和 Rocha(1988)猜测一些 *Melipona* 无刺蜂种也会在食物源附近留下气味标记,但是他们没有提供试验性的证据(Nieh and Roubik,1995 ;Jarau,2009)。

仍对 *Melipona* 属无刺蜂采用了引导飞行进行招募的证据持谨慎态度。研究者反对 *M. panamica* 无刺蜂采用引导飞行进行招募定位的，其中一个论据就是研究者发现即使已经抓住被标记的招募蜂，新被招募同巢蜂几分钟后也会到达食物源，而且抵达的新被招募蜂往往不是及时地一起抵达食物源的（Nieh and Roubik,1995, 1998 b）。另外，这种招募中似乎也必须有积极招募的工蜂存在。如果捕捉了训练到达喂食器的招募工蜂，那么与另一方向上的对照喂食器相比，新被招募工蜂就不会偏好于这个招募工蜂访问过的喂食器。因此，*M. panamica* 无刺蜂的采集过程中显然涉及了招募工蜂某种形式的引导作用，但是新被招募蜂所利用的信息的本质仍有待探索。

研究发现 *Melipona quadrifasciata* 无刺蜂采集蜂还会表现出局部"之"字形飞行的行为（Esch et al.,1965; Esch,1967），但 Hrncir 等（2000）在他们的研究中发现，新被招募蜂没有跟随招募蜂表现出"之"字形飞行（Jarau et al., 2003）。相反，被招募工蜂会径直飞向食物源，而且新被招募蜂很少与招募工蜂一起到达食物源（Jarau et al.,2003）。

Partamona orizabaensis 无刺蜂的招募交流也同样令人费解（Flaig et al., 2016）。这种大规模招募型蜂种的采集蜂能够在没有任何气味轨迹信号的情况下引导同伴到达食物源的位置，而由于其栖息地光线条件昏暗，它们应该也不太可能使用了引导飞行（Flaig et al.,2016）。一种解释是采集蜂在飞行中释放了气味，从而形成了空中气味轨迹（Kerr,1960; Flaig et al.,2016）。那么它们是否采用了引导飞行，还是研究者猜想的空中气味轨迹，还需要进一步具体研究，进而探索无刺蜂是否普遍采用了这些招募形式（Jarau,2009）。

10.3 无刺蜂的"窃听"行为

无刺蜂对一个食物源和蜂巢之间路径的标记或标记食物源本身都会吸引竞争者的注意，竞争者可以"偷听"这种交流，这一过程被称为"窃听"[13]。例如，Lindauer 和 Kerr（1958）发现，*Scaptotrigona postica* 采集蜂会尾随附近同种蜂群的气味轨迹。*Agelaia vicina* 胡蜂因其蜂群规模能庞大到超过一百万只而闻名（Zucchi et al.,1995），它们会尾随食肉性 *Trigona hypogea* 无刺蜂的气味标记（Noll,1997）。研究表明，*Nannotrigona testaceicornis*（Schmidt

---

[13] 笔者追溯了窃听的一个广义的定义，该定义里将社会性的线索与信号都作为窃听者的社会性信息来源（Johnstone,2001; Danchin et al.,2004; Valone,2007）。其他定义则将窃听限制在信号的应用上（Slaa et al.,2003; Slaa and Hughes,2009）。

et al.,2005）、*Tetragonisca angustula*（Villa and Weiss,1990） 和 *Trigona corvina*（Boogert et al.,2006）还会被其他同一蜂种的蜂群工蜂所留下的气味标记所吸引，而 *Tetragonisca angustula* 和 *Trigona corvina* 的采集蜂会忽视不同蜂种工蜂留下的气味标记（Boogert et al.,2006）。另一种 *Trigona* 属无刺蜂蜂种——大规模招募型的 *T. Spinipes* 则会被 *Melipona rufiventris* 无刺蜂留下的气味所吸引，*T. Spinipes* 蜂群会通过赶走 *Melipona* 采集蜂来占据食物源（Nieh et al.,2004 a）。

如前所述，无刺蜂是否依赖气味标记找寻食物源通常取决于特定环境，这种依赖性可以随着无刺蜂的学习情况而有所变化（Nieh et al.,2004 a; Solórzano-Gordillo et al.,2018）。对于新发现的食物源，*Trigona spinipes* 采集蜂更喜欢 *M. Rufiventris* 无刺蜂的气味标记，偏好程度超过自己的气味标记；对于已被 *T. spinipes* 采集蜂开采过的食物源，*Trigona spinipes* 采集蜂则喜欢自己的气味标记（Nieh et al.,2004 a）胜过 *M. Rufiventris* 无刺蜂的气味标记。相反，*Melipona rufiventris* 采集蜂会避开 *T. spinipes* 的气味标记。尽管人们认为 *Trigona hyalinata* 无刺蜂在竞争食物源上比 *T.spinipes* 更有优势，但实际上 *Trigona hyalinata* 采集蜂通常会避开 *T. spinipes* 无刺蜂的招募信息素（Lichtenberg et al.,2011）。这可能是因为两个蜂种都具有攻击性，避开另一种好斗蜂种的招募信息素的行为能够降低在争抢食物源时斗争的风险投入（Lichtenberg et al.,2011）。*T. spinipes* 采集蜂反过来也不会对 *T. hyalinata* 无刺蜂的招募信息素做出反应（Lichtenberg et al.,2011）。"窃听"也要取决于信息素轨迹的强度。一个信息素轨迹如果信号较弱则意味着这个轨迹所指食物源可能防守也较弱，这就将降低了无刺蜂去尝试占领食物源的潜在成本。因此，当 *T. hyalinata* 采集蜂遇到的信号较弱的 *T. spinipes* 无刺蜂信息素轨迹时，它们就会由回避转为被吸引（Lichtenberg et al.,2014）。

不同无刺蜂的蜂种特异性的信号策略能够决定"窃听"者付出的代价（Lichtenberg et al.,2014）。例如，为什么无刺蜂的信息素轨迹通常在食物源不远处（见上文）也可以用这个来解释，因为这样减少了竞争者或捕食者顺着这些轨迹发现食物源位置或蜂群本身的风险（Nieh et al.,2004 b; Lichtenberg et al., 2011）。有趣的是，竞争者未必会选择窃听不明显的交流。如果信息素能向窃听者传递其需付出代价的信号，如攻击性对抗等，那么竞争者也会偏好窃听明显的信号（Lichtenberg et al.,2014）。

10.4 局部增强学习和刺激增强学习

当一个食物源上或食物源附近存在其他蜂（或它们的气味），这种情况会被无刺蜂当作社会性线索进行采集决策，例如一朵花上存在其他蜂时，可能就会刺激无刺蜂采集蜂降落在这朵花上（见上文）。然后采集蜂可以学习这一食物源的特征，这是所谓的局部增强学习（local enhancement）的一种社会性学习形式（Heyes,1994）[14]。而在刺激增强学习（stimulus enhancement）的过程中，采集蜂不仅学会认识到这个局部的花区是有奖励的，而且会产生对同类花的普遍偏好（Heyes,1994; Avargues-Weber and Chittka,2014）。局部增强学习和刺激增强学习由两个连续部分构成：首先是社会线索引起吸引性组分，而后吸引性组分为采集蜂创造了一个学习食物源特征（如颜色、气味、地标）与奖励回馈之间的关系的机会。"窃听"和"局部增强"是相关的过程，但根据无刺蜂应用这两种模式的方式，二者在4个方面有所不同。第一，"窃听"指的是利用非同巢同伴提供的信息，而"局部增强"学习则并不对复制和被复制个体之间的关系做出假设（Heyes,1994）。第二，"窃听"主要指的是利用社会信息（信号和线索）从而产生的最终过程（Lichtenberg et al.,2014），而"局部增强"则重点强调新行为习得的近似基础（Heyes,1994）。第三，"窃听"通常不是必须要学习的，而"局部增强"是一个社会学习的过程。最后，"窃听"可能涉及利用信号或社会线索，而"局部增强"通常局限于社会线索的使用（Slaa and Hughes,2009）。

如果花朵已被其他工蜂占领，那么这种情况会阻止采集蜂降落在此处，这种现象称为"局部抑制"（Slaa et al., 2003）。与驱避性气味标记（一种局部抑制）的情况一样，由于被占领花朵的资源可能已经耗尽，因此避开这些花朵可降低采集成本。研究发现，在一些无刺蜂物种中存在基于视觉线索的局部增强和局部抑制（Slaa and Hughes,2009）。一只采集蜂是会被花朵上其他蜂的存在所吸引还是所阻退可能也取决于其先前的经验（第10.2.2节）（Slaa et al., 2003; Sanchez et al., 2008; Dawson et al., 2013; Avargues-Weber and

---

[14] 不同的研究者以不同的方式定义了"局部增强"（Hoppitt and Laland,2008）。一些作者定义的局部增强学习中不包括学习需要，而是主要将吸引部分作为局部增强的关键准则（Slaa et al.,2003; Hoppitt and Laland,2008）。笔者追溯了 Heyes（1994）的研究并将局部增强学习定义为一个有学习行为参与的过程。考虑采集蜂确实有学习食物源相关线索的倾向（第8章），那么采集蜂基本不太可能在没有后续学习食物源特征的情况下被食物源所吸引。

Chittka, 2014）。例如，对于 *Trigona amalthea* 无刺蜂来说，缺乏采集经验的采集蜂会表现出局部增强，而经验丰富的采集蜂则会被花朵上的同巢蜂阻退（Slaa et al., 2003）。对食物源处的其他采集蜂做出什么样的反应可能取决于食物源上的采集蜂是否具有攻击性。正如研究者猜想的那样，在花朵选择实验中，无刺蜂采集蜂避开了体型较大且更有攻击性的异种蜂（Slaa et al., 2003）。因此，花上的异种蜂是否能诱发局部增强可能取决于参与竞争的蜂种的竞争等级（Slaa and Hughes, 2009）。

无刺蜂对食物源上存在的其他蜂的反应具有物种特异性，这些物种特异性差异会影响不同蜂种采集蜂在花区上的分布：局部抑制导致采集蜂更加分散，而局部增强则可以促进采集蜂群的形成（第 8 章）（Slaa et al., 2003; Slaa and Hughes, 2009）。反过来，采集蜂群的形成也会进一步保护采集蜂，抵御竞争者和天敌。

10.5 招募交流的进化

那么无刺蜂不同的招募行为是何时和以什么顺序进化而来的？这是一个十分有趣的问题。研究发现，蜜蜂（von Frisch, 1967）和熊蜂（Dornhaus and Chittka, 1999）在巢内也会兴奋地飞行（如摇晃、"之"字形和推挤式前进）。因此，巢内兴奋性前进很可能代表一种原始的祖先行为，这种行为能够提醒其他采集蜂某种特定食物类型的存在，并激励它们寻找这种食物。在无刺蜂和蜜蜂蜂群中，采集蜂通常会中断其兴奋前行状态来进行交哺行为，这使其他工蜂能够品尝食物并学习它的气味（见上文和第 8 章）。奇怪的是，熊蜂（*Bombini*）不进行交哺行为，可能是因为它们蜂群规模相对较小，这使得寻找交哺同伴会消耗太多时间（Anderson and Ratnieks, 1999）。兴奋前行行为和交哺行为这两种同时发生的行为可能代表了无刺蜂祖先招募交流演化的一个早期阶段（Kerr, 1960; Nieh, 2004）。大多数乃至所有无刺蜂兴奋前进过程中普遍会产生嗡嗡声（Kerr, 1960），无刺蜂和蜜蜂中这种嗡嗡前进行为可能也产生了独立进化（Hrncir and Barth, 2014; I'Anson Price and Grüter, 2015）。

Jarau（2009）基于对 *Scaptotrigona* 属和 *Trigona* 属无刺蜂的观察提出信息素轨迹并不是无刺蜂的祖先状态，这两个属无刺蜂是相对派生属，因其会形成信息素轨迹而闻名（图 2.2）。研究者提出了一种假设，即无刺蜂的信息素轨迹代表了一种派生特征（图 10.11），而一项祖先状况重建研究支持这一假设。该研究分析表明，铺设信息素轨迹的蜂种亲缘关系相对紧密，在

3 500万—3 000万年前无刺蜂就曾出现铺设信息素轨迹的情况。而 *Tetragona* 的祖先中信息素轨迹交流方式似乎已经消失了（图 10.11）。有趣的是，Jarau 等（2003）观察到，*Tetragona clavipes* 采集蜂会表现出类似于在食物源上做气味标记的行为，但他们并没有发现其采集蜂铺设的气味轨迹。该 *Tetragona* 蜂种的招募能力较弱（Jarau et al.,2003），研究者尚未对其采集交流进行详细研究，对于大多数其他无刺蜂属来说也是如此。

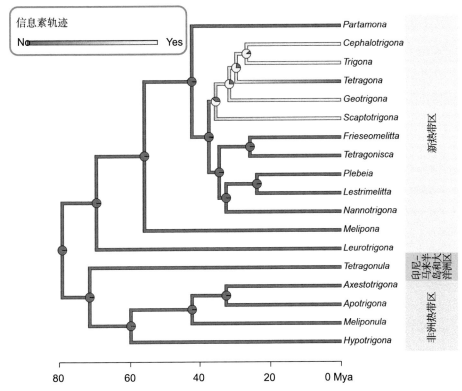

图 10.11 （是 / 否）铺设轨迹追踪信息素到食物源的无刺蜂蜂种的祖先状态的建树情况（表 10.1）

注：图中树枝的颜色是基于 MCMC 法计算而来（Paradis,2011）。饼图提供了基于连续时间马尔可夫链模型的估值［关于如何进行这种分析的更多细节，见 Grüter（2018）］。图中记录的蜂种时间则是基于 Rasmussen 和 Cameron（2010）。所有方法都表明铺设轨迹追踪信息素到食物源是无刺蜂在 3 500万—3 000万年前出现的一种衍生状态。

参考文献

Aguilar I, Briceño D（2002）Sounds in *Melipona costaricensis*（Apidae: Meliponini）:

effect of sugar concentration and nectar source distance. Apidologie 33:375–388

Aguilar I, Fonseca A, Biesmeijer JC (2005) Recruitment and communication of food source location in three species of stingless bees (Hymenoptera, Apidae, Meliponini). Apidologie 36:313–324

Aguilar I, Sommeijer MJ (2001) The deposition of anal excretions by *Melipona favosa* foragers (Apidae: Meliponinae): behavioural observations concerning the location of food sources. Apidologie 32:37–48

Alavez-Rosas D, Malo EA, Guzmán MA, Sánchez-Guillén D, Villanueva-Gutiérrez R, Cruz-López L (2017) The Stingless Bee *Melipona solani* Deposits a Signature Mixture and Methyl Oleate to Mark Valuable Food Sources. Journal of Chemical Ecology 43:945–954

Al Toufailia HM, Grüter C, Ratnieks FLW (2013) Persistence to unrewarding feeding locations by honeybee foragers (*Apis mellifera*): the effects of experience, resource profitability and season. Ethology 119:1096–1106

Anderson C, Ratnieks FLW (1999) Task partitioning in insect societies. I. Effect of colony size on queueing delay and colony ergonomic efficiency. The American Naturalist 154:521–535

Avargues-Weber A, Chittka L (2014) Local enhancement or stimulus enhancement? Bumblebee social learning results in a specific pattern of flower preference. Animal Behaviour 97:185–191

Barth FG, Hrncir M, Jarau S (2008) Signals and cues in the recruitment behavior of stingless bees (Meliponini). Journal of Comparative Physiology A 194:313–327

Beckers R, Deneubourg JL, Goss S, Pasteels JM (1990) Collective decision making through food recruitment. Insectes Sociaux 37:258–267

Beekman M, Dussutour A (2009) How to tell your mates: costs and benefits of different recruitment mechanisms. In: Jarau S, Hrncir M (eds) Food Exploitation by Social Insects: Ecological, Behavioral, and Theoretical Approaches. CRC Press, Taylor & Francis Group, Boca Raton, pp 115–134

Beekman M, Sumpter DJT, Ratnieks FLW (2001) Phase transition between disordered and ordered foraging in Pharaoh's ants. Proceedings of the National Academy of Sciences of the United States of America 98:9703–9706

Biesmeijer JC, de Vries H (2001) Exploration and exploitation of food sources by social insect colonies: a revision of the scout-recruit concept. Behavioral Ecology and Sociobiology 49:89–99

Biesmeijer JC, van Nieuwstadt MGL, Lukacs S, Sommeijer MJ（1998）The role of internal and external information in foraging decisions of *Melipona* workers（Hymenoptera : Meliponinae）. Behavioral Ecology and Sociobiology 42:107–116

Blum MS, Crewe RM, Kerr WE, Keith LH, Garrison AW, Walker MM（1970）Citral in stingless bees: isolation and functions in trail-laying and robbing. Journal of Insect Physiology 16:1637–1648

Boogert NJ, Hofstede FE, Monge IA（2006）The use of food source scent marks by the stingless bee *Trigona corvina*（Hymenoptera: Apidae）: the importance of the depositor's identity. Apidologie 37:366–375

Camazine S, Deneubourg JL, Franks NR, Sneyd J, Theraulaz G, Bonabeau E（2001）Self-Organization in Biological Systems. Princeton University Press, Princeton, NJ

Contrera FAL, Nieh JC（2007）Effect of forager-deposited odors on the intra-patch accuracy of recruitment of the stingless bees *Melipona panamica* and *Partamona peckolti*（Apidae, Meliponini）. Apidologie 38:584–594

Czaczkes TJ, Salmane AK, Klampfleuthner FA, Heinze J（2016）Private information alone can trigger trapping of ant colonies in local feeding optima. Journal of Experimental Biology 219:744–751

Dall SRX, Giraldeau LA, Olsson O, McNamara JM, Stephens DW（2005）Information and its use by animals in evolutionary ecology. Trends in Ecology & Evolution 20:187–193

Danchin E, Giraldeau LA, Valone TJ, Wagner RH（2004）Public information: from nosy neighbors to cultural evolution. Science 305:487–491

Darchen R（1969）Sur la biologie de *Trigona*（*Apotrigona*）*nebulata* komiensis Cock. I. Biologia Gabonica 5:151–183

Dawson EH, Avargues-Weber A, Chittka L, Leadbeater E（2013）Learning by observation emerges from simple associations in an insect model. Current Biology 23:1–4

De Marco RJ（2006）How bees tune their dancing according to their colony's nectar influx: re-examining the role of the food-receivers' "eagerness." Journal of Experimental Biology 209:421–432

Dechaume-Moncharmont FX, Dornhaus A, Houston AI, McNamara JM, Collins EJ, Franks NR (2005) The hidden cost of information in collective foraging. Proceedings of the Royal Society of London Series B-Biological Sciences 272:1689–1695

Dornhaus A, Chittka L (1999) Evolutionary origins of bee dances. Nature 401:38–38

Dornhaus A, Chittka L (2004) Why do honey bees dance? Behavioral Ecology and Sociobiology 55:395–401

Esch H (1967) Die Bedeutung der Lauterzeugung für die Verständigung der stachellosen Bienen. Zeitschrift für vergleichende Physiologie 56:199–220

Esch H, Esch I, Kerr WE (1965) Sound: An element common to communication of stingless bees and to dances of the honey bee. Science 149:320–321

Farina WM (2000) The interplay between dancing and trophallactic behavior in the honey bee *Apis mellifera*. Journal of Comparative Physiology A 186:239–245

Farina WM, Grüter C, Diaz PC (2005) Social learning of floral odours within the honeybee hive. Proceedings of the Royal Society of London Series B-Biological Sciences 272:1923–1928

Flaig IC, Aguilar I, Schmitt T, Jarau S (2016) An unusual recruitment strategy in a mass-recruiting stingless bee, *Partamona orizabaensis*. Journal of Comparative Physiology A 202:679–690

Franklin EL (2014) The journey of tandem running: the twists, turns and what we have learned. Insectes Sociaux 61:1–8

Gil M, De Marco RJ (2005) Olfactory learning by means of trophallaxis in *Apis mellifera*. Journal of Experimental Biology 208:671–680

Gilley DC, Kuzora JM, Thom C (2012) Hydrocarbons emitted by waggle-dancing honey bees stimulate colony foraging activity by causing experienced foragers to exploit known food sources. Apidologie 43:85–94

Giurfa M, Núñez JA (1992) Honeybees mark with scent and reject recently visited flowers. Oecologia 89:113–117

Goulson D, Chapman JW, Hughes WOH (2001) Discrimination of unrewarding flowers by bees: direct detection of rewards and use of repellent scent marks. Journal of Insect Behavior 14:669–678

Granovskiy B, Latty T, Duncan M, Sumpter DJT, Beekman M (2012) How dancing honey bees keep track of changes: the role of inspector bees. Behavioral Ecology 23:588–596

Grüter C, Acosta LE, Farina WM（2006）Propagation of olfactory information within the honeybee hive. Behavioral Ecology and Sociobiology 60:707–715

Grüter C, Balbuena MS, Farina WM（2008）Informational conflicts created by the waggle dance. Proceedings of the Royal Society of London Series B-Biological Sciences 275:1321–1327

Grüter C, Farina WM（2009 a）Past experiences affect interaction patterns among foragers and hive-mates in honeybees. Ethology 115:790–797

Grüter C, Farina WM（2009 b）The honeybee waggle dance: can we follow the steps? Trends in Ecology & Evolution 24:242–247

Grüter C, Czaczkes TJ（2019）Communication in social insects and how it is shaped by individual experience. Animal Behaviour 151:207–215

Grüter C, Balbuena MS, Farina WM（2009）Retention of long-term memories in different age groups of honeybee（*Apis mellifera*）workers. Insectes Sociaux 56:385–387

Grüter C, Wüst M, Cipriano AP, Nascimento FS（2018）Tandem recruitment and foraging in the ponerine ant *Pachycondyla harpax*（Fabricius）. Neotropical Entomology 47:742–749

Heyes CM（1994）Social learning in animals: categories and mechanisms. Biological Reviews of the Cambridge Philosophical Society 69:207–231

Hölldobler B, Wilson EO（1990）The Ants. The Belknap Press of Harward University, Cambridge, MA

Hoppitt W, Laland KN（2008）Social processes influencing learning in animals: A review of the evidence. Advances in the Study of Behavior 38:105–165

Hrncir M（2009）Mobilizing the Foraging Force Mechanical Signals in Stingless Bee Recruitment. In: Jarau S, Hrncir M（eds）Food Exploitation by Social Insects: Ecological, Behavioral, and Theoretical Approaches. CRC Press, Taylor & Francis Group, Boca Raton, FL

Hrncir M, Barth FG（2014）Vibratory Communication in Stingless Bees（Meliponini）: The Challenge of Interpreting the Signals. In: Cocroft RB, Gogala M, Hill PSM, Wessel A（eds）Studying Vibrational Communication. Springer Berlin Heidelberg,

Berlin, Heidelberg, pp 349–374

Hrncir M, Gravel AI, Schorkopf DLP, Schmidt VM, Zucchi R, Barth FG（2008 a）Thoracic vibrations in stingless bees（*Melipona seminigra*）: resonances of the thorax influence vibrations associated with flight but not those associated with sound production. Journal of Experimental Biology 211:678–685

Hrncir M, Jarau S, Zucchi R, Barth FG（2000）Recruitment behavior in stingless bees, *Melipona scutellaris* and *M. quadrifasciata*. II. Possible mechanisms of communication. Apidologie 31:93–113

Hrncir M, Jarau S, Zucchi R, Barth FG（2004）On the origin and properties of scent marks deposited at the food source by a stingless bee, *Melipona seminigra*. Apidologie 35:3–13

Hrncir M, Maia-Silva C（2013）The fast versus the furious - On competition, morphological foraging traits, and foraging strategies in stingless bees. In: Vit P, Roubik DW（eds）Stingless bees process honey and pollen in cerumen pots. Universidad de Los Andes, Mérida, Chapter 13. Universidad de Los Andes, Mérida

Hrncir M, Schmidt VM, Schorkopf DLP, Jarau S, Zucchi R, Barth FG（2006）Vibrating the food receivers: a direct way of signal transmission in stingless bees（*Melipona seminigra*）. Journal of Comparative Physiology A 192:879–887

Hrncir M, Schorkopf DLP, Schmidt VM, Zucchi R, Barth FG（2008 b）The sound field generated by tethered stingless bees（*Melipona scutellaris*）: inferences on its potential as a recruitment mechanism inside the hive. Journal of Experimental Biology 211:686–698

Hubbell SP, Johnson LK（1978）Comparative foraging behavior of six stingless bee species exploiting a standardized resource. Ecology 59:1123–1136

Hubbell SP, Johnson LK（1977）Competition and nest spacing in a tropical stingless bee community. Ecology 58:949–963

I'Anson Price R, Grüter C（2015）Why, when and where did honey bee dance communication evolve? Frontiers in Ecology and Evolution 3:1–7

I'Anson Price R, Dulex N, Vial N, Vincent C, Grüter C（2019）Honeybees forage more successfully without the "dance language" in challenging environments. Science Advances 5:eaat0450

Jarau S（2009）Chemical Communication during Food Exploitation in Stingless

Bees. In: Jarau S, Hrncir M (eds) Food Exploitation by Social Insects: Ecological, Behavioral, and Theoretical Approaches. CRC University Press, Boca Raton, FL

Jarau S, Dambacher J, Twele R, Aguilar I, Francke W, Ayasse M (2010) The trail pheromone of a stingless bee, *Trigona corvina* (Hymenoptera, Apidae, Meliponini), varies between populations. Chemical Senses 35:593–601

Jarau S, Hrncir M, Ayasse M, Schulz C, Francke W, Zucchi R, Barth FG (2004a) A Stingless Bee (*Melipona seminigra*) Marks Food Sources with a Pheromone from Its Claw Retractor Tendons. Journal of Chemical Ecology 30:793–804

Jarau S, Hrncir M, Schmidt VM, Zucchi R, Barth FG (2003) Effectiveness of recruitment behavior in stingless bees (Apidae, Meliponini). Insectes Sociaux 50:365–374

Jarau S, Hrncir M, Zucchi R, Barth FG (2004b) A stingless bee uses labial gland secretions for scent trail communication (*Trigona recursa* Smith 1863). Journal of Comparative Physiology A 190:233–239

Jarau S, Hrncir M, Zucchi R, Barth FG (2000) Recruitment behavior in stingless bees, *Melipona scutellaris* and *M. quadrifasciata*. I. Foraging at food sources differing in direction and distance. Apidologie 31:81–91

Jarau S, Schulz CM, Hrncir M, Francke W, Zucchi R, Barth FG, Ayasse M (2006) Hexyl decanoate, the first trail pheromone compound identified in a stingless bee, *Trigona recursa*. Journal of Chemical Ecology 32:1555–1564

Jesus TNCS, Venturieri GC, Contrera FAL (2014) Time–place learning in the bee *Melipona fasciculata* (Apidae, Meliponini). Apidologie 45:257–265

John L, Aguilar I, Ayasse M, Jarau S (2012) Nest-specific composition of the trail pheromone of the stingless bee *Trigona corvina* within populations. Insectes Sociaux 59:527–532

Johnson DL (1967) Communication among honey bees with field experience. Animal Behaviour 15:487–492

Johnson LK (1987) The pyrrhic victory of nest-robbing bees: did they use the wrong pheromone? Biotropica 19:188–189

Johnstone RA (2001) Eavesdropping and animal conflict. Proceedings of the National Acadamy of Sciences USA 98:9177–9180

Kaschef AH (1957) Über die Einwirkung von Duftstoffen auf die Bienentänze.

Zeitschrift für vergleichende Physiologie 39:562–576

Kerr WE（1960）Evolution of communication in bees and its role in speciation. Evolution 14:386–387

Kerr WE（1994）Communication among *Melipona* workers（Hymenoptera: Apidae）. Journal of Insect Behavior 7:123–128

Kerr WE, Ferreira A, Mattos NS de（1963）Communication among Stingless Bees-Additional Data（Hymenoptera: Apidae）. Journal of the New York Entomological Society 71:80–90

Kerr WE, Rocha FH（1988）Communicação em *Melipona rufiventris* e *Melipona compressipes*. Ciencia e Cultura 40:1200–1202

Koethe S, Fischbach V, Banysch S, Reinartz L, Hrncir M, Lunau K（2020）A Comparative Study of Food Source Selection in Stingless Bees and Honeybees: Scent Marks, Location, or Color. Frontiers in Plant Sciences 11:516

Krausa K, Hager FA, Kiatoko N, Kirchner WH（2017a）Vibrational signals of African stingless bees. Insectes Sociaux 64:415–424

Krausa K, Hager FA, Kirchner WH（2017 b）The effect of food profitability on foraging behaviors and vibrational signals in the African stingless bee *Plebeina hildebrandti*. Insectes Sociaux 64:567–578

Leonhardt SD（2017）Chemical ecology of stingless bees. Journal of Chemical Ecology 43:385–402

Lichtenberg EM, Graff Zivin J, Hrncir M, Nieh JC（2014）Eavesdropping selects for conspicuous signals. Current Biology 24:R598–R599

Lichtenberg EM, Hrncir M, Turatti IC, Nieh JC（2011）Olfactory eavesdropping between two competing stingless bee species. Behavioral Ecology and Sociobiology 65:763–774

Lindauer M（1956）Über die Verständigung bei indischen Bienen. Zeitschrift für vergleichende Physiologie 38:521–557

Lindauer M（1957）Communication Among the Honeybees and Stingless Bees of India. Bee World 38:3–14

Lindauer M（1948）Über die Einwirkung von Duft- und Geschmacksstoffen sowie anderer Faktoren auf die Tänze der Bienen. Zeitschrift für vergleichende Physiologie 31:348–412

Lindauer M, Kerr WE (1960) Communication between the workers of stingless bees. Bee World 41:29–71

Lindauer M, Kerr WE (1958) Die gegenseitige Verständigung bei den stachellosen Bienen. Journal of Comparative Physiology A 41:405–434

Martinez A, Farina WM (2008) Honeybees modify gustatory responsiveness after receiving nectar from foragers within the hive. Behavioral Ecology and Sociobiology 62:529–535

Mc Cabe SI, Hrncir M, Farina WM (2015) Vibrating donor-partners during trophallaxis modulate associative learning ability of food receivers in the stingless bee *Melipona quadrifasciata*. Learning and Motivation 50:11–21

Nieh JC (2004) Recruitment communication in stingless bees (Hymenoptera, Apidae, Meliponini). Apidologie 35:159–182

Nieh JC (1998 a) The food recruitment dance of the stingless bee, *Melipona panamica*. Behavioral Ecology and Sociobiology 43:133–145

Nieh JC (1998 b) The role of a scent beacon in the communication of food location by the stingless bee, *Melipona panamica*. Behavioral Ecology and Sociobiology 43:47–58

Nieh JC, Barreto LS, Contrera FAL, Imperatriz-Fonseca VL (2004 a) Olfactory eavesdropping by a competitively foraging stingless bee, *Trigona spinipes*. Proceedings of the Royal Society of London B: Biological Sciences 271:1633–1640

Nieh JC, Contrera FA, Yoon RR, Barreto LS, Imperatriz-Fonseca VL (2004 b) Polarized short odor-trail recruitment communication by a stingless bee, *Trigona spinipes*. Behavioral Ecology and Sociobiology 56:435–448

Nieh JC, Contrera FAL, Nogueira-Neto P (2003 a) Pulsed mass recruitment by a stingless bee, *Trigona hyalinata*. Proceedings of the Royal Society of London B: Biological Sciences 270:2191–2196

Nieh JC, Contrera FAL, Rangel J, Imperatriz-Fonseca VL (2003 b) Effect of food location and quality on recruitment sounds and success in two stingless bees, *Melipona mandacaia* and *Melipona bicolor*. Behavioral Ecology and Sociobiology 55:87–94

Nieh JC, Ramírez S, Nogueira-Neto P (2003 c) Multi-source odor-marking of food by a stingless bee, *Melipona mandacaia*. Behavioral Ecology and Sociobiology 54:578–586

Nieh JC, Roubik DW (1998 a) Potential mechanisms for the communication of height and distance by a stingless bee, *Melipona panamica*. Behavioral Ecology and Sociobiology 43:387–399

Nieh JC, Roubik DW（1995）A stingless bee（*Melipona panamica*）indicates food location without using a scent trail. Behavioral Ecology and Sociobiology 37:63–70

Nieh JC, Roubik DW（1998 b）Potential mechanisms for the communication of height and distance by a stingless bee, *Melipona panamica*. Behavioral Ecology and Sociobiology 43:387–399

Nieh JC, Sánchez D（2005）Effect of food quality, distance and height on thoracic temperature in the stingless bee *Melipona panamica*. Journal of Experimental Biology 208:3933–3943

Nieh JC, Tautz J, Spaethe J, Bartareau T（2000）The communication of food location by a primitive stingless bee, *Trigona carbonaria*. Zoology 102:238–246

Noll FB（1997）Foraging behavior on carcasses in the nectrophagic bee *Trigona hypogea*（Hymenoptera: Apidae）. Journal of Insect Behavior 10:463–467

Pankiw T, Nelson M, Page RE, Fondrk MK（2004）The communal crop: modulation of sucrose response thresholds of pre-foraging honey bees with incoming nectar quality. Behavioral Ecology and Sociobiology 55:286–292

Peng T, Pedrosa J, Batista JE, Nascimento FS, Grüter C（in press）Foragers of the stingless bee *Plebeia droryana* inform nestmates about the direction, but not the distance to food sources. Ecol Entomol

Reichle C, Aguilar I, Ayasse M, Jarau S（2011）Stingless bees（*Scaptotrigona pectoralis*）learn foreign trail pheromones and use them to find food. Journal of Comparative Physiology A 197:243–249

Reichle C, Aguilar I, Ayasse M, Twele R, Francke W, Jarau S（2013）Learnt information in species-specific "trail pheromone" communication in stingless bees. Animal Behaviour 85:225–232

Reinhard J, Srinivasan MV, Zhang SW（2004）Olfaction: Scent-triggered navigation in honeybees. Nature 427:411–411

Roselino AC, Rodrigues AV, Hrncir M（2016）Stingless bees（*Melipona scutellaris*）learn to associate footprint cues at food sources with a specific reward context. Journal of Comparative Physiology A 202:657–666

Roubik DW（1989）Ecology and Natural History of Tropical Bees. Cambridge

University Press, New York

Saleh N, Chittka L (2006) The importance of experience in the interpretation of conspecific chemicals signals. Behavioral Ecology and Sociobiology 61:215–220

Sánchez D, Kraus FB, Hernández MJ, Vandame R (2007) Experience, but not distance, influences the recruitment precision in the stingless bee *Scaptotrigona mexicana*. Naturwissenschaften 94:567–573

Sánchez D, Nieh JC, Hénaut Y, Cruz L, Vandame R (2004) High precision during food recruitment of experienced (reactivated) foragers in the stingless bee *Scaptotrigona mexicana* (Apidae, Meliponini). Naturwissenschaften 91:346–349

Sánchez D, Nieh JC, Vandame R (2008) Experience-based interpretation of visual and chemical information at food sources in the stingless bee *Scaptotrigona mexicana*. Animal Behaviour 76:407–414

Schmidt VM, Hrncir M, Schorkopf DLP, Mateus S, Zucchi R, Barth FG (2008) Food profitability affects intranidal recruitment behaviour in the stingless bee *Nannotrigona testaceicornis*. Apidologie 39:260–272

Schmidt VM, Schorkopf DLP, Hrncir M, Zucchi R, Barth FG (2006a) Collective foraging in a stingless bee: dependence on food profitability and sequence of discovery. Animal Behaviour 72:1309–1317

Schmidt VM, Zucchi R, Barth FG (2006b) Recruitment in a scent trail laying stingless bee (*Scaptotrigona* aff. *depilis*): Changes with reduction but not with increase of the energy gain. Apidologie 37:487–500

Schmidt VM, Zucchi R, Barth FG (2005) Scent marks left by *Nannotrigona testaceicornis* at the feeding site: cues rather than signals. Apidologie 36:285–291

Schmidt VM, Zucchi R, Barth FG (2003) A stingless bee marks the feeding site in addition to the scent path (*Scaptotrigona* aff. *depilis*). Apidologie 34:237–248

Schorkopf DLP, Jarau S, Francke W, Twele R, Zucchi R, Hrncir M, Schmidt VM, Ayasse M, Barth FG (2007) Spitting out information: *Trigona* bees deposit saliva to signal resource locations. Proceedings of the Royal Society of London Series B 274:895–898

Schorkopf DLP, Morawetz L, Bento JMS, Zucchi R, Barth FG (2011) Pheromone paths attached to the substrate in meliponine bees: helpful but not obligatory for recruitment success. Journal of Comparative Physiology A 197:755–764

Schorkopf DLP, de Sá Filho GF, Maia-Silva C, Schorkopf M, Hrncir M, Barth FG(2016) Nectar profitability, not empty honey stores, stimulate recruitment and foraging in *Melipona scutellaris*（Apidae, Meliponini）. Journal of Comparative Physiology A 202:709–722

Seeley TD（1995）The wisdom of the hive: The social physiology of honey bee colonies. Harward University Press, Cambridge, MA

Seeley TD（1986）Social foraging by honeybees - How colonies allocate foragers among patches of flowers. Behavioral Ecology and Sociobiology 19:343–354

Sherman G, Visscher PK（2002）Honeybee colonies achieve fitness through dancing. Nature 419:920–922

Slaa EJ, Hughes WOH（2009）Local Enhancement, Local Inhibition, Eavesdropping, and the Parasitism of Social Insect Communication. In: Jarau S, Hrncir M（eds）Food exploitation by Social Insects: Ecological, Behavioral, and Theoretical Approaches. CRC Press, Taylor & Francis Group, Boca Raton, FL

Slaa EJ, Wassenberg J, Biesmeijer JC（2003）The use of field-based social information in eusocial foragers: local enhancement among nestmates and heterospecifics in stingless bees. Ecological Entomology 28:369–379

Solórzano-Gordillo E, Rojas JC, Cruz-López L, Sánchez D（2018）Associative learning of non-nestmate odor marks between colonies of the stingless bee *Scaptotrigona mexicana* Guérin（Apidae, Meliponini）during foraging. Insectes Sociaux 65:393–400

Stangler ES, Jarau S, Hrncir M, Zucchi R, Ayasse M（2009）Identification of trail pheromone compounds from the labial glands of the stingless bee *Geotrigona mombuca*. Chemoecology 19:13–19

Stout JC, Goulson D, Allen JA（1998）Repellent scent-marking of flowers by a guild of foraging bumblebees（*Bombus* spp.）. Behavioral Ecology and Sociobiology 43:317–326

Tanner DA, Visscher PK（2008）Do honey bees average directions in the waggle dance to determine a flight direction? Behavioral Ecology and Sociobiology 62:1891–1898

Valone TJ（2007）From eavesdropping on performance to copying the behavior of others: a review of public information use. Behavioral Ecology and Sociobiology 62:1–14

Villa JD, Weiss MR (1990) Observations on the use of visual and olfactory cues by *Trigona* spp foragers. Apidologie 21:541–545

von Frisch K (1919) Über den Geruchsinn der Biene und seine blütenbiologische Bedeutung. Zoologisches Jahrbuch (Zoologie und Physiologie) 37:1–238

von Frisch K (1923) Über die "Sprache" der Bienen. Zoologisches Jahrbuch (Zoologie und Physiologie) 40:1–186

von Frisch K (1967) The dance language and orientation of bees. Harvard University Press, Cambridge, MA

von Frisch K, Jander R (1957) Über den Schwänzeltanz der Bienen. Zeitschrift für vergleichende Physiologie 40:239–263

Wilson EO (1971) The insect societies. Harvard University Press, Cambridge, MA

Wyatt TD (2010) Pheromones and signature mixtures: defining species-wide signals and variable cues for identity in both invertebrates and vertebrates. Journal of Comparative Physiology A 196:685–700

Zucchi R, Sakagami SF, Noll FB, Mechi MR, Mateus S, Baio MV, Shima SN (1995) *Agelaia vicina*, a Swarm-Founding Polistine with the Largest Colony Size among Wasps and Bees (Hymenoptera: Vespidae). Journal of the New York Entomological Society 103:129–137